Springer Series in Reliability Engineering

Series Editor

Hoang Pham, Department of Industrial and Systems Engineering, Rutgers University, Piscataway, NJ, USA

Today's modern systems have become increasingly complex to design and build, while the demand for reliability and cost effective development continues. Reliability is one of the most important attributes in all these systems, including aerospace applications, real-time control, medical applications, defense systems, human decision-making, and home-security products. Growing international competition has increased the need for all designers, managers, practitioners, scientists and engineers to ensure a level of reliability of their product before release at the lowest cost. The interest in reliability has been growing in recent years and this trend will continue during the next decade and beyond.

The Springer Series in Reliability Engineering publishes books, monographs and edited volumes in important areas of current theoretical research development in reliability and in areas that attempt to bridge the gap between theory and application in areas of interest to practitioners in industry, laboratories, business, and government.

Indexed in Scopus

Interested authors should contact the series editor, Hoang Pham, Deptartment of Industrial and Systems Engineering, Rutgers University, Piscataway, NJ 08854, USA. Email: hopham@rci.rutgers.edu, or Anthony Doyle, Executive Editor, Springer, London. Email: anthony.doyle@springer.com.

More information about this series at http://www.springer.com/series/6917

Durga Rao Karanki · Gopika Vinod ·
Srividya Ajit
Editors

Advances in RAMS Engineering

In Honor of Professor Ajit Kumar Verma on His 60th Birthday

Springer

Editors
Durga Rao Karanki
Siemens Mobility AG
Wallisellen, Switzerland

Gopika Vinod
Bhabha Atomic Research Centre
Trombay, Mumbai, India

Srividya Ajit
Luleå University of Technology
Luleå, Sweden

ISSN 1614-7839 ISSN 2196-999X (electronic)
Springer Series in Reliability Engineering
ISBN 978-3-030-36517-2 ISBN 978-3-030-36518-9 (eBook)
https://doi.org/10.1007/978-3-030-36518-9

This Springer imprint is published by the registered company Springer Nature Switzerland AG
The registered company address is: Gewerbestrasse 11, 6330 Cham, Switzerland

Dedication

To

Professor Ajit Kumar Verma, our dear teacher, a philosopher, friend, and guide to his students on the occasion of his sixtieth birthday the 10th December, 2019.

Brief Profile

He was educated at IIT Kharagpur and he taught at IIT Bombay as a Professor in Reliability Engineering/Electrical Engineering for around 15 years and is now with the Western Norway University of Applied Sciences as Professor (Technical Safety) for almost 8 years. He has co-supervised/supervised 39 Ph.D. students and around 100 Master's students. He has been the founding EIC of International Journal of System Assurance Engineering and Management, the EIC of Life Cycle Reliability and Safety Engineering, and has been in editorial capacity with several journals including being the EIC of Opsearch several years back, all published by Springer, besides being the Series Editor of three-book series published by Springer and associated with numerous books either as an author or an editor. His senior collaborators include Prof. Lotfi Zadeh, Prof. Roy Billinton, Prof. Osaki, Prof. Uday Kumar, and Prof. P. K. Kapur among others. He learnt Yoga in his early student days from the Bihar School of Yoga and taught Yoga through various workshops in India and abroad. His interests were greatly influenced by his spiritual mentors Paramhamsa Swami Satyananda Saraswati, a legendary Yogi and Bhagwan Sri Sathya Sai Baba of Puttaparthi and felt his life to be spiritually connected and touched by them. A significant part of his spare time in his life at IIT Bombay was spent in learning along with his wife, Prof. A. Srividya, the esoteric practices of Srichakra upasana from his Guru, Sri Jairamanji who had dedicated his life in teaching and initiating these mystical traditions for its continuity. An avid trekker in the Himalayas, he has led and trekked at high altitudes during the past three decades in the Himalayas in India, Nepal, and Tibet and has lectured extensively on Indian mysticism and Kundalini Yoga. He had been closely associated and coordinated various Devi Padmavathy temple activities and was a coordinator of NSS activities for many years at IIT Bombay. His mother was fond of him but his parents are no more. His son Amardeep and his two brothers Ranjit and Pradeep work and live in India. His wife Srividya accompanies him in Norway.

Foreword

I am pleased to write the foreword for this well-edited book on Advances in Reliability, Availability, Maintainability, and Safety (RAMS) Engineering. Contributed chapters in the book cover latest trends as well as applications to various fields including nuclear engineering, software engineering, power systems engineering, and mechanical engineering. The chapters have succeeded in achieving a fine balance between the theory and practice.

I am indeed delighted to know that this book was brought on a special occasion of 60th birthday of my dear friend Prof. Ajit Kumar Verma, who has contributed immensely to the field of reliability and safety. He truly deserves this special honor and tribute by some of his former Ph.D. students and collaborators.

Advances in safety assessment of nuclear power plants, maintenance aspects of complex engineering systems, and reliability of power systems are some of the highlights of the book. I am sure that students, research scholars, scientists, engineers, and practitioners of safety and reliability engineering will greatly benefit from this book.

I strongly recommend this book for its comprehensive coverage on the advances of reliability and safety engineering and their practical applications.

Many best wishes to you, Prof. Ajit Kumar Verma, on your 60th birthday!

Hoang Pham
IEEE Fellow, IIE Fellow, Distinguished Professor
Department of Industrial and Systems Engineering
Rutgers University
Piscataway, NJ, USA

Preface

Reliability, Availability, Maintainability, and Safety (RAMS) engineering is attracting ever-increased attention in the era of Industry 4.0, fuelled by digitalization, Internet of things, and artificial intelligence. RAMS tasks inevitably appear throughout life cycle of engineering systems, ranging from concept phase to decommissioning. Typical RAMS activities include apportionment of RAMS requirements, evaluation of design alternatives, hazard analysis, maintenance planning, operator support systems, RAM demonstration, safety assessment, etc. These activities significantly contribute to efficient and economical design and operation of systems as well as supports homologation and regulation.

Ever-increasing complexity of today's engineering systems requires multidisciplinary expertise. This book presents advances in RAMS analysis for several engineering disciplines including software engineering, electrical and electronics engineering, civil and mechanical engineering, chemical engineering, and nuclear engineering. This book presents practical applications from the various industries including software, electrical and electronic, nuclear power plants, process and chemical plants, railways, etc. The book is organized as follows:

Chapters "DC and AC Contingency Solvers in Composite Power System Adequacy Assessment" to "Reliability Considerations in Analysis of Tunnels in Squeezing Rock" cover advances in reliability and availability of engineering systems; chapters "Integrated RAMS, LCC and Risk Assessment for Maintenance Planning for Railways" to "Fuzzy Logic Based Analysis of Dissolved Decay Contents in Transformer Oil" focus on maintainability; and chapters "Probabilistic Safety Assessment in Nuclear and Non-nuclear Facilities: In a Glimpse" to "Integrated Deterministic and Probabilistic Safety Assessment" are on safety engineering. Power systems and electrical/electronic systems reliability are addressed in chapters "DC and AC Contingency Solvers in Composite Power System Adequacy Assessment" to "Fatigue Life Model for PLCC Solder Joints Under Thermal Cycling Stress". Chapters "An Integrated Approach of BOCR Modeling Framework for Decision Tool Evaluation" to "The Unpopularity of the Software Tester Role Among Software Practitioners: A Case Study" emphasize software reliability, followed by focus on mechanical and civil engineering in chapters "A Study on Reliability of Rotors Using XLrotor" to "Reliability Considerations in Analysis of Tunnels in Squeezing Rock". An overview of these chapters is presented below.

Reliability and Availability

Application of probabilistic methods in power system reliability studies—both at the generation level and at the composite (generation plus transmission) level—is a highly developed field with plenty of available literature. Adequacy assessment studies deal with the evaluation of whether the generation capacity of a system is sufficient to supply the load requirement of the system. Chapter "DC and AC Contingency Solvers in Composite Power System Adequacy Assessment" elaborates on both DC and AC contingency solvers used in composite power system adequacy studies.

Reliability analysis is crucial for the design and maintenance of a microgrid system. In chapter "Reliability Analysis of Microgrid Systems Using Hybrid Approaches", few hybrid techniques are proposed to assess failure probability and reliability of microgrid system. Hybrid approaches are presented using both fault tree analysis (FTA) and binary decision diagram (BDD) to evaluate performance of a microgrid system.

Chapter "Reliability Prediction of Instrumentation and Control Cables for NPP Applications" deals with the methods and models developed toward determination of the reliability of instrumentation and control (I&C) cables used in nuclear power plants (NPPs). Several reliability prediction methodologies based on performance indicators were developed from analytical and experimental approaches. Chapter "Reliability Prediction of Instrumentation and Control Cables for NPP Applications" demonstrates these methodologies with the accelerated life testing data obtained under thermal and radiation aging, and also from the data from literature.

Solder joints are inevitable part of assembly of an electronic system. Life cycle stresses affect the integrity of these contacts and lead to jeopardizing of its functions and in turn to failure of system. Life estimation of solder joints is important to predict the time to failure and take countermeasures. Chapter "Fatigue Life Model for PLCC Solder Joints Under Thermal Cycling Stress" addresses this issue by specializing an established empirical model for fatigue failure, Coffin-Manson, for a PLCC solder joint using experimental data.

Chapter "An Integrated Approach of BOCR Modeling Framework for Decision Tool Evaluation" illustrates an analytical methodology representing benefits & opportunities (BO), and latter costs & risks (CR) BOCR models. The modeling has been carried out in the combination of appropriate analytical models and suitable data aggregation techniques. The proposed framework is illustrated by two case studies. The first case study illustrates a holistic model in the context of prototype dependability assessment of software at the prototype level. The second case study demonstrates a model validation quantitatively for quality of services (QoS) for real-world SOA-based applications.

IT is backbone of modern business in the digital world, which is increasingly becoming complex, due to disruptive technologies and co-existing with legacy applications. Chapter "DevOps for IT Service Reliability and Availability" discusses DevOps framework, practices across the software development life cycle. The impact of DevOps practices to improve software reliability and availability will be discussed with metrics that are available. The chapter will provide practices for reliability and availability improvement.

Software testing is one of the crucial supporting processes in software development. Unfortunately, the role is stigmatized partly due to misperception and partly due to the treatment of the role in the industry. Chapter "The Unpopularity of the Software Tester Role Among Software Practitioners: A Case Study" aims to analyze the situation exploring what limits an individual from taking up a testing career, in a way that exposes actual reactions to this role in the software industry.

Chapter "A Study on Reliability of Rotors Using XLrotor" is focused on computation of rotating machines in XLrotor in order to predict the failure due to disk offset and to check reliability. Reliability of rotary machines is not only dependent on static design stress but also dynamic forces generated during operating speeds. To check reliability of rotating machines, the XLrotor computational tool is considered optimum in incorporating the effect of mass, stiffness, inertia & imbalance effectively. The simulation-based methodology is used for modeling and analysis for the simple & complex rotors. The vibration results determine the impact of disk offset on rotor model and its performance.

Most of the advanced nuclear reactors implement passive systems for better safety and availability in order to reduce human error and active component malfunctions. Reliability of the passive systems degrades with time due to aging, random/stochastic loading or strength degradation. Hence, it is very important to evaluate the passive system reliability by considering both static and time variant analysis. Chapter "Time Variant Reliability Analysis of Passive Systems" deals with these aspects.

Underground openings and excavations are increasingly being used for civilian and strategic purposes all over the world. Evaluation of safety of a tunnel, especially a non-conforming one, in soft ground or poor rock mass needs an interactive analysis. Interaction is an issue of great relevance in the case of tunnels in lower Himalayas. Chapter "Reliability Considerations in Analysis of Tunnels in Squeezing Rock" evaluates tunnel safety and stability through reliability analysis under squeezing conditions.

Maintainability

Chapters "Integrated RAMS, LCC and Risk Assessment for Maintenance Planning for Railways" to "Fuzzy Logic Based Analysis of Dissolved Decay Contents in Transformer Oil" cover advances in maintainability engineering.

As of today, about 70% of the transportation infrastructure has already been built for the needs of customers, business, and society, where Railways is the major infrastructure. Due to the hierarchical nature of Railways, it is necessary for railway infrastructure managers to design a generic framework for the decision-making process when planning maintenance and interventions, which is an important functional block of asset management in railway infrastructures. Chapter "Integrated RAMS, LCC and Risk Assessment for Maintenance Planning for Railways" proposes an integrated methodology to perform maintenance decision-making using definitive "building blocks", namely, Reliability, Availability, Maintainability, and Safety (RAMS), Life Cycle Costing (LCC), and risk assessment.

Rapid developments in technologies such as robotics, digital automation, Internet of things, and AI have heralded the Fourth Industrial Revolution, commonly referred to as Industry 4.0 (i4.0). Industrial operations and products have since become more competitive and hence more demanding. Chapter "Implementation of Predictive Maintenance Systems in Remotely Located Process Plants under Industry 4.0 Scenario" highlights the need for identifying the needs of condition monitoring preparedness of process plants located in remote places, especially in a logistic sense. Issues related to assessment of the need for the new paradigm in condition monitoring, challenges faced by such plants in the transition from legacy systems to a new system, and customization and optimization of predictive maintenance under Industry 4.0 (PdM 4.0) have been discussed.

Unexpected failure of any component of the system may increase the maintenance and downtime cost due to unavailability of the system. A methodology using mathematical modeling facility of fuzzy set theory is presented in chapter "Application of Fuzzy Sets in Reliability and in Optimal Condition Monitoring Technique Selection in Equipment Maintenance", which is effective in situations wherein the data available is mostly subjective and it is difficult to get precise quantitative data. Multi-attribute decision-making methods with application to ranking and optimal condition monitoring technique selection from maintenance engineering domain is highlighted.

Chapter "Fuzzy Logic Based Analysis of Dissolved Decay Contents in Transformer Oil" presents a fuzzy logic methodology based on statistical techniques to monitor, diagnose, and predict the health index of electric transformer using furanic contents. This is to address transformer condition monitoring, ensuring good health, safety of operation, and maintenance.

Safety

Chapters "Probabilistic Safety Assessment in Nuclear and Non-nuclear Facilities: In a Glimpse" to "Integrated Deterministic and Probabilistic Safety Assessment" address advances and applications in safety engineering.

Chapter "Probabilistic Safety Assessment in Nuclear and Non-nuclear Facilities: In a Glimpse" presents a glimpse of probabilistic safety assessment (PSA) of nuclear and chemical facilities. An overview of PSA methodology in nuclear industry is given, followed by several applications in safety informed decision-making. PSA process steps in chemical industry are explained as well as its application in risk-based inspection of a chemical plant is demonstrated.

Chapter "Passive System Reliability Assessment and Its Integration into PSA" describes most widely used methods for Passive system reliability analysis and present an approach for its integration into PSA by modeling one of the operational transients, normally expected to occur in a typical nuclear power plant.

Chapter "Project Stage Considerations for an Inherently Safe and Reliable Chemical Plant" examines the factors that influence eventual safety-operability-reliability of a chemical manufacturing unit, right at the inception and execution stage. The relevant tools and practices used in the chapter are Engineering Project Process, Process Risk Assessment and management methods like Hazard and Operability (HAZOP) methodology, Layers of Protection Analysis (LOPA), Safety Integrity Level (SIL), and management of change.

Chapter "Integrated Deterministic and Probabilistic Safety Assessment" introduces the concept of integrated deterministic and probabilistic safety assessment (IDPSA), highlights benefits as well as its limitations. Challenges to these approaches include modeling of dynamic interactions among physical process, safety systems, and operator actions as well as propagation of these model uncertainties. Case study on a medium loss of coolant accident in a nuclear power plant is presented, which focuses on a comparison between IDPSA and traditional PSA considering impact of accident dynamics.

This book is useful for advanced undergraduate and postgraduate engineering students of electrical, electronic, information technology, mechanical, nuclear, and chemical disciplines. It will provide a good reference for research scholars of reliability and safety engineering, industrial engineering, and system engineering. Practicing engineers and safety managers will get an overview of latest trends in RAMS engineering.

Wallisellen, Switzerland — Durga Rao Karanki
Mumbai, India — Gopika Vinod
Luleå, Sweden — Srividya Ajit

Acknowledgements

It is a great privilege to honor Prof. Ajit Kumar Verma who immensely contributed to shape careers of several students, research scholars, and colleagues. We are sincerely grateful to him for allowing us to bestow a tribute to him.

We have received excellent support from the authors of the chapters in timely preparations as well as revisions. We are grateful for their commitment and dedication.

We are thankful to reviewers for their constructive criticism and useful suggestions during the review of the chapters. The reviewers of special mention are Santosh, Hari Prasad, Suresh, Vijay Venu, Manoj, Krishna Mohan, Adhitya, Solanki, Suraj, Ramesh, and Bimal.

We express our sincere thanks to Dr. Sebastian Klabes of Siemens Mobility AG and Dr. J. Chattopadhyay of Bhabha Atomic Research Centre for their encouragement and full moral support.

Special thanks to Praveena, Harshi, and Madhav for their love and cooperation.

We are grateful to Prof. Hoang Pham for his suggestions and encouragement. We thank Ms. Vidyaa Shri K., Ms. Bhagyalakkshme S., and Mr. P. Clarie for their great support in managing the production of this book.

October 2019

Durga Rao Karanki
Gopika Vinod
Srividya Ajit

Contents

Reliability and Availability Engineering

DC and AC Contingency Solvers in Composite Power System Adequacy Assessment 3
Øystein Stake Laengen and Vijay Venu Vadlamudi

Reliability Analysis of Microgrid Systems Using Hybrid Approaches 49
Moushumi Patowary, Bimal C. Deka and Gayadhar Panda

Reliability Prediction of Instrumentation and Control Cables for NPP Applications 67
T. V. Santhosh, V. Gopika, A. K. Ghosh, B. G. Fernandes and J. Chattopadhyay

Fatigue Life Model for PLCC Solder Joints Under Thermal Cycling Stress 85
Rohit Khatri, Diana Denice and Manoj Kumar

An Integrated Approach of BOCR Modeling Framework for Decision Tool Evaluation 109
Krishna Mohan Kovur and Ravi Kumar Gedela

DevOps for IT Service Reliability and Availability 149
Mahesh Kuruba

The Unpopularity of the Software Tester Role Among Software Practitioners: A Case Study 185
Yadira Lizama, Daniel Varona, Pradeep Waychal and Luiz Fernando Capretz

A Study on Reliability of Rotors Using XLrotor 199
H. N. Suresh, N. Madhusudan, D. Saravana Bhavan and S. B. Murgayya

Time Variant Reliability Analysis of Passive Systems 217
Hari Prasad Muruva

Reliability Considerations in Analysis of Tunnels in Squeezing Rock 237
G. Venkatachalam, Mallika Sreedharan and A. Srividya

Maintainability Engineering

Integrated RAMS, LCC and Risk Assessment for Maintenance Planning for Railways 261
Adithya Thaduri and Uday Kumar

Implementation of Predictive Maintenance Systems in Remotely Located Process Plants under Industry 4.0 Scenario 293
P. G. Ramesh, Saurav Jyoti Dutta, Subhas Sarma Neog, Prandip Baishya and Indrani Bezbaruah

Application of Fuzzy Sets in Reliability and in Optimal Condition Monitoring Technique Selection in Equipment Maintenance 327
Rajesh S. Prabhu Gaonkar

Fuzzy Logic Based Analysis of Dissolved Decay Contents in Transformer Oil 361
Nitika Ghosh, Vikas Singh Bhadoria, Ram Naresh Sharma and Vivek Shrivastava

Safety Engineering

Probabilistic Safety Assessment in Nuclear and Non-nuclear Facilities: In a Glimpse 389
Gopika Vinod

Passive System Reliability Assessment and Its Integration Into PSA 417
R. B. Solanki

Project Stage Considerations for an Inherently Safe and Reliable Chemical Plant 435
Raghavendra Pai, Suraj Rane and Anusha Pai

Integrated Deterministic and Probabilistic Safety Assessment 455
Durga Rao Karanki

Contributors

Prandip Baishya Mechanical Maintenance Department and Area-in-Charge of Diesel Hydro-Treater Unit of Numaligarh Refinery Limited, Numaligarh, Assam, India

Indrani Bezbaruah Mechanical Engineering Department, Kaziranga University, Jorhat, Assam, India

Vikas Singh Bhadoria Electrical and Electronics Engineering Department, ABES Engineering College, Ghaziabad, India

D. Saravana Bhavan School of Mechanical Engineering, Dayananda Sagar University, Bengaluru, India

Luiz Fernando Capretz Department of Electrical and Computer Engineering, Western University, London, ON, Canada;
Department of Computer Science, New York University, Abu Dhabi, United Arab Emirates

J. Chattopadhyay Reactor Safety Division, Bhabha Atomic Research Centre, Trombay, Mumbai, India

Bimal C. Deka Department of Electrical Engineering, Assam Engineering College, Guwahati, Assam, India

Diana Denice Control Instrumentation Division, Bhabha Atomic Research Centre, Trombay, Mumbai, India

Saurav Jyoti Dutta Mechanical Maintenance Department, Numaligarh Refinery Ltd., Numaligarh, Assam, India

B. G. Fernandes Indian Institute of Technology Bombay, Mumbai, India

Ravi Kumar Gedela Banking Labs, Toronto, Canada

A. K. Ghosh Reactor Safety Division, Bhabha Atomic Research Centre, Trombay, Mumbai, India

Nitika Ghosh Electrical and Electronics Engineering Department, ABES Engineering College, Ghaziabad, India

V. Gopika Reactor Safety Division, Bhabha Atomic Research Centre, Trombay, Mumbai, India

Durga Rao Karanki Siemens Mobility AG, Wallisellen, Switzerland

Rohit Khatri Homi Bhabha National Institute, Mumbai, India

Krishna Mohan Kovur Banking Labs, Toronto, Canada

Manoj Kumar Control Instrumentation Division, Bhabha Atomic Research Centre, Trombay, Mumbai, India

Uday Kumar Division of Operation and Maintenance Engineering, Luleå University of Technology, Luleå, Sweden

Mahesh Kuruba Tata Consultancy Services (TCS), Tokyo, Japan

Øystein Stake Laengen Department of Electric Power Engineering, Norwegian University of Science and Technology (NTNU), Trondheim, Norway

Yadira Lizama Cultureplex Laboratories, Western University, London, ON, Canada

N. Madhusudan Diagnostic Engineers, Bengaluru, India

S. B. Murgayya Department of Automobile Engineering, Dayananda Sagar College of Engineering, Bengaluru, India

Hari Prasad Muruva Bhabha Atomic Research Centre, Trombay, Mumbai, India

Subhas Sarma Neog Predictive Maintenance Department, Numaligarh Refinery Limited, Numaligarh, Assam, India

Anusha Pai Computer Engineering Department, Padre Conceicao College of Engineering, Verna, Goa, India

Raghavendra Pai Project Management Office Lead (Asia Pacific), Syngenta, Corlim, Ilhas, Goa, India

Gayadhar Panda Department of Electrical Engineering, NIT Meghalaya, Shillong, Meghalaya, India

Moushumi Patowary Department of Electrical Engineering, NIT Meghalaya, Shillong, Meghalaya, India

Rajesh S. Prabhu Gaonkar School of Mechanical Sciences, Indian Institute of Technology Goa (IIT Goa), Ponda, Goa, India

P. G. Ramesh Mechanical Engineering Department, The Assam Kaziranga University, Jorhat, Assam, India

Suraj Rane Mechanical Engineering Department, Goa College of Engineering, Farmagudi, Ponda, Goa, India

T. V. Santhosh Reactor Safety Division, Bhabha Atomic Research Centre, Trombay, Mumbai, India

Ram Naresh Sharma Electrical Engineering Department, National Institute of Technology, Hamirpur, Himachal Pradesh, India

Vivek Shrivastava Electrical Engineering Department, National Institute of Technology, Delhi, India

R. B. Solanki Atomic Energy Regulatory Board (AERB), Mumbai, India

Mallika Sreedharan Dar, Pune, India

A. Srividya Luleå University of Technology, Luleå, Sweden;
Western Norway University of Applied Sciences, Haugesund, Norway

H. N. Suresh Department of Automobile Engineering, Dayananda Sagar College of Engineering, Bengaluru, India

Adithya Thaduri Division of Operation and Maintenance Engineering, Luleå University of Technology, Luleå, Sweden

Vijay Venu Vadlamudi Department of Electric Power Engineering, Norwegian University of Science and Technology (NTNU), Trondheim, Norway

Daniel Varona Cultureplex Laboratories, Western University, London, ON, Canada

G. Venkatachalam Mukesh Patel School of Technology Management and Engineering, Mumbai, India

Gopika Vinod Reactor Safety Division, Bhabha Atomic Research Center, Trombay, Mumbai, India

Pradeep Waychal CSIBER (Chhatrapati Shahu Institute of Business Education and Research), Kolhapur, India

Reliability and Availability Engineering

DC and AC Contingency Solvers in Composite Power System Adequacy Assessment

Øystein Stake Laengen and Vijay Venu Vadlamudi

Application of probabilistic methods in Power System Reliability (PSR) studies—both at the generation level (also known as Hierarchical Level (HL) I) and at the composite (generation plus transmission) level (also known as HL II)—is a highly developed field with plenty of available literature. Such application has widespread applications in the planning, operation and management of power systems. Adequacy assessment studies, a subset of PSR studies dealing exclusively with static conditions of a power system, in their simplest form, deal with the evaluation of whether the generation capacity of a system is sufficient to supply the load requirement of the system. Other considerations can also be included in the assessment, such as whether the transmission and distribution facilities of the system can provide sufficient energy transportation from the generating facilities to the end consumers. Methodologies—both analytical and simulation-based—for the quantification of power system adequacy through indices such as Loss of Load Expectation (LOLE) and Expected Energy Not Served (EENS), are well established. However, there is a much felt need for more transparency and pedagogical clarity in the exposition of methodologies for assessing power system adequacy at the composite level. More importantly, details surrounding the contingency solvers for composite system state evaluation need elaboration so that it is easier to replicate the results of research works related to the assessment of power system reliability. This chapter elaborates on both DC and AC contingency solvers used in composite power system adequacy studies.

Ø. S. Laengen · V. V. Vadlamudi (✉)
Department of Electric Power Engineering, Norwegian University of Science and Technology (NTNU), Trondheim, Norway
e-mail: vijay.vadlamudi@ntnu.no

D. R. Karanki et al. (eds.), *Advances in RAMS Engineering*,
Springer Series in Reliability Engineering,
https://doi.org/10.1007/978-3-030-36518-9_1

1 Composite System Adequacy

In the adequacy assessment studies at the composite level of a power system, simply termed as composite power system adequacy assessment [1–7], there is a need for a model representing the network topology in addition to the load and generation models. The evaluation at HLI is limited to a comparison of the generation capacity against the load requirement, while the composite system assessment depends on a load flow based analysis. Therefore, a choice should be made regarding the desired accuracy of the load flow analysis. A simplified approach is to use a DC-based load flow analysis, but if more accurate results are desired, a full AC-based load flow analysis should be used. Whichever the analysis of choice, additional network data[1] must be supplied as input. More specific generation and load data are also needed where allocation of generation capacity and load requirement among the buses is specified.

Both analytical methods and Monte Carlo Simulation (MCS) methods include the step of the evaluation of system states; the selection of system states is done by some suitable contingency selection criterion [1–4] in the analytical methods, whereas in the case of MCS methods, a suitable sampling method is used. If considering the MCS, the two popularly used methods are state sampling method and state transition sampling method.

The system states are evaluated by the use of two different contingency solvers, one based on DC-load flow analysis and another based on AC-load flow analysis (Note that these are all easily implementable in MATLAB). The proposed methodology of constructing the solvers is presented in the following sections, with a section describing the general parts that are similar for both the DC and AC solvers. Thereafter, both contingency solvers are presented in detail. The explanation for solvers is given with the consideration of the selection of system states using MCS sampling methods. For fundamental details of the two popular sampling methods, the reader is referred to [3, 7].

1.1 General Elements

There are some general elements common for the two contingency solver approaches, AC and DC, proposed in this chapter. Both methods depend on input data in a specific format. For example, through MCS sampling the system state of a given time increment is obtained and handed over to the contingency solver for further evaluation. Thus, a specific format of input data to solver, e.g., the system state is needed.

[1]Network topology, impedances and current limits of the lines.

1.1.1 Input Data

A representation of the network data format used by the MCS state sampling method for full AC analysis is shown in Table 1. Per unit (p.u.) system of units is used for line parameters and line limits. If the load flow analysis is based on DC, the columns of the resistances and shunt susceptances are left out. The MCS state transition method uses a similar format except that the Forced Outage Rate (FOR) column is replaced by two columns, one for the failure rates and one for the repair rates.

For the AC based analysis to be conducted, reactive power capabilities of the generators must be added to the generator input data. Thus, the specified format used by the MCS state sampling method is presented in Table 2, where the minimum and maximum values of the reactive power capability of each generator are specified. If the MCS state transition method is used, the FOR column is replaced by two transition rate columns. The DC based analysis uses similar input, without the columns of reactive power capabilities.

An additional table specifying the specific bus data of the system is also included as input. Both the AC and the DC solvers need one column specifying the allocation of loads in the system and one column where the cost of load curtailments at each bus are specified. For the AC- analysis, the minimum and maximum voltage limits of the buses are included as well. The format used by the AC solver is presented in Table 3.

1.1.2 System State

A system state of a time increment is sampled through MCS sampling, giving the states of the generators and lines as two vectors. Each component state is given by a

Table 1 Line input data, state sampling AC-solver

Line	From Bus	To Bus	FOR	Resistance [p.u.]	Reactance [p.u.]	Half of shunt susceptance [p.u.]	Current limit [p. u.]
1	1	2	FOR_1	R_{12}	X_{12}	y_{10}	$Ilim_1$
2	2	3	FOR_2	R_{23}	X_{23}	y_{20}	$Ilim_2$
n	5	6	FOR_n	R_{56}	X_{56}	y_{50}	$Ilim_n$

Table 2 Generator input data, state sampling AC-solver

Generator	Capacity [MW]	Bus #	Min reactive [MVAr]	Max reactive [MVAr]	FOR
1	$P_{cap,1}$	N_1	$Qmin_1$	$Qmax_1$	FOR_1
2	$P_{cap,2}$	N_2	$Qmin_2$	$Qmax_2$	FOR_2
n	$P_{cap,n}$	N_n	$Qmin_n$	$Qmax_n$	FOR_n

Table 3 Bus specification, AC-solver

Bus	Share of load	Vmin [pu]	Vmax [pu]	Interruption cost [$/kWh]
1	$Load_1$	$Vmin_1$	$Vmax_1$	$Cost_1$
2	$Load_2$	$Vmin_2$	$Vmax_2$	$Cost_2$
n	$Load_n$	$Vmin_n$	$Vmax_n$	$Cost_n$

binary value [0, 1], termed X_i, where a value of zero denotes an available component and a value of one denotes that the component is unavailable. Thus, the vectors giving the states of the n generators and the m lines are of formats presented in the following.

$$P_g = [X_1 \quad X_2 \quad \ldots \quad X_n]^T \tag{1}$$

$$L = [X_1 \quad X_2 \quad \ldots \quad X_m]^T \tag{2}$$

Due to the allocation of generators at different buses of the system, the generator capacities at each of the k buses can be combined, according to their state given from (1) and their rated capacity, to give the bus generation capacities. Thus, the system's generation capacity can be represented by a generation capacity vector with k elements for the DC approach.

$$P_{g,lim} = [P_{cap,1} \quad P_{cap,2} \quad \ldots \quad P_{cap,k}]^T \tag{3}$$

For the AC approach, the generation capacity vector is extended to a matrix with three columns, which give the respective active power, minimum reactive power and maximum reactive power capabilities of the buses.

$$G_{lim} = \begin{bmatrix} P_{cap,1} & Q_{min,1} & Q_{max,1} \\ P_{cap,2} & Q_{min,2} & Q_{max,2} \\ \vdots & \vdots & \vdots \\ P_{cap,k} & Q_{min,k} & Q_{max,k} \end{bmatrix} \tag{4}$$

In addition, the load requirement allocation among the k buses of a system is needed. Hence, the active power load requirement is represented by a load vector of k elements for the DC approach.

$$P_{load} = [P_{load,1} \quad P_{load,2} \quad \ldots \quad P_{load,k}]^T \tag{5}$$

For the AC approach, the load requirement vector is extended with an additional column giving the reactive power requirement of the loads at the buses.

$$P_{load} = \begin{bmatrix} P_{load,1} & P_{load,2} & \cdots & P_{load,k} \\ Q_{load,1} & Q_{load,2} & \cdots & Q_{load,k} \end{bmatrix}^T \quad (6)$$

1.1.3 Isolated Buses

A possible situation that might arise during the selection of system states is the occurrence of multiple lines on outage at the same time. This might lead to isolation of one or more buses or parts of the system being islanded, depending on the number of outages and where they occur. If the developed load flow analysis tools lack a part that detects and handles the isolation of buses properly, the load flow problem becomes infeasible, due to the inclusion of isolated buses in the matrix representing the system, i.e., bus admittance matrix, Ybus. Among the problems encountered can be the nonexistence of an inverse admittance matrix. A point worth noting regarding the development of a suitable algorithm, is that a system of small size does not necessarily reveal the limitations of a proposed algorithm, which might become evident only when the approach is applied to a test system of larger size. The decision strategy on how to handle possible isolation of buses used by the *contingency solvers* presented is presented in the following:

Step 1: When the state of a line is sampled as a failure, i.e., outage, the admittance of the line is set to zero.

Step 2: When an isolated bus is identified, the elements corresponding to that bus are removed from the matrices and vectors representing the system. This step ensures that the optimal power flow (OPF) problem remains solvable.

Step 3: After the identification of isolated buses and the subsequent matrix modifications, the load curtailments due to the isolation of buses are given according to the following criteria:

(a) The slack bus of the system, i.e. bus 1[2] is the only bus able to operate in islanded mode.

    (i) If the slack bus is isolated from the rest of the system, all loads are shed in the system except from the loads at the slack bus.
    (ii) The generators at an isolated bus are not able to provide the load requirement at the bus.

[2]The slack bus of the system must always be bus number 1 for the approach to be valid. If the slack bus of the system has a different original number, the buses of the system must be given new numbers.

Table 4 The susceptance matrix of the RBTS

11	0	−11	0	0	0
0	3	0	−3	0	0
−11	0	19	0	−8	0
0	−3	0	3	0	0
0	0	−8	0	17	−8
0	0	0	0	−8	8

1.1.4 Identification of Isolated Buses

The Roy Billinton Test System (RBTS) [8], is used as a simple example to show how isolated buses and possible islands can be identified through inspection of either the conductance or susceptance matrix. A case of islanding, i.e., isolation of bus 2 and bus 4 from the rest of the system, occurs if lines L3, L4 and L8 of the RBTS are on outage. The isolation can be identified by looking at the system's susceptance matrix presented in Table 4.

When investigating row two of the susceptance matrix, it can be identified that bus 2 has no connection to bus 1, but only a connection to bus 4. Further investigation of row 4 in the matrix reveals that bus 4 has a connection to bus 2 only. The other four buses of the system are interconnected. Based on the above observation, a simple approach is presented:

Step 1: Check for left-connectivity, i.e., examine whether the bus under consideration is connected to a bus with a lower number.

(a) Iterate from bus number 2 to the last bus [2, k].

  (i) For each bus under consideration, examine whether it is connected to a bus with a lower number.

  - If no, flag the bus under consideration as 'isolated'.
  - If yes, also check if the bus to which the bus under consideration is connected has already been marked as 'isolated'.
    - If yes, flag the bus under consideration as 'isolated'.

Step 2: Check the right-connectivity, i.e., examine whether the bus under consideration is connected to a bus with a higher number.

(a) Iterate from the last bus to bus number 2 [k, 2].

  (i) For each bus under consideration flagged as isolated from Step 1, examine whether it is connected to a bus with a higher number that is not flagged as isolated. If yes, the 'isolated' flag is removed for the bus under consideration. If no, the 'isolated' flag is retained.

However, the approach is found to be insufficient for more complicated system configurations, thus representing a possible pitfall. Even for the RBTS system, a

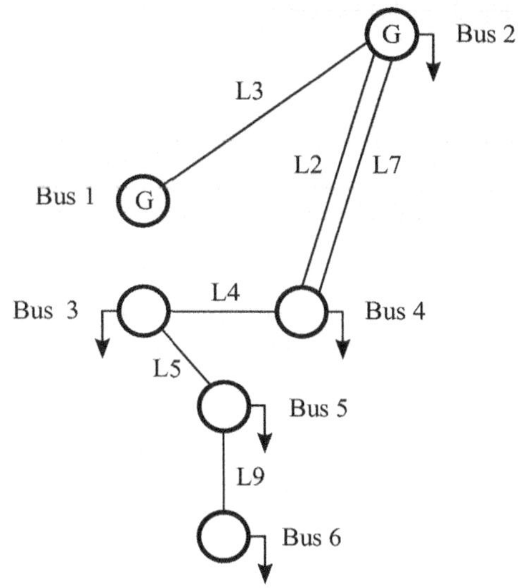

Fig. 1 The RBTS special case with outages

Table 5 Intermediate flags, RBTS case

Bus	Flag after step 1	Flag after step 2	Result
1	0	0	Not isolated
2	0	0	Not isolated
3	1	0	Not isolated
4	0	0	Not isolated
5	1	1	Isolated
6	1	1	Isolated

system of relatively small size, an error occurs if the approach is applied to the system configuration of Fig. 1 where lines L1, L6 and L8 are on outage.

If a thorough inspection of the algorithm's steps applied to the case of Fig. 1 is performed, it becomes clear that the algorithm marks buses 5 and 6 incorrectly as isolated. The intermediate results obtained by applying the algorithm step by step are presented in Table 5, where a one denotes an isolation flag.

When lines L1, L6 and L8 are on outage, a visual inspection of Fig. 1 shows that none of the buses is isolated. However, the algorithm has resulted in the incorrect isolation of buses 5 and 6. Thus, a limitation of the above presented bus isolation algorithm is revealed. Such a limitation is encountered when the outages of lines lead to the creation of new radials containing buses numbered in no particular sequence. An illustration of such a configuration where the bus isolation algorithm might encounter difficulties, is presented in Fig. 2, which is a sample illustration of a 14 bus system with buses (indicated as dots) numbered as B1, B2, B3,...... B14.

Based on the special configuration of the RBTS and the radial example of Fig. 2, a new more detailed algorithmic approach is proposed, with basis in the first suggested algorithm:

Fig. 2 Radial example

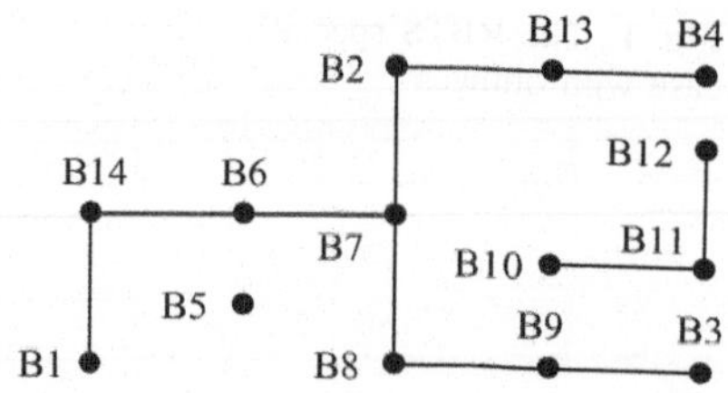

Step 1: Check the left-connectivity, i.e., examine whether the bus under consideration is connected to a bus with a lower number.

(a) Iterate from bus number 2 to the last bus [2, k].

  (i) For each bus under consideration, examine whether it is connected to a bus with a lower number.

   - If no, flag the bus under consideration as 'isolated'.
   - If yes, also check if the bus to which the bus under consideration is connected has already been marked as 'isolated'.

     – If yes, flag the bus under consideration as 'isolated'.
     – If no, check whether the bus under consideration is directly connected to any of the other buses with lower numbers that have already been flagged as isolated.

     If yes, clear the flags of all the buses that are directly connected to the bus under consideration.

Step 2: Examine whether the bus under consideration is connected to any other another bus. To ensure that none of the buses are incorrectly marked as isolated after finalization of step 2, the step is started over again if a special combination occurs; the bus is cleared from its isolation flag and leads to the clearing of additional buses' flags.

(a) Iterate from the last bus to bus number 2 [k, 2], iterator m.

  (i) For each bus under consideration flagged as isolated, examine whether it is connected to another bus.
  (ii) If a connection is found, clear the flag of bus m.

   (1) Iterate from bus number 2 to bus k [2, k], iterator n.

    - If bus n is flagged as isolated and a connection to bus m exists, clear the flag of bus n.

   (2) If one or more flags are cleared during the loop by iterator n, Step 2 is restarted from the beginning with iterator m starting from the last bus.

The importance of including a restarting of the algorithm's Step 2 becomes clear by applying the new suggested algorithm step by step on the radial example of

Table 6 Radial example for identification of isolated buses

Bus	Step 1	Step 2 # stage 1	Step 2 # stage 2	Step 2 # stage 3	Step 2 # stage 4
1	0				0
2	1	(0)			0
3	1			(0)	0
4	1		(0)		0
5	1				1
6	1 (0)				0
7	1	0			0
8	1	1 (0)			0
9	1	1		0	0
10	1	1		1	1
11	1	1		1	1
12	1	1		1	1
13	1	1	0	0	0
14	0	0	0	0	0

Fig. 2. In Table 6, the obtained intermediate results are presented. The parentheses surrounding some of the numbers indicate a clearing of a flag in the inner loop of the algorithm, i.e., point a.(ii).(1) in Step 2 described above. Under the column 'Step 1' in Table 6, if an entry is '1', it denotes an isolation flag; if one starts with an entry 1 for a bus, and ends with entry '0' in the last stage (i.e., stage 4) of Step 2 due to the application of the algorithm above, it means that the bus that was flagged as isolated in Step 1 is deemed as 'not isolated' by the time Step 2 is finished.

1.2 *DC—Contingency Solver*

The considerations of the DC contingency solver are presented in this section. First, an introduction to the approximations and equations used to represent the system is made, before the OPF problem is formulated. The solver is then tested on a selection of system states, before a final example illustrating the details is given.

1.2.1 Network Model

In the DC based approach, the network is represented by the DC power flow formulations and approximations found in most available power system analysis text books, for example [9]. By using the assumptions of DC-power flow, it is possible to formulate the power flows through the lines as linear functions of the net power injections at the buses. The assumptions of the DC-power flow formulations are listed in the following:

(i) The resistance of a line is much smaller than its reactance ($r_{ij} << x_{ij}$).
(ii) The difference in voltage phasor angle between two interconnected buses is small. Thus, two reasonable approximations are to set the *sin* of the difference in phasor angle equal to the difference and the *cos* of the same difference equal to one[3] ($\sin \delta_{ij} = \delta_{ij}$ and $\cos \delta_{ij} = 1$).
(iii) The lines' susceptances to earth are neglected ($b_{io} = 0$ and $b_{jo} = 0$).
(iv) The voltages are fixed at a magnitude of one p.u., ($V_i = 1$).

By using the stated assumptions, the power flow equations are simplified to expressions in terms of the lines' susceptances and the differences in voltage phasor angles between the buses.

$$P_i = \sum_{j=1}^{k} B_{ij}\delta_{ij} \text{ where } \delta_{ij} = \delta_i - \delta_j \tag{7}$$

where the susceptance elements, B_{ij} and $B_{ii,}$ are defined according to.

$$B_{ij} = b_{ij} = -\frac{1}{X_{ij}} \text{ and } B_{ii} = -\sum_{j=1, j\neq i}^{k} b_{ij} \tag{8}$$

The formulation of (7) is rewritten to matrix notation in (9), where the net power injection vector is expressed in terms of the susceptance matrix and the column vector of voltage phasor angles δ.

$$[P] = [B] \cdot [\delta] \tag{9}$$

A general view of the elements in the susceptance matric, B, is given in (10).

$$B = \begin{bmatrix} B_{11} & B_{12} & \cdot & B_{1k} \\ B_{21} & B_{22} & \cdot & B_{2k} \\ \cdot & \cdot & \cdot & \cdot \\ B_{k1} & B_{k2} & \cdot & B_{kk} \end{bmatrix} \tag{10}$$

The set of linear equations is singular, since one of the rows could be expressed as the linear combination of the other rows. To overcome the problem, the concept of a slack bus is introduced. The implication is that the row and vector elements corresponding to the slack bus, are removed from the susceptance matrix, giving a sub matrix $\mathbf{B'}$. If the first bus is chosen as the slack bus, the corresponding sub matrix is defined by (11).

[3]The angles must be expressed in radians.

$$B' = \begin{bmatrix} B_{22} & \cdot & B_{2k} \\ \cdot & \cdot & \cdot \\ B_{k2} & \cdot & B_{kk} \end{bmatrix} \tag{11}$$

Since the net power injection of the slack bus could be expressed as the linear combination of power injections at the other buses of the system, a new set of equations becomes:

$$[P] = [B'] \cdot [\delta] \tag{12}$$

$$P_{slack} = \sum_{j=1, j \neq slack}^{k} P_j \tag{13}$$

The implication is that the slack bus compensates the surplus or deficit of generation in the system. A further development of (12) give the voltage phasor angles expressed in terms of the inverse of the sub matrix $\mathbf{B'}$ and the net power injections. To include the slack bus into the expression again, the row and the column of the slack bus are reintroduced in the inverse sub matrix B' with values of zero, naming the new matrix $\mathbf{Z}$.

$$[\delta] = \begin{bmatrix} 0 & 0 & \cdots & 0 \\ 0 & & & \\ \vdots & & [B']^{-1} & \\ 0 & & & \end{bmatrix} \cdot [P] = [Z] \cdot [P] \tag{14}$$

The active power flow through a line between bus i and j is further given as the difference in voltage phasor angles, between the two connected buses, divided by the reactance of the line.

$$P_{ij} = \frac{\delta_i - \delta_j}{X_{ij}} \tag{15}$$

If the above notations of (14) are used, each voltage phasor angle of (15) could be expressed as a row of the $\mathbf{Z}$ matrix times the net power injection vector. Thus, the power flow through the line between bus i and j could be reformulated:

$$P_{ij} = \frac{Z_{i1} - Z_{j1}}{X_{ij}} P_1 + \frac{Z_{i2} - Z_{j2}}{X_{ij}} P_2 + \ldots + \frac{Z_{ik} - Z_{jk}}{X_{ij}} P_k \tag{16}$$

Rewriting the expression of (16) to an expression in terms of the power distribution factors, denoted $a_{ij,k}$, gives the simplified format of the line flow:

$$P_{ij} = a_{ij,1}P_1 + a_{ij,2}P_2 + \ldots + a_{ij,k}P_k \tag{17}$$

When the rest of the system's line flow equations are developed similarly, the line flows are expressed in terms of the sensitivity matrix, **A**, times the net bus injection vector, **P**.

$$[T_P] = [A] \cdot [P] \tag{18}$$

The net injection vector, **P**, is defined in terms of the vector of load requirements subtracted from the vector of generation. The load requirements are considered as constants in the analysis. If the loads need to be reduced to maintain the power balance, it is done by curtailing loads at the load buses. Thus, an additional load curtailment vector, C_p, is introduced into the net injection vector of (19).

$$[P] = [P_g] + [C_P] - [P_{load}] \tag{19}$$

1.2.2 Contingency Solver Description

The network model of the system by a DC power flow representation simplifies the constraints of the load flow problem. In DC power flow, only the active power is accounted for while the losses are ignored. Hence, the list of the constraints that are considered by the contingency solver is presented in the following:

(i) The system's generation capacity must be larger than or equal to the load requirement.
(ii) The power flow through the lines are limited by their power ratings.
(iii) The load curtailment at a bus cannot exceed the load requirement of the bus.
(iv) The actual generation at a bus cannot exceed the generation capacity of the bus.
(v) The load curtailment at a bus cannot be negative.
(vi) The actual generation at a bus cannot be negative.

When a system state is handed over to the "contingency solver", the solver tries to find a feasible operating state without violating any of the above listed constraints. If a violation of a constraint is present, a measure or a combination of measures is taken to restore the system back into a feasible operating point. First, the solver tries to reschedule the generation, but if the action is insufficient in removing the violation, it will be necessary to try load curtailments at one or more buses. The order of the possible actions is controlled by introducing an objective function with differing costs of rescheduling of generation and load curtailments. If the costs of load curtailments are set higher than the costs of rescheduling of generation, the solver will ensure that rescheduling of generation is tried before load curtailments are considered. By differing the costs of load curtailments at the buses, it is possible to make a prioritized list of the loads, where the loads at the bus with the lowest cost of curtailment are curtailed first [10]. Hence, the costs of the possible actions can be expressed by a row vector of 2 times k elements, where k is given as the number of buses in the system.

$$W = \begin{bmatrix} w_1 & w_2 & \dots & w_{2k} \end{bmatrix} \tag{20}$$

The first k elements of the cost vector correspond to the cost of rescheduling of generation, while the next k elements correspond to the cost of load curtailments. In this chapter, the costs of rescheduling of generation are set equal to zero, while the costs of load curtailments are set according to the specification in the input data. Each cost element has a corresponding decision variable, which is optimized with the goal of minimizing the overall cost. The set of decision variables are represented by a column vector, where the first k elements give the generation, P_{gi}, at the buses, while the next k elements give the load curtailments, C_i, at the buses.

$$X = \begin{bmatrix} P_{g1} & \dots & P_{gk} & C_1 & \dots & C_k \end{bmatrix}^T \tag{21}$$

1.2.3 Optimal Power Flow Formulation

After representing the system by matrices and vectors, the final step is to formulate the problem as an OPF problem. A general representation of such an OPF problem is given below in accordance with the formulation presented in [3] where the objective is to minimize the load curtailments.

$$\text{Min}\, f = \sum_{i=1}^{k} C_i \tag{22}$$

$$\sum_{i=1}^{k} P_{gi} + \sum_{i=1}^{k} C_i = \sum_{i=1}^{k} P_{load,i} \tag{23}$$

$$|[A] \cdot [P]| \leq [T_{Lim}] \tag{24}$$

$$0 \leq P_{gi} \leq P_{cap,i} \tag{25}$$

$$0 \leq C_i \leq P_{load,i} \tag{26}$$

The formulation needs some minor modifications to be in a format suited for solving with the dual simplex method of the built-in "linprog" function of MATLAB. Using the above presented vector notations, the objective function of (22) is defined as the cost vector times the vector of decision variables.

$$\text{Min f} = [W] \cdot [X] \tag{27}$$

The equality constraint of (23), which stipulates that the sum of the system's power generation and load curtailment must equal the load requirement, is modified to (28). In the equation, K is a row vector of 2 times k elements with values of one. Thus, an expression in terms of the decision variables is obtained.

$$[K] \cdot [X] = \sum_{i=1}^{k} P_{load,i} \tag{28}$$

The inequality constraint of (24), limiting the power flows through the lines, must be converted to two inequality constraints to remove the absolute value sign from the equation. The absolute value sign in the equation is needed because the multiplication of A and P can give negative values. When using the fact that the load requirements are constant, the sensitivity matrix times the load requirement vector could be moved over to the constant side of the inequality. Thus the new inequalities are given in terms of the decision variables as shown below.

$$[A \quad A] \cdot [X] \le [T_{lim}] + [A] \cdot [P_{load}] = [T_{lim1}] \tag{29}$$

$$-[A \quad A] \cdot [X] \le [T_{lim}] - [A] \cdot [P_{load}] = [T_{lim2}] \tag{30}$$

1.2.4 Contingency Solver Verification

A few selected system states, using the RBTS test system at peak load of 185 MW, are tested in the contingency solver, to verify that the proposed OPF methodology of the DC-contingency solver gives valid results. The load requirement of the system is the same and divided according to Table 7 in all the test cases.

Case 1

The first test is performed to check that there are no erroneous load curtailments when there are no outages of the components. When this is the case, the generation capacity of buses one and two are given by their installed capacity, cf. Table 8 As

Table 7 The RBTS load distribution at a peak load of 185 MW

Bus	Load demand [MW]
1	0
2	20
3	85
4	40
5	20
6	20
Sum:	185

Table 8 The installed generation capacity of the RBTS

Bus	Active power [MW]
1	110
2	130

Table 9 The results of case 1—DC solver

Bus	Generation [MW]	Curtailment [MW]
1	55.00	0.00
2	130.00	0.00
3	0.00	0.00
4	0.00	0.00
5	0.00	0.00
6	0.00	0.00
Sum:	185.00	0.00

Table 10 The generation capacity of case 2—DC solver

Bus	Active power [MW]
1	100
2	50

Table 11 The results of case 2—DC solver

Bus	Generation [MW]	Curtailment [MW]
1	100.00	0.00
2	50.00	0.00
3	0.00	35.00
4	0.00	0.00
5	0.00	0.00
6	0.00	0.00
Sum:	150.00	35.00

can be seen in Table 9, the contingency solver returns no erroneous load curtailments.

Case 2

The solver is also tested to check whether the load curtailments are correct when the generation capacity of the system is insufficient. A specification of the generation capacity is presented in Table 10. All the lines are available during the test. In Table 11, a total load curtailment of 35 MW is the result, making up for the deficit of generation capacity.

Table 12 The results of case 3—DC solver

Bus	Generation [MW]	Curtailment [MW]
1	110.00	0.00
2	0.00	20.00
3	0.00	15.00
4	0.00	40.00
5	0.00	0.00
6	0.00	0.00
Sum:	110.00	75.00

Table 13 The results of case 4—DC solver

Bus	Generation [MW]	Curtailment [MW]
1	15.00	0.00
2	130.00	0.00
3	0.00	0.00
4	0.00	0.00
5	0.00	20.00
6	0.00	20.00
Sum:	145.00	40.00

Table 14 The generation capacity of case 5—DC solver

Bus	Active power [MW]
1	20
2	130

Case 3

The generation capacity of the system is equal to the installed capacity, specified in Table 8. Buses 2 and 4 are isolated from the rest of the system, due to outages of lines L3, L4 and L8. The stated assumption that only the slack bus can operate in islanded mode, gives load curtailment equal to the load requirements at buses 2 and 4. In addition, the generation capacity at bus 1 is insufficient to meet the total load requirement of buses 3, 5 and 6. The complete load curtailments are presented in Table 12.

Case 4

The generation capacity of the system is still equal to the installed capacity, given Table 8, but this time there are outages of lines L5 and L8, leading to the islanding of buses 5 and 6. The load curtailments of this case are presented Table 13, and are found to be equal to the load requirements at the isolated buses.

Case 5

In this case, a combination of generator and line outages occurs at the same time. The generation capacity is reduced according to the specification in Table 14, while line L2 and L7 are on outage. Here, the power transfer limit through line L3 is the constraint that limits the active power generation at bus 2. The resulting load

Table 15 The results of case 5—DC solver

Bus	Generation [MW]	Curtailment [MW]
1	20.00	0.00
2	91.00	0.00
3	0.00	34.00
4	0.00	0.00
5	0.00	20.00
6	0.00	20.00
Sum:	111.00	74.00

curtailments of 74 MW are given in Table 15. It can be observed that the generation of bus 2 is limited to 91 MW for this case, which shows that the line flow constraint of L3 is binding.

1.2.5 Illustrative Example

The following example is presented to illustrate the basic principles of the composite system assessment methodology in a pedagogical way, where the step by step numerical calculations are included. The test system of the example is shown in Fig. 3. It consists of 3 buses, where bus 1 is a generator bus with two connected generators, and buses 2 and 3 are load buses with constant load requirements. A network of three lines, all with equal transfer limits but different reactances, interconnect the buses. The system data is summarized in Tables 16, 17 and 18.

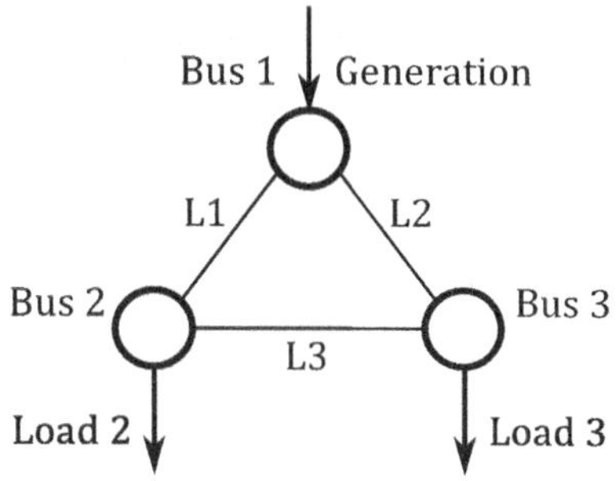

Fig. 3 The 3-bus test system

Table 16 The generator data of the 3-bus test system

Generator	Bus	Capacity [MW]	FOR
G1	1	50	0.01
G2	1	50	0

Table 17 The loads of the 3-bus test system

Bus	Load [MW]	Cost of load curtailment [$/kWh]
1	0	–
2	30	1
3	40	1

Table 18 The network data of the 3-bus test system

Line	Reactance [pu]	Transfer limit [MW]	FOR
L1	0.1	50	0.01
L2	0.2	50	0
L3	0.2	50	0

The system is configured to have a limited number of possible contingencies, to have an example of limited size. It is accomplished by having only two components, a generator and a line, with FOR values larger than zero. Thus, the other components are assumed to be available 100% of the time. The power base of the system is 100 MVA.

Step by Step Calculations

The step by step numerical calculations performed when the equations are set up according to the presented methodology, are shown for two system states. In this section, the base case where all the components are available, is shown.

Step 1: Obtain the elements of the sub matrix, $\mathbf{B'}$, of the susceptance matrix using (8).

(a) Calculate the susceptance of the three lines:

$$\begin{aligned} b_{12} = b_{21} &= -\frac{1}{0.1} = -10 \\ b_{13} = b_{31} &= -\frac{1}{0.2} = -5 \\ b_{23} = b_{32} &= -\frac{1}{0.2} = -5 \end{aligned} \tag{31}$$

(b) Calculate the elements of the $\mathbf{B}$ matrix:

$$
\begin{aligned}
B_{11} &= -(b_{12}+b_{13}) = 10+5 = 15\\
B_{12} &= B_{21} = b_{12} = -10\\
B_{13} &= B_{31} = b_{13} = -5\\
B_{22} &= -(b_{21}+b_{23}) = 10+5 = 15\\
B_{23} &= B_{32} = b_{23} = -5\\
B_{33} &= -(b_{31}+b_{32}) = 5+5 = 10
\end{aligned} \tag{32}
$$

(c) Create the **B** matrix:

$$
B = \begin{bmatrix} 15 & -10 & -5 \\ -10 & 15 & -5 \\ -5 & -5 & 10 \end{bmatrix} \tag{33}
$$

(d) Bus 1 is chosen as the slack bus, giving the sub matrix **B′**:

$$
B' = \begin{bmatrix} 15 & -5 \\ -5 & 10 \end{bmatrix} \tag{34}
$$

Step 2: When the sub matrix **B′** is established, the next step is to obtain the **Z** matrix of (14)

(a) Calculate the inverse of the **B′** matrix

$$
[B']^{-1} = \begin{bmatrix} 0.08 & 0.04 \\ 0.04 & 0.12 \end{bmatrix} \tag{35}
$$

(b) Use the notation of to obtain **Z** with 3 × 3 elements:

$$
Z = \begin{bmatrix} 0 & 0 & 0 \\ 0 & 0.08 & 0.04 \\ 0 & 0.04 & 0.12 \end{bmatrix} \tag{36}
$$

Step 3: Establish the power transfer distribution factor matrix, **A**, of (18).

(a) The factors are obtained for each line by use of (16).

(i) First, the factors of L1 are calculated:

$$\begin{aligned} P_{12} &= \frac{z_{11} - z_{21}}{X_{12}} \cdot P_1 + \frac{z_{12} - z_{22}}{X_{12}} \cdot P_2 + \frac{z_{13} - z_{23}}{X_{12}} \cdot P_3 \\ &= \frac{0 - 0}{0.1} \cdot P_1 + \frac{0 - 0.08}{0.1} \cdot P_2 + \frac{0 - 0.04}{0.1} \cdot P_3 \\ &= 0 \cdot P_1 - 0.8 \cdot P_2 - 0.4 \cdot P_3 \end{aligned} \tag{37}$$

(ii) Similarly, the factors of L2 are calculated:

$$\begin{aligned} P_{13} &= \frac{z_{11} - z_{31}}{X_{13}} \cdot P_1 + \frac{z_{12} - z_{32}}{X_{13}} \cdot P_2 + \frac{z_{13} - z_{33}}{X_{13}} \cdot P_3 \\ &= \frac{0 - 0}{0.2} \cdot P_1 + \frac{0 - 0.04}{0.2} \cdot P_2 + \frac{0 - 0.12}{0.2} \cdot P_3 \\ &= 0 \cdot P_1 - 0.2 \cdot P_2 - 0.6 \cdot P_3 \end{aligned} \tag{38}$$

(iii) Finally, the factors of L3 are calculated:

$$\begin{aligned} P_{23} &= \frac{z_{21} - z_{31}}{X_{23}} \cdot P_1 + \frac{z_{22} - z_{32}}{X_{23}} \cdot P_2 + \frac{z_{23} - z_{33}}{X_{23}} \cdot P_3 \\ &= \frac{0 - 0}{0.2} \cdot P_1 + \frac{0.08 - 0.04}{0.2} \cdot P_2 + \frac{0.04 - 0.12}{0.2} \cdot P_3 \\ &= 0 \cdot P_1 + 0.2 \cdot P_2 - 0.4 \cdot P_3 \end{aligned} \tag{39}$$

(b) Then, the sensitivity matrix is built with the power transfer distribution factors:

$$A = \begin{bmatrix} 0 & -0.8 & -0.4 \\ 0 & -0.2 & -0.6 \\ 0 & 0.2 & -04 \end{bmatrix} \tag{40}$$

Step 4: Modify the line's power flow limits according to (29) and (30), by using the load data in Table 17 and the line limits in Table 18.

(a) Obtain the elements of the $\mathbf{T_{lim1}}$ vector:

$$\begin{aligned} T_{Line1} &= T_{lim} + 0 \cdot P_{L1} - 0.8 \cdot P_{L2} - 0.4 \cdot P_{L3} \\ &= 0.5 + 0 \cdot 0 - 0.8 \cdot 0.3 - 0.4 \cdot 0.4 = 0.1 \\ T_{Line2} &= T_{lim} + 0 \cdot P_{L1} - 0.2 \cdot P_{L2} - 0.6 \cdot P_{L3} \\ &= 0.5 + 0 \cdot 0 - 0.2 \cdot 0.3 - 0.6 \cdot 0.4 = 0.2 \\ T_{Line3} &= T_{lim} + 0 \cdot P_{L1} + 0.2 \cdot P_{L2} - 0.4 \cdot P_{L3} \\ &= 0.5 + 0 \cdot 0 + 0.2 \cdot 0.3 - 0.4 \cdot 0.4 = 0.4 \end{aligned} \tag{41}$$

(b) Obtain the elements of the $\mathbf{T_{lim2}}$ vector:

$$\begin{aligned} T_{Line1} &= T_{lim} - 0 \cdot P_{L1} + 0.8 \cdot P_{L2} + 0.4 \cdot P_{L3} \\ &= 0.5 - 0 \cdot 0 + 0.8 \cdot 0.3 + 0.4 \cdot 0.4 = 0.9 \\ T_{Line2} &= T_{lim} - 0 \cdot P_{L1} + 0.2 \cdot P_{L2} + 0.6 \cdot P_{L3} \\ &= 0.5 - 0 \cdot 0 + 0.2 \cdot 0.3 + 0.6 \cdot 0.4 = 0.8 \\ T_{Line3} &= T_{lim} - 0 \cdot P_{L1} - 0.2 \cdot P_{L2} + 0.4 \cdot P_{L3} \\ &= 0.5 - 0 \cdot 0 - 0.2 \cdot 0.3 + 0.4 \cdot 0.4 = 0.6 \end{aligned} \tag{42}$$

(c) The new limits in vector form are given by:

$$T_{lim,1} = \begin{bmatrix} 0.1 \\ 0.2 \\ 0.4 \end{bmatrix} \text{ and } T_{lim,2} = \begin{bmatrix} 0.9 \\ 0.8 \\ 0.6 \end{bmatrix} \tag{43}$$

Step 5: The final formulation of the OPF problem is formulated by reducing the number of decision variables from 6 to 3, recognizing the non-existence of generation at buses 2 and 3 and load requirement at bus 1. Thus, the final OPF formulation becomes:

$$\text{Min f} = 0 \cdot P_{g1} + 1 \cdot C_2 + 1 \cdot C_3 \tag{44}$$

$$1 \cdot P_{g1} + 1 \cdot C_2 + 1 \cdot C_3 = 0.7 \tag{45}$$

$$\begin{aligned} 0 \cdot P_{g1} - 0.8 \cdot C_2 - 0.4 \cdot C_3 &\leq 0.1 \\ 0 \cdot P_{g1} - 0.2 \cdot C_2 - 0.6 \cdot C_3 &\leq 0.2 \\ 0 \cdot P_{g1} + 0.2 \cdot C_2 - 0.4 \cdot C_3 &\leq 0.4 \end{aligned} \tag{46}$$

$$\begin{aligned} 0 \cdot P_{g1} + 0.8 \cdot C_2 + 0.4 \cdot C_3 &\leq 0.9 \\ 0 \cdot P_{g1} + 0.2 \cdot C_2 + 0.6 \cdot C_3 &\leq 0.8 \\ 0 \cdot P_{g1} - 0.2 \cdot C_2 + 0.4 \cdot C_3 &\leq 0.6 \end{aligned} \tag{47}$$

$$\begin{aligned} P_{g1} &\leq 1.0 \\ C_2 &\leq 0.3 \\ C_2 &\leq 0.4 \end{aligned} \tag{48}$$

$$P_{g1}, C_2, C_3 \geq 0 \tag{49}$$

Note that for the case with a line outage, the line's susceptance is set to zero, consequently changing the system susceptance matrix, and the steps outlined above are repeated.

Composite System Adequacy Assessment

The possible system states of the 3-bus example are presented in Table 19 Since only two of the system's five components have forced outage rates larger than zero, the possible system states are given by combining the possible states of the two components with non-zero FOR. An underline of a component denotes a component on outage. The load curtailments are obtained by solving the OPF problem corresponding to each system state.

The probability of each system state, is calculated in (50) by multiplication of the state probabilities of G1 and L1, using the component data of Tables 16, 17 and 18.

$$\begin{aligned} P(N) &= 0.99 \cdot 0.99 = 0.9801 \\ P(A) &= 0.01 \cdot 0.99 = 0.0099 \\ P(B) &= 0.99 \cdot 0.01 = 0.0099 \\ P(C) &= 0.01 \cdot 0.01 = 0.0001 \end{aligned} \tag{50}$$

The four system states and the resulting power flows are illustrated in Figs. 4, 5 and 6.

Table 19 The system states with probability and severity

Event	State of the components	Probability	Load curtailment [MW]
P(N)	G1, L1	0.9801	0
P(A)	$\underline{\text{G1}}$, L1	0.0099	20
P(B)	G1, $\underline{\text{L1}}$	0.0099	20
P(C)	$\underline{\text{G1}}$, $\underline{\text{L1}}$	0.0001	20

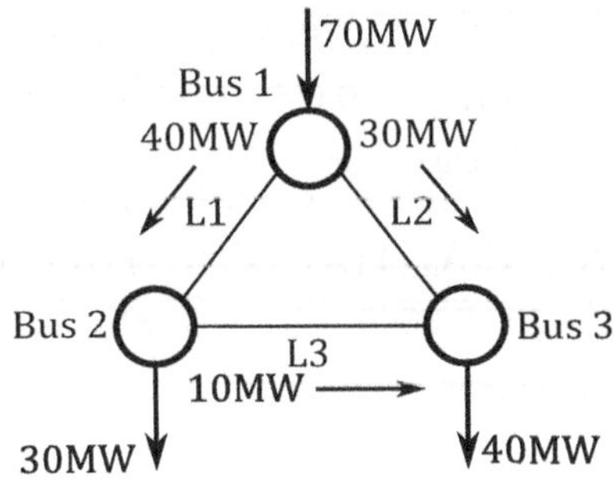

Fig. 4 Event P(N)

Fig. 5 Event P(A)

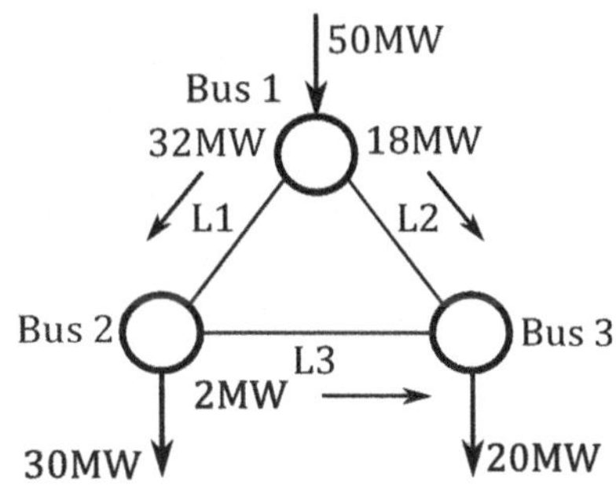

Fig. 6 Events P(B) and P(C)

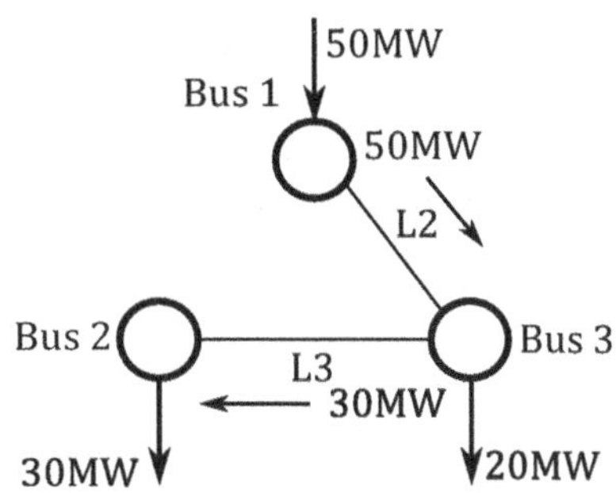

Calculation of the Reliability Indices

The reliability indices used in the assessment are the LOLE and EENS indices [4]. A LOLE of 0.0199 years in one year and an EENS of 3486 MWh in one year are obtained, when assuming a constant yearly peak load (CYPL). The numerical calculations are shown below with the corresponding formulae:

$$\begin{aligned} \text{LOLE} &= X_N \cdot P(N) + X_A \cdot P(A) + X_B \cdot P(B) + X_C \cdot P(C) \\ &= 0 \cdot 0.9801 + 1 \cdot 0.0099 + 1 \cdot 0.0099 + 1 \cdot 0.0001 \\ &= 0.0199\,\text{years/year} \end{aligned} \tag{51}$$

$$\begin{aligned} \text{EENS} &= [C(N) \cdot P(N) + C(A) \cdot P(A) + C(B) \cdot P(B) + C(C) \cdot P(C)] \cdot T \\ &= (0 \cdot 0.9801 + 20 \cdot 0.0099 + 20 \cdot 0.0099 + 20 \cdot 0.0001) \cdot 8760 \\ &= 3486.48\,\text{MWh/year} \end{aligned} \tag{52}$$

1.3 AC—Contingency Solver

The considerations of the AC contingency solver are presented in this section. First, an introduction to the equations and assumptions that are used is made, before the general form of the AC OPF problem is formulated. Then tests of the solver are performed on the same selection of system states as that of the DC contingency solver. A final illustrative example is given, to highlight the intermediate steps of the presented methodology.

1.3.1 Network Model

A choice must be made regarding how the network model is represented in the OPF problem of the AC contingency solver. There exist two network models that are commonly used, namely the bus injection model and the branch flow model [11]. The bus injection model represents a compact form of the AC power flow equations, where the system is represented in terms of nodal variables at each system bus, e.g. active and reactive power injections, voltage phasors and current injections. It has been the most widely used network model in OPF problems since the first presented papers on OPF [11]. The branch flow model represents the system in terms of power flows and currents through each branch instead. This chapter uses the AC power flow equations derived from the bus injection model to represent the network. Derivations of the AC power flow equations can be found in most power system analysis textbooks, such as [9].

An important part of the AC power flow equations is how the various branch elements of the network are represented. Most of these, including transmission lines, cables and nominal transformers, can simply be represented by their π-equivalent model, cf. Fig. 7.

For transmission lines and cables, the branch characteristics are normally specified in terms of a series impedance and a shunt susceptance; it is common to neglect the shunt susceptance for transformers. As shown in Fig. 7, the shunt susceptance is equally divided between the two buses at each end of the branch,

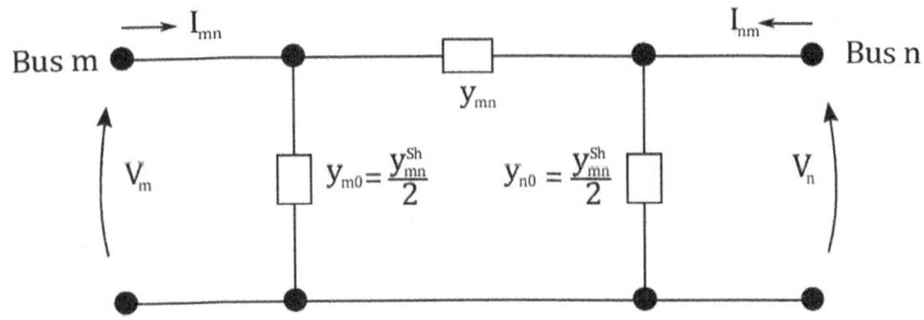

Fig. 7 π-equivalent model

while the series admittance of the figure, y_{mn}, is calculated from the series impedance as follows.

$$y_{mn} = \frac{1}{r_{mn} + jx_{mn}} = \frac{r_{mn}}{r_{mn}^2 + x_{mn}^2} - j \cdot \frac{x_{mn}}{r_{mn}^2 + x_{mn}^2} = g_{mn} + jb_{mn} \tag{53}$$

The net injections of currents at the two buses Fig. 7 can be expressed in terms of the system's admittances and voltages. Derivation from Kirchhoff's Current Law, expressing that the sum of currents flowing into a node must equal the sum of currents flowing out of the node, gives the net injection of current at each bus expressed in terms of a bus admittance matrix, Y_{bus}, times the nodal voltages [9].

$$\begin{bmatrix} I_m \\ I_n \end{bmatrix} = \begin{bmatrix} Y_{mm} & Y_{mn} \\ Y_{nm} & Y_{nn} \end{bmatrix} \cdot \begin{bmatrix} V_m \\ V_n \end{bmatrix} \tag{54}$$

The elements of the Y_{bus} matrix are mounted according to the following scheme. Each row and column corresponds to a bus, e.g., the elements of row 2 and column 2 correspond to bus 2. The diagonal elements of the matrix are mounted by summing the following for each of the lines that are connected to the bus: the series admittance and half of the shunt susceptance. Off-diagonal elements are set to zero if there are no branch elements between the corresponding buses, otherwise the elements are mounted by adding the negative of the branch element's series admittance. If two or more lines are connected in parallel between the two buses, the negatives of the series admittances are added together. Thus, the Y_{bus} of the two-bus system in Fig. 7 is given by:

$$Y_{bus} = \begin{bmatrix} y_{mn} + y_{m0} & -y_{mn} \\ -y_{nm} & y_{nm} + y_{n0} \end{bmatrix} \tag{55}$$

If the transformer is off-nominal, a more complex branch model must be used by introducing additional variables into the equations. A tap changing transformer has a real turns ratio, a:1, while a phase shifting transformer has a complex turns ratio. This turns ratio can be represented in polar coordinates with magnitude T_{mn} and phase shift φ_{mn}. If the turns ratio is real, the phase shift is set to zero.

Fig. 8 Off-nominal transformer

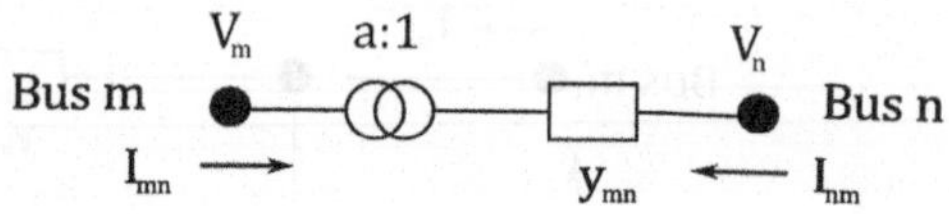

$$a_{mn} = T_{mn} \cdot e^{j\varphi_{mn}} \tag{56}$$

A single line representation of an off-nominal transformer is shown in Fig. 8. As previously stated it is common to neglect the shunt susceptance of transformers. Thus, the Y_{bus} elements of an off-nominal transformer branch are given in. An inspection of the matrix elements shows that the two off-diagonal elements, Y_{mn} and Y_{nm}, differ by opposite signs of the phase shift variables, giving an unsymmetrical Y_{bus}.

$$Y_{bus} = \begin{bmatrix} \frac{y_{mn}}{T_{mn}^2} & -\frac{y_{mn}}{T_{mn} \cdot e^{-j\varphi_{mn}}} \\ -\frac{y_{mn}}{T_{mn} \cdot e^{j\varphi_{mn}}} & y_{mn} \end{bmatrix} \tag{57}$$

The system's bus admittance matrix can be mounted by adding the elements of each branch together. It is worth noting that the Y_{bus} of a large system would be sparse, consisting of mostly zeros in the off-diagonal elements, due to each bus being directly connected to only one or a few other buses.

$$Y_{bus} = \begin{bmatrix} Y_{11} & Y_{12} & \cdots & Y_{1k} \\ Y_{21} & Y_{22} & \cdots & Y_{2k} \\ \vdots & \vdots & \ddots & \vdots \\ Y_{k1} & Y_{k2} & \cdots & Y_{kk} \end{bmatrix} \tag{58}$$

This chapter uses a separation of the Y_{bus} matrix into its real and imaginary parts, namely a separation into a conductance matrix and a susceptance matrix, to avoid having complex numbers in the AC power flow equations.

$$[Y_{bus}] = [G_{bus}] + j \cdot [B_{bus}] \tag{59}$$

It is common to separate the AC power flow equations into active and reactive power injections at each bus. Thus, the equations where the voltages are expressed in polar coordinates and the bus admittance matrix elements are expressed in rectangular coordinates, are given by (60) and (61) for each bus:

$$P_i(V,\delta) = V_i \cdot \sum_{j=1}^{k} V_j \left[G_{ij} \cdot \cos(\delta_i - \delta_j) + B_{ij} \cdot \sin(\delta_i - \delta_j)\right] \tag{60}$$

$$Q_i(V,\delta) = V_i \cdot \sum_{j=1}^{k} V_j \left[G_{ij} \cdot \sin(\delta_i - \delta_j) - B_{ij} \cdot \cos(\delta_i - \delta_j)\right] \tag{61}$$

The vectors giving the net injection of powers at each bus, both active and reactive, are defined similarly as in the DC approach. Thus, (60) gives the net injection of active power vector, while (62) gives the net injection of reactive power vector. Both give the net injections as the load requirements subtracted from the sum of actual power generations and load curtailments. It is worth noting that the load requirements at each bus are treated as a constant in the OPF problem, while a vector of load curtailments are introduced to have the option of reducing the loads to reach a feasible solution to the OPF problem.

$$[Q] = [Q_g] + [C_Q] - [Q_{load}] \tag{62}$$

Another important part of the network model are the constraints limiting the flow of current through the branches. The magnitude of the current flowing through a branch is given by the magnitude of the voltage drop over the branch, times the magnitude of the series admittance:

$$|I_{mn}| = |V_m - V_n| \cdot |y_{mn}| \leq I_{mn}^{max} \tag{63}$$

1.3.2 Contingency Solver Description

A representation of the system by the AC power equations leads to a larger of number of considered constraints in the OPF problem than for the DC based approach. The list of constraints that are considered by the AC contingency solver is presented in the following:

(a) The system's generation capacity must be larger than or equal to the sum of the system's load requirements and losses, where both active and reactive power are considered.
(b) The power flow through the lines are limited by the maximum current rating of the lines.
(c) The load curtailment at a bus cannot exceed the load requirement of the bus, where both active and reactive power are considered.
(d) The actual generation at a bus cannot exceed the generation capacity of the bus, where both active and reactive power are considered.
(e) The actual active power generation at a bus cannot be negative.

(f) The actual reactive power generation at a bus cannot be lower than the minimum reactive power capability of the bus.
(g) The load curtailment at a bus cannot be negative, when both active and reactive power are considered.[4]
(h) The voltage magnitude at each bus must be inside the specified limits.

The possible actions of the AC contingency solver are similar to the ones of the DC based solver, e.g., rescheduling of generation and load curtailment. When a system state is handed to the AC contingency solver, the solver tries to find a feasible operating point that does not violate any of the above listed constraints. Rescheduling of generation is tried first, before load curtailments are considered. Due to the inclusion of reactive power considerations in the analysis, there is an increased number of system states that require actions compared to when the DC approximations are applied. An example of such a situation is when the voltage at one or more buses drops below the specified voltage limit, due to voltage drops in the transmission lines in spite of sufficient generation capacity in the system. In a similar fashion as for the DC based approach, the control actions are controlled by the objective function of the OPF problem. The objective function consists of a vector of decision variables that is multiplied by a cost vector. Each element of the cost vector corresponds to the cost of increasing a decision variable. In the approach suggested in this thesis, there are only costs associated with load curtailments. The cost of load curtailments at each bus are specified according to the input data of the test system.

$$W = \begin{bmatrix} w_1 & w_2 & \cdots & w_{6k} \end{bmatrix}^T \tag{64}$$

Compared to the OPF problem of the DC contingency solver, the number of decision variables is increased. It is common to partition the decision variables into two sets [11]: a set of control variables and a set of state variables. Control variables are the independent variables that are controllable, typically active and reactive power generation at each bus. The voltage magnitudes and voltage angles at the buses form the set of state variables that are dependent. Load requirements at the buses are fixed parameters for each system state. If it is necessary to reduce the load of the system to overcome a constraint violation, it is handled by introducing load curtailment variables at each bus to the set of control variables. Thus, the decision variables used in this work are:

$$X = \begin{bmatrix} P_{g1} & \cdots & Q_{g1} & \cdots & C_P & \cdots & C_Q & \cdots & V_1 & \cdots & \delta_1 & \cdots \end{bmatrix} \tag{65}$$

The vector of decision variables consists of 6 k elements, where k elements[5] of each decision variable type are ordered in sequence: active power generation,

[4]The reactive power part of the load requirement is limited to positive values. A negative load requirement value is defined as a generation capacity instead.

[5]Each element corresponds to one of the system's k buses.

reactive power generation, active power load curtailment, reactive power load curtailment, voltage magnitude and voltage angle. It is also possible to include additional control variables in the set of decision variables. Among them are more advanced controls, such as tap changing and phase shifting of off-nominal transformers, thus increasing the number of variables considerably and complicating the problem [11]. If more control variables are added to the set of decision variables, the vectors of (64) and (65) must be updated accordingly.

1.3.3 Optimal Power Flow Formulation

The OPF problem formulation used in this chapter has its basis in the classic formulation presented in [11]. It is a variant based on the classic form presented in [12], which has the typical objective of reducing the total cost of generation. For the considered OPF problems of this thesis, the desired objective is to minimize the total cost of load curtailments. In the following, the general OPF problem is formulated using the notations and matrices of the previous sections. It is important to note that all quantities must be in per unit notation, and angles must be expressed in radians, for the equations to be applicable.

$$\min f = [X] \cdot [W] \tag{66}$$

$$P_{gi} + C_{Pi} - P_{load,i} - P_i(V, \delta) = 0 \tag{67}$$

$$Q_{gi} + C_{Qi} - Q_{load,i} - Q_i(V, \delta) = 0 \tag{68}$$

$$I_{mn} \leq I_{mn}^{max} \tag{69}$$

$$H_{Load,i} \cdot C_{Pi} - C_{Qi} = 0 \tag{70}$$

$$0 \leq P_{gi} \leq P_{gi}^{max} \tag{71}$$

$$Q_{gi}^{min} \leq Q_{gi} \leq Q_{gi}^{max} \tag{72}$$

$$0 \leq C_{Pi} \leq P_{load,i} \tag{73}$$

$$0 \leq C_{Qi} \leq Q_{load,i} \tag{74}$$

$$V_i^{min} \leq V_i \leq V_i^{max} \tag{75}$$

$$-\pi \leq \delta_i \leq \pi \tag{76}$$

In the approach used in this work, bus number 1 of the system is selected as the slack bus. Thus, the constraint of (76) for this bus is limited to a fixed angle of 0 radians. Equations (67) and (68) are extended to complete form by rewriting them in terms of the decision variables by use of (60) and (61). The inequalities in the form of (69) must also be rewritten in a format suitable for the nonlinear solver. A suitable form is given in (77) by rewriting (63) in terms of the decision variables, where the current limit is moved to the left side of the inequality.

$$(V_m \cdot \cos\delta_m - V_n \cdot \cos\delta_n)^2 + (V_m \cdot \sin\delta_m - V_n \cdot \sin\delta_n)^2 - \left(\frac{I_{mn}^{max}}{y_{mn}}\right)^2 \leq 0 \quad (77)$$

An equation set in the form of (70) is needed to keep the power factors of the loads fixed, when loads are curtailed. $H_{load,i}$ gives the specified relation between active and reactive power load requirement at each bus, thus ensuring a constant power factor. The equation set represents a set of linear equality constraints, which could be expressed by an **A** matrix times the vector of decision variables in (78), where the **A** matrix is filled to comply with (70).

$$[A] \cdot [X] = \begin{bmatrix} 0 \\ \vdots \\ 0 \end{bmatrix} \quad (78)$$

If additional control variables, such as tap changing and phase shifting, are included in the set of decision variables, additional constraints that limit the range of these controls must be added to the problem.

$$T_{mn}^{min} \leq T_{mn} \leq T_{mn}^{max} \quad (79)$$

$$\varphi_{mn}^{min} \leq \varphi_{mn} \leq \varphi_{mn}^{max} \quad (80)$$

1.3.4 Initial Starting Point

In nonlinear optimization, the choice of starting point is important for the solution of the OPF problem, e.g., the initialization of the system. Two common approaches are typical: either use a "flat start" where voltage magnitudes are set to 1.0 p.u. and voltage angles to zero, or a "warm start" where the voltage magnitudes and voltage angles are initialized according to a pre-solved load flow [11]. The convergence of the OPF problem's power flow equations rely on the choice of starting point.

During the work leading to this chapter, different choices of starting points have been tested to ensure that the contingency solver returned valid solutions for all possible system states. The nonlinear solver used in this work, is the *fmincon* solver of MATLAB with the interior point algorithm [13]. An inbuilt feature of the solver is that an exitflag can be returned along with the OPF solution. The exitflag provides additional information regarding the solution according to the following:

(a) Exitflag +2: Might be an optimal solution, but the solution should be tested further. The solver ends, because the change in the solution is smaller than the

Table 20 Tuning of the fmincon solver

Test system setting	Maximum function evaluations
Standard	3000
Specific for RBTS	50,000

step tolerance of 1E-10, i.e., the solver tries to take a step that is smaller than this value.

(b) Exitflag +1: A local optimum is found.

(c) Exitflag 0: The solver returns a solution due to reaching the maximum number of iterations or function evaluations. The standard settings of the *fmincon* solver, interior point algorithm, are 3000 for the MaxFunctionEvalutaions and 10000 for the MaxIterations settings.

(d) Exitflag −1: The solver is interrupted and ended by an output or plot function.

(e) Exitflag −2: No feasible solution is found.

Thus, a solution should only be accepted if an exitflag of +1 is returned along with the solution. By practical experience during the work leading to this chapter, it has been discovered that it is advantageous to increase the number of maximum function evaluations according to Table 20.

It has also been discovered through testing on the test systems, that different starting points give correct solutions for some system states, but fail to deliver correct solutions for other system states. When such is the case, the solver also returns an exitflag different from +1. Thus, a scheme that is presented below is created by taking advantage of the returned exitflags. Common for the tries of starting points presented in the following are that all use a "flat start" for the state variables, e.g., voltage magnitudes of 1.0 p.u. and angles set to zero. Load curtailments, both active and reactive, are also initialized to zero.

Try 1: The starting point, x0, is initialized with active power generation set equal to the available capacity at each bus, while the reactive power generation at each bus is set to zero.

(a) If the solver returns an exitflag of +2 or 0, the solver is run again with the new solution as starting point.

  (i) If an exitflag of +1 is returned, the solution is accepted.
  (ii) Any other exitflag, leads to try number 2.

(b) If an exitflag of +1 is returned, the solution is accepted.

(c) Any other exitflag, leads to try number 2.

Try 2: The starting point, x0, is initialized with both active and reactive power generation set equal to the available capacity at each bus.

(a) If the solver returns an exitflag of +2 or 0, the solver is run again with the new solution as starting point.

  (i) If an exitflag of +1 is returned, the solution is accepted.

(ii) Any other exitflag, leads to try number 3.

(b) If an exitflag of +1 is returned, the solution is accepted.
(c) Any other exitflag, leads to try number 3.

Try 3: The starting point, x0, is initialized with both active and reactive power generation set to zero.

(a) If the solver returns an exitflag of +2 or 0, the solver is run again with the new solution as starting point.

    (i) If an exitflag of +1 is returned, the solution is accepted.

(b) If an exitflag of +1 is returned, the solution is accepted.

To capture cases where the final solution is received with an exitflag different from +1, an error script needs to be written, which writes a text file with system details for system states that have already been solved by more than one tries of initial starting points.

1.3.5 Contingency Solver Verification

The contingency solver has to be tested to verify the suggested OPF methodology on a similar selection of system states as the ones used for the DC contingency solver, i.e., on the RBTS test system at peak load of 185 MW with various configurations. In Table 21, the load requirements at each bus are specified.

Case 1

In the first test, all components are assumed to be working. Thus, the generation capacities at buses 1 and 2 are equal to the installed capacities, cf. Table 22. The test is performed to verify that no loads are curtailed erroneously (Table 23).

Case 2

Another test is performed to verify that the contingency solver curtails loads when the system's generation capacity is insufficient. During the test, all lines of the system are available. The generation capacity of the case is specified in Table 24,

Table 21 The RBTS load distribution at a peak load of 185 MW

Bus	Active load demand [MW]	Reactive load demand [MVAr]
1	0	0
2	20	4
3	85	17
4	40	8
5	20	4
6	20	4
Sum:	185	37

Table 22 The installed generation capacity of the RBTS

Bus	Active power [MW]	Min reactive power [MVAr]	Max reactive power [MVAr]
1	110	−37	53
2	130	−43	75

Table 23 The results of case 1—AC solver

Bus	Generation [MW]	Generation [MVAr]	Curtailment [MW]	Voltage [p.u.]
1	94.00	31.30	0.00	1.04
2	95.50	−1.30	0.00	1.04
3	0.00	0.00	0.00	1.00
4	0.00	0.00	0.00	1.00
5	0.00	0.00	0.00	0.99
6	0.00	0.00	0.00	0.98
Sum:	189.50	–	0.00	–

Table 24 The generation capacity of case 2—AC solver

Bus	Active power [MW]	Min reactive power [MVAr]	Max reactive power [MVAr]
1	100	−37	46
2	50	−14	34

Table 25 The results of case 2—AC solver

Bus	Generation [MW]	Generation [MVAr]	Curtailment [MW]	Voltage [p.u.]
1	100.00	12.30	0.00	1.05
2	50.00	−1.20	0.00	1.05
3	0.00	0.00	37.40	1.02
4	0.00	0.00	0.00	1.02
5	0.00	0.00	0.00	1.01
6	0.00	0.00	0.10	1.00
Sum:	150.00	–	37.50	–

with generators 7, 8 and 11 on outage. A total load curtailment of 37.5 MW can be seen in Table 25, which is 2.5 MW higher than the result obtained by the DC contingency solver in Table 11. The higher load curtailment is reasonable, because the losses of the lines are included in the AC solution.

Case 3

Another test is performed to verify that the contingency solver handles the isolation of buses properly. The system's generation capacity is equal to the installed capacity, cf. Table 22. Lines L3, L4 and L8 are on outage, leading to the islanding

Table 26 The results of case 3—AC solver

Bus	Generation [MW]	Generation [MVAr]	Curtailment [MW]	Voltage [p.u.]
1	110.00	27.30	0.00	1.05
2	0.00	0.00	20.00	0.00
3	0.00	0.00	17.40	1.01
4	0.00	0.00	40.00	0.00
5	0.00	0.00	0.00	0.99
6	0.00	0.00	0.10	0.98
Sum:	110.00	–	77.50	–

of buses 2 and 4. A total load curtailment of 77.5 MW is observed in Table 26, where the loads at the isolated buses are curtailed correctly although the isolated part of the system has connected generators. There are also load curtailments at buses 3 and 6, due to bus 1 being unable to supply the total load demand of the connected part. It can be observed that the AC based solver returns a higher load curtailment than the DC based solver does for the same case, as it takes the transfer losses into account.

Case 4

The system's generation capacity is still equal to the installed capacity, cf. Table 22. Lines L5 and L8 are on outage, leading to the islanding of buses 5 and 6. A total load curtailment of 40 MW can be seen in Table 27, where the loads at the isolated buses are curtailed correctly. For this case, the generation capacity is sufficient to handle the sum of load demands and losses in the connected part, and thus the AC contingency solver gives an equal load curtailment as the DC contingency solver does for this case.

Case 5

Another test is performed on a case where both generators and lines are on outage. The generation capacity is specified in Table 28 where generators 1, 3 and 4 are on outage. Lines L2 and L7 are on outage, and thus the transfer of power that is generated at bus 2, is limited to line L3. A total load curtailment of 100.1 MW can

Table 27 The results of case 4—AC solver

Bus	Generation [MW]	Generation [MVAr]	Curtailment [MW]	Voltage [p.u.]
1	71.00	18.50	0.00	1.02
2	76.40	−3.00	0.00	1.03
3	0.00	0.00	0.00	1.00
4	0.00	0.00	0.00	1.00
5	0.00	0.00	20.00	0.00
6	0.00	0.00	20.00	0.00
Sum:	147.40	–	40.00	–

Table 28 The generation capacity of case 5—AC solver

Bus	Active power [MW]	Min reactive power [MVAr]	Max reactive power [MVAr]
1	20	−7	12
2	130	−43	75

Table 29 The results of case 5—AC solver

Bus	Generation [MW]	Generation [MVAr]	Curtailment [MW]	Voltage [p.u.]
1	20.00	12.00	0.00	1.01
2	68.10	6.40	0.00	1.05
3	0.00	0.00	84.80	0.98
4	0.00	0.00	0.00	0.97
5	0.00	0.00	0.00	0.97
6	0.00	0.00	15.30	0.97
Sum:	88.10		100.10	

be seen in Table 29. The higher load curtailment, 26.1 MW, than the one of the DC contingency solver in Table 15, is due the large voltage drop associated with power transfers trough the remaining lines. For the DC case, the current rating of line L3 is the binding constraint, but for the AC case the voltage limit at bus 6 is the binding constraint.

1.3.6 Illustrative Example

A small example is presented to illustrate the basic principles of the AC based composite system assessment methodology in a pedagogical way. Numerical calculations that are performed when the OPF problem is formulated, are included. The test system is the same 3-bus test system that has been presented in the DC contingency solver segment of this chapter, cf. Figure 3. Additional system details are included in Tables 30, 31 and 32, due to the need to consider reactive power and voltages. The system power base is 100 MVA. Voltages at the buses are limited to a range from 0.95 to 1.05 p.u. The power factors of the loads are fixed at 0.98, implying that a curtailment of 1 p.u. active load must be met by a curtailment of 0.2 p.u. reactive load to maintain a constant power factor.

Table 30 The generator data of the 3-bus test system

Generator	Bus	Active power [MW]	Min reactive power [MVAr]	Max reactive power [MVAr]	FOR
G1	1	50	−10	20	0.01
G2	1	50	−10	20	0

Table 31 The loads of the 3-bus test system

Bus	Active load [MW]	Reactive load [MVAr]	Cost of load curtailment [$/kWh]
1	0	0	–
2	30	6	1
3	40	8	1

Table 32 The network data of the 3-bus test system

Line	From	To	Resistance [p.u.]	Reactance [p.u.]	Half of shunt susceptance [p. u.]	Transfer limit [p. u]	FOR
L1	1	2	0.02	0.1	0.01	0.5	0.01
L2	1	3	0.04	0.2	0.02	0.5	0
L3	2	3	0.04	0.2	0.02	0.5	0

The number of possible contingencies for the system are limited, due to only considering the outages of G1 and L1 as possible. FOR values of the components are similar to those of the DC based example, yielding equal system states and corresponding possibilities.

Step by Step Calculations

A step by step approach with numerical details, where the OPF problem is formulated according to the suggested methodology, is presented in this section. The base case is presented here.

Step 1: Obtain the conductance and susceptance matrices of (59):

(a) Calculate the series admittance of the lines by (53):

$$
\begin{aligned}
y_{12} &= \frac{1}{0.02+j0.1} = \frac{0.02}{0.02^2+0.1^2} - j\frac{0.1}{0.02^2+0.1^2} = 1.9231 - j9.6154 \\
y_{13} &= \frac{1}{0.04+j0.2} = \frac{0.04}{0.04^2+0.2^2} - j\frac{0.2}{0.04^2+0.2^2} = 0.9615 - j4.8077 \\
y_{23} &= y_{13}
\end{aligned}
\tag{81}
$$

(b) Calculate the elements of the conductance matrix, $\mathbf{G_{bus}}$:

$$\begin{aligned} G_{11} &= g_{12} + g_{13} = 1.9231 + 0.9615 = 2.8846 \\ G_{22} &= g_{12} + g_{23} = 1.9231 + 0.9615 = 2.8846 \\ G_{33} &= g_{13} + g_{23} = 0.9615 + 0.9615 = 1.9231 \\ G_{12} &= G_{21} = -g_{12} = -1.9231 \\ G_{13} &= G_{31} = -g_{13} = -0.9615 \\ G_{23} &= G_{32} = -g_{23} = -0.9615 \end{aligned} \tag{82}$$

(c) Calculate the elements of the susceptance matrix, $\mathbf{B_{bus}}$:

$$\begin{aligned} B_{11} &= b_{12} + b_{13} + b_{10}^{L1} + b_{10}^{L2} = -9.6154 - 4.8077 + 0.01 + 0.02 = -14.3931 \\ B_{22} &= b_{12} + b_{23} + b_{20}^{L1} + b_{20}^{L3} = -9.6154 - 4.8077 + 0.01 + 0.02 = -14.3931 \\ B_{33} &= b_{13} + b_{23} + b_{30}^{L2} + b_{30}^{L3} = -4.8077 - 4.8077 + 0.02 + 0.02 = -9.5754 \\ B_{12} &= B_{21} = -b_{12} = 9.6154 \\ B_{13} &= B_{31} = -b_{13} = 4.8077 \\ B_{23} &= B_{32} = -b_{23} = 4.8077 \end{aligned} \tag{83}$$

(d) Create the conductance matrix, $\mathbf{G_{bus}}$:

$$G_{bus} = \begin{bmatrix} 2.88 & -1.92 & -0.96 \\ -1.92 & 2.88 & -0.96 \\ -0.96 & -0.96 & 1.92 \end{bmatrix} \tag{84}$$

(e) Create the susceptance matrix, $\mathbf{B_{bus}}$:

$$B_{bus} = \begin{bmatrix} -14.39 & 9.62 & 4.81 \\ 9.62 & -14.39 & 4.81 \\ 4.81 & 4.81 & -9.58 \end{bmatrix} \tag{85}$$

Step 2: Set up the AC power flow equations (60) and (61) for each of the three buses:

(a) Active power:

$$\begin{aligned}P_1(V,\delta) &= V_1V_1[G_{11}\cos(\delta_1-\delta_1)+B_{11}\sin(\delta_1-\delta_1)]\\&\quad+V_1V_2[G_{12}\cos(\delta_1-\delta_2)+B_{12}\sin(\delta_1-\delta_2)]\\&\quad+V_1V_3[G_{13}\cos(\delta_1-\delta_3)+B_{13}\sin(\delta_1-\delta_3)]\\&=2.88\cdot V_1^2-1.92\cdot V_1V_2\cos(\delta_1-\delta_2)+9.62\cdot V_1V_2\sin(\delta_1-\delta_2)\\&\quad-0.96\cdot V_1V_3\cos(\delta_1-\delta_3)+4.81\cdot V_1V_3\sin(\delta_1-\delta_3)\end{aligned}\tag{86}$$

$$\begin{aligned}P_2(V,\delta) &= V_2V_1[G_{21}\cos(\delta_2-\delta_1)+B_{21}\sin(\delta_2-\delta_1)]\\&\quad+V_2V_2[G_{22}\cos(\delta_2-\delta_2)+B_{22}\sin(\delta_2-\delta_2)]\\&\quad+V_2V_3[G_{23}\cos(\delta_2-\delta_3)+B_{23}\sin(\delta_2-\delta_3)]\\&=-1.92\cdot V_2V_1\cos(\delta_2-\delta_1)+9.62\cdot V_2V_1\sin(\delta_2-\delta_1)+2.88\cdot V_2^2\\&\quad-0.96\cdot V_2V_3\cos(\delta_2-\delta_3)+4.81\cdot V_2V_3\sin(\delta_2-\delta_3)\end{aligned}\tag{87}$$

$$\begin{aligned}P_3(V,\delta) &= V_3V_1[G_{31}\cos(\delta_3-\delta_1)+B_{31}\sin(\delta_3-\delta_1)]\\&\quad+V_3V_2[G_{32}\cos(\delta_3-\delta_2)+B_{32}\sin(\delta_3-\delta_2)]\\&\quad+V_3V_3[G_{33}\cos(\delta_3-\delta_3)+B_{33}\sin(\delta_3-\delta_3)]\\&=-0.96\cdot V_3V_1\cos(\delta_3-\delta_1)+4.81\cdot V_3V_1\sin(\delta_3-\delta_1)\\&\quad-0.96\cdot V_3V_2\cos(\delta_3-\delta_2)+4.81\cdot V_3V_2\sin(\delta_3-\delta_2)+1.92\cdot V_3^2\end{aligned}\tag{88}$$

(b) Reactive Power:

$$\begin{aligned}Q_1(V,\delta) &= V_1V_1[G_{11}\sin(\delta_1-\delta_1)-B_{11}\cos(\delta_1-\delta_1)]\\&\quad+V_1V_2[G_{12}\sin(\delta_1-\delta_2)-B_{12}\cos(\delta_1-\delta_2)]\\&\quad+V_1V_3[G_{13}\sin(\delta_1-\delta_3)-B_{13}\cos(\delta_1-\delta_3)]\\&=14.39\cdot V_1^2-1.92\cdot V_1V_2\sin(\delta_1-\delta_2)-9.62\cdot V_1V_2\cos(\delta_1-\delta_2)\\&\quad-0.96\cdot V_1V_3\sin(\delta_1-\delta_3)-4.81\cdot V_1V_3\cos(\delta_1-\delta_3)\end{aligned}\tag{89}$$

$$\begin{aligned} Q_2(V,\delta) &= V_2V_1[G_{21}\sin(\delta_2-\delta_1) - B_{21}\cos(\delta_2-\delta_1)] \\ &\quad + V_2V_2[G_{22}\sin(\delta_2-\delta_2) - B_{22}\cos(\delta_2-\delta_2)] \\ &\quad + V_2V_3[G_{23}\sin(\delta_2-\delta_3) - B_{23}\cos(\delta_2-\delta_3)] \\ &= -1.92\cdot V_2V_1\sin(\delta_2-\delta_1) - 9.62\cdot V_2V_1\cos(\delta_2-\delta_1) + 14.39\cdot V_2^2 \\ &\quad - 0.96\cdot V_2V_3\sin(\delta_2-\delta_3) - 4.81\cdot V_2V_3\cos(\delta_2-\delta_3) \end{aligned} \tag{90}$$

$$\begin{aligned} Q_3(V,\delta) &= V_3V_1[G_{31}\sin(\delta_3-\delta_1) - B_{31}\cos(\delta_3-\delta_1)] \\ &\quad + V_3V_2[G_{32}\sin(\delta_3-\delta_2) - B_{32}\cos(\delta_3-\delta_2)] \\ &\quad + V_3V_3[G_{33}\sin(\delta_3-\delta_3) - B_{33}\cos(\delta_3-\delta_3)] \\ &= -0.96\cdot V_3V_1\sin(\delta_3-\delta_1) - 4.81\cdot V_3V_1\cos(\delta_3-\delta_1) \\ &\quad - 0.96\cdot V_3V_2\sin(\delta_3-\delta_2) - 4.81\cdot V_3V_2\cos(\delta_3-\delta_2) + 9.58\cdot V_3^2 \end{aligned} \tag{91}$$

Step 3: Set up the power balance equations for each bus, by using the AC power flow equations from step 2 and the net injection equations of (19) and (62), and the nonlinear line current constraints for each line using (77):

(a) The power balance at each bus:

$$\begin{aligned} P_{g1} + C_{P1} - P_{load,1} - P_1(V,\delta) &= P_{g1} + C_{P1} - P_1(V,\delta) = 0 \\ P_{g2} + C_{P2} - P_{load,2} - P_2(V,\delta) &= P_{g2} + C_{P2} - P_2(V,\delta) - 0.3 = 0 \\ P_{g3} + C_{P3} - P_{load,3} - P_3(V,\delta) &= P_{g3} + C_{P3} - P_3(V,\delta) - 0.4 = 0 \\ Q_{g1} + C_{Q1} - Q_{load,1} - Q_1(V,\delta) &= Q_{g1} + C_{Q1} - Q_1(V,\delta) = 0 \\ Q_{g2} + C_{Q2} - Q_{load,2} - Q_2(V,\delta) &= Q_{g2} + C_{Q2} - Q_2(V,\delta) - 0.06 = 0 \\ Q_{g3} + C_{Q3} - Q_{load,3} - Q_3(V,\delta) &= Q_{g3} + C_{Q3} - Q_3(V,\delta) - 0.08 = 0 \end{aligned} \tag{92}$$

(b) The current constraints for the three lines:

(i) Line L1:

$$\begin{aligned} (V_1\cos\delta_1 - V_2\cos\delta_2)^2 + (V_1\sin\delta_1 - V_2\sin\delta_2)^2 - \left(\frac{I_{12}^{max}}{\sqrt{g_{12}^2 + b_{12}^2}}\right)^2 &\le 0 \\ (V_1\cos\delta_1 - V_2\cos\delta_2)^2 + (V_1\sin\delta_1 - V_2\sin\delta_2)^2 - \frac{0.5^2}{1.92^2 + 9.62^2} &\le 0 \\ (V_1\cos\delta_1 - V_2\cos\delta_2)^2 + (V_1\sin\delta_1 - V_2\sin\delta_2)^2 - 0.002598 &\le 0 \end{aligned} \tag{93}$$

(ii) Line L2:

$$\begin{aligned}
&(V_1 \cos \delta_1 - V_3 \cos \delta_3)^2 + (V_1 \sin \delta_1 - V_3 \sin \delta_3)^2 - \left(\frac{I_{13}^{max}}{\sqrt{g_{13}^2 + b_{13}^2}} \right)^2 \leq 0 \\
&(V_1 \cos \delta_1 - V_3 \cos \delta_3)^2 + (V_1 \sin \delta_1 - V_3 \sin \delta_3)^2 - \frac{0.5^2}{0.96^2 + 4.81^2} \leq 0 \\
&(V_1 \cos \delta_1 - V_3 \cos \delta_3)^2 + (V_1 \sin \delta_1 - V_3 \sin \delta_3)^2 - 0.010392 \leq 0
\end{aligned} \tag{94}$$

(iii) Line L3:

$$\begin{aligned}
&(V_2 \cos \delta_2 - V_3 \cos \delta_3)^2 + (V_2 \sin \delta_2 - V_3 \sin \delta_3)^2 - \left(\frac{I_{23}^{max}}{\sqrt{g_{23}^2 + b_{23}^2}} \right)^2 \leq 0 \\
&(V_2 \cos \delta_2 - V_3 \cos \delta_3)^2 + (V_2 \sin \delta_2 - V_3 \sin \delta_3)^2 - \frac{0.5^2}{0.96^2 + 4.81^2} \leq 0 \\
&(V_2 \cos \delta_2 - V_3 \cos \delta_3)^2 + (V_2 \sin \delta_2 - V_3 \sin \delta_3)^2 - 0.010392 \leq 0
\end{aligned} \tag{95}$$

Step 4: Set up the equations that maintain the power factor of the loads at each load bus:

$$\begin{aligned}
0.2 \cdot C_{P2} - C_{Q2} = 0 \\
0.2 \cdot C_{P3} - C_{Q3} = 0
\end{aligned} \tag{96}$$

Step 5: Set up the final OPF problem, recognizing that bus 1 is a generator bus without connected loads and that buses 2 and 3 have no connected generators. Thus, the number of decision variables could be reduced. The angle of the slack bus is the reference angle of the system, and it is fixed at zero radians.

$$\begin{aligned}
\text{Min} f = 0 \cdot P_{g1} + 0 \cdot Q_{g1} + 1 \cdot C_{P2} + 1 \cdot C_{P3} + 1 \cdot C_{Q2} + 1 \cdot C_{Q3} \\
+ 0 \cdot V_1 + 0 \cdot V_2 + 0 \cdot V_3 + 0 \cdot \delta_1 + 0 \cdot \delta_2 + 0 \cdot \delta_3
\end{aligned} \tag{97}$$

$$\begin{aligned}
P_{g1} - 2.88 \cdot V_1^2 + 1.92 \cdot V_1 V_2 \cos(\delta_1 - \delta_2) - 9.62 \cdot V_1 V_2 \sin(\delta_1 - \delta_2) \\
+ 0.96 \cdot V_1 V_3 \cos(\delta_1 - \delta_3) - 4.81 \cdot V_1 V_3 \sin(\delta_1 - \delta_3) = 0
\end{aligned} \tag{98}$$

$$\begin{aligned}
C_{P2} + 1.92 \cdot V_2 V_1 \cos(\delta_2 - \delta_1) - 9.62 \cdot V_2 V_1 \sin(\delta_2 - \delta_1) - 2.88 \cdot V_2^2 \\
+ 0.96 \cdot V_2 V_3 \cos(\delta_2 - \delta_3) - 4.81 \cdot V_2 V_3 \sin(\delta_2 - \delta_3) - 0.3 = 0
\end{aligned} \tag{99}$$

$$\begin{aligned}
C_{P3} + 0.96 \cdot V_3 V_1 \cos(\delta_3 - \delta_1) - 4.81 \cdot V_3 V_1 \sin(\delta_3 - \delta_1) \\
+ 0.96 \cdot V_3 V_2 \cos(\delta_3 - \delta_2) - 4.81 \cdot V_3 V_2 \sin(\delta_3 - \delta_2) - 1.92 \cdot V_3^2 - 0.4 = 0
\end{aligned} \tag{100}$$

$$\begin{aligned}
Q_{g1} - 14.39 \cdot V_1^2 + 1.92 \cdot V_1 V_2 \sin(\delta_1 - \delta_2) + 9.62 \cdot V_1 V_2 \cos(\delta_1 - \delta_2) \\
+ 0.96 \cdot V_1 V_3 \sin(\delta_1 - \delta_3) + 4.81 \cdot V_1 V_3 \cos(\delta_1 - \delta_3) = 0
\end{aligned} \tag{101}$$

$$\begin{aligned} & C_{Q2} + 1.92 \cdot V_2V_1 \sin(\delta_2 - \delta_1) + 9.62 \cdot V_2V_1 \cos(\delta_2 - \delta_1) - 14.39 \cdot V_2^2 \\ & \quad + 0.96 \cdot V_2V_3 \sin(\delta_2 - \delta_3) + 4.81 \cdot V_2V_3 \cos(\delta_2 - \delta_3) - 0.06 = 0 \end{aligned} \quad (102)$$

$$\begin{aligned} & C_{Q3} + 0.96 \cdot V_3V_1 \sin(\delta_3 - \delta_1) + 4.81 \cdot V_3V_1 \cos(\delta_3 - \delta_1) \\ & + 0.96 \cdot V_3V_2 \sin(\delta_3 - \delta_2) + 4.81 \cdot V_3V_2 \cos(\delta_3 - \delta_2) - 9.58 \cdot V_3^2 - 0.08 = 0 \end{aligned} \quad (103)$$

$$\begin{aligned} & (V_1 \cos \delta_1 - V_2 \cos \delta_2)^2 + (V_1 \sin \delta_1 - V_2 \sin \delta_2)^2 - 0.002598 \le 0 \\ & (V_1 \cos \delta_1 - V_3 \cos \delta_3)^2 + (V_1 \sin \delta_1 - V_3 \sin \delta_3)^2 - 0.010392 \le 0 \\ & (V_2 \cos \delta_2 - V_3 \cos \delta_3)^2 + (V_2 \sin \delta_2 - V_3 \sin \delta_3)^2 - 0.010392 \le 0 \end{aligned} \quad (104)$$

$$\begin{aligned} 0.2 \cdot C_{P2} - C_{Q2} = 0 \\ 0.2 \cdot C_{P3} - C_{Q3} = 0 \end{aligned} \quad (105)$$

$$\begin{aligned} 0 \le P_{g1} \le 1.0 \\ -0.2 \le Q_{g1} \le 0.4 \\ 0 \le C_{P2} \le 0.3 \\ 0 \le C_{P3} \le 0.4 \\ 0 \le C_{Q2} \le 0.06 \\ 0 \le C_{Q3} \le 0.08 \end{aligned} \quad (106)$$

$$\begin{aligned} 0.95 \le V_1 \le 1.05 \\ 0.95 \le V_2 \le 1.05 \\ 0.95 \le V_3 \le 1.05 \\ 0 \le \delta_1 \le 0 \\ -\pi \le \delta_2 \le \pi \\ -\pi \le \delta_3 \le \pi \end{aligned} \quad (107)$$

Note that for the case with a line element on outage, only the system conductance and susceptance matrices (i.e., the system bus admittance matrix) will change, and the steps outlined above are repeated.

Composite System Adequacy Assessment

The system states and their probability of occurrence are presented in Table 33. State probabilities are equal to the ones of the DC based example, which are

Table 33 The system states with probability and severity

Event	State of the components	Probability	Load curtailment [MW]
P(N)	G1, L1	0.9801	0
P(A)	$\underline{\text{G1}}$, L1	0.0099	20.3
P(B)	G1, $\underline{\text{L1}}$	0.0099	19.3
P(C)	$\underline{\text{G1}}$, $\underline{\text{L1}}$	0.0001	21.0

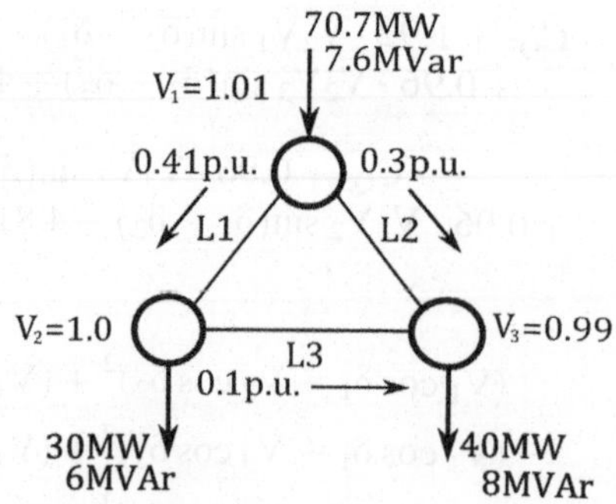

Fig. 9 Event P(N)

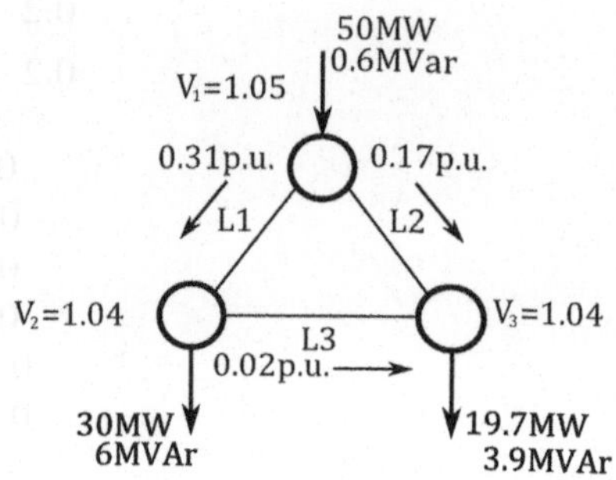

Fig. 10 Event P(A)

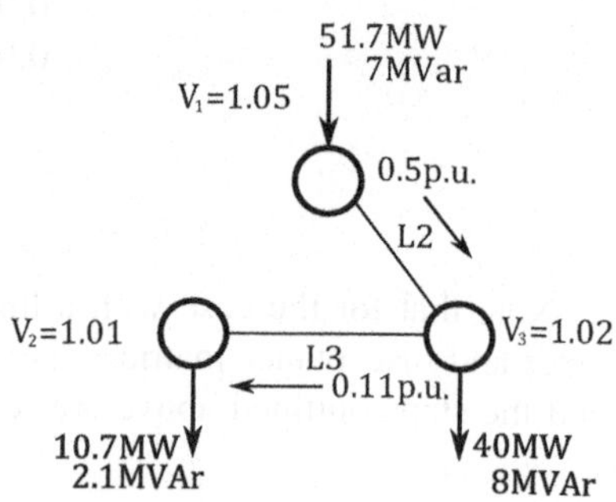

Fig. 11 Event P(B)

calculated in (50). The severity of each state, e.g., the active power load curtailment of each system state, is calculated by solving the corresponding OPF problem.

The following illustrations in Figs. 9, 10, 11 and 12 show the system states where the flow of currents are included along with some additional details.

The reliability indices used in the assessment are the LOLE and EENS indices. A LOLE of 0.0199 years in one year, same as the LOLE of the DC based example, is obtained, since load curtailments occur for the same system states. However, the

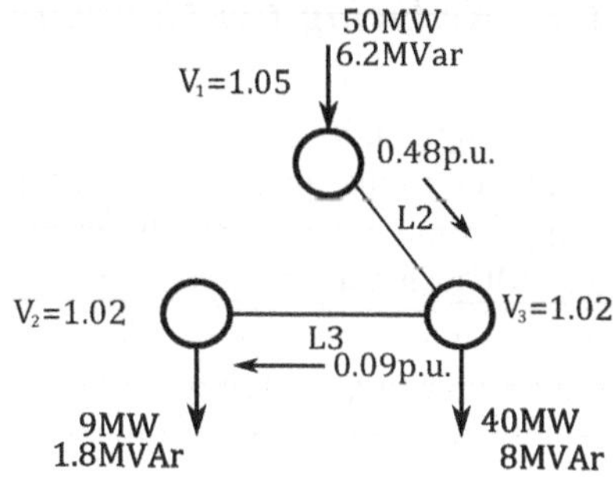

Fig. 12 Event P(C)

Table 34 Additional example

Event	State of the components	Probability	Load curtailment [MW]
P(N)	G1, L1	0.9801	0
P(A)	G1, L1	0.0099	20.3
P(B)	G1, L1	0.0099	21.9
P(C)	G1, L1	0.0001	21.9

calculation for the EENS index yields a different result of 3452 MWh in one year, which is slightly lower than the DC based result. The higher EENS result of the DC based solution is due to the approximation of voltages at 1.0 p.u., not accounting for the increased power transfer capability of a transmission line when the voltages are increased.

$$\begin{aligned}\mathrm{EENS} &= [\mathrm{C(N)}\cdot\mathrm{P(N)}+\mathrm{C(A)}\cdot\mathrm{P(A)}+\mathrm{C(B)}\cdot\mathrm{P(B)}+\mathrm{C(C)}\cdot\mathrm{P(C)}]\cdot\mathrm{T}\\ &= (0\cdot 0.9801+20.3\cdot 0.0099+19.3\cdot 0.0099+21\cdot 0.0001)\cdot 8760\\ &= 3452.67\mathrm{MWh/year}\end{aligned} \quad (108)$$

To highlight the impact of reducing the maximum voltage limit, the severities of each system state when the maximum voltage limit is changed to 1 p.u., are provided in Table 34.

The new severities give an expected EENS of 3679 MWh per year as shown below, which is higher than the expected EENS of the DC based approach.

$$\begin{aligned}\mathrm{EENS} &= [\mathrm{C(N)}\cdot\mathrm{P(N)}+\mathrm{C(A)}\cdot\mathrm{P(A)}+\mathrm{C(B)}\cdot\mathrm{P(B)}+\mathrm{C(C)}\cdot\mathrm{P(C)}]\cdot\mathrm{T}\\ &= (0\cdot 0.9801+20.3\cdot 0.0099+21.9\cdot 0.0099+21.9\cdot 0.0001)\cdot 8760\\ &= 3678.94\mathrm{MWh/year}\end{aligned} \quad (109)$$

1.4 *Reducing the Computation Time of HLII Assessment*

The computation time associated with HLII evaluation of system states increases compared to the time spent on evaluation of system states at HLI, because the evaluation involves solving OPF problems instead of simple algebraic equations. When the OPF problems are based on an AC representation of the network, the computation time increases even more. Thus, it can be inferred that measures have to be taken to reduce the computation time spent on the evaluation of system states. Criteria can be established which separate the system states into states that need to be solved by a contingency solver and states that are certain to have no load curtailments, by performing some simple algebraic calculations. A reduction in the number of system states that need to be run through a contingency solver ought to be achieved. Another measure taken to improve the speed of the simulations is by implementing a parallel simulation scheme. For more details in this regard, the reader is referred to [7].

1.4.1 Parallel Computation

MATLAB has an inbuilt toolbox [14] that enables processing of loops in parallel by different *workers*. MATLAB handles the computation in parallel threads automatically when the standard "for" loop command is replaced with a "parfor" loop. The requirement that needs to be in place is the independence between iterations, e.g., an iteration cannot depend on the calculations performed in the previous iteration. When a "parfor" loop runs, the iterations are run in a random sequence that is different from the deterministic sequence when a "for" loop is executed. Each worker has its own unique random number stream, ensuring that the streams are independent. The advantage of using parallel processing increases with the number of available CPU cores, where parallel processing in two cores can reduce the computation time by half if effective parallelization of the code into threads is achieved. A further reduction in computation time can be achieved if more CPU cores are available, but it can be limited by overhead from communication between the threads.

2 Concluding Remarks

In this chapter, the small but important details required to successfully implement the analytical methods or MCS methods for power system adequacy assessment from the perspective of contingency solvers have been discussed. Using the contingency solvers described in this chapter, it has been further verified that the adequacy indices pertaining to standard reliability test systems are similar to the ones obtained in corresponding benchmark research papers. It is observed that the adequacy indices at the various load points in a system are dependent upon the load curtailment philosophy. It is also observed that the estimates obtained by the DC-based approach are lower than those that include all the AC considerations into the assessment.

The AC based approach, however, has a large computation time associated with the evaluation of system states, due to the AC OPF solutions. Thus, it would be interesting to look at a decoupled approach, where voltage limits and reactive power generation are considered, to see what gains could be achieved in terms of reduced computation time, and how estimates of the indices are affected. It should, however, be noted that a decoupled OPF approach is different from a decoupled power flow approach, because decoupled OPF is approximate. It could also be interesting to examine the effect of using the branch flow model instead of the bus injection model to represent the system in the AC OPF problems.

References

1. Bhavajaru MP, Billinton R, Reppen ND, Ringlee RJ, Albrecht PF (1988) Requirements for composite system reliability evaluation models. IEEE Trans Power Syst 3(1):149–157
2. Billinton R, Khan E (1992) A security based approach to composite power system reliability evaluation. IEEE Trans Power Syst 7(1):65–72
3. Billinton R, Li W (1994) Reliability assessment of electric power systems using Monte Carlo methods. Plenum Press, New York
4. Billinton R, Allan RN (1996) Reliability evaluation of power systems, 2nd edn. Plenum Press, New York
5. Jirutitijaroen P, Singh C (2008) Comparison of simulation methods for power system reliability indexes and their distributions. IEEE Trans Power Syst 23(2):486–493
6. Kadhem AA, Wahab NIA, Aris I, Jasni J, Abdalla AN (2017) Computational techniques for assessing the reliability and sustainability of electrical power systems: a review. Renew Sustain Energy Rev 80:1175–7786
7. Laengen ØS (2018) Application of Monte Carlo simulation to power system adequacy assessment. Masters thesis, Norwegian University of Science and Technology
8. Billinton R et al (1989) A reliability test system for educational purposes—basic data. IEEE Trans Power Syst 4(3):1238–1244
9. Glover JD, Sarma MS, Overbye TJ (2011) Power system analysis and design, 5th edn. Nelson Education, Stamford
10. Wangdee W, Billinton R (2006) Impact of load shedding philosophies on bulk electric system reliability analysis using sequential Monte Carlo simulation. Electr Power Components Syst 34(3):355–368
11. Frank S, Rebennack S (2016) An introduction to optimal power flow: theory, formulation, and examples. IIE Trans 48(12):1172–1197
12. Dommel H, Tinney W (1968) Optimal power flow solutions. IEEE Trans Power Appar Syst PAS-87(10):1866–1876

13. Mathworks (2018) Optimization toolbox user's guide R2018a. The MathWorks, Inc, Natick, Massachusetts
14. Mathworks (2018) Parallel computing toolbox user's guide R2018a. The MathWorks, Inc, Natick, Massachusetts

Øystein Stake Laengen was born in Oslo, Norway, in 1989. He graduated from the five-year integrated Masters programme in Energy and Environmental Engineering, with specialization in Electric Power Systems in 2018 at the Norwegian University of Science and Technology (NTNU), Trondheim. In 2019, he was chosen as the recipient of the Norwegian Electro Technical Association's prestigious prize for Best Masters Student in Electric Power Engineering at NTNU. He is presently working at Statnett, the Transmission System Operator of the Norwegian Power System, as Trainee.

Dr. Vijay Venu Vadlamudi is currently Associate Professor and Deputy Head of the Department at the Department of Electric Power Engineering, NTNU. He obtained his M. Tech. and Ph.D. degrees from the Indian Institute of Technology (IIT) Bombay, India, in 2006 and 2011, respectively. He was a Postdoctoral Research Fellow at the Department of Electric Power Engineering, NTNU, from 2011–2015. Prior to this, he was a visiting researcher at the Power Systems Research Group of the University of Saskatchewan, Canada, from 2010–2011. His areas of research interest include:

- Reliability and Risk—based Power System Operation and Planning Practices
- Probabilistic Methods Applied to Power System Analysis
- Reliability—based Appraisal of Smart Grid Challenges and Realisation
- Resource Adequacy in Electricity Markets
- Electrical Energy Informatics.

Reliability Analysis of Microgrid Systems Using Hybrid Approaches

Moushumi Patowary, Bimal C. Deka and Gayadhar Panda

Abstract The reliability analysis is a crucial phenomenon for the design and maintenance of a microgrid system. In this Chapter, few hybrid techniques are proposed to assess the failure probability and reliability of the microgrid system. An inverter dominated photovoltaic (PV) system interfacing to IEEE 5-bus microgrid system is developed and system performances are evaluated in terms of its failure probability. Considering different design topologies, four different possible cases are presented to analyse mean-time-to-failure (MTTF), sectional and overall failure rates of the microgrid system using reliability block diagram (RBD) technique. Another two hybrid approaches are presented using both fault tree analysis (FTA) and binary decision diagram (BDD) to evaluate the performance of the microgrid system in terms of reliability. Moreover, a comparative study is also executed in order to inspect the effectiveness of the proposed hybrid approaches.

Keywords Microgrid · Reliability block diagram (RBD) · Fault tree analysis (FTA) · Binary decision diagram (BDD) · Reduced BDD (RBDD)

M. Patowary (✉) · G. Panda
Department of Electrical Engineering, NIT Meghalaya, Bijni Complex, Laitumkhrah, Shillong 793003, Meghalaya, India
e-mail: moushumi.patowary@nitm.ac.in

G. Panda
e-mail: gayadhar.panda@nitm.ac.in

B. C. Deka
Department of Electrical Engineering, Assam Engineering College, Jalukbari, Kamrup Metro, Guwahati 781013, Assam, India
e-mail: deka_bimal@rediffmail.com

D. R. Karanki et al. (eds.), *Advances in RAMS Engineering*, Springer Series in Reliability Engineering,
https://doi.org/10.1007/978-3-030-36518-9_2

1 Introduction

The present scenario of power system utility is more prone towards massive penetration of renewable energy sources (RESs) to sustain a healthy atmospheric condition. To address the climate changes and depletion of fossil fuels, RESs on a massive scale are playing an important role in transitioning the existing power market to a low carbon feeding modern power infrastructure. Power system blackouts have turned out to be a turning point for the power industries that causes a large series of consequences. Customers are thus, eyeing for a resilient utility followed by well-configured backup systems which have the enough endurances to promise the demand of the customers with utmost reliability. A microgrid is nothing but the amalgamation of RES, loads, conventional power plants, power electronic interfacing devices and the existing utility grid, which is the biggest revolution and solution to the modern power industry. A microgrid can be categorized into DC microgrid, AC microgrid, and hybrid microgrid. Though the availability of photovoltaic sources (PVs) is abundant, yet, it is intermittent in nature which affects quality and reliability of the supplied power. When a fault or short circuit happens in a particular section of a microgrid system, the other section of the system might turn out to be overloaded or isolated through the tripping action of the switchgear for load-redistribution. This load-redistribution frequently leads to cascading occurrences and is propagated throughout the microgrid system. This could be the reason of component or system level failures causing power disruptions in the system. Its negative influences on socio-economic front requires that a microgrid has to be very reliable and sustainable from the design time itself. As the percentage of penetration of RESs is rapidly growing high, reliability challenges and its innovative solutions are thus analysed beforehand in order to sustain system availability. In evaluating the reliability of a microgrid system, factors such as failure rates and mean-time-to-failure (MTTF) plays the significant role in defining lifetime of the system. However, reliability assessment and maintenance of a microgrid system is the fundamental procedure to be carried out for a system designer. This can provide the designers, operators and customers probable failure modes of a system/sub-system/component.

By definition, reliability is the ability of an item or a system to perform a mandatory objective under the given environment and operational conditions for a certain period of time (ISO8402) [22]. The arrival of intelligent power networks that allow efficient exploitation of the energy resources, decrease earth's temperature, carbon emissions, increases sustainability of microgrid and promises of a greener atmosphere [30]. It provides better flexibility in the management, control and operation of the system [31]. One of the prime issues regarding the uses of distributed generations (DGs), such as photovoltaic (PV) or wind energy resources to the existing utility grid is their consequences on system reliability and availability [8], Spinato et al. [27]. As the percentage of DG capacity is growing highly, reliability challenges and its innovative solutions are also discovered. In evaluating reliability of microgrid system, numerous factors such as the component failure, system configuration, ambient conditions, environmental disaster etc. are mostly affected [18]. Some of these factors may not be the ideal cause for total system failure, but have the de-rating impact on power generation and distribution. As the solar irradiance reaching to horizontal earth surface is irregular in nature, therefore, production of power from a PV is unpredictable [29]. Other ambient factors with de-rating effects on the microgrid systems are dust, temperature, snow accrual, shading and climate condition etc. [7, 20, 24]. The essential parts of a microgrid system subject to its failure include PV modules, DC-DC converterS, DC-AC inverterS, transformers, distribution lines, bus-bars, point of common coupling (PCC) etc. where PVs and inverters are the most vulnerable [10].

The reliability assessment of a microgrid system is very crucial for the system design and its maintenance. It can provide the designers, operators and customers about probable failure modes of a system or sub-system or component [12]. The reliability assessment methods contain, FMECA-failure modes, effect and criticality analysis, BBN-Bayesian belief networks, ETA-event tree analysis, PRA-probabilistic risk assessment, RBD-reliability block diagrams, FTA-fault tree analysis, MRM-Markov reliability modelling. The FMECA is an additional scope of failure mode and effects analysis (FMEA). FMEA is a bottom-up, inductive logical technique which is performed either at the component or system level. The FMECA analysis on PV plants considers all the failure and de-gradation modes from both the security and performance perspectives and based on this results, reliability analysis is made for each PV plant to decide the main failure-modes through grading and prioritizing the failure-modes [26]. A number of approaches with fuzzy theories have also been suggested in literature to overcome the drawbacks of FMECA method. Two different approaches for handling the shortcomings of traditional FMEA are listed in Gargama et al. [11]. The BBN's facility to allow impacts of a classification decision to classify the practical components into hardware and software is shown in Olson et al. [21].

Another reliability assessment method, ETA which is an inductive method, checks occurrence of an event and their potential deviation. The discrepancies of ETA method and a way out to this is also suggested for a simple system in Andrews et al. [2]. FTA is a top-down deductive fault or failure investigation methodology that is used to establish root-cause of a failure in a component or system. The failures and its possible modes are inter-connected to logical gates [1, 17]. ETA and FTA methods are complimentary to each other. In ETA, the focus is on the penalty after the occurrence of an event, while in FTA, the focus is on the causes that lead to the event. BDD-Binary Decision Diagram is again another tool for decision-making used in the analysis of ETA and FTA. It assists by developing qualitative and quantitative evaluation of reliability [15, 16]. BDD tells about the failure logic of a component or a system in disjoint form which gives the benefit of less computational burden issue. In BDD structure, first a fault tree is constructed and then it is converted to BDD by following the best ordering rule that provides efficient representation of failure logic. Once the ordering is done, Rauzy approach is utilized repeatedly to generate the BDD by employing an **if-then-else** (***ite***) structure from Shannon's formula [5, 23]. Another method, SBDD-Sequential Binary Decision

Diagram, for the evaluation of cold-standby systems is reported in Xing et al. [32]. A reduced BDD (RBDD) method for the evaluation of defect-tolerant systems-on-chip is discussed in Carrasco et al. [6], Moeinzadeh et al. [19]. PRA is employed to demonstrate how a relevant approach to blackout probabilistic risk assessment allows conducting such studies [13]. The aim of PRA in cascading outages is to recognize cascading outage scenarios, and to assess their unfavourable penalties, mentioned in Henneaux et al. [14]. RBD is an another popular probabilistic analysis technique for the assessment of reliability and this can be applied to any realistic model [3], Rausand et al. 2009. MRM is based on probabilistic transitions relating to vigorous, intermediary, and failed states of a component or system. A complete structure for assessment of reliability of an induction motor under indirect field-control using MRM is presented in Bazzi et al. [4].

Although, FMEA and its extended version FMECA, FTA and its expansion ETA are used in mechanical and electrical industries, yet, all these risk analysis tools are very time overwhelming processes, and thus, the end result may not be able to thoroughly recognize the possible failure modes. This shortcoming may conquer through the combined effect of FMEA and FTA. Again, BBN learning is enormously expensive in terms of computation. It performs well if the network configuration is predefined, otherwise it is hard to implement for a large dimensional network. Moreover, it is an un-automated learning risk analysis method. In contrast, RBD is the most popular tool to assess reliability but it can't guarantee the correctness owing to their innate limitations. On the other hand, the MRM method, which can be applied to both repairable and non-repairable systems, needs full attention in model building phase. It requires exponential distribution function in order to represent time to failure and repair date providing further constraints. Therefore, MRM model can become unnatural if all these constraints are not being satisfied. However, it cannot produce accurate answer as it mostly depends upon the knowledge based experience and decision of the modeller.

In this Chapter, reliability modelling and quantitative assessment of the IEEE 5-bus (G. Sybille (Hydro-Quebec), Initializing A 5-Bus Network with the Load Flow Tool of Powergui) microgrid system are presented considering different system design topologies. Overall system failure probability and reliability calculations are also performed using few hybrid assessment techniques. RBD, FTA, FTA-BDD and FTA-RBDD approaches are applied to grid-connected PV integrated IEEE 5-bus microgrid system for its probabilistic assessment. Sectional and overall system failure rates along with its MTTFs are calculated based on its design strategies. FTA-BDD and FTA-RBDD which does not necessarily require exponential distribution for the time to failure analysis can be used to side-step the shortcomings of traditional risk analysis methodologies. Moreover, comparative estimation is also computed among the RBD, FTA, FTA-BDD and FTA-RBDD hybrid techniques in order to demonstrate effeteness and easy adaptability of FTA-RBDD method in reliability evaluation of the microgrid system.

2 System Descriptions

The circuit diagram shown in Fig. 1 mainly consists of PV modules, incremental condenser (InC.)-MPPT based parallel-boost converters, DC-link capacitors (C_{dc}) that connects to the output of boost converter, centralized DC-AC inverter and passive filter/PFC bank that connects to the output of inverter. The inverter dominated parallel-PV system is then connected to the multi-string, multi-bus microgrid system with an IEEE-5 bus radial system where 300 kW parallel-PV systems are connected in parallel to the 150 MW diesel generator (DG). The whole power plant is then integrated to the utility grid. A dual-stage power conversion technology is being incorporated for the integration of parallel-PV system to the utility grid through the inverter control operation. The first stage is the boost converter operation with InC. based MPPT logic and second stage is the DC-AC inverter operation which is used for fetching the PV generated DC-power to the loads and the surplus power is fed back to the utility grid. A stable DC-link voltage is usually applied at the inverter input terminal which is controlled by two control loop, viz. (i) outer-voltage-control loops (i.e. using PIs) (ii) inner-current-control loop (i.e. using adaptive current control). This multi-string, multi-bus microgrid system is considered for the evaluation of system reliability based on the following reliability assessment techniques.

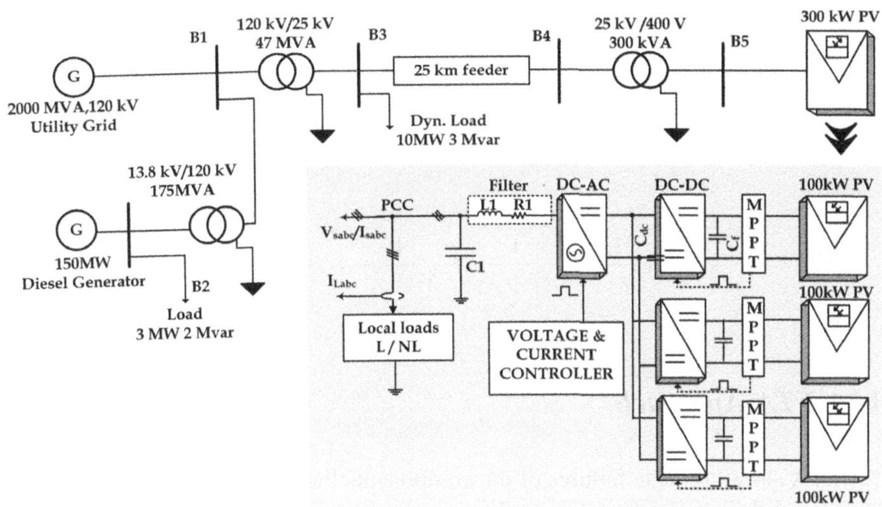

Fig. 1 IEEE-5 bus radial system with grid-connected 300 kW-PVs and 150 MW-diesel generator —a multi-string, multi-bus microgrid system

3 Hybrid Reliability Assessment Approaches

3.1 RBD Approach

The RBD is a system-level reliability evaluation tool which can be expressed with the block diagrams by taking into consideration the function of each and every components or sub-systems. RBDs are appropriate for those systems which have non-repairable components and where the order of failures does not matter. The quantitative measures for the reliability evaluation include time to failure, reliability function R(t), failure rate function, mean time to failure (MTTF) etc. It is assumed that all the failure distributions are supposed to have the exponential distributions. Let us consider two non-repairable systems viz. system1 and system2. In system1, components A and B are connected in parallel and the whole system is connected in series with the component C and in system2 all the components A, B and C are connected in parallel. If λ_1, λ_2 and, λ_3 be the failure rates of the components A, B and C respectively, then R(t) and MTTF for the systems1 and system2 are derived as follows.

For system1,

$$R(t) = e^{-(\lambda_1+\lambda_2)t} + e^{-(\lambda_2+\lambda_3)t} - e^{-(\lambda_1+\lambda_2+\lambda_3)t} \tag{1}$$

$$MTTF = \frac{1}{\lambda_1+\lambda_2} + \frac{1}{\lambda_2+\lambda_3} - \frac{1}{\lambda_1+\lambda_2+\lambda_3} \tag{2}$$

For system2,

$$R(t) = e^{-\lambda_1 t} + e^{-\lambda_2 t} + e^{-\lambda_3 t} - (e^{-(\lambda_1+\lambda_2)t} + e^{-(\lambda_2+\lambda_3)t} + e^{-(\lambda_1+\lambda_3)t}) + e^{-(\lambda_1+\lambda_2+\lambda_3)t} \tag{3}$$

$$MTTF = \frac{1}{\lambda_1} + \frac{1}{\lambda_2} + \frac{1}{\lambda_3} - \left(\frac{1}{\lambda_1+\lambda_2} + \frac{1}{\lambda_2+\lambda_3} + \frac{1}{\lambda_1+\lambda_3}\right) - \frac{1}{\lambda_1+\lambda_2+\lambda_3} \tag{4}$$

3.2 FTA Approach

The FTA shows diverse failures of the components that are essential to result in the TOP-event failure. It gives the qualitative and quantitative assessment on system unavailability. Figure 2 reflects one of the basic FTA diagrams. The evaluation of the complex system may produce numerous combination of events, which are known as the cut-sets that leads to TOP-event failure. The determination of the minimal cut sets for a complex system is also a time-consuming process even on high speed computers. If a fault tree has many minimal cut sets, then the exact TOP-event failure probability calculation would be a very lengthy process which is

Fig. 2 A FT model

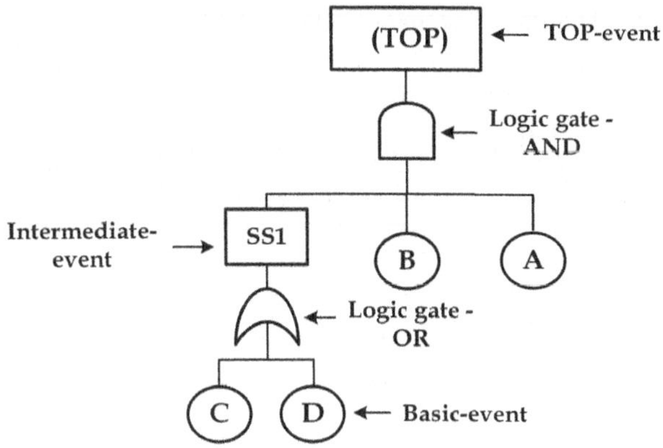

beyond the capability of computers. As a consequence, approximation techniques have been introduced with a loss of accuracy.

3.3 *FTA-BDD and FTA-RBDD Approaches*

The BDD method is an alternative method to the conventional one in evaluating the FTA. BDD has the benefits both in accurateness and competence over the conventional methods. The BDD finds out the failure logics in a mutually exclusive way that gives the benefits from the computational burden issue. In BDD structure, first a fault tree is constructed and then it is converted to a BDD by following the best variable ordering rule that provides the efficient representation of failure logic. Once ordering is done, Rauzy approach is utilized repeatedly to generate the BDD by applying an ***ite*** structure used in Shannon's formula. Let R and S be any two nodes in BDD, where, R = $ite(a, f_1, f_2)$, S = $ite(\mathrm{b}, g_1, g_2)$ and, $< bop >$ corresponds to a boolean operation.

if $a <$ b (i.e. a performs before b in variable ordering) then,

$$R < bop > S = ite(a, f_1 < bop > S, f_2 < bop > S) \tag{5}$$

if a = b then,

$$R < bop > S = ite(a, f_1 < bop > g_1, f_2 < bop > g_2) \tag{6}$$

A BDD has to be in its minimal form to generate the minimal cut sets, but this is not an ideal case always. Therefore, in order to get minimal cut sets from a BDD, it has to undergo some minimization techniques which leads to the generation of RBDD. The reduction rules are—*Rule#1*: eliminate duplicate terminals. If BDD contains more than one terminal 0-node, then we redirect all the edges which point to such 0-nodes to just one of them. A similar technique is used for terminal 1-node too. Therefore, the new BDD model has only two terminal nodes. *Rule#2*: eliminate redundant nodes. If both the edges of non-terminal nodes are pointing to the same nodes, then one can be eliminated. *Rule#3*: merge duplicate nodes, i.e., nodes must be the unique one.

4 Results and Assessment

4.1 Reliability Assessment of Microgrid System Employing RBD Approach

The RBD representation of PV interfaced grid-connected IEEE-5 bus microgrid system is drawn in Fig. 3. The different part names of the PV module, DC-DC and DC-AC systems along with their failure rates (λ_i) are tabulated in Tables 1 and 2 respectively. The failure rates for different components of the microgrid system are listed in Table 3 [9, 25]. To assist the quantitative reliability assessment, emphasis has been given on structural design of microgrid system. Depending upon proper component connection method or design strategies, the lifetime of a system can be figured out. Therefore, four such different design strategies have been considered as case studies, which are—(i) one PV and one inverter, (ii) parallel PVs and one inverter, (iii) one PV and parallel inverters, and (iv) parallel PVs and parallel inverters. By using Eqs. (3), (4), and (7)–(10), a comparative assessment is performed based on sectional, overall system failure rates and mean time to failure which is shown in Table 4. The failure rates of one PV and one inverter (INV) with three leg system are calculated as follows.

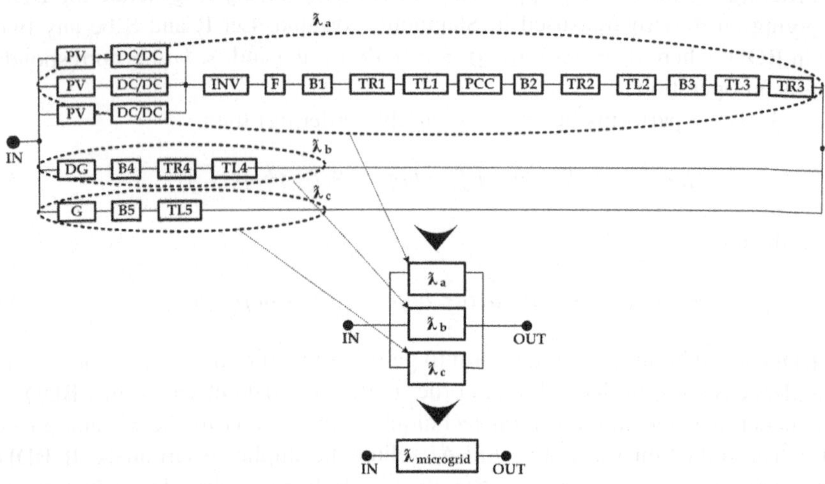

Fig. 3 RBD for the microgrid system

Table 1 Part name and failure rates (f/hr) for the PV panel and DC-DC (boost) converter

Part name		Failure rate (λ_i), f/hr
PV Panel (λ_1)		4.566×10^{-6}
DC-DC (boost) converter	IGBT (λ_2)	0.9×10^{-6}
	Inductance (λ_3)	0.038×10^{-6}
	Capacitance (λ_4)	0.87×10^{-6}
	Diode (λ_5)	0.8×10^{-6}
Voltage sensor (λ_6)		0.57×10^{-6}
Current sensor (λ_7)		0.5×10^{-6}
Filter capacitor (λ_8)		3.03×10^{-6}

Table 2 Part name and failure rates (f/hr) for DC-AC converter (inverter)

Part name	Failure rate (λ_i), f/hr
IGBT (λ_2)	0.9×10^{-6}
Diode (λ_5)	0.8×10^{-6}
Snubber (λ_9)	0.001×10^{-6}
Gate drive (λ_{10})	1.0×10^{-6}
DC-link capacitor (λ_8)	3.03×10^{-6}
Voltage sensor (λ_6)	0.56×10^{-6}
Current sensors (λ_7)	0.5×10^{-6}

Table 3 Component failure rates in failure/hour for the microgrid system

Components	Failure rate (λ_i), f/hr
PV system (λ_{PV})	11.274×10^{-6}
DC-AC (inverter) system (λ_{INV})	20.306×10^{-6}
Filter (λ_F)	14.2692×10^{-6}
Busbar (λ_B)	0.205×10^{-6}
Transmission line (λ_{TL})	0.0913×10^{-6}
Transformer (λ_{TR})	2.283×10^{-6}
Circuit breaker/PCC (λ_{PCC})	2.2831×10^{-6}
Diesel generator (DG) (λ_{DG})	11.4155×10^{-6}
Grid (λ_G)	2.283×10^{-6}

$$\lambda_{PV} = \lambda_1 + (\lambda_2 + \lambda_3 + \lambda_4 + \lambda_5 + \lambda_8) + (\lambda_6 + \lambda_7) \quad (7)$$

$$\lambda_{INV} = 6 \times (\lambda_2 + \lambda_5 + \lambda_9 + \lambda_{10}) + \lambda_8 + (\lambda_6 + \lambda_7) \quad (8)$$

The failure rates of parallel PVs and parallel INVs (i.e. in operation N = 2, redundant X = 1) are calculated as follows [33].

Table 4 Comparison of sectional, overall system failure rates (λ_i) (f/year) and mean time to failure ($MTTF_i$) (years)

Cases	Sectional failure rates (λ_i)			Sectional MTTFs ($MTTF_i$)			Overall failure rates ($\lambda_{microgrid}$)	Overall MTTFs ($MTTF_{microgrid}$)
	(λ_a)	(λ_b)	(λ_c)	($MTTF_a$)	($MTTF_b$)	($MTTF_c$)		
Case (i)	0.48944	0.14261	0.02260	2.04322	7.01286	44.25652	0.022098	45.251316
Case (ii)	0.51202			1.95306			0.022101	45.247340
Case (iii)	0.52501			1.90476			0.022103	45.245322
Case (iv)	0.54476			1.83560			0.022106	45.242570

$$\lambda_{N+X} = \frac{1}{MTTF_{N+X}} = \frac{1}{\frac{1}{\lambda_{PV}} \sum_{j=N}^{N+1} \frac{1}{j}} \quad (9)$$

The analysis shown in Table 4 says that grid is always considered in-operation i.e. with rare failure cases, and its MTTF is calculated as 44.256514 years. That means it takes 44.256514 years for its failure as the failure rate is very low, i.e. 0.022596 f/year. Again, the section with diesel generator has also showing lower failure rate i.e. 0.142596 f/year. Therefore, its MTTF will also be high which is calculated as 7.012855 years. The failure rates and MTTFs for the Sect. 1 have different values corresponding to four different possible combination of inverter dominated PV system. The combination of one PV and one inverter system shown in case (i) is considered as the normal case whose sectional failure rate/year, overall system failure rate/year and MTTF in years are calculated as 0.489424 f/year, 0.022098 f/year and 45.251316 years respectively. But in case (ii), where the parallel PVs and one inverter system is considered, i.e. the increase in power generation with redundant units, we are also increasing the sectional failure rate. At the same time, sectional MTTF is going down. Same are the cases with overall failure rate and MTTF. In case (iii), i.e. same generation but with the redundant inverter units, we are increasing the sectional failure rate and decreasing the sectional MTTF more than in case (ii). Again, in case (iv), which is the combination of case (ii) and case (iii), will deteriorate sectional, overall failure rates and MTTFs more than case (iii). Therefore, with the proposed system design complexities, system's overall failure rate increases and overall MTTF decreases from the point of reliability. But from the point of generation, this much of reliability evaluation risk is worth for consideration. This may not be the case with different system design constraints. Thus, reliability evaluation of a particular system is more of design specific.

4.2 *Reliability Assessment of Microgrid System Employing FTA-BDD and FTA-RBDD Approaches*

The following microgrid system shown in Fig. 4 has been consideration for reliability assessment using FTA-BDD and FTA-RBDD. The equivalent RBD and FT models are depicted in Figs. 5 and 6 respectively. The reliability evaluation of above shown FT model using BDD and RBDD concept has been applied as below. In ***ite***-structure, selection of the sub-node sharing not only reduces the requirement of computer memory allocation, but it also increases system efficiency. Actually, once an ***ite***-structure is calculated, there has no need of repetition. One of the predominant concepts in ***ite***-structure is ordering of variables. Let us consider, each component connected in IEEE-5 bus microgrid model shown in Fig. 4 is represented by a variable as described in Table 6.

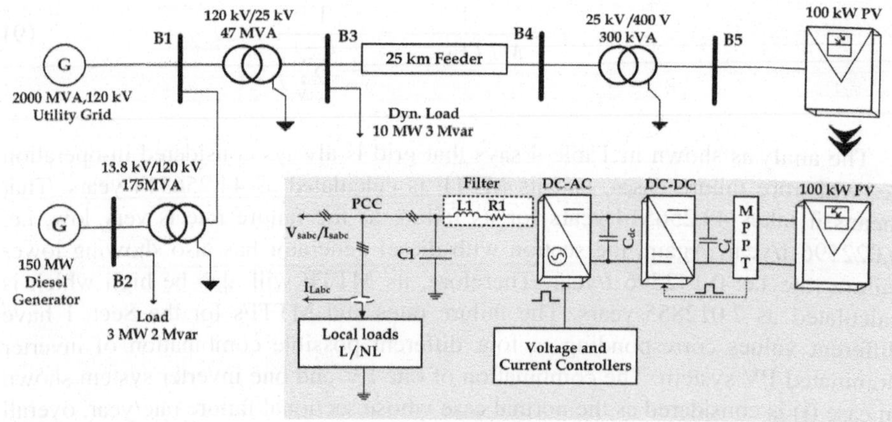

Fig. 4 IEEE-5 bus microgrid system with grid-connected 100 kW-PV and 150 MW—diesel generator

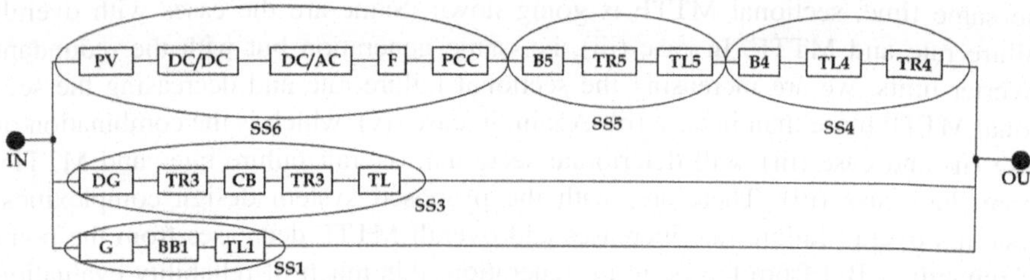

Fig. 5 RBD representation of microgrid model

The ordering of the variables is considered from the point of connection of the components in the system. Therefore, for the variable ordering, a < b < c < d < e < f < g < h < i < j < k < l < m < n < o < p < q < r < s, microgrid failure is calculated as follows.

$$microgridfailure = ite(a, ite(l, 1, ite(m, 1, ite(n, 1, ite(o, 1, ite(p, ite(q, 1, ite(r, 1, ite(s, 1, 0))), 0))))) \cdot ite(b, 1, ite(c, 1, ite(d, 1, ite(e, 1, ite(f, 1, ite(g, 1, ite(h, 1, ite(i, 1, ite(j, ite(q, 1, ite(r, 1, ite(s, 1, 0))), ite(k, ite(l, 1, ite(m, 1, ite(n, 1, ite(o, 1, ite(p, ite(q, 1, ite(r, 1, ite(s, 1, 0))), 0))))), 0))))))))))$$

The generation of FTA-BDD and FTA-RBDD models from the ***ite*** structure are drawn in Fig. 7a, b respectively. The microgrid failure probability calculated from the RBDD model is equal to the summation of failure probabilities of the disjoint paths. From Fig. 7b, possible mutually exclusive or disjoint paths are, $\bar{a}\bar{b}\bar{c}\bar{d}\bar{e}\bar{f}\bar{g}\bar{h}\bar{i}\bar{j}k\bar{l}\bar{m}\bar{n}\bar{o}p\bar{q}\bar{r}\bar{s}$ and al. Therefore,

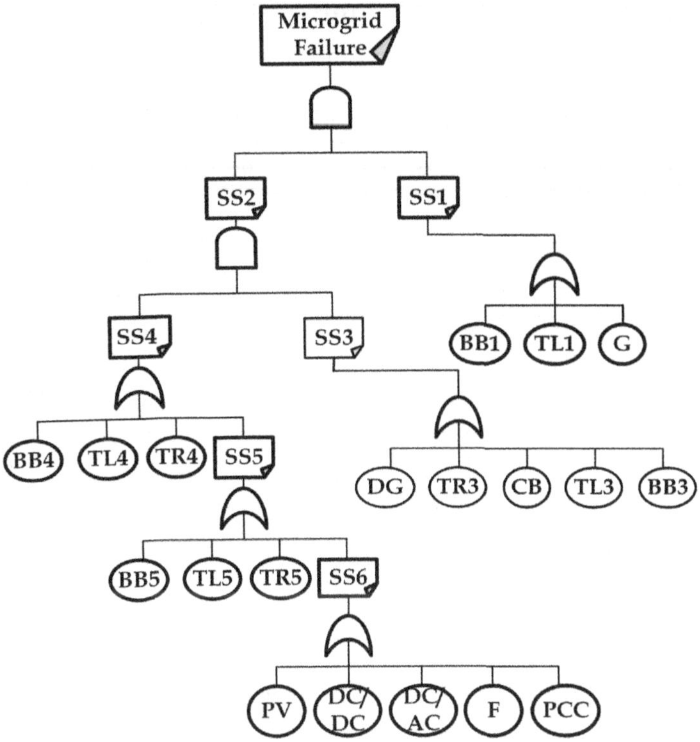

Fig. 6 FTA representation of microgrid model

Table 6 Variables assigned corresponding to components of microgrid system

Component name	Variable name	Component name	Variable name
PV panel (PV)	a	DC/DC converter (DC/DC)	b
DC/AC converter (DC/AC)	c	Filter (F)	d
PCC	e	Transformer-5 (TR5)	f
Transmission line-5 (TL5)	g	Busbar-5 (BB5)	h
Transformer-4 (TR4)	i	Transmission line-4 (TL4)	j
Busbar-4 (BB4)	k	Diesel generator (DG)	l
Transformer-3 (TR3)	m	Circuit breaker (CB)	n
Transmission line-3 (TL3)	o	Busbar-3 (BB3)	p
Busbar-1 (BB1)	q	Transmission line-1 (TL1)	r
Grid (G)	s	Sub-Station-1-6 (SS1-SS6)	

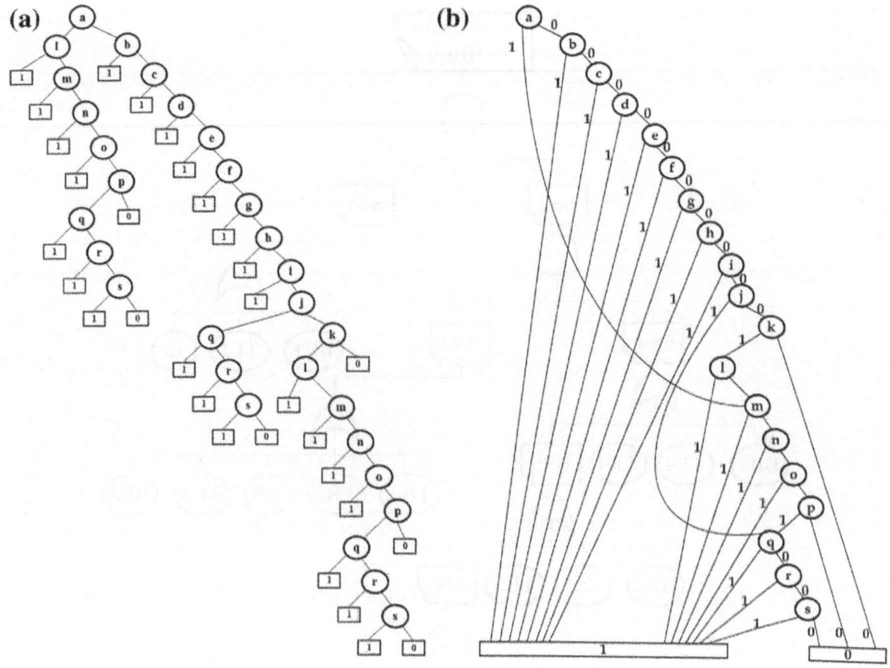

Fig. 7 **a** Creation of FTA-BDD model and **b** Creation of FTA-RBDD model

$$\begin{aligned} P(microgridfailure) &= P(\bar{a}\bar{b}\bar{c}\bar{d}\bar{e}\bar{f}\bar{g}\bar{h}\bar{i}\bar{j}k\bar{l}\bar{m}\bar{n}\bar{o}p\bar{q}\bar{r}s + al) \\ &= P(\bar{a}\bar{b}\bar{c}\bar{d}\bar{e}\bar{f}\bar{g}\bar{h}\bar{i}\bar{j}k\bar{l}\bar{m}\bar{n}\bar{o}p\bar{q}\bar{r}s) + P(al) \end{aligned} \quad (10)$$

4.3 Comparative Assessment

In order to verify the superiority of the FTA-RBDD model in the estimation of system failure probability over the other above mentioned approaches, let us conceder the availability and unavailability of each component present in the microgrid model as 0.98 and 0.02 respectively. From Table 7, it is observed that the FTA-RBDD performs superior than the RBD, FTA and FTA-BDD in accessing probabilistic assessment of the microgrid system in terms of its failure probability or reliability. And, FTA-BDD performs well as compared to the RBD and FTA. RBD and FTA are the most conventional approaches for evaluating the reliability of a system. In FTA-BDD, there are multiple terminal-1 s and terminal-0 s, which is the main drawback in this approach as it produces overlapped cut-sets. To determine minimal cut-set, this approach needs further efficient techniques that will eventually consume more time for the calculation and increases the computational complexity. Literature says that, RBDD technique is the most drill-down technique

Table 7 Comparative assessment on microgrid failure probability analysis among hybrid approaches

Probability/ Methods	RBD	FTA	FTA-BDD	FTA-RBDD
Microgrid failure probability	0.001126	0.001126	0.001046	0.0004511
Microgrid reliability	0.998774	0.998774	0.998954	0.9996489
Remarks	• conventional method of evaluating reliability	• conventional method of evaluating reliability • requires long probabilistic formulas and derivations to solve top event probability • time consuming process • computational burden issue etc.	• requires additional methods to figure out minimal cut-sets • overlapping of cut-sets • computational burden issues • time consuming • incorrect top event probability calculation etc.	• easy and efficient to figure out disjoint paths • very less computational burden issue or no burden • correct top event probability calculation • less time consuming etc.

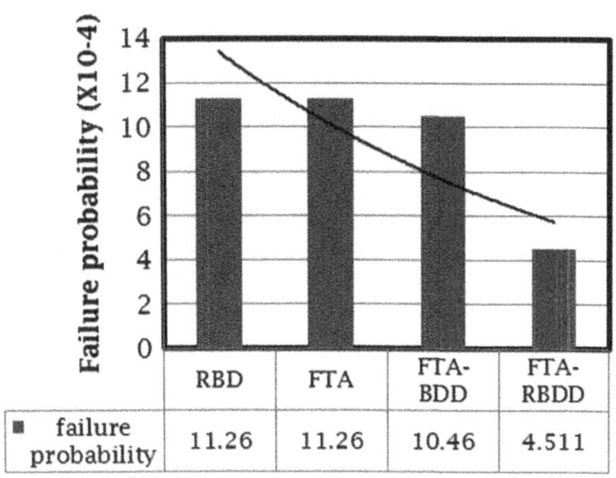

Fig. 8 Microgrid failure probability distribution

in formulating the accurate disjoint paths and it is validated through the comparative assessments of the proposed hybrid approaches. Formulation of accurate minimal cut-sets or disjoint paths will generate accurate calculation of Top-event failure probability. Thus, Table 7 and Fig. 8 validate that the FTA-RBDD method can provide the best result for probabilistic analysis in an application of the microgrid system.

5 Conclusions

This Chapter highlights hybrid reliability assessment approaches and its comparative verification performed in an IEEE-5 bus microgrid system. The proposed hybrid techniques are RBD, FTA, FTA-BDD and FTA-RBDD that are applied in the analysis of failure probability of microgrid system. Considering different design topologies, four different possible cases are presented in order to analyse the sectional, overall failure rates and MTTFs of the microgrid systems using RBD. Again, FTA is initiated for IEEE-5 bus microgrid system using computer-science dominated BDD and RBDD techniques. Later, a comparative study has also been performed based on the overall system failure probability among all the hybrid approaches. In FTA-BDD, as there are multiple terminal-1 s and terminal-0 s, it is not being a worth technique to figure out the minimal cut-sets. Additionally, it requires efficient techniques for minimal cut-set determination that will omit the overlapped paths completely. The comparative assessment proves that FTA-RBDD technique is the most drill-down technique in order to find out the minimal cut-sets or disjoint paths. It can provide accurate calculation of Top-event failure probability, and thus establishes that FTA-RBDD approach is comparatively superior and easy to implement than the others reliability evaluation approaches in calculating the system failure probability.

References

1. Aliee H, Zarandi H (2013) A fast and accurate fault tree analysis based on stochastic logic implemented on field-programmable gate arrays. IEEE Trans Reliab 62:13–22
2. Andrews JD, Dunnett SJ (2000) Event-tree analysis using binary decision diagrams. IEEE Trans Reliab 49:230–238
3. Anthony M, Arno R, Dowling N, Schuerger R (2013) Reliability analysis for power to fire pump using fault tree and RBD. IEEE Trans Ind Appl 49:997–1003
4. Bazzi AM, Dominguez-Garcia A, Krein PT (2012) Markov reliability modeling for induction motor drives under field-oriented control. IEEE Trans Power Electron 27:534–546
5. Bryant RE (1986) Graph-based algorithms for boolean function manipulation. IEEE Trans Comput C-35:677–690
6. Carrasco JA, Sune V (2009) An ROBDD-based combinatorial method for the evaluation of yield of defect-tolerant systems-on-chip. IEEE Trans Very Large Scale Integ (VLSI) Syst 17:207–220
7. Deline C, Dobos A, Janzoua S, Meydbray J, Donovan M (2013) A simplified model of uniform shading in large photovoltaic arrays. Sol Energy 96:274–282
8. Denholma P, Margolis RM (2007) Evaluating the limits of solar photovoltaics (PV) in traditional electric power systems. Energy Policy 35:2852–2861
9. Department of US Defense (2005) MIL-HDBK-21F Notice-1 and 2. Military Handbook-Reliability Prediction of Electronic Equipment, Washington, DC
10. Eltawil MA, Zhao Z (2010) Grid-connected photovoltaic power systems: technical and potential problems-a review. Renew Sustain Energy Rev 14:112–129
11. Gargama H, Chaturvedi SK (2011) Criticality assessment models for failure mode effects and criticality analysis using fuzzy logic. IEEE Trans Reliab 60:102–110
12. Hare J, Shi X, Gupta S, Bazzi A (2014) A review of faults and fault diagnosis in micro-grids electrical energy infrastructure. In: Energy conversion congress and exposition (ECCE), pp 3325–3332

13. Henneaux P, Member P, Labeau P, Maun J (2013) Blackout probabilistic risk assessment and thermal effects: impacts of changes in generation. IEEE Trans Power Syst 28:4722–4731
14. Henneaux P, Labeau PE, Maun JC, Haarla H (2016) A two-level probabilistic risk assessment of cascading outages. IEEE Trans Power Syst 31:2393–2403
15. Lin YH, Li YF, Zio E (2016) A reliability assessment framework for systems with degradation dependency by combining binary decision diagrams and monte carlo simulation. IEEE Trans Syst Man Cybern Syst 46:1556–1564
16. Mahdi L, Nadji B (2013) Application of the binary decision diagram (BDD) in the analysis of the reliability of the inverters. In: International conference on power engineering, energy and electrical drives. pp 1265–1271
17. Mao C, Canavero F (2016) System-level vulnerability assessment for EME: from fault tree analysis to bayesian networks #x2014;Part I: methodology framework. IEEE Trans Electromag Compat 58:180–187
18. Meyer EL, Dyk EEV (2004) Assessing the reliability and degradation of photovoltaic module performance parameters. IEEE Trans Reliab 53:83–92
19. Moeinzadeh H, Mohammadi M, Pazhoumand-dar H, Mehrbakhsh A, Kheibar N, Mozayani N (2009) Evolutionary-reduced ordered binary decision diagram. In: International conference on modelling and simulation, pp 25–29
20. Nimnuan P, Janjai S (2012) An approach for estimating average daily global solar radiation from cloud cover in Thailand. Procedia Eng 32:399–406
21. Olson JT, Rozenblit JW, Talarico C, Jacak W (2007) Hardware/Software partitioning using bayesian belief networks. IEEE Trans Syst Man Cybernet Part A Syst Humans 37:655–668
22. Rausand M, Høyland A (2005) System reliability theory: models, statistical methods, and applications, 2nd edn. Wiley, Hoboken, NJ
23. Rauzy A (1993) New algorithms for fault tree analysis. Reliab Eng Syst Saf 40:203–211
24. Sarver T, Al-Qaraghuli A, Kazmerski LL (2013) A comprehensive review of the impact of dust on the use of solar energy: History, investigations, results, literature, and mitigation approaches. Renew Sustain Energy Rev 22:698–733
25. Shi X, Bazzi AM (2015) Reliability modeling and analysis of a micro-grid with significant clean energy penetration, In: International conference on power electronics and ECCE Asia (ICPE-ECCE Asia), pp 202–207
26. Shrestha SM, Mallineni JK, Yedidi KR, Knisely B, Tatapudi S, Kuitche J, Mani GT (2015) Determination of dominant failure modes using FMECA on the field deployed c-Si modules under hot-dry desert climate. IEEE J Photovolt 5:174–182
27. Spinato F, Tavner PJ, Bussel GJWV, Koutoulakos E (2009) Reliability of wind turbine subassemblies. IET Renew Power Gener 3:387–401
28. Sybille G (Hydro-Quebec): https://www.mathworks.com/help/physmod/sps/examples/initializing-a-5-bus-network-with-the-load-flow-tool-ofpowergui.html
29. Tyagi VV, Rahim NAA, Jeyraj A, Selvaraj L (2013) Progress in solar PV technology: research and achievement. Renew Sustain Energy Rev 20:443–461
30. U.S. Department of Energy (2003) Grid 2030: A National Vision for Electricity's Second 100 Years
31. Wencong Su and Jianhui Wang (2012) Energy management systems in microgrid operations. Electr J 25:45–60
32. Xing L, Tannous O, Dugan JB (2012) 'Reliability analysis of nonreplicable cold-standby systems using sequential binary decision diagrams. IEEE Trans Syst Man Cybern Part A: Syst Humans 42:715–726
33. Yu X, Khambadkone AM (2012) Reliability analysis and cost optimization of parallel-inverter system. IEEE Trans Ind Electron 59:3881–3889

Moushumi Patowary received her Ph.D. degree in Electrical Engineering from the National Institute of Technology Meghalaya (NIT Meghalaya) Shillong, India, in 2018. With teaching plus industry experiences for more than 12 years, she is currently a Guest Faculty with the Department of Electrical Engineering, National Institute of Technology, Arunachal Pradesh, India. Her research interests include distributed generation, modeling and control of microgrid, power quality assessment, microgrid reliability evaluation, artificial intelligence (AI) techniques, power electronics (PE), and drives-based power systems.

Bimal C. Deka received his Ph.D. degree in Power System Reliability from IIT Bombay, Mumbai, India, in 2004. With teaching experience of more than 26 years, he is currently an Associate Professor with the Department of Electrical Engineering, Assam Engineering College, Guwahati, India. His research interests include power system reliability and renewable energy systems.

Gayadhar Panda (M'16–SM'19) received his B.E. degree in Electrical Engineering from the Institute of Engineers, Kolkata, India, in 1996; master's degree in Power Electronics from Bengal Engineering College (presently IIEST), Shibpur, India, in 1998; and Ph.D. degree in Electrical Engineering from Utkal University, Bhubaneswar, India, 2007. He is currently a Professor with the Electrical Engineering Department, National Institute of Technology, Meghalaya, India. He has served as head of the department and chairman of various committees at institute level. He is also currently looking after dean academic affairs (AA) and Chief Vigilance Officer at NIT Meghalaya. He has published more than 80 technical papers in national and international conferences proceedings or journals. He has more than 20 years of teaching experience. His work involves design, implementation, and operation of ac/dc microgrid with interfacing converters that use digital signal processing, artificial intelligence techniques, and other novel control methods. His current research interests include automatic generation control, stability improvements using flexible alternating current transmission system devices, power quality, power electronic converters, and distributed power generation. Dr. Panda has received Institution Medal for obtaining the highest marks in graduation and Power Medal for his one of the research papers.

Reliability Prediction of Instrumentation and Control Cables for NPP Applications

T. V. Santhosh, V. Gopika, A. K. Ghosh, B. G. Fernandes and J. Chattopadhyay

1 Background

Instrumentation and control (I and C) cables are one of the most important components in nuclear power plants (NPPs) because they provide power to safety-related equipment and also to transmit signals to and from various controllers to perform safety and control functions. A nuclear power plant may contain more than 50,000 electric cable circuits, of which about 60% are control circuits, 20% are instrumentation, 13% are AC power, 1% are DC power, and the remainder are miscellaneous communications circuits [1]. Depending on their location and application, these cables are exposed to a wide range of stressors such as temperature, radiation, humidity etc. The polymeric materials such as polyvinyl chloride (PVC), cross-linked polyethylene (XLPE), ethylene propylene rubber (EPR), ethylene propylene diene monomer (EPDM), poly ether ether ketone (PEEK), etc. used for insulation and jacket (sheath) in I and C cables are subjected to ageing and degradation mechanisms caused by these stressors during service conditions. If an accident such as a loss of coolant accident (LOCA) occurs, the aged cable must have sufficient properties remaining after normal ageing so that it will remain viable during design basis accident (DBA) conditions [2]. The ageing of components in NPPs is an important concern since the degradation caused by ageing can impact the performance of susceptible equipment. Operating experience from the existing nuclear reactors shows that the number of cable failures increases with plant age

T. V. Santhosh (✉) · V. Gopika · A. K. Ghosh · J. Chattopadhyay
Reactor Safety Division, Bhabha Atomic Research Centre, Trombay, India
e-mail: santutv@barc.gov.in

B. G. Fernandes
Indian Institute of Technology Bombay, Mumbai, India

D. R. Karanki et al. (eds.), *Advances in RAMS Engineering*,
Springer Series in Reliability Engineering,
https://doi.org/10.1007/978-3-030-36518-9_3

resulting in plant transients, shutdowns, and in some cases, the loss of safety functions. As a result, cable condition monitoring (CM) and remaining useful life (RUL) estimation have become increasingly important in recent years [3].

2 Ageing and Condition Monitoring

The electrical cables in NPPs are exposed to a variety of environmental and operational stressors. Over time, these stressors and combinations of these stressors can cause ageing and degradation mechanisms that will result in a gradual degradation of the cable insulation and jacket materials. Much of the degradation due to ageing is controlled through periodic maintenance and/or component replacement. However, I and C cables do not receive periodic maintenance or monitoring once they are installed. Moreover, replacing a cable in an NPP is complex and expensive task [2]. The integrity and function of electrical cables are monitored indirectly through the performance of in-service testing of the safety-related systems and components. However, while these tests can demonstrate the function of the cables under test conditions, they do not verify their continued successful performance for extended periods as they would under anticipated normal service operating conditions or under design basis conditions. Most of the work on degradation of cable insulation and life assessment discusses about the traditional methods of monitoring the cable degradation through parameters such as elongation at break (EAB), insulation resistance (IR), oxidation induction time (OIT), time domain reflectometry (TDR), Fourier transform infrared (FTIR) spectroscopy, etc. [4]. In many countries, qualification of safety-related I and C cables is based on compliance with IEEE-323 (1983) and IEEE-383 (1974) standards, which detail testing procedures aimed at demonstrating the ability to survive a design basis event (DBE) after a 40 year service life. However, since these standards were written, a better understanding of the degradation behavior of cable materials has been reached.

There have been some cases where cables which had initially been qualified for a 40 year design life failed a design basis event test after removal from the NPP after less than 10 years [5]. Also, despite the development of several accelerated ageing tests, there is still no simple ageing test that can reliably assess and/or predict the performance of cables for use in reliability assessment. In life estimation studies, the impact of non-polymeric materials such as conductor and shield on insulation ageing has never been addressed and there are no guidelines to account such effect. The present acceptance criterion used for installed cables is 50% absolute elongation-at-break and is based on years of experience in testing and analysis of accelerated and field aged cables. It is assumed, generally, that this elongation value will provide sufficient margin to ensure that the insulation maintains its electrical properties during a design basis event. However, it has been noted in the literature that, in certain cases, an elongation of only 5% of the original value may be sufficient for the cable to function electrically during a design basis event [6].

3 The Proposed Approach

Though several condition monitoring and life estimation techniques are available, currently there is no standard methodology or an approach towards estimating the time dependent reliability of I and C cables. The state-of-the-art for incorporating cable ageing effects into probabilistic safety assessment (PSA) is still evolving and the current assumptions that need to be made on the failure rates and common cause effects are based on sparse data. Therefore, identification and quantification of ageing of electrical cables is very much essential for an accurate prediction of system reliability for use in PSA of NPPs. The objective of this study is to develop a methodology to assess the susceptibility of polymeric cable insulation to various ageing mechanisms; and to predict the state of the insulation at any chosen time in order to evaluate the remaining life-time of operating cables. This work aims to develop the background and technical basis for incorporating the ageing effects of I and C cables in the reliability assessment of reactor systems in nuclear power plants for PSA applications.

The proposed methodology is shown in Fig. 1. The chemical changes in the polymers occur due to thermal, radiation and moisture intrusion. Because of these chemical changes in the polymer structure, there could be change in the physical, mechanical and electrical properties which are indicative of the ageing and degradation. The parameters thus measured are used for degradation assessment and reliability prediction. The key performance indicators such as IR, EAB, and OIT are determined from analytical, simulation and experimental approaches by subjecting the cable samples to thermal and radiation ageing. The time dependent reliability is estimated from the reliability prediction models developed based on stress-strength interference theory and Weibull reliability concepts.

4 Reliability Prediction by Stress Strength Interference Theory

Development of remaining life models for cable insulation consists of looking for adequate relationships between insulation life and the applied stresses from both the life testing data and the physics of failure considering the time dependent degradation. The main activities are identification of failure modes, characterization of the ageing process, derivation of life expressions and validating the models. The approach is mainly based on the study of physical-chemical properties of insulating materials subjected to various stressors that are present in nuclear power plants. The model, which will be function of one or more stresses, will eventually describe the state of the insulation with the help of time-dependent degradation of insulating materials. In order to validate the developed models based on the physical-chemical properties, experiments will be conducted under accelerated ageing conditions. The

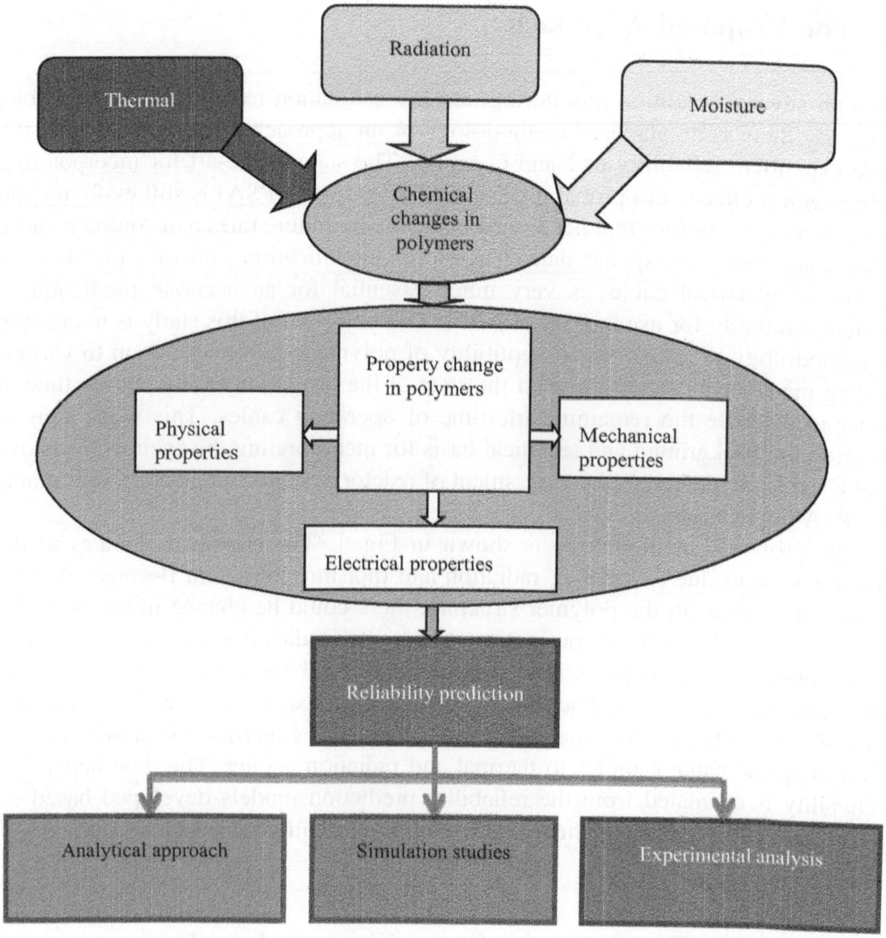

Fig. 1 Proposed approach of reliability prediction

experimental data available in the literature may also be used to validate (if required) the proposed remaining life prediction models.

Although, the stress-strength interference (SSI) theory has well been recognized and successfully applied to various fields of risk and reliability assessment, it has not yet been applied to remaining life estimation of electrical cables in NPPs. In this section a framework for estimating the life time of I and C cable is discussed. The main aim of developing this framework is to perform reliability evaluation of safety-related and safety critical system taking into account the reliability or probability of failure of associated I and C cables for PSA of NPP. The XLPE cables are extensively being used in safety systems of Indian NPPs and their failure will not only interrupt the plant performance but also causes the malfunctioning of associated systems essential during emergency shutdown. Therefore, the prior knowledge of their remaining life time is very much essential to maintain the adequate plant safety.

4.1 Methodology

The performance of an electrical cable depends on its insulation resistance (conductor to ground or shield, conductor to conductor) and dielectric strength. High temperature, humidity, radiation and voltage can degrade the insulation resistance beyond a point which may not be acceptable for I and C circuits to maintain the accuracy [7]. A cable failure occurs when the stress is higher than the strength or capacity. For a random loading or strength, the probability that the strength is always greater than the load in mission duration provides product reliability for that time period. From the stress-strength interference (SSI) theory, a cable failure will occur when the current insulation resistance is less than a specified threshold value. The SSI model for stress (S_1) and strength (S_2) relationship is as follows [8]:

$$R = P(S_2 > S_1) \tag{1}$$

In regard to the cable life assessment, the insulation resistance of the cable represents the strength of the material and the temperature, humidity, voltage etc. represent various stresses that act on the cable insulation to degrade and result in its eventual failure. Furthermore, if both stress and strength are treated as continuous random variables and their probability density functions denoted by $f_1(S_1)$ and $f_2(S_2)$ respectively, then Eq. (1) can be rewritten as the following [9]:

$$R = \int_{-\infty}^{\infty} f_2(S_2)\left[\int_{-\infty}^{S_2} f_1(S_1)dS_1\right]dS_2 \tag{2}$$

The strength, in this case the insulation resistance of cable material, is a function of one or more stresses. The insulation system of a cable consists of a primary insulation, a secondary insulation and a jacket material. All these materials are usually organic compounds typically made from a family of polymers such as polyethylene, polyvinyl chloride; ethylene propylene rubber etc. and they are subjected to various stresses. Therefore, the insulation resistance of a cable can be a time dependent function of stresses such as temperature, humidity, voltage etc. and can be formulated as the following:

$$IR = f(T, H, V, t) \tag{3}$$

where, IR is the insulation resistance of the cable material, T is the temperature H is the humidity, V is the applied voltage, t is the time.

Since, temperature is one of the dominant stressors present in an NPP; the proposed methodology has considered temperature as the ageing stress. However, the other stressors such as humidity, radiation, etc. can also be modeled accordingly. The insulation resistance of the insulating system depends upon the number of free electrons available to conduct electricity between the conductor to ground/shield or between conductor to conductor. The number of free electrons in an organic compound depends upon temperature. Keeping the applied voltage V constant at some fixed value V_0, the leakage resistance of the cable insulation system is given by the following equation [7]:

$$IR = R_0 e^{\alpha T} \tag{4}$$

where,

R_0 constant depending upon material property (Ω),
IR resistance at temperature T (Ω),
α voltage dependent constant (1/K),
T Ambient temperature (K).

At some ambient temperature T_1, Eq. (4) becomes:

$$IR_1 = R_0 e^{\alpha T_1} \tag{5}$$

Similarly at an ambient temperature of T_2, Eq. (4) becomes:

$$IR_2 = R_0 e^{\alpha T_2} \tag{6}$$

Dividing Eq. (5) by Eq. (6) yields,

$$\frac{IR_1}{IR_2} = e^{\alpha[T_1 - T_2]} \tag{7}$$

Taking natural logarithm on both sides of Eq. (7) and rearranging for α,

$$\alpha = \frac{\ln \frac{IR_1}{IR_2}}{[T_1 - T_2]} \tag{8}$$

Similarly at an ambient temperature of T_3, Eq. (7) becomes:

$$\frac{IR_1}{IR_3} = e^{\alpha[T_1 - T_3]} \tag{9}$$

Substituting for α and rearranging,

$$IR_3 = IR_1 e^{\left(\ln \frac{IR_1}{IR_2}\right)\left(\frac{T_1 - T_3}{T_2 - T_1}\right)} \tag{10}$$

Due to the upgradation of I and C systems in NPPs, a periodic PSA is carried out to ensure that the adequate safety levels are maintained. Therefore, a time dependent reliability of cables and other equipment taking into account the degradation needs to be evaluated. In order to account for a time dependent degradation of the insulation resistance, let IR_3 be the initial insulation resistance at a given temperature, T and λ be the linear degradation rate due to normal ageing over a time period t, then IR at a given temperature for a period of time t can be formulated as:

$$IR(T,t) = IR_3(1 - \lambda t) \tag{11}$$

If the degradation process is assumed to be exponential, then Eq. (11) becomes,

$$IR(T,t) = IR_3 e^{-\lambda t} \tag{12}$$

Using Eq. (11) or Eq. (12), the remaining insulation resistance of a cable at time t can be estimated. Now, IR_3 becomes the remaining strength of the cable insulation material. From Eq. (2), the reliability of an electrical cable can be estimated at any temperature T as stress. This is further formulated as a structural reliability problem as follows.

A cable is said to be failed when the remaining insulation resistance is less than or equal to a specified threshold value IR_{th} when subjected to one or more stresses. Therefore, the probability of a cable failure is then formulated by the following equation:

$$P_f = P(IR \leq IR_{th}) \tag{13}$$

Now, this is in the form of a load-resistance relationship and can be solved by using the structural reliability concepts through first or second order reliability methods or Monte Carlo simulation method. However, in this study this problem has been solved by using structural reliability methods through limit state concepts [10].

The limit state function for structural reliability problem can be formulated as:

$$g(IR, IR_{th}) = IR_3 - IR_{th} \tag{14}$$

Substituting for IR_3, Eq. (14) becomes,

$$g(IR, IR_{th}) = IR_1 e^{\left(\ln\frac{IR_1}{IR_2}\right)\left(\frac{T_1 - T_3}{T_2 - T_1}\right)} - IR_{th} \tag{15}$$

From the structural reliability methods, the probability of failure is computed from the following relation:

$$P_f = P(g(IR, IR_{th}) = IR_3 - IR_{th} \leq 0) \tag{16}$$

Substituting for IR_3, Eq. (16) becomes,

$$P_f = P\left(g(IR, IR_{th}) = IR_1 e^{\left(\ln\frac{IR_1}{IR_2}\right)\left(\frac{T_1-T_3}{T_2-T_1}\right)} - IR_{th} \le 0\right) \tag{17}$$

From structural reliability,

$$P_f = \Phi(-\beta) \tag{18}$$

where, P_f is the probability of failure, $\Phi(.)$ is the standard normal cumulative distribution function (CDF) and β is the reliability index, which is defined as the minimum distance from the origin to the limit state surface in the standard normal space [11]. Accordingly, the reliability is given by:

$$R = 1 - P_f \tag{19}$$

Thus, reliability of an electrical cable can be estimated from Eq. (19) for use in PSA applications as a part of ageing management programs.

In general, when cables are exposed to extreme ambient temperatures their insulation resistance typically drops several orders of magnitude following an exponential behaviour. The acceptable IR level depends on the application; therefore, it is different for a cable used in control circuit, than for a cable used in a power supply circuit. Based on the insulation resistance approach, the USNRC has established a minimum acceptance criterion of $10^6\ \Omega$ over a 1000 foot length of a conductor or cable for applications using less than 1000 volts [1].

This is based on the assumption that for most modern cable insulations such as XLPE, PVC, EPR, etc. the IR drops exponentially as temperature increases and most of the cable insulation materials commonly used in nuclear power plants behave in similar manner; so the IR drops by order of magnitude at the temperature surrounding the cable increases. Since, most of the I and C cables in Indian nuclear power plants operate within the range of 1000 volts the minimum criterion suggested by NUREG/CR-7000 can be adopted as failure criterion in evaluating the life time of cables in this study.

4.2 *Analysis and Results*

In order to validate the proposed methodology and to show the usefulness in assessing the cable life and subsequent application in PSA studies of NPPs, an accelerated life testing data on a typical XLPE I and C cable has been taken from the literature and the probability of failure has been estimated using structural reliability method. The data obtained from the literature is shown in Table 1 [12].

Table 1 Insulation resistance data on XLPE cable

Sl. no.	Temperature (K)	IR (Ω)
1	375	1.0E+06
2	475	1.0E+05
3	575	1.0E+04
4	675	1.0E+03

Table 2 Insulation resistance values from proposed methodology

Sl. no.	Temperature (K)	IR model (Ω)	IR expt. (Ω)
1	273	1.05E+07	–
2	313	4.17E+06	–
3	333	2.63E+06	–
4	353	1.66E+06	–
5	375	1.0E+06	1.0E+06
6	475	1.0E+05	1.0E+05
7	575	1.0E+04	1.0E+04
8	675	1.0E+03	1.0E+03

From Table 1,

IR_1 1.0E+06Ω, corresponding to $T_1 = 375$ K,
IR_2 1.0E+05Ω, corresponding to $T_2 = 475$ K.

From Eq. (10), the new IR can be estimated corresponding to a given temperature T. Table 2 shows the predicted values of insulation resistance corresponding to various temperatures from Eq. (10).

Figure 2 shows the plot of temperature verses predicted insulation resistance from the IR model.

Considering the degradation rate of 0.02Ω/yr the behaviour of insulation resistance of a typical XLPE cable when exposed to temperature with the degradation rate being either linear or exponential is shown in Fig. 3.

The degradation rate considered in this study is only for illustration. However, the actual degradation rate should be considered when performing reliability analysis of specific cable systems of NPP. A periodic insulation resistance measurement yields results on the state of insulation. However, it is practically not possible to carry out an IR measurement on an installed cable in the field. Other alternate methods may include storing the similar cables within the installed environment and carrying out periodic experiments for determining the actual degradation rate.

Now, using the data from Table 1, the probability of failure of a typical cable corresponding to a temperature of 40 °C (313 K) can be calculated from the structural reliability method. This temperature is chosen under the assumption that most of these I and C cables in a typical nuclear power plant are operating at this temperature. The input data considered for calculating the reliability of an XLPE cable from the structural reliability method is shown in Table 3. The coefficient of

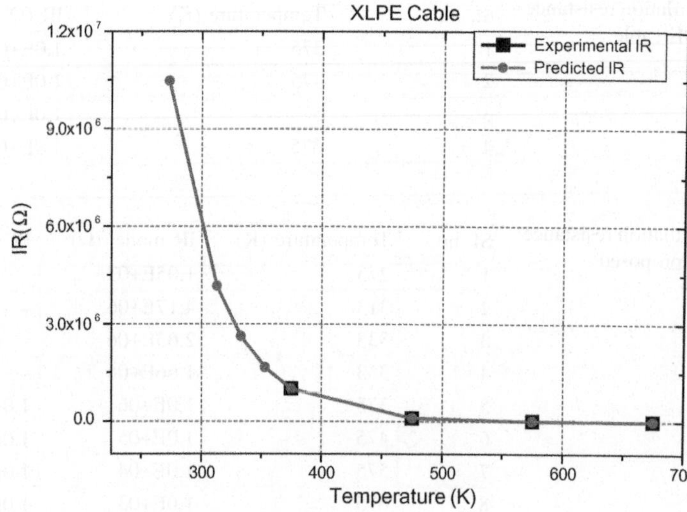

Fig. 2 Temperature versus insulation resistance

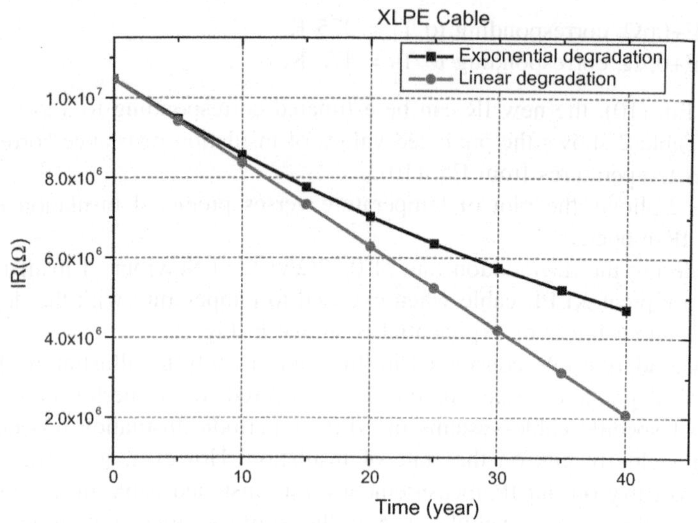

Fig. 3 Insulation resistance versus time

variation (COV) values for basic random variables have been taken from the literature.

This problem has been solved by using the first order reliability method from the COMREL [13] tool, which is a standard software tool for solving structural reliability problems. COMREL stands for COMponent RELiability, a module of

Table 3 Input data for structural reliability problem

Parameter	Distribution	Mean	COV
IR_1	Normal	1.0E+06 Ω	0.1
IR_2	Normal	1.0E+05 Ω	0.1
IR_{th}	Constant	1.0E+06 Ω	–
T_1	Constant	375 K	–
T_2	Constant	475 K	–
T_3	Constant	313 K	–

STRUREL [14] which stands for STRUctural RELiability developed by Reliability Consulting Programs, Germany. Both time-variant and time-invariant reliability evaluations can be performed using COMREL software. COMREL comprises the several structural reliability methods such as first order reliability method (FORM), second order reliability method (SORM), and Monte Carlo simulation (MCS) techniques. The reliability index β corresponding to a temperature of 313 K is found to be 5.789 and consequently the probability of failure is 3.56e−9. This is without the consideration of time dependent degradation.

Now, considering the degradation rate of 0.02Ω/yr, the time dependent probability of failure of an XLPE cable corresponding to a temperature of 40 °C (313 K) is shown in Fig. 4. It is evident from Fig. 4 that P_f increases suddenly after about 30 years of service when the degradation process is linear. This shows that the safety margin is significantly less after about 30 years of service. However, when the degradation is due to temperature the insulation resistance behaviour will

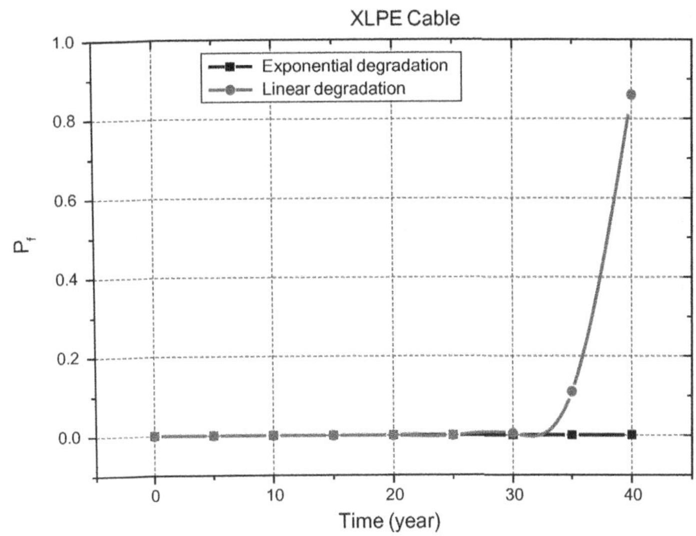

Fig. 4 Probability of failure versus time

usually be exponential. From Fig. 4, a time dependent probability of failure value obtained can be directly used in the reliability analysis of cable systems for PSA applications as a part of ageing management of NPPs.

4.3 Summary

A framework for time dependent reliability prediction of I and C cables for use in PSA of NPP has been developed by considering the thermal aging of XLPE cable as a part of ageing management. The proposed methodology has been illustrated with the data obtained from the literature on a typical XLPE I and C cable. It is observed from the results that with a few sets of accelerated life testing data on a typical XLPE cable the remaining life can be estimated from the developed methodology. The behaviour of insulation resistance when the degradation process is linear or exponential has been modeled. Also, the time dependent probability of failure model has been developed in order to account for time dependent degradation. The degradation rate considered in this study is only for illustration. However, the actual degradation rate should be considered when performing reliability analysis of specific cable systems. The reliability index or probability of failure obtained from this framework is used in system reliability analysis to account for cable ageing. The proposed methodology can be extended to other degradation mechanisms such as humidity, voltage etc.

5 Reliability Prediction from Experimentally Determined Performance Indicators

Experimental techniques provide means for evaluating the level of ageing and degradation of electrical cables. Condition monitoring for electric cable systems involves inspection and measurement of one or more indicators, which can be correlated to the condition or functional performance of the electrical cable on which it is applied [15–18]. Furthermore, it is desirable to link the measured indicators such as elongation-at-break, insulation resistance, etc. with an independent parameter, such as time or cycles, in order to identify trends in the condition of the cable. Ideally, condition monitoring data and trends in performance indicators guide the cable engineer's decisions to effectively manage the ageing and degradation in electrical cables, cable splices, or other accessories in a cable system before they reach the end of life or degraded performance that may adversely affect the safe and reliable operation of the associated components and systems.

Nuclear power plant receives large number of low voltage cables of various manufacturers for I and C applications. Although, all the manufacturers meet the required specifications there can be a significant variation in the performance of cables from manufacturer to manufacturer. This is due to many factors such as design, material processing, purity of the raw materials, workmanship, etc. Also, a few manufacturers add plasticizers and other additives for enhanced performance and ease of manufacturing. The experimental techniques such as insulation resistance testing, tensile elongation, differential scanning calorimetry, Fourier transform infrared spectroscopy, etc. will provide strong basis on the adequate performance of these cables. These experimental techniques will also help in establishing correlation against standard benchmark characterization techniques for lifetime prediction of insulating materials. Prior to accelerated ageing the insulation resistance measurement was carried out and the key electrical parameters were assessed for the cable samples under study.

The performance parameters thus obtained from various condition assessment techniques can be used for reliability prediction. Based on the property such as EAB, OIT, etc. an appropriate reliability technique can be applied to estimate the reliability. Since EAB is related to material resistance to the applied load, the stress-strength interference theory is employed to determine the cable reliability. Similarly, the OIT is related to characteristic life or mean time to reach end-of-life, the Weibull reliability model is employed to determine the cable reliability.

5.1 *Reliability Prediction from EAB*

The exponential model obtained from the experimental data for insulation aged at 110 °C is given in Eq. (20) and the model parameters are shown in Table 4.

$$EAB_{model} = A + Bexp^{-ct} \tag{20}$$

From the structural reliability methods, the probability of failure is computed from the following relation:

$$P_f = P(g(EAB_{model}, EAB_{th}) = EAB_{model} - EAB_{th} \leq 0) \tag{21}$$

where, g(.) is the limit state function, EAB_{model} is the model derived from the thermal ageing data and EAB_{th} is the 50% absolute elongation which is the threshold value considered as an accepted criteria.

Table 4 Model parameters for insulation aged at 110 °C

Parameter	Value	Standard error
A	22.72	51.97
B	485.79	57.58
C	0.025	0.007

Table 5 Input data for structural reliability problem

Parameter	Distribution	Mean	COV
A	Lognormal	22.72	0.1
B	Constant	485.79	–
C	Constant	0.025	–
t	Constant	0	–

Substituting Eq. (20) in Eq. (21) we get,

$$P_f = P(g(EAB_{model}, EAB_{th}) = (A + Bexp^{-ct}) - EAB_{th} \leq 0) \tag{22}$$

The distribution and coefficient of variation (COV) for basic parameters for structural reliability are shown in Table 5. Since parameter A is contributing to the EAB (measured in %) and is related to strength of the material, the lognormal distribution is considered to account for uncertainty in the random variable. Unit of B is in %, C is in days^{-1} and t is in days.

Equation (22) has been solved by using the first order reliability method from the COMREL [13], which is one of the modules of STRUREL [14].

5.2 Results with Lognormal Distribution of Parameter A

Analysis was carried out by varying the time from 0 to 100 days considering lognormal distribution. The probability of failure is shown in Fig. 5.

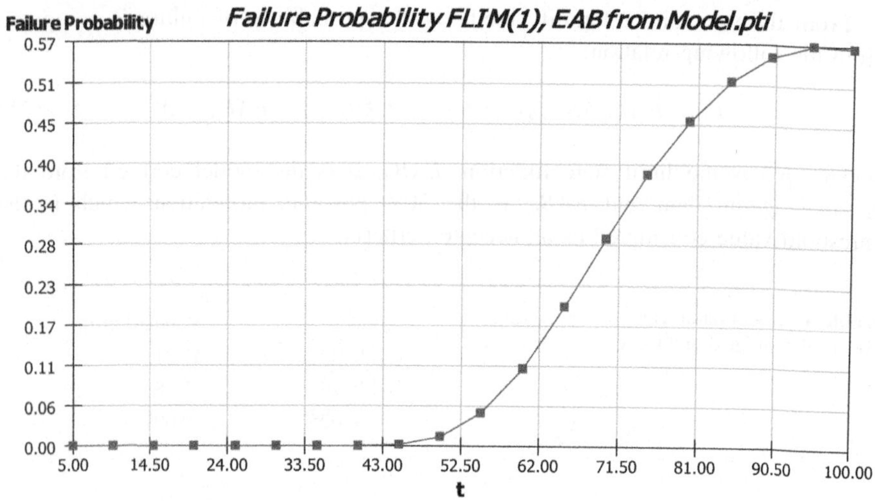

Fig. 5 Probability of failure as a function of time at 110 °C of ageing

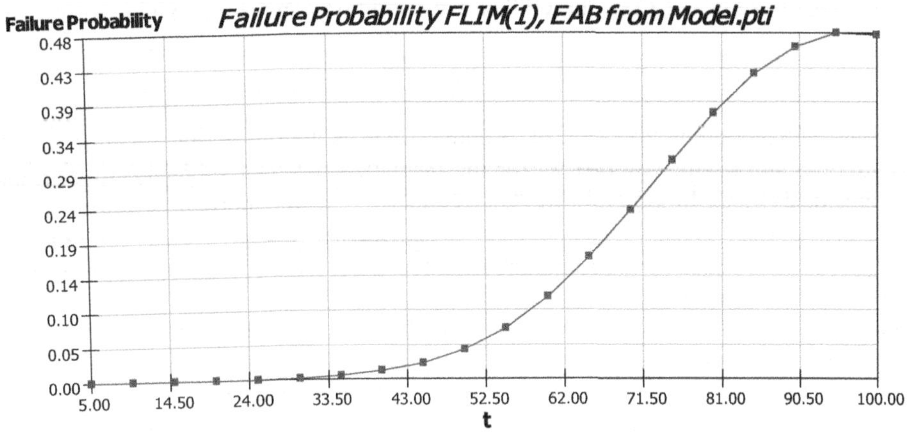

Fig. 6 Probability of failure as a function of time at 110 °C of ageing

5.3 *Results with Gumbel (Min) Distribution of Parameter A*

Analysis was carried out by varying the time from 0 to 100 days with Gumbel distribution. Gumbel distribution is a Type I extreme value distribution which has two forms. One is based on the smallest extreme and the other is based on the largest extreme. Gumbel distribution has two parameters namely, location parameter α and scale parameter β [12, 17]. The probability of failure is shown in Fig. 6.

6 Reliability Prediction from OIT

Generally, OIT decreases exponentially with increase in the isothermal temperature used for the test and this relationship follows Arrhenius behaviour. Therefore, the OIT under isothermal condition can be related to Arrhenius equation as shown in Eq. (23).

$$OIT = Ae^{\frac{\Phi}{kT_{OIT}}} \tag{23}$$

where, T_{OIT} is the isothermal temperature, A is constant, Φ is the activation energy and k is the Boltzmann constant. OIT in Eq. (23) can be considered as characteristic life of insulation to reach end-of-life criterion. Now considering the Arrhenius acceleration factor given by:

$$AF = \frac{t_{use}}{t_{acc}} = \frac{OIT_{use}}{OIT_{acc}}. \tag{24}$$

Using Eq. (23) and (24), the OIT at use condition can be estimated from:

$$OIT_{use} = OIT_{acc} e^{\frac{\Phi}{k}\left[\frac{1}{T_{use}} - \frac{1}{T_{acc}}\right]} \tag{25}$$

OIT corresponding to 50% of original EAB is found to be 8.4 min under isothermal temperature of 200 °C for the insulation material subjected to thermal ageing of 110 °C. In general, 40 °C is considered as the normal operating temperature in NPPs. Hence, *OIT* under 40 °C considering Φ = 1.15 eV is found to be 29 years from Eq. (25). In order for the reliability prediction, an appropriate life distribution is chosen to describe the failure characteristics. The Weibull distribution is commonly employed in reliability studies and it is well suited to fitting the 'weakest-link' properties of typical lifetime data. Different mechanisms of failure can sometimes be distinguished by the Weibull parameters needed to fit the results [18–22]. The reliability function from Weibull distribution is given by:

$$R(t) = e^{-\left(\frac{t}{\eta}\right)^{\beta}} \tag{26}$$

where, η is the scale parameter or characteristic life, β is the shape parameter and t is the time. Since OIT is equivalent to characteristic life, the time dependent reliability can be determined from Eq. (26). Now with η = OIT = 29 years and varying t from 0 to 40 years, the time dependent reliabilities have been calculated from Eq. (26). The time dependent reliabilities of the thermally aged cable insulation for different values of β are shown in Fig. 7.

It is observed from Fig. 7 that when times-to-failure of the insulation material follows an exponential distribution, the reliability is substantially high as failures are treated as random failures. Study also demonstrates that under thermal and radiation ageing both insulation and sheath materials exhibited an exponential

Fig. 7 Reliability versus time

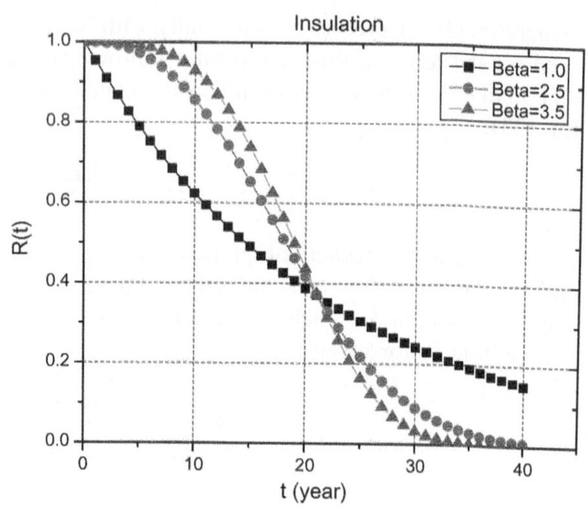

degradation process; hence the exponential failure rate for modelling cables is found to be appropriate. As samples were subjected to accelerated ageing other failure distributions were also considered in this study to demonstrate the reliabilities from non-exponential failures. The predicted reliability is used in PSA of NPP to account for the cable failures.

7 Summary

Several reliability prediction methodologies based on performance indicators have been developed from analytical and experimental approaches. The proposed methodologies have been demonstrated with the accelerated life testing data obtained under thermal and radiation ageing, and also from the data taken from literature. It was apparent from the study that with a few sets of initial accelerated life testing data the reliability of cable can be estimated from the proposed methods.

References

1. Villaran M, Lofaro R (2010) Essential elements of an electric cable condition monitoring program, Brookhaven National Laboratory, BNL-NUREG-90318-2009
2. Lofaro RJ, Grove E, Soo P (2000) An evaluation of condition monitoring techniques for low-voltage electric cables, Brookhaven National Laboratory, BNL-NUREG-67389
3. Shumaker Brent D, Ledlow Jonathan B, O'Hagan Ryan D, McCarter Dan E, Hashemian Hashem M (2013) Remaining useful life estimation of electrical cables in nuclear power plants. Chem Eng Trans 33:877–882
4. Simmons KL, Fifield LS, Westman MP, Ramuhalli P, Pardini AF, Tedeschi JR, Jones AM (2013) Determining remaining useful life of ageing cables in nuclear power plants-Interim study FY13, PNNL 22812, US Department of Energy
5. IAEA-TECDOC-932 (1997) Pilot study on the management of ageing of instrumentation and control cables, International Atomic Energy Agency
6. Anandkumaran A, Sedding HG (2012) Ageing management of cables in nuclear power plants, D1-301, Cigre
7. Alsammarae AJ, Behera AK, Akhtar S (1990) An approach to calculate insulation resistance (IR) at different temperatures and voltages. IEEE Trans Nucl Sci 37(2)
8. Santosh TV, Saraf RK, Ghosh AK, Kushwaha HS (2006) Optimum step length selection rule in modified HL-RF method for structural reliability. Int J Press Vessels Pip 83:742–748
9. Haldar A, Mahadevan S (2000) Reliability assessment using stochastic finite element analysis. Wiley
10. Wang L, Grandhi RV (1996) Safety index calculations using intervening variables for structural reliability. Comput Struct 59(6), 1139–1148
11. Liu Pei-Ling, Der Kiureghian Armen (1991) Optimization algorithms for structural reliability. Struct Saf 9:161–177
12. Steven N (1990) An investigation of the effects of thermal aging on the fire damageability of electric cables, Sandia National Laboratories, Albuquerque, NM, SAND-1990-0696, United States Nuclear Regulatory Commission, Washington, DC, NUREG/CR-5546
13. COMREL TI+TV, version 7.10. RCP Consulting Software, 1987–2000

14. STRUREL TI+TV, version 7.10. RCP Consulting Software, 1987–2000
15. Kröhnke 2012 Polymer science: a comprehensive reference, vol 8.14-Polymer Additives, Elsevier
16. Kudo H, Shimada A, Idesaki A, Ohshima T, Tamura K, Seguchi T (2010) Degradation mechanisms of cable insulation materials during radiation thermal ageing in radiation environment. In: Proceedings of the international symposium on the ageing management and maintenance of nuclear power plants, pp 84–91
17. NUREG/CR-6704 (2001) Assessment of environmental qualification practices and condition monitoring techniques for low voltage electrical cables, Brookhaven National Laboratory
18. Tavner PJ, Ran L, Penman J, Sedding H (2008) Condition monitoring of rotating electrical machines. In: Power and energy series, no. 56, IET
19. Mason LR, Reynolds AB (1998) Comparison of oxidation induction time measurements with values derived from oxidation induction temperature measurements for EPDM and XLPE polymers. Polym Eng Sci 38(7):1149–1153
20. Bell RM, Doyle TE, Reynolds AB, Correlation of oxidative induction time with thermal and radiation aging of EPR and XPLE cable insulation. In: IEEE international symposium on electrical insulation, Toronto, Canada, June 3–6, 1990
21. Dissado LA (2002) Predicting electrical breakdown in polymeric insulators from deterministic mechanisms to failure statistics. IEEE Trans Dielectr Electr Insul 5:860–875
22. Fothergill JC, Dissado LA (1992) Electrical degradation and breakdown in polymers. P. Peregrinus for IEE, London

Dr. Santhosh joined Reactor Safety Division of Bhabha Atomic Research Centre in 2001 after completion of Post Graduate Diploma in Nuclear Science and Engineering. He received his B.E. in Electrical and Electronics from Bangalore University, M. Tech in Reliability Engineering from IIT Bombay and Ph.D. in Engineering from Homi Bhabha National Institute, Mumbai. He is faculty of BARC Training School and Member of Atomic Energy Regulatory Board for Power Plant Safety Review. He has over 90 publications in refereed journals and conferences. He is Referee to several Elsevier and IEEE journals. His areas of research include artificial intelligence, operator support system, ageing studies, risk and reliability assessment of nuclear and other industrial facilities.

Fatigue Life Model for PLCC Solder Joints Under Thermal Cycling Stress

Rohit Khatri, Diana Denice and Manoj Kumar

Abstract Solder joints are inevitable part of assembly of an electronic system. These joints are mainly for establishing electrical connection between components, in addition these are also responsible for mechanical bonding and act as heat paths. Life cycle stresses affect the integrity of these contacts and lead to jeopardizing of its functions and in turn to failure of system. The reducing size and changing geometry of solder joints coupled with high power density of the components has increased the contribution of solder joint failure in total system failure. Cycle stresses lead to fatigue failure of solder joints. Thermal cycling is reported to be the major cause of failures. Life estimation of solder joints is important to predict the time to failure and take counter measures. The chapter addresses this issue by specializing an established empirical model for fatigue failure, Coffin-Manson, for a PLCC solder joint using experimental data.

Keywords Solder joints · Thermal cycle · Fatigue life · Coffin-Manson · Finite element method

1 Introduction

Solder joints form a vital part of electronic systems. Their role is to provide (i) electrical inter-connection (ii) mechanical bond and (iii) thermal conduit for removal of heat. Over the years, component reliability has improved because of improvement in manufacturing technologies. However, solder joint reliability has decreased due to shrinking of size, change in technology from PTH (Plated through Hole) to SMD (Surface Mount Device) joints, migration from leaded components to leadless components etc. and hence can be a limiting case and hamper system

R. Khatri
Homi Bhabha National Institute, Mumbai, India

D. Denice · M. Kumar (✉)
Control Instrumentation Division, Bhabha Atomic Research Centre, Mumbai, India
e-mail: kmanoj@barc.gov.in

D. R. Karanki et al. (eds.), *Advances in RAMS Engineering*,
Springer Series in Reliability Engineering,
https://doi.org/10.1007/978-3-030-36518-9_4

reliability. Therefore, estimation of solder joint life is critical to ensure system reliability goals for given operating conditions. Also, extensive studies have shown that 15% of all hardware failures in electronic systems are due to solder joint failures [1] as shown in Fig. 1.

Electronic components come in various packages such as DIP (Dual in-line package), BGA (Ball Grid Array), PLCC (Plastic Leaded Chip Carrier) etc. For each of these packages, solder joints take a different geometry. Solder joint life depends on solder material, their microstructure and geometry, in addition to environmental factors [2]. As a result, a universal model for life estimation of each type of solder joint is not possible.

Plastic Leaded Chip Carrier (PLCC) package components have been extensively used in contemporary designs. These components come with J-leads and the solder joint geometry is quite unique. The solder material used in older designs was Sn-37Pb, i.e. lead solder, while with RoHS (Restriction of Hazardous Substances) directive in force for new designs lead solders are restricted. In literature solder joint life models are available for older packages (i.e. packages in use for quite some time) with lead solder joints. With RoHS, most of the research has focused on leadless solder joints. As for the design under study, the solder material is lead based, so this has created a gap area for PLCC solder (lead) joints.

Environmental factors that greatly influence the life of solder joints include temperature, thermal cycling, vibration, shock etc. [3]. Each of these accelerates a failure mechanism in the solder material causing them to fail.

In this work, thermal cycling stress has been chosen to estimate life of PLCC solder joints due to its predominance in the actual use environment. Since, fatigue is the dominant failure mechanism under thermal cycling, it has been considered for estimation of solder joint life.

Section 2 gives a background on solder joints of electronic components, fatigue failure and fatigue mechanism in solder joint due to thermal cycling. Section 3

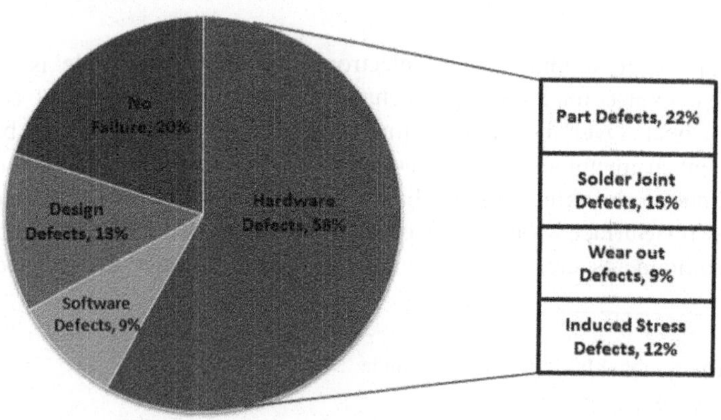

Fig. 1 Distribution of failure causes in an electronic system

introduces the package and solder joint taken for life estimation. Finite Element (FE) model of the PLCC solder joint is given in Sect. 4. Thermal cycle stress and results of FE analysis is given in Sect. 5. In Sect. 6, details of the experimentation, defect recording systems and test setup is given. Section 7 shows the estimation of fatigue model parameters and finally summary and discussion is presented in Sect. 8.

2 Background

2.1 Solder Joints

Solder joints microstructure and geometry coupled with environmental conditions significantly affect their life [2]. The geometry of solder joints varies with different component packages. For BGA packages, it takes a spherical shape while for PLCC packages, it takes an M shape. Since solder joint life greatly depends on geometry, it is imperative to evaluate life of individual type of solder joints.

Solders are classified into lead and leadless ones based on the presence or absence of lead (Pb). Sn-37Pb and Sn-3.5Ag-3Bi are examples of lead and leadless materials, respectively. Implementation of RoHS (Restriction of Hazardous Substances) directive has imposed restriction on the use of Pb in the industry. Ever since, leadless solders gained the market and current research is also in line with that. As a result, life data is not available for solder joints with lead solder material, for concurrent package.

The electronics for critical applications, such as NPP, avionics and space, still use lead solder joints due to their well-known behavior and established reliability. The chapter attempts to fill this gap for lead solder joints of PLCC packages.

2.2 Fatigue Failure

It is known that all materials are subjected to degradation and ageing with time and environment. A solder joint can be termed as failed if cracks have developed forming discontinuities. The three main causes of solder joint failure [2] are:

i. Fracture: Mechanical overloading due to tensile rupture
ii. Creep: Damage due to long–lasting permanent loading at elevated temperature
iii. Fatigue: Damage due to cyclic loading

Fatigue failure is defined as a process causing localized, progressive and permanent micro-structural damage when the material is subjected to fluctuating stresses and strains at some material point or points [4]. It occurs under alternating stresses, causes failure when the maximum operating stress is of sufficiently high value, or there is enough variation of the applied stress or even if there is a sufficiently large number of stress cycles. Sooner or later, it leads to crack initiation and propagation and finally failure [2].

To quantify fatigue life, empirical fatigue life models are available in literature. Widely used models are Coffin-Manson, Total strain, Solomon, Engel Maier etc. [5]. These models are based on plastic strain due to thermal cycling and/or vibration loading for a given environmental conditions.

2.3 *Fatigue Mechanism Due to Thermal Cycling*

Fatigue caused by thermal cycling is categorized as a low cycle fatigue (LCF) phenomenon. That is, the number of cycles to failure is less than 10^4 cycles [1]. The fatigue occurs due to mismatch in coefficients of thermal expansion (CTE) between the component leads, solder and the PCB [5]. On repeated exposure to thermal cycling, these materials expand and contract differently due to their CTE mismatch. This induces plastic strain or deformation in the solder joint, culminating in crack initiation and further growth [5].

For fatigue life due to thermal cycling, Coffin-Manson model is well accepted. Therefore, Coffin-Manson is adopted here which considers plastic strain as the main contributor for failure in case of thermal cycling.

2.3.1 Fatigue Model

In the Coffin-Manson fatigue model, the total number of cycles to failure (N_f) is dependent on the plastic strain range ($\Delta\varepsilon_p$), the fatigue ductility exponent (c) and the fatigue ductility coefficient (ε'_f). It is expressed by,

$$\frac{\Delta\varepsilon_p}{2} = \varepsilon'_f (2N_f)^c \tag{1}$$

where:

- $\Delta\varepsilon_p$ plastic strain range
- ε'_f fatigue ductility coefficient, defined as the failure strain for a single reversal ($2N_f = 1$)
- $2N_f$ number of reversals to failure
- c an empirical constant also known as the fatigue ductility exponent. It depends on the material and its geometry

From (1), it can be seen that $\Delta\varepsilon_p$, ε'_f and c need to be obtained for concerned solder joints to estimate N_f.

Due to the small size of solder joints, traditional methods of strain measurements such as strain gauges cannot be used. Literature suggests the use of Finite Element method for solder joints to estimate strain ranges. To estimate other model parameters, it is also necessary to conduct accelerated life test experiments.

3 Description of Sample

44 pin PLCC component on a 1–layer PCB (with 1.6 mm thickness) has been taken for the study. The package has a pitch of 1.27 mm (0.05 inches). The solder joint is 'J' shaped and solder material is Sn-37Pb.

Each PCB has 4 such packages, so for study PCB around the PLCC package in centre is cut in a size 48 mm × 48 mm. Four samples have been cut out from a single PCB. The PCB and the sample cut from it are shown in Fig. 2.

4 Estimation of Stress/Strain Relationship Using Finite Element Model (FEM)

As discussed earlier, models for fatigue require information related to stress or strain in element (solder joint) to predict the fatigue life. The stress or strain data can be collected experimentally using strain gauges but the decreasing solder joint size makes measurement impractical. This is where FEA (finite element analysis) is the more practical route for obtaining stress/strain relationships. But, it does not discount the requirement of carrying out experiment of actual parts in order to verify life predictions.

Fig. 2 44–pin PLCC test sample

FEA aids in predicting essential information about the internal states of the solder joints i.e. displacement, inelastic strains and stresses etc. FEA procedure comprises of following five steps to obtain satisfactory estimates:

- Geometric and material definition
- Definition of connection between bodies
- Meshing the model
- Definition of load and boundary conditions
- Verification of the results

The cyclic thermal stress is going to induce plastic strain [6]. So, FEA shall give the states of stress and plastic strain (for thermal loading) for different levels of loadings. A non-linear analysis is needed for this purpose.

4.1 Non-Linear FEA

The assumptions for non-linear FEA of the given solder joints are as follows (Table 1):

- All the parts in the package are assumed to be bonded to each other.
- Temperature change and displacement is assumed to be uniform throughout.
- Except solder joints, all other materials are assumed to behave as linear elastic.
- All material properties are time invariant.
- A quarter model is taken for analysis, as it possesses symmetry in boundary conditions and applied loadings. The quarter FE model has added advantage of reducing the total unknown degrees of freedom for the problem compared to the full model, thus reducing the cost of computations.

Thus a quarter model, nonlinear and static analysis is carried out with finite element solver. The dimension of the model is given in Table 3 in Chap. 3.

The dimension details have been extracted using datasheet of the package [7]. The geometry of the package as developed in FEA software is presented in Fig. 3.

As discussed earlier, Copper (Cu) pads and J–lead, plastic package encapsulation and PCB are modeled as linearly elastic materials. The Sn–37Pb solder is

Table 1 Dimensions of FEM model geometry

Component	Dimension (mm)
PCB	48 × 48 × 1.6
PLCC package	16.5 × 16.5 × 4.3
Copper pads	0.65 × 2
Solder joint height	0.762
Stencil size	0.60 × 1.95

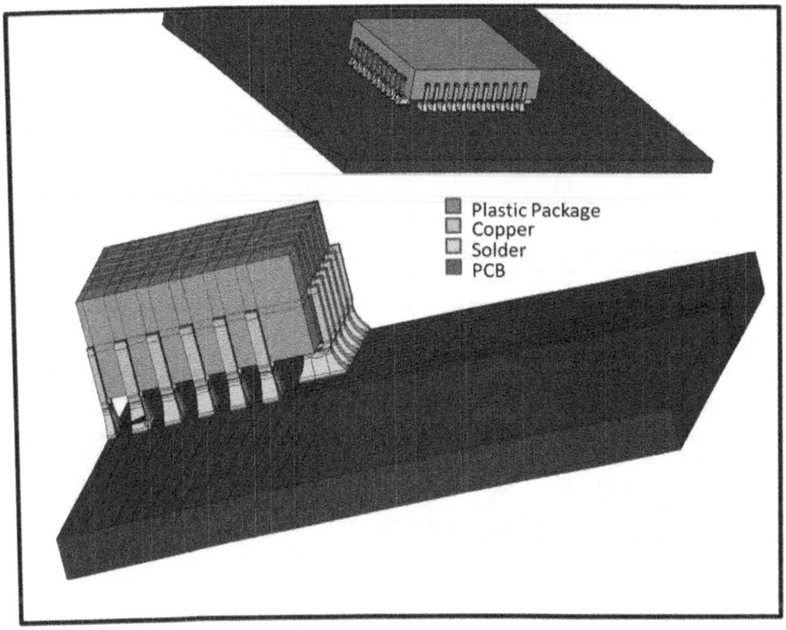

Fig. 3 Full and quarter geometry model of PLCC

modeled as elastic-plastic material with yield strength of 27.23 MPa. The mass of materials of sample package and their properties [3, 8] taken for analysis is given in Tables 2 and 3, respectively.

The size of elements and the degree of interpolation function chosen for the elements affect the accuracy of FE calculations of field variables, particularly in regions where the gradient of the variables is high. While fine sized element mesh could ensure accuracy of the FE results [9], the large number of associated elements renders the computational time long and expensive.

The meshed geometry of the quarter model is presented in Fig. 4. Mesh generation is meticulously planned and executed using mesh generation tool of the FEA software. Mesh density is more in solder joint volumes for accurate results and intentionally rarer at other areas so as to save on computational time.

The quarter model consists of 739965 nodes and 140694 elements with 3D SOLID 186 elements. The meshed geometry of the solder joint and J–lead is presented in Fig. 5. The Fig. 5 shows the meshed geometry of J–lead with solder joint on right and on left depicts the meshed solder joint volume.

Table 2 Mass of materials of sample package

Material	Mass (gms)
Plastic package	2.44 (with J–leads)
PCB	8.43

Table 3 Material properties used in FEA [4, 23]

Material	CTE (ppm/°C)	Elastic modulus (GPa)	Poisson ratio	Density (kg/m^3)
Copper pad and leads	16.7	117	0.35	8940
Plastic package	22	13.79	0.30	1800
PCB	18	22	0.28	2200
Sn–37Pb solder	24.7	15.7	0.4	8500

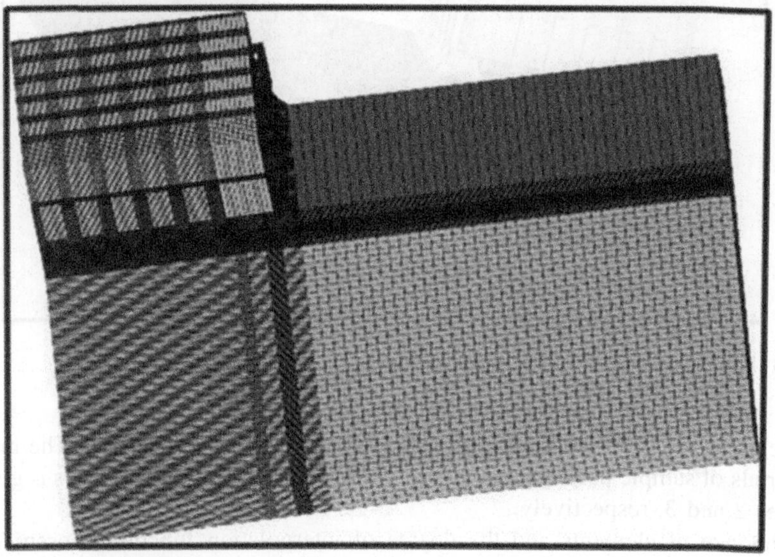

Fig. 4 Meshed quarter geometry model

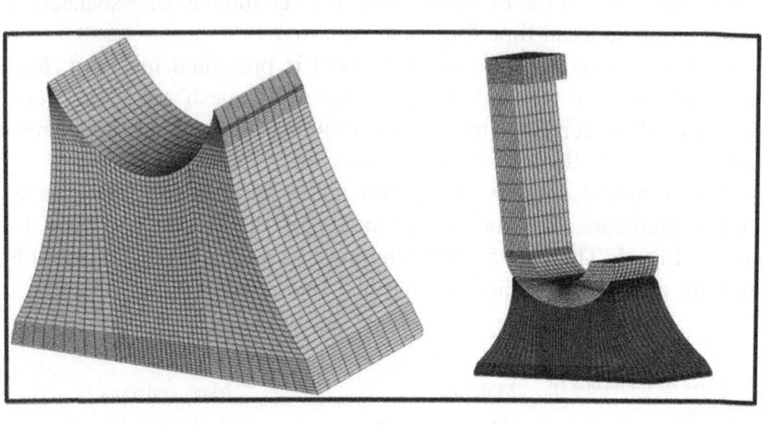

Fig. 5 Meshed solder joint and J–lead

5 Thermal Cycle Stress and Results

For analysis two different test thermal cycle levels namely ALT I (for 5 to 165 °C) and ALT II (for 5 to 145 °C) are considered. The temperature load is applied to all the nodes of quarter geometry model for the specified two test levels. The thermal load cycle for FEA is depicted in Fig. 6.

As shown in Fig. 6, the total cycle duration is 40 min with a dwell/hold period of 10 min, at the maximum and minimum temperature during the thermal cycles. The ramp rate is 16 °C/min and 14 °C/min for ALT I and ALT II, respectively.

5.1 *FEA Results*

Post FEA and processing the stress/strain relationship, the plastic strain information of solder joints for load conditions ALT I and ALT II are shown in Fig. 7 through Fig. 10.

The colour bar indicates the distribution of plastic strain in the solder volume. As shown in Figs. 7, 8 and 9, the maximum strains occur at the outer edges of the solder volume, at the interface of the J–lead and solder.

The strain is greater on the outer side of solder than inner side (end of J bend) due to higher stiffness towards the outer side as the lead is bounded at top and bottom as compared to the inner side where it is free from top.

Figure 10 shows the distribution of strain at the bottom surface of solder joints. The strain is lesser at bottom surface due to greater surface area and a gradual shape of discontinuity.

The inferences from the plastic strain information are as follows:

- Plastic strain range as seen from Figs. 8 and 9 is maximum at the top surface of the solder joints on the outer edges.
- There is similarity in distribution of plastic strain range in both cases with only change in magnitude of maximum strain value.

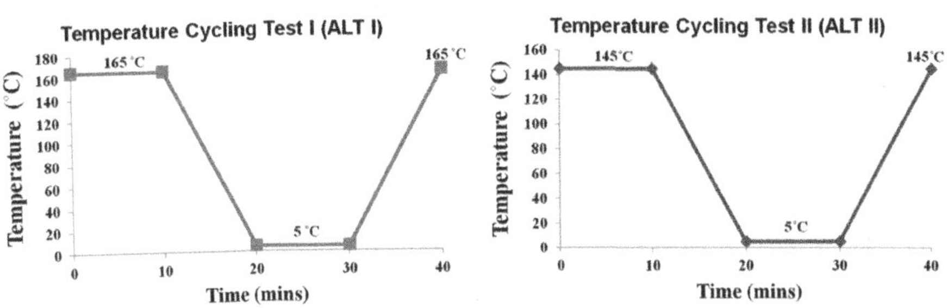

Fig. 6 Thermal cycle environment for FEA

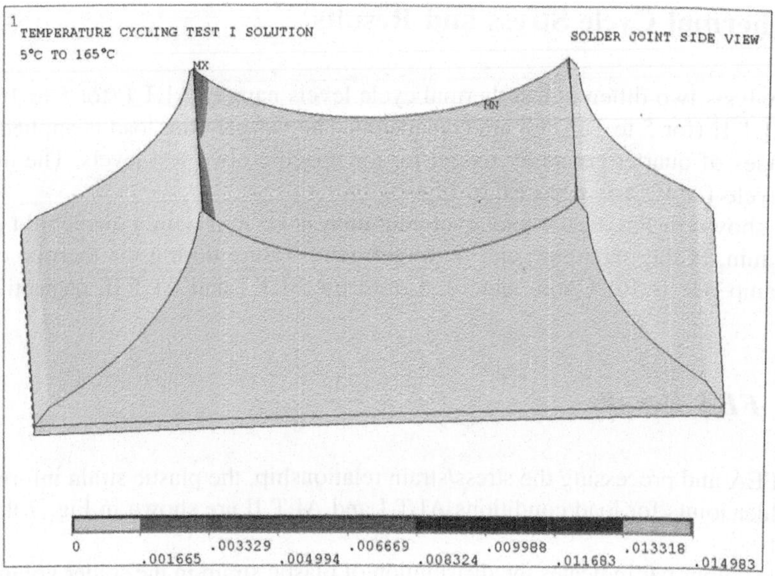

Fig. 7 Plastic strain in solder joints for ALT I (Side view)

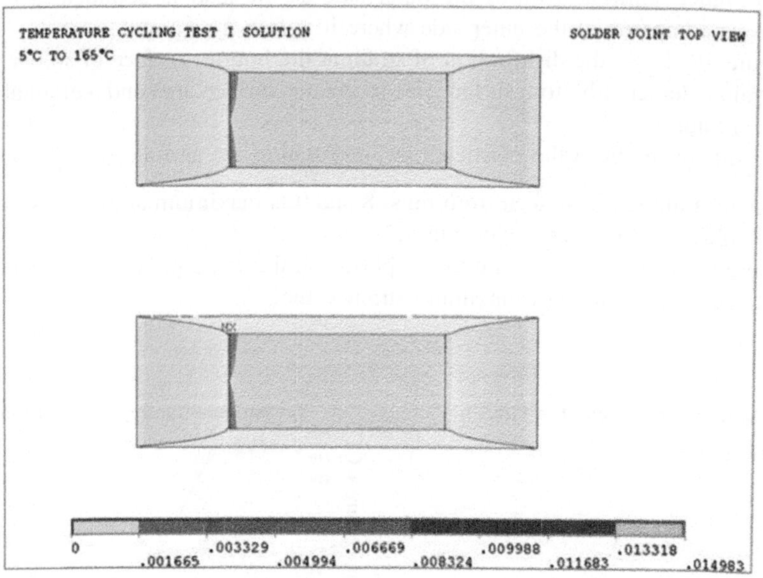

Fig. 8 Plastic strain in solder joints for ALT I (Top view)

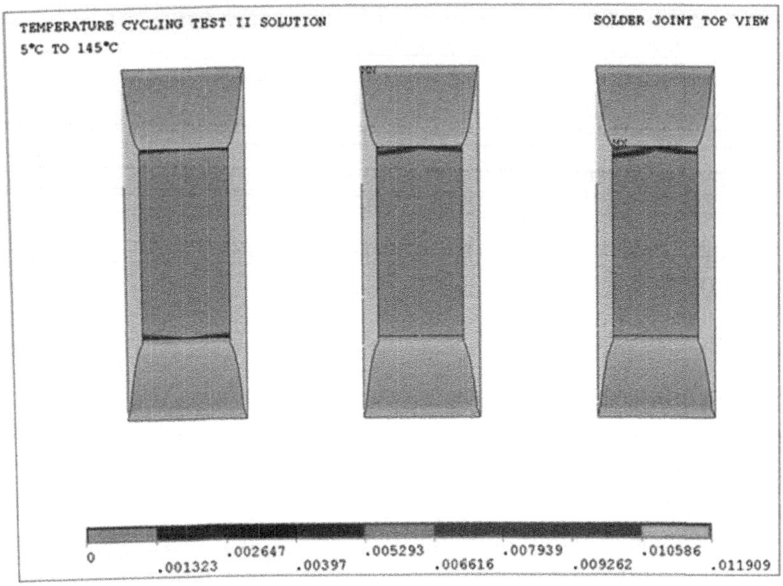

Fig. 9 Plastic strain in solder joints for ALT II (Top view)

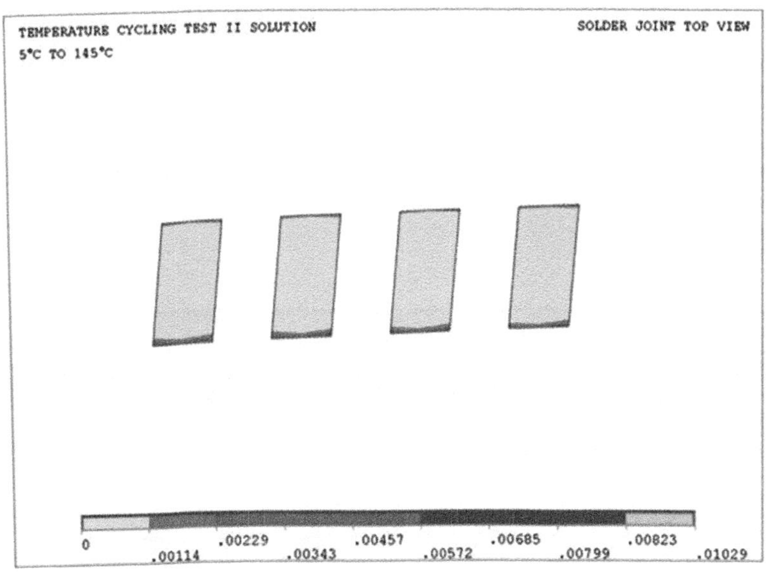

Fig. 10 Plastic Strain in Solder Joints for ALT II (Bottom View)

- The plastic strain range at the bottom surface of solder joints in Fig. 10 is on the higher side but lesser than that at the top surface.
- The regions of maximum plastic strain are the interface of the solder joint with J–lead at top and copper pad at bottom. This is in consonance with literature, since these are the areas of IMC (inter-metallic compounds) formation.
- The plastic strain range is observed to be maximum for the corner solder joints and thus can be deemed as the critical solder joints in the sample under analysis.

The plastic strain range for ALT I is 0.014983. The plastic strain range for ALT II is 0.011909.

6 Experimentation

6.1 Pre-Experiment Defect Record

Visual inspection aided by magnification between 4X to 10X and use of additional magnification to resolve suspected defect is a reliable and widely used method [6]. Therefore, following three methods have been employed for presence of cracks and thus identification of failure of solder joints:

i. Stereo microscope imaging of the samples.
ii. Scanning Electron Microscopy (SEM) of the sample.
iii. Resistance measurement of solder joints using Low Resistance measurement facility (100–200 Ω)

The SEM and Stereo Microscope images of the samples prior to commencement of experiment are shown in Figs. 11, 12, 13, and 14. As visible from the images of the individual solder joints from both the sources, no pre–existing cracks are present.

Fig. 11 Pre experiment SEM imaging of samples (45X Mag)

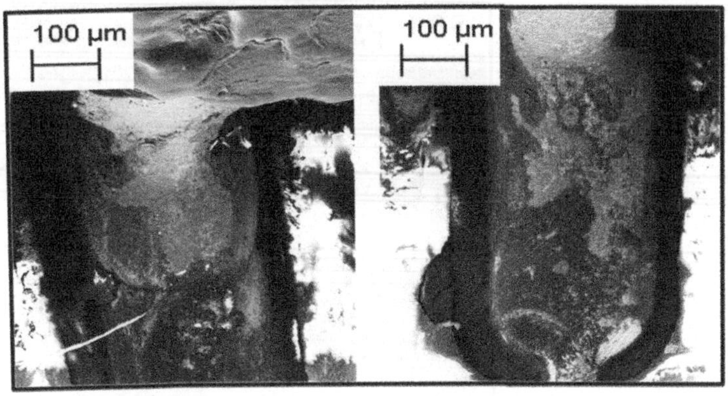

Fig. 12 Pre experiment SEM imaging of samples (90X Mag)

Fig. 13 Pre experiment stereo microscope imaging of samples (20X Mag)

The resistances of the individual solders joints at the start of the experiment were also recorded. The range of resistance for healthy solder joints (including component leads) was found to be in the range 105–120 mΩ. The complete listings of individual solder joint resistances are available in author's dissertation [10].

All of the above being offline techniques warrant removing samples from the test benches at regular intervals during duration of fatigue life tests. The samples prior to being put for fatigue life testing were checked for presence of defects and data was recorded using the above three methods.

Fig. 14 Pre experiment stereo microscope imaging of samples (80X Mag)

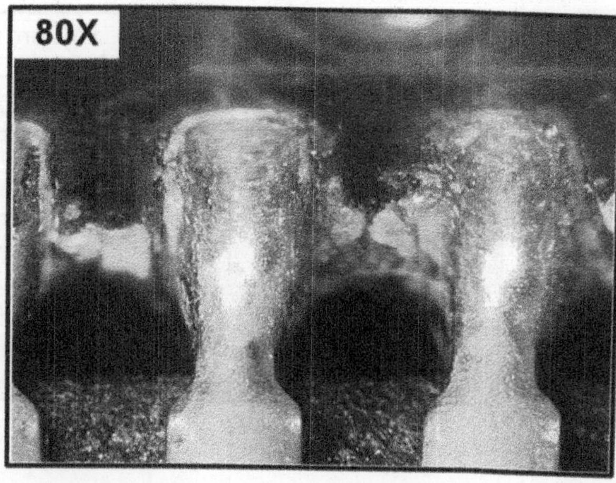

6.2 Planning of Test Profile

The IPC–SM–785 [3] provides the worst case use environments and recommended accelerated testing levels for surface mounted electronics. The standard also prescribes that the rate of change of temperature should be less than 20 °C/min in order to avoid thermal shock. Thermal shock may give rise to other failure modes in solder joints. The solder joints shall be kept at the extreme temperature for quite some time so that every point of the bulk is at same temperature.

Two stress levels chosen for carrying out temperature cycling. ALT I is cycling between 5 °C and 165 °C, while ALT II is cycling between 5 °C and 145 °C. The cycle duration is 40 min in both cases. Also, the ideal ramp and dwell time is 10 min each. However, in view of non-availability of ramp rate control in chambers, the cycle duration has been modified. The modified cycle is reflected in Fig. 15.

Steps are visible in the thermal cycle above in Fig. 15. However the solder joints will not follow the step due to thermal inertia and the temperature change would nearly follow the ramp.

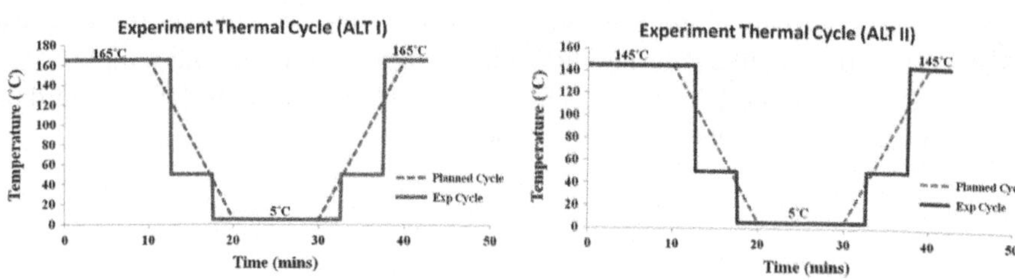

Fig. 15 Thermal cycling levels used in experiment

6.3 Acceleration Factor (AF) Estimation

Norris–Landzberg proposed that the plastic strain range in thermal fatigue of solder joints is proportional to the thermal range of the cyclic loading (ΔT) [11]. They also accounted for effects of thermal cycling frequency (f) and maximum temperature (T_{max}).

The model provides a basis for test design and estimating the duration of the testing suitable to replicate the field lifetime duration of interest. The suffix field and test are used for field and test conditions respectively in the model equation for calculating AF, given by:

$$AF = \left(\frac{N_{field}}{N_{test}}\right)^{-m}\left(\frac{\Delta T_{field}}{\Delta T_{test}}\right)^{-n} e^{\left(\frac{E_a}{K}\left(\frac{1}{T_{field}}-\frac{1}{T_{test}}\right)\right)} \quad (2)$$

where,

E_a Activation Energy
K Boltzmann constant
m, n Parameters for material

The values above for fitted parameters m and n have been determined for Sn–37Pb eutectic solder [11]. Also, the use environment as provided in [4] for equipment is 5–45 °C at one cycle per day. Pecht [12] also provides the use life of compliant leaded solder joints at use temperature environment as 35000 cycles. Therefore from these field conditions the estimated acceleration factors and thus approximate number of cycles required at test levels ALT I and ALT II are estimated in Table 4 in Chap. 4.

6.4 Test Setup

The test set up as shown in Fig. 16, consists of two thermal chambers maintained at 165 and 145 °C respectively. The cold temperature of 5 °C is achieved using a water bath. The samples are manually cycled between the two chambers in accordance with the duration depicted in Fig. 15.

6.5 Results

The samples were temperature cycled in the set up shown in Fig. 16 simultaneously at ALT I and ALT II levels with an average 12 cycles per day.

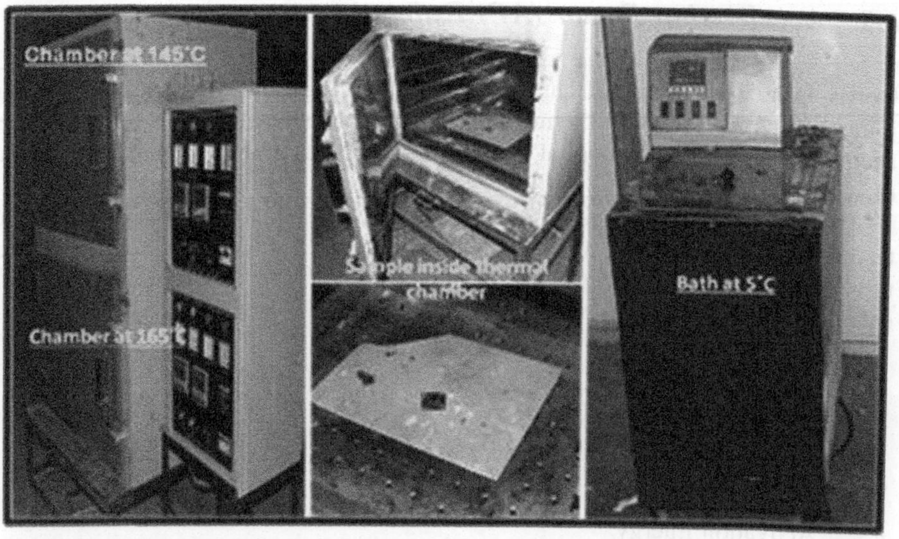

Fig. 16 Temperature cycling test set–up

The schedule of the conduct of test was:

- The samples were examined for defects after every 50 cycles till identification of first failure and every 25 cycles thereafter.
- Stereo imaging and resistance measurement were taken at these regular intervals.
- The first defect was observed at 200 for ALT I conditions.
- The first defect was observed at 325 cycles for ALT II conditions.
- The test was further continued and stopped at 255 cycles for ALT I and at 375 cycles for ALT II conditions (for failure of 50% of solder joints).

Number of cycles required to observe first defect during both test levels match with the prediction/plan given in Table 4.

Figures 17 and 18 present the stereo microscopic images of the first cracks occurring in ALT I and ALT II test conditions. While Figs. 19 and 20 reflect the solder joints with defects at the end of two thermal cycling test conditions.

Table 4 Acceleration factor calculations

Parameter	ALT I	ALT II
f (cycles per day)	36	36
ΔT	160	140
Tmax (K)	433	413
Acceleration factor	155.50	105.87
Ntest cycles	225	330

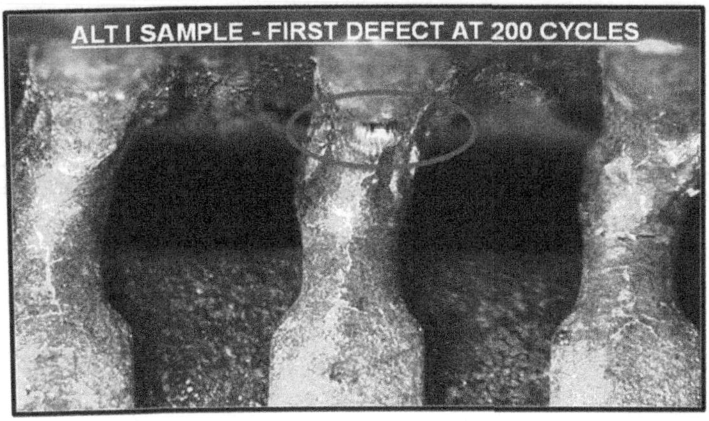

Fig. 17 Stereo microscope image of first crack at 200 cycles for ALT I

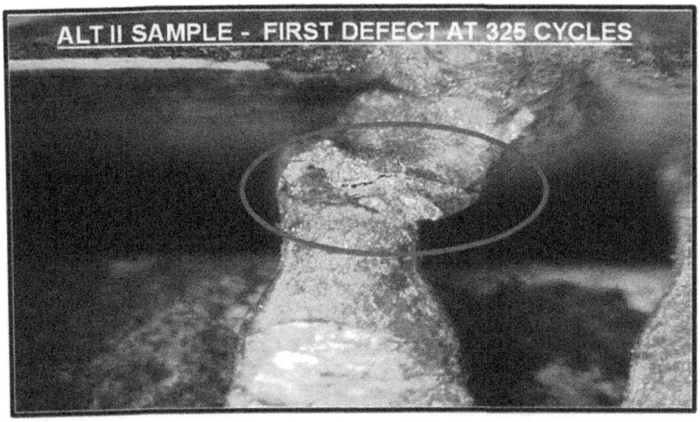

Fig. 18 Stereo microscope image of first crack at 325 cycles for ALT II

The location of crack as seen in Figs. 17, 18, 19, and 20 is on the outside of the interface between solder joint and J–lead. This is clearly in consonance with the results of FEA, which indicated the very same region with maximum plastic strain.

Figures 21 and 22 present the SEM images of the first cracks observed in ALT I and ALT II test conditions while Figs. 23 and 24 show solder joints with defects at the end of two thermal cycling test conditions.

The resistance of individual solder joints with defects has been observed to vary from 245 mΩ (in case of small cracks) to 2 kΩ (for cracks spread throughout the volume).

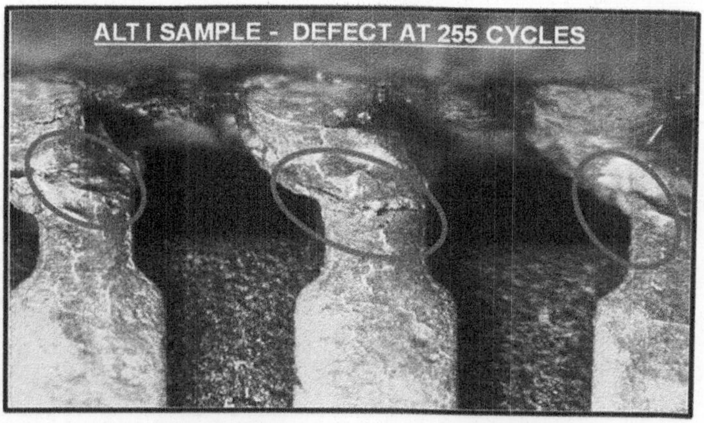

Fig. 19 Stereo microscope image of crack after 255 cycles for ALT I

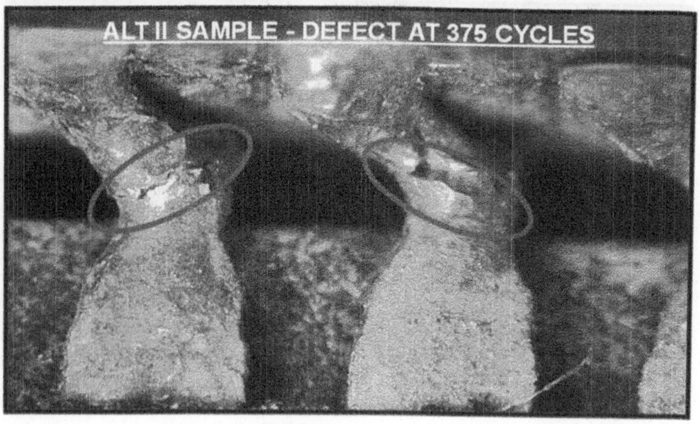

Fig. 20 Stereo microscope image of crack after 375 cycles for ALT II

It is observed that the change in resistance of the corner solder joints (Pin no. 1, 11, 12, 22, 23, 33, 34 and 44) is the greatest (greater than 2Ω). This observation is in agreement with FEA results, where maximum plastic strain range was observed in the corner solder joints. Thus, the corner solder joints for a PLCC package are the critical solder joints.

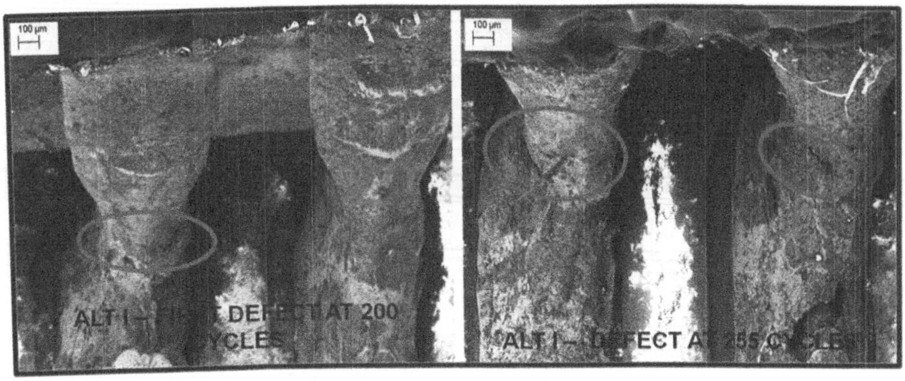

Fig. 21 SEM image of first crack at 200 cycles for ALT I

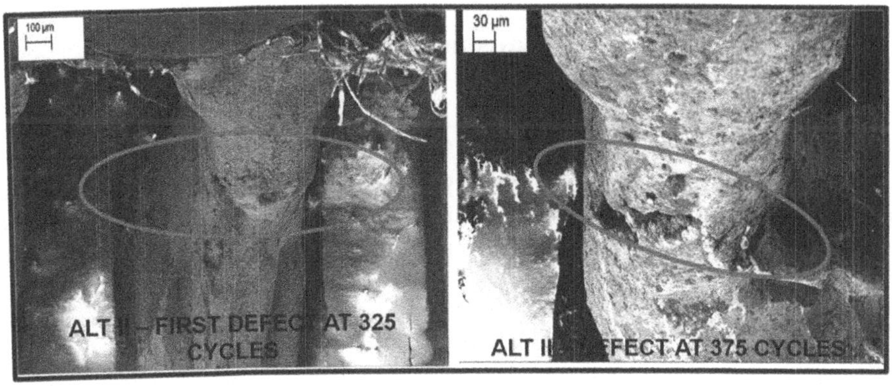

Fig. 22 SEM image of first crack at 325 cycles for ALT II

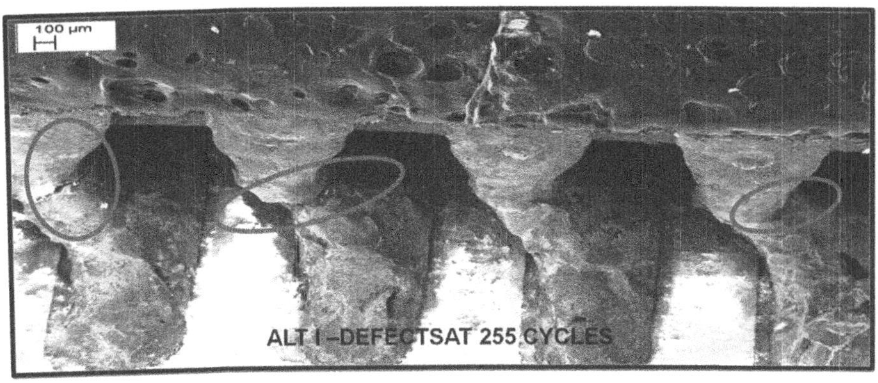

Fig. 23 SEM image of crack after 255 cycles for ALT I

Fig. 24 SEM image of crack after 375 cycles for ALT II

7 Estimation of Coffin-Manson Model Parameters

The results from experiment and FEA, are used for estimation of fatigue model parameters. The results are summarized Table 5.

As discussed, the total number of cycles to failure (N_f) is dependent on the plastic strain range ($\Delta\varepsilon_p$), the fatigue ductility coefficient (c) and the fatigue ductility exponent (ε_f). As per [3, 11] ε_f is taken as 0.325 for Sn–37Pb solder material. Therefore (1) can be rewritten as:

$$\log(\Delta\varepsilon_p) = \log(0.65) + c\log(2N_f) \tag{3}$$

Equation (3) clearly indicates a linear relationship between logarithm of $2N_f$ and $\Delta\varepsilon_p$. Table 5 lists the values of N_f and $\Delta\varepsilon_p$ for two ALT levels.

The results from FEA and experiments are used in (3) to obtain the following data points.

The regression analysis using the data points shown in Table 6 has been shown in Fig. 25. The slope of best fit line gives the value of c. As seen from Fig. 25, c is estimated to be –0.604. The value of c lies within the specified range (–0.5 to –0.7) [5]. Thus, the fatigue model for PLCC (with J-lead and Sn-37Pb solder) solder joints under temperature cycling can be expressed as.

$$\Delta\varepsilon_p = (0.65)(2N_f)^{-0.604} \tag{4}$$

Table 5 Temperature cycling loading results

Experiment	Nf (cycles)	$\Delta\varepsilon p$(%)
ALT I (5 to 165 °C)	255	1.4983
ALT I (5 to 165 °C)	375	1.1909

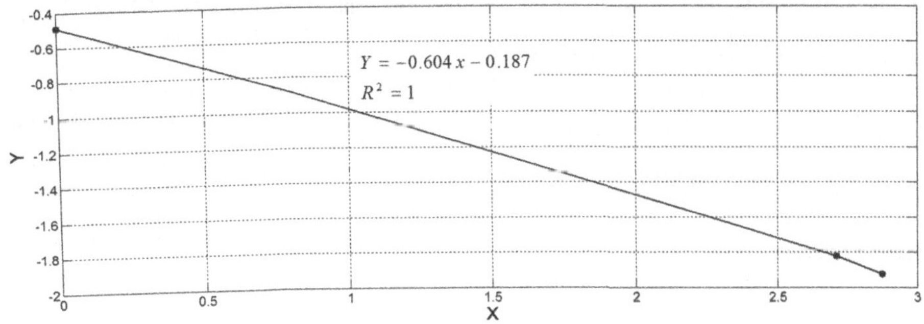

Fig. 25 Regression analysis for estimation of *c*

Table 6 Data points for regression analysis for estimation of *c*

Data point	x log (2*Nf*)	y log (Δεp)
A	0	–0.4881
B	2.7076	–1.8244
C	2.8751	–1.9241

8 Summary and Discussion

General empirical models for estimation of fatigue life are available in literature. These models have parameters which are function of geometry and material. In this study, parameters of Coffin-Manson model for PLCC solder joint with Sn-37Pb solder material under thermal cycle loading are estimated. FE models were used to estimate the strain range. The results of strain analysis indicate highest strain at the outer edge of the solder joint and the corner pins of the package.

Limitation of the present model is that it does not consider the accumulated damage in the solder joints. This is the gap area for the future work, to access the accumulated damage or N_f based on the observable (physical or micro-structural) parameters of solder-joint.

The specific fatigue model for PLCC solder joint can be used for predicting its life in use environment as part of overall system reliability estimation. It will also be useful in deciding the need of environmental control and estimation of Remaining Useful Life (RUL).

Acknowledgements The work presented here requires support of a number of tools and test setups. We got active support from experts of various divisions of BARC for the experiments. To name a few—Shri Deependra Singh, RRDPD for FE analysis; Shri Sanjay Kumar, Shri N. Saikathik, Dr. R. Tiwari and Dr. V. Kain of MSD for SEM and stereo imaging; Shri N.B. Shrestra and Shri P.K. Ramateke of RSD for thermal cycling experiments; and Shri P. Trivedi, ACnD for resistance measurement. We thank all for their kind support and making this work possible.

References

1. Ohring M (2014) Reliability and failure of electronic materials and devices. Academic
2. Liu J (2010) Reliability of microtechnology: interconnects devices and systems. Springer, Berlin
3. IPC SM 785, Guidelines for Accelerated Reliability Testing of Surface Mount Solder Attachments, IPC, 1992
4. Anand L, Park DM (2004) Mechanics and materials: supplementary notes. Massachusetts Institute of Technology
5. Lee WW, Nguyen LT, Selvaduray GS (2000) Solder joint fatigue models: review and applicability to chip scale packages. Microelectron Reliab 40(2):231–244
6. Fox L, Sofia J, Shine M (1985) Investigation of solder fatigue acceleration factors. IEEE Trans Compon. Hybrids Manuf Technol 8(2):275–282
7. Zilog Product Specification Z08536 Counter/Timer and Parallel I/O Unit. http://www.zilog.com
8. Siewert T, Stephen L, Smith DR, Madeni JC (2002) Database for solder properties with emphasis on new lead-free solders. NIST and Colorado School of Mines, Release 4
9. Tamin M, Norhashimah S (2014) Solder joint reliability assessment: finite element simulation methodology. Springer, Berlin
10. Rohit K (2016) Estimation of Fatigue life of PLCC solder joints due to vibration and thermal cycling, M. Tech. Thesis, HBNI
11. Norris KC, Landzberg AH (1969) Reliability of controlled collapse interconnections. IBM J Res Dev 13(3):266–271
12. Pecht M (1999) Integrated circuit hybrid and multichip module package design guidelines: a focus on reliability. Wiley, New York

Rohit Khatri is a commissioned Electrical Officer in the Indian Navy since 2008 and specializes in submarines. He has rich an experience in Operations and Maintenances of electrical & electronic equipment onboard submarines. He has completed his Masters in Electrical Engineering with specialization in Nuclear Engineering from Homi Bhabha National Institute, Mumbai. It is during this M. Tech phase that he developed an inclination towards research in field of Reliability Prediction, physics-of-failure (PoF) and remaining useful life (RUL) of Electronic Systems. He completed his Master's Dissertation with Control Instrumentation Division of BARC, Mumbai.

Diana Denice is working as a Scientific Officer in BARC, Mumbai. She has completed her Masters in Electronics Engineering in 2012 at Homi Bhabha National Institute, Mumbai and currently pursuing her Ph. D. in Prognostics of high-density electronic components. For the past 8 years, she has been working in the field of Design, Development and Reliability evaluation of Computer based systems for Nuclear Power Plant. Her research interests and field of expertise include reliability prediction of electronic systems, accelerated life testing and prognostics of electronic components.

Manoj Kumar received his B. Tech. (Electronics and Communication) from JMI, N. Delhi (1998) and Ph. D. (Engg.) from IIT Bombay (2008). He has been with Control Instrumentation Division of Bhabha Atomic Research Centre (BARC), Mumbai as Scientific Officer since 1998 and working in the area of area development and dependability analysis of computer based systems for safety applications. He has over 30 research papers to his credit and has supervised ten M. Tech. dissertations. He has authored two books in the area of dependability modeling and accelerated life testing of electronic systems. His current research interests include prognostics and application of machine learning in dependability prediction.

An Integrated Approach of BOCR Modeling Framework for Decision Tool Evaluation

Krishna Mohan Kovur and Ravi Kumar Gedela

Abstract In the present era of high standards competitive market and business globalization, it becomes difficult to balance financial uncertainties based on market requirements that are influenced by multiple attributes (criteria) involving multiple stakeholders. Further, we encounter many challenges arising from multiple input data sets of vague and imprecise nature that reflect the final options and outcomes (alternatives). Moreover, the prevailing models lack proper applicability of aggregation techniques to resolve the ambiguity of multiple data sets or where more than one experts are involved. Further, do not focus to quantify the uncertainty of complex decision problems representing positive (acceptance) and negative (renunciation) attributes. Those attributes, signifying benefits, and opportunities (BO), and latter costs, and risks (CR) respectively. To address these issues, this book chapter illustrates an analytical methodology representing BOCR models. The modeling has been carried out in the combination of appropriate analytical models and suitable data aggregation techniques. The proposed framework is illustrated by considered two case studies. The first case study illustrates a holistic model in the context of prototype dependability assessment of software at the prototype level. The second case study demonstrates a model validation quantitatively for quality of services (QoS) for real-world SOA based applications.

Keywords BOCR analysis · AHP · ANP · Priority vectors · Fuzzy logic · Arithmetic interval operations · Decision-making

K. M. Kovur (✉) · R. K. Gedela
Banking Labs, Toronto, Canada
e-mail: krishna.kovur@bankinglabs.com

R. K. Gedela
e-mail: Ravi.Gedela@bankinglabs.com

D. R. Karanki et al. (eds.), *Advances in RAMS Engineering*,
Springer Series in Reliability Engineering,
https://doi.org/10.1007/978-3-030-36518-9_5

1 Introduction

This book chapter finds some of the key findings of an extensive literature survey that was relied on to arrive at the decision analysis framework. Beginning with issues on prototyping and validation of real word applications, the book chapter covers the significant aspects in the realm of feasibility analysis, model validation and resource planning that complement the study of quantitative decision assessment engineering management. A prelude to this work in this article needs to be tackled in its own right before initiating dedicated research into the core premise of furthering the performance analysis of software systems. Collating the literature on data discrepancy from multiple inputs in group decision making is the first and the foremost step towards thoroughly understanding the potential hurdle one faces while dealing with data aggregation. Therefore, quantitative estimation and prediction of feasibility assessment along with model validation in critical software systems and factors associated on these issues are presented.

Although BOCR evaluation in all likelihood is the most complex challenging task of the decision-making process, it is actually the most worthwhile to investigate its relevance. The drawbacks of the prevailing tools, concepts, and theories show a strong need to improve and evaluate the various possible outcomes (alternatives) quantitatively. Based on our comprehensive literature review, we accomplish an analysis of economic feasibility that impacts the considered application problem to assess the decision making process on the different project objectives. The traditional economic study exploration is every so often to handle the influence of negative and positive financial impacts of application simultaneously based on qualitative analysis. However, this undervalues its adaptability in considering intangible values. This book chapter encompasses the proposed BOCR analysis—a framework aimed at decision analysis modeling of considered problems comprising with a multi-level objective; multiple attributes; various uncertainties; and numerous actors or stakeholders. Further, emphasis on the data collected from various sources with a focus on multiple inputs analysis by means appropriate aggregation techniques. Figure 1 depicts the proposed BOCR framework schematic in this book chapter.

Therefore, BOCR analysis [1–4] is further defined as a tool for effective decision-making aids through

- assessing yet-to-decide outcomes (alternatives) based on various policies and strategies of the application problem
- allowing two or more different alternatives with multiple influencing factors at the same time
- considering micro and macro details of external and internal aspects that affect the application attributes
- measuring positive facets of a project representing benefits and opportunities and while negative consequences in lieu of risks and costs
- quantification and evaluation of benefits, opportunities, costs, and risks (BOCR) offer a numerical operational framework of a decision problem

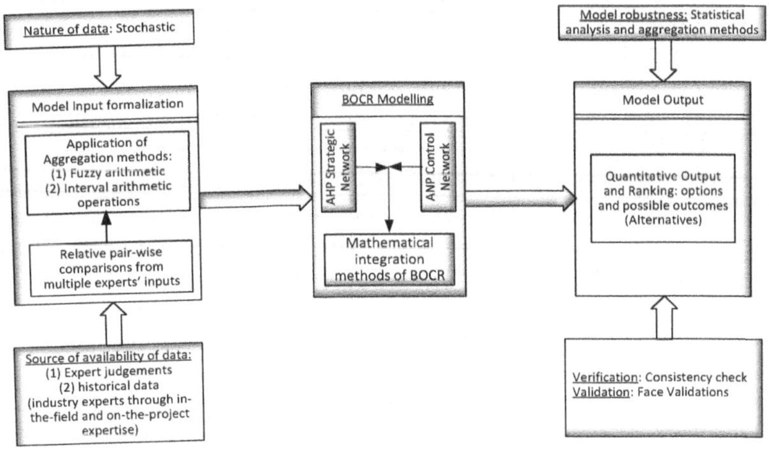

Fig. 1 Proposed BOCR framework schematic

- the multifaceted nature of the actors (cluster of persons or establishments); objective (to accomplish, improve, and realize by decision makers); alternative (to achieve their objectives); and attributes (a feature of a potential alternative used to assess w.r.t. stated objective)
- the validation of the suitability of real-world applications such (as performance, QoS, and reliability merits) based on the data profile and expert judgments proficiency to endorse the acceptability of the outcomes that are associated with a project.

2 Related Background and Literature Review

In this Sect. 2, we deal with associated procedures in the background of this book chapter and have been reviewed in respect to the following: (a) Quantitative Framework in Decision Analysis (b) Definition and Prototyping Evolution (c) Feasibility Analysis in Software Industry (d) Fuzzy Set Theory (e) Interval Arithmetic Basic Operations (f) BOCR Quantitative Approach.

2.1 Quantitative Framework in Decision Analysis

In decision making science, one of the most significant issues is about group decision making [5–10]. The aim of this book chapter is to aggregate multiple inputs provided by the different expert stakeholder of same maturity levels. The inputs are intuition-based and as well as interpreted through historical data.

Therefore, uncertainties that are governed by the events which are not at the disposal of the decision makers. In specific, the decision experts provide the input information which is characterized through a statistical relevance. As a result, quantitative decision analysis [11–15] methods are suited for those applications with a well-defined objective with more than one outcomes of possible options, i.e. alternatives. Thereby gauging combined benefits and opportunities over mutual costs and risks associated with various alternatives takes the driver's seat in the decision framework. The framework combines analytical models and aggregation methods in the context of multi-criteria decision making.

The book chapter integrates perceptions, concepts, tools, and techniques from computer science, analytical mathematics, and applied statistics intended for modeling and assessment aimed at multifaceted decision-making methodology for feasibility and model validations at the prototype level. The underlying importance of proposed framework to control complex decision making which is beyond the disposal of the decision making expertise to

- demonstrate a scientific approach for various current practices prevailing to carry out volatile business transformation with the combination of appropriate models
- develop tools via. core concepts from analytic mathematical models; and quantitative techniques through various economic perceptions
- provide a quantitative solution for several business applications by exploring and identifying factors, sub-factors, etc. affecting resource availability
- strike balance between production/demand ratio to yield feasible strategies in order to obtain maximum profits
- apply proper fitting of data distributions like exponential, normal, and etc. for data analysis for the model input parameters setting
- optimum distribution usage, selection, deployment of resources and strategies to acquire maximum output with minimum adverse effects/events
- deliver pointers for current market needs; meet business competition, and successful on-time product/project delivery
- expedite and as well streamline the decision-making process using granular analysis at varying degrees of application complexities associated with various uncertainties at different levels of abstraction
- smooth transition between various stockholders with improved communication gaps; effective quick review process; and insignificant bias
- permit the applicability of prevailing quantitative models and aim to develop a tailor-made model framework based on the 'source of availability of data', 'characteristics of uncertainty' and 'customer's requirements' to capture influences of decision-making.

2.2 Definition and Prototyping Evaluation

One of the main tasks of decision making is about the assessment of prototype in any industry connecting several stockholder's accountabilities. The Prototyping Evaluation offers a sense of assurance and confidence prior to a decision to accept or modify or reject as the indicators of an idea before further taking it up in a large scale [16–23]. Therefore, a prototype characterizes a thought process or idea based on past experience and historical data into a pragmatic representation or a real implementation. The prototype evaluation progresses through a rapid feedback process from the end-users to access its feasibility in regards to marketability abilities. Further, assists through the reduction of functional and non-functional conflicts at the early stages to maximize the acceptance rate from the customer. The book chapter focusses on the various aspects of the internal and external factors to balance between benefits and opportunities set against costs and risks in order to investigate the best alternative quantitatively. The main aim of this study is to examine the viability of project/product development via economic, social, technical, political, and operational feasibility of a new proposal or idea of a proposed model. Thereby, to benefit innovators to verify and quickly validate the ideas beforehand of its potentiality to further undertake the development process. The framework proposed works majorly on quantitative perspective concerned with an overall numerical evaluation in respect to economic (non-functional), technical (functional) and costs associated with the prototype assessment.

2.3 Feasibility Analysis in Software Industry

As in many engineering disciplines, prototyping in software industry uses several methods and techniques to evaluate the calculation of risks and costs of new addressed projects or products. Hence in software engineering prototyping has also been implemented as a technique to access feasibility before proceeding for full-blown development. In recent years, there a need to address the difficulties in software development prototyping and the criticism aimed at traditional life cycle become has been increasingly observed to improve planning and success of software projects [23]. Software Prototyping encompasses all phases of software development as a whole and normally viewed as evolutionary vision producing early working versions ("prototypes") of the future application system. Thus, prototyping [23] delivers a communication base for deliberations amongst of all groups involved especially between users and developers to track project improvement that

- aid us to recognize when to archive an anticipated degree of quality
- aims at approximations and estimates
- predict and analyze different types of defects

- determine design complexity module and system levels
- validate using best practices using various techniques and simulations.

There are several software prototyping practices currently underway at the various industries. For instance, like exploratory prototyping, experimental Prototyping, evolutionary prototyping, and horizontal prototyping. In the exploratory prototyping are considered when the application problem under consideration is not clear. Here, it focusses on the preliminary ideas to bridge the gaps between management necessities with the users to proceed with the prototyping process. Whereas, in experimental prototyping works differently with well-defined ideas and a focus on technical implementation towards the developmental goal or objective. Hence it maintains a balance between ideas and infrastructure support such as system requirements. Thereby, the emphasis given for appraisal in regards to feasibility and correctness of application which is considered an extension for exploratory prototyping. In case of evolutionary prototyping, it is a process-oriented technique with a continuous procedure to cater quickly to varying organizational constraints. Therefore, it continuously tracks the application within its process framework. Normally, prototyping methods covers application semantics and software architecture with a focus to ease system design. Lastly, we discuss horizontal and vertical prototyping to build over various layers via. a user interface to an operating system to data management. In horizontal prototyping merely particular layers of the system, for example, the menu form through the user interface layer with functional layers and whereas, in vertical prototyping, it works on a part of the systems incorporating all the layers. Our Proposed framework quantifies the best feasibility decision alternatives (options of possible outcomes) analytically by multi-factor analysis and assessment. Also, addresses various data discrepancies from multiple data inputs obtained from several sources and as well from experts' opinion judgments.

2.4 *Data Discrepancy from Multiple Inputs in Group Decision Making*

One of the major challenges to be tackled is when multiple input data conflicts happen in the group decision making results in a breakdown affecting the final outcome adversely. The other challenge that surfaces is about when to aggregate [24–26] the multiple inputs i.e. 'before' or 'after' modeling execution. In the former case, different experts' inputs are aggregated into a single value and provide a source of input to the model. In the latter case, where each expert evaluates the model independently to produce corresponding outputs; subsequently, aggregate multiple outputs to obtained into a single aggregated output. These application problems which are under consideration have been conventionally resolved by means of aggregation [24–26], but selecting the precise level of aggregation at different perspectives needs determining the stability between cumulative or aggregation error with the sampling error. Here, in this book chapter, we address two types of aggregation techniques, the first one been fuzzy set theory and whereas, second technique been interval arithmetic operations. We use the appropriate technique as per the situation that fits the aggregation approximation to produce near the optimal and unambiguous output.

2.4.1 Fuzzy *Logic Set Theory*

In one of the case studies presented in this book chapter, fuzzy logic set theory has been applied based on customer's requirement and input data prerequisite. In this section, reviewed some of the basic perceptions related to the fuzzy concept [14, 27–31]. The fuzzy logic set theory was initially proposed by Zadeh [36] to evaluate uncertainty and vagueness (fuzziness) associated with application problems in the various domain areas of a real-world environment. Here, we briefly encapsulate some of the underlying concepts associated with fuzzy sets such as fuzzification, fuzzy numbers and along with associated membership functions. The fuzzification is the technique of translating of crisp values into different levels of membership. A fuzzy number is a generalization of a real number, meaning it doesn't associate with a particular single number but instead bring together a group of probable values so that each probable value will have a range of its own values between with 0 and 1 which is termed as the membership function [14, 27–31]. A Fuzzy set is well-defined by its uncertain limits or boundaries. Its properties are defined or described by its membership function $\mu(x)$. Membership function aids to alter or change the categorize set into a fuzzy set. In our proposed model framework, the fuzzy logic is applied over into the dataset governed by various aspects of the considered decision application problem over statistical data during the software development phases [8]. Triangular and trapezoidal membership functions have been used in this book chapter as per the requirements of the considered case study application problems.

Here, we discuss the basic concept of triangular membership curve which is defined as a vector function of x and governed by scalar parameters [36], a_1, a_2, and a_3, shown in Fig. 2. The parameters a_1 and a_3 identify the triangle base whereas, the parameter a_2 points peak of the triangle. Therefore, let triangular member M be

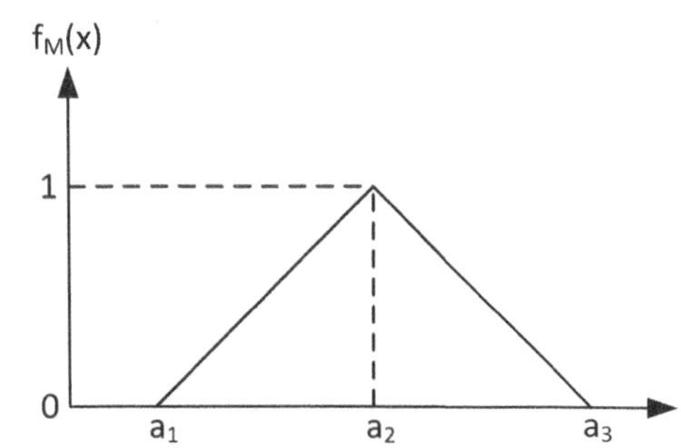

Fig. 2 Triangular membership function

(a_1, a_2, a_3), Fig. 2. The corresponding membership function is defined in Eq. 1 as follows

$$a_\mu(X : a_1, a_2, a_3) = \left\{ \begin{array}{ll} \frac{X-a_1}{a_2-a_1} & a_1 < X <= a_2 \\ \frac{X-a_2}{a_3-a_2} & a_2 < X <= a_3 \\ 0 & \text{otherwise} \end{array} \right\} \quad (1)$$

Fuzzy arithmetic: Next we briefly review basic fuzzy arithmetic operations. Zahed [36] proposed the fuzzy extension rules for the two triangular fuzzy numbers S_1 and S_2 in respect to addition, subtraction, multiplication, and division defined by $S_1(a_{11}, a_{12}, a_{13})$ and $S_2(a_{21}, a_{22}, a_{23})$ as follows through the Eqs. 2–5.

$$s_1 \oplus s_2 = (a_{11} + a_{21}, a_{12} + a_{22}, a_{13} + a_{23}) \quad (2)$$

$$s_1 - s_2 = (a_{11} - a_{23}, a_{12} - a_{22}, a_{13} - a_{21}) \quad (3)$$

$$s_1 \otimes s_2 = (a_{11} \times a_{21}, a_{12} \times a_{22}, a_{13} \times a_{23}) \quad (4)$$

$$s_1/s_2 = (a_{11}/a_{23}, a_{12}/a_{22}, a_{13}/a_{21}) \quad (5)$$

Likewise, we revisit the MCDM based application problem for the data sets governed by trapezoidal fuzzy sets. Figure 3 displays a trapezoidal membership M be (a_1, a_2, a_3, a_4) and corresponding membership function through Eq. 6. Both triangular and trapezoidal membership functions are illustrated here for the purpose of comparison to deal with their applicability with the same application problem under consideration.

$$a_\mu(X : a_1, a_2, a_3, a_4) = \left\{ \begin{array}{ll} 0 & (X < a_1) \text{ or } (X > a_4) \\ \frac{X-a_1}{a_2-a_1} & a_1 < X <= a_2 \\ 1 & a_2 < X <= a_3 \\ \frac{a_4-X}{a_4-a_3} & a_3 < X <= a_4 \end{array} \right\} \quad (6)$$

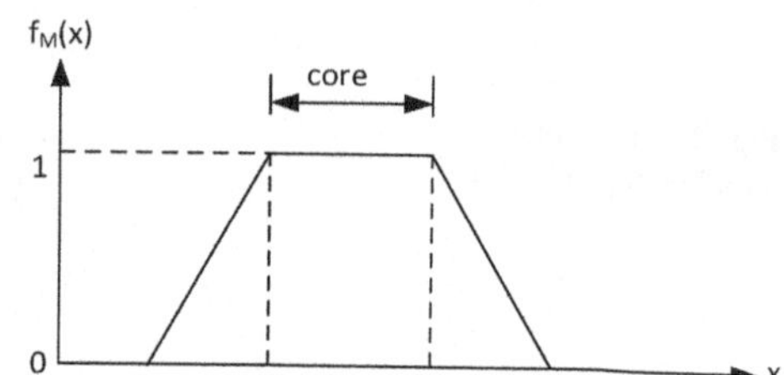

Fig. 3 Trapezoidal membership function

2.4.2 Arithmetic Interval Operations

Here, in this section, other aggregation techniques, the arithmetic interval operations have been revisited. Based on the literature review and illustrated application case study illustrated in the book chapter, interval operations have been considered to perform aggregation of multiple outputs of BOCR (AHP/ANP) model. The basic underlying procedures in the interval arithmetic analysis [32–35] between two intervals is administered by addition, subtraction, multiplication and division operations. For example, if two intervals are defined by (α_1, β_1) and (α_2, β_2), then preliminary athematic operations are defined via. following Eq. 7 through Eq. 10 shown in Table 1. In the book chapter, applied these derived equations on the outcome results of the BOCR framework methodology evaluated by three different experts. The purpose of this aggregation analysis is to compute the output in a spectrum of values, so as to incorporate inconsistency among different experts involved in group decision making. Therefore, we have introduced the interval analysis on benefits, opportunities, costs and risks models assessed through multiple data sets assesses by multiple experts.

2.5 *BOCR Quantitative Approach*

Finally, put forward the underlying BOCR mathematical derived Eqs. 11–15 proposed by Saaty [2] applied for the proposed framework assessment in the book chapter. Due to prerequisite need, appropriate data aggregations techniques along with suitable analytical models AHP and ANP are used in BOCR framework. Here, we review basic model-driven concepts of BOCR merits usage with regard to the proposed framework analysis and assessment. Through literature review and expert judgment, priory weights are evaluated for each of the factor elements (criterion) and alternatives under each BOCR-ANP model merits via. preliminary unweighted

Table 1 Arithmetic interval operations

Arithmetic operations	Interval	Min and max definition	Equation number
Addition:	$(\alpha_1, \beta_1) + (\alpha_2, \beta_2)$	$\min(\alpha_1 + \alpha_2, \alpha_1 + \beta_2, \beta_1 + \beta_2, \alpha_1 + \beta_2)$ $\max(\alpha_1 + \alpha_2, \alpha_1 + \beta_2, \beta_1 + \alpha_1, \beta_1 + \beta_2)$	(7)
Subtraction:	$(\alpha_1, \beta_1) - (\alpha_2, \beta_2)$	$\min(\alpha_1 - \alpha_2, \alpha_1 - \beta_2, \beta_1 - \alpha_1, \beta_1 - \beta_2)$ $\max(\alpha_1 - \alpha_2, \alpha_1 - \beta_2, \alpha_2 - \beta_1, \beta_1 - \beta_2)$	(8)
Multiplication:	$(\alpha_1, \beta_1) \times (\alpha_2, \beta_2)$	$\min(\alpha_1 \times \alpha_2, \alpha_1 \times \beta_2, \beta_1 \times \alpha_2, \beta_1 \times \beta_2)$ $\max(\alpha_1 \times \alpha_2, \alpha_1 \times \beta_2, \beta_1 \times x2, \beta_1 \times \beta_2)$	(9)
Division:	$(\alpha_1, \beta_1) / (\alpha_2, \beta_2)$	$\min(\alpha_1 \div \alpha_2, \alpha_1 \div \beta_2, \beta_1 \div \alpha_1, \beta_1 \div \beta_2)$ $\max(\alpha_1 \div \alpha_2, \alpha_1 \div \beta_2, \beta_1 \div \alpha_2, \beta_1 \div \beta_2)$	(10)

to intermediary weighted super-matrix towards final limiting super-matrix. Similarly, the importance of b-o-c-r to each strategic criterion are accessed through BOCR-AHP. Thus, the following BOCR equations are implemented in conjunction with ANP and AHP models.

$$\text{Additive}\, P_i = bB_i + oO_i + c[1/C_i]_{normalized} + r][1/R_i]_{normalized} \tag{11}$$

$$\text{Probabilistic additive}\, P_i = bB_i + oO_i + c(1 - C_i)] + r(1 - R_i) \tag{12}$$

$$\text{Subtractive}\, P_i = bBi + oO_i - cC_i - rR_i \tag{13}$$

$$\text{Multiplicative priority powers}\, P_i = B_i^b O_i^o [(1/C_i)_{normalized}]^c [(1/R_i)_{normalized}]^r \tag{14}$$

$$\text{Multiplicative}\, P_i = B_i O_i / C_i R_i \tag{15}$$

3 Customized Research Methodology and Modified Methods Used in the Proposed Framework

In the book chapter, we have customized the research methodology based on the considered case studies and the customer's requirements. The Fuzzy ANP (FANP); Fuzzy AHP (FAHP); statistical based fuzzy slope evaluation; and athematic interval operations; are combined with BOCR models as and when required. Therefore, these methods are used suitably employed in the proposed BOCR framework, for instance, triangular fuzzy numbers are established for linguistic terms of importance in the pair-wise comparison matrix, shown in Table 2.

Similarly, athematic interval operations are used for aggregation of multiple BOCR models output evaluated from different experts with modified BOCR calculations. The modified BOCR mathematical equations through the application of arithmetic interval operations are shown through Eqs. 16 and 17. The revised Eq. 16 is a modified formula of Eq. 13. Similarly, Eq. 17 is a modified formula of Eq. 15.

Table 2 Scale of the fuzzy linguistic variable

Linguistic performance variable	Linguistic relations of importance	Triangular Fuzzy Numbers (TFN)
Very Low (VL)	Equivalent	[1,1,3]
Low (L)	A slight strong	[1,3,5]
Medium (M)	Strong	[3,5,7]
High (H)	Very strong	[5,7,9]
Very High (VH)	Extremely strong	[9,9,9]

Subtractive P_i:

$$[b_L b_H].[B_L B_H] + [o_L o_H].[O_L O_H] - [c_L c_H].[C_L C_H] - [r_L r_H].[R_L R_H] \quad (16)$$

Multiplicative P_i:

$$[B_L B_H].[O_L O_H]/[C_L C_H].[R_L R_H] \quad (17)$$

Finally, statistical based fuzzy slope methodology evaluation is derived for strategic hierarchy (BOCR-AHP model) assessment depicted in Fig. 4. An illustration 'mean' of effort in person-hours where r is the effort estimate, r_j with jth effort time, and n is the no. of data sets considered. Here, used t-distribution function, degree of freedom to indicate lower and upper bounds obtained from sample data sets. Equation 18 illustrates mean and membership bounds between r_n and r_{n-1}

$$\left.\begin{array}{l} r = \frac{1}{n}\sum_{j}^{1} r_j \\ -(t\alpha/2(n-1) \leq \frac{r-\mu}{s\sqrt{n}} t\alpha/2(n-1) \\ r_{(n-1)} = (r - t\alpha/2(n-1)\frac{s}{\sqrt{n}} r + t\alpha/2(n-1)\frac{s}{\sqrt{n}} = r_{(n)} \end{array}\right\} \quad (18)$$

4 Case Studies Illustrations

In this chapter, we have considered two wide case studies for the proposed integrated approach of BOCR modeling framework for decision tool evaluation. The analytically based decision making analysis is performed against three possible viable options in association with suitable and selected data aggregation techniques. The first case study is focused on the decision model with regards to the feasibility aspect of software development at prototype level prior to actual or real implementation. Whereas, the second case study dedicated to model validation quantitatively to ascertain cross authentication of a proposed SOA model developed for

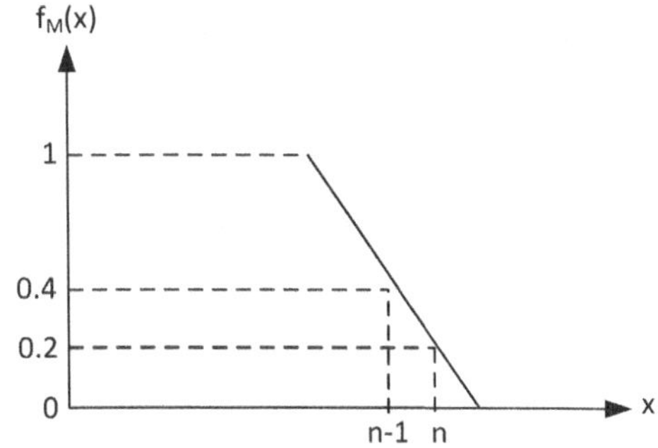

Fig. 4 Fuzzy slope calculations

QoS considerations. Both the case studies demonstrate for a wider range of applicability that exemplifies as holistic feasibility models. Therefore, identified various economic, social, technical and trade influencing factors (criteria) and sub-factors (sub-criteria) under respective merit of BOCR models [3]. These identified factors have an impact on the consequences with respect to practical adaptability of a commercial project in the software industry. Therefore, in the book chapter, multiple data inputs from different experts are been considered for both case studies. All influencing factor elements are modeled according to the problem characteristics and customer's wide requirements. For example, the relationships among the factor elements, such as dependency and independence on each other are examined through various associations in a realistic manner. Accordingly, multi-criteria based AHP [37–42] or ANP models [43–48]—that governs the underlying assessment of BOCR framework are reflected in the modeling design. In other words, AHP works through hierarchy relationship as a linear architecture and whereas, ANP is accomplished via. non-linear architecture through various clusters relationships (i.e. network) that allows several relations among connecting influencing factors, sub-factors, and alternatives. Further, in ANP, we can broadly classify various relations (influences) as the inner (self) and outer dependencies across the network. On the top this, suitable aggregation techniques have been deployed to address the multiple user inputs and expert judgments.

4.1 *Case Study 1*

4.1.1 Introduction

The feasibility of real-time applications in the software industry is used as an illustration. Here, the software application projects which have been developed over a period of the last four years and have varied maturity levels of various complexities are considered. Therefore, carefully chosen 15 different projects are compared and aggregated through an appropriate technique (athematic interval operations) in conjunction with BOCR models. The feasibility quantification is performed through 'BOCR (ANP/AHP)—arithmetic interval operations' decision model with the consensus of multiple experts based on 15 different projects. The purpose of arithmetic interval analysis is to minimize the discrepancy among the expert's experience and preferences involved in a group decision-making progression. Here, in the case study, three expert's inputs are considered to execute BOCR (ANP/AHP) models independently to yield a respective set of possible outputs. Consequently, upon application of arithmetic interval operations (namely addition, subtractions, multiplication, and division) on the obtained outputs to evaluate the final output through a spectrum bounds of lower and upper values. In other terms, we combine different sets of output from BOCR (ANP/AHP), in our case we obtain three sets of outputs after the modeling process. The analysis and modeling flow are depicted in Fig. 5.

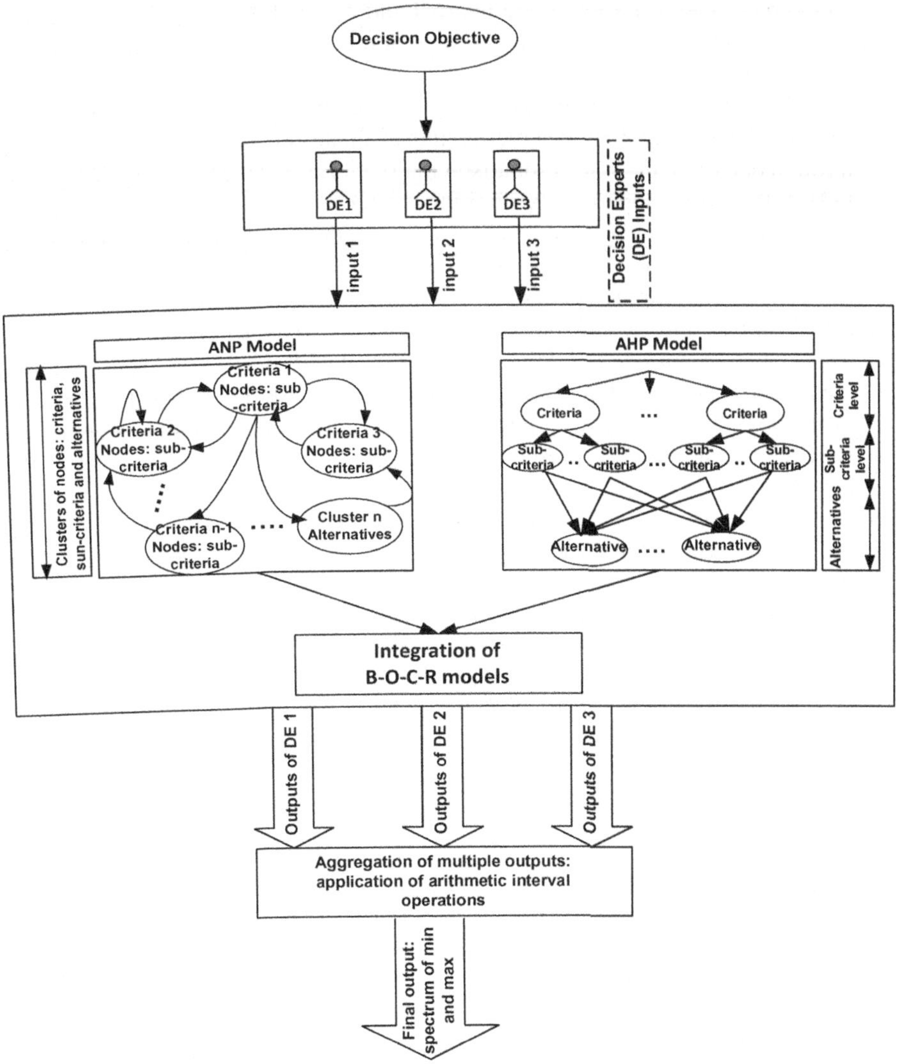

Fig. 5 BOCR(ANP/AHP) with arithmetic interval analysis

4.1.2 Proposed Steps

Steps in BOCR-(ANP/AHP) with Arithmetic Interval Analysis: modeling and assessment

(1) *Formulation of the problem statement and solution strategy:* defining the decision problem objective and identification of different domain experts. Formulate the BOCR-ANP control network and BOCR-AHP strategic

hierarchy models—based on the internal and external aspects of the considered problem. Determine the number of three domain experts needed to provide the inputs to the model.

(2) *BOCR-ANP control network assessment:* comprehensive modeling of ANP for each of the B-O-C-R model merits and evaluate the priorities of alternatives through derived pair-wise comparison matrices via 1–9 scale proposed by Saaty [43]; followed by supermatrix; limiting matrix as per the BOCR-ANP models.
(3) *BOCR-AHP strategic hierarchy evaluation*: modeling of strategic hierarchy and calculate the priorities—here, in this case, it is the—importance of BOCR (b, o, c, r) based on the computed pair-wise strategic matrix and the weights derived from statistical based fuzzy slope evaluation (ref. Eq. 18) as per BOCR-AHP model.
(4) *Application of arithmetic interval analysis on BOCR-ANP*: select and segregate the min (L) and Max (H)—(L and H represent lower and higher values) priority evaluations from three different experts for alternatives under each of the BOCR control models represented by (B_L, B_H); (O_L, O_H); (C_L, C_H); (R_L, R_H) acquired from step 2.
(5) *Application of arithmetic interval analysis on BOCR-ANP*: similarly, select the priority values obtained from importance's from three different experts as (b_L, b_H); (o_L, o_H); (c_L, c_H); (r_L,r_H) from step 3.
(6) *Final result assessment of BOCR-arithmetic analysis:* apply, arithmetic interval operations (ref. to Eq. 7 through 10) to evaluate final priority values of alternatives in a spectrum of upper and lower limits via revised formulations (ref. Eqs. 16 and 17).
(7) *Interpretation of results outcome:* finally, interpret results for the considered decision problem.

4.1.3 Problem Formulation

The objective of the case study is to achieve the best possible alternative (options and outcome) which is to *evaluate prototype dependability assessment of software at the prototype level.* The underlying concept that represents best alternative by highest quantified value which have maximum benefits and offers more opportunities; and at the same time with lowest quantified values that offer lower risks and lesser costs obtained by BOCR (ANP/AHP) models. Here, in this case, the output is obtained in a spectrum of intervals (lower and upper limits) from the proposed BOCR(AHP/ANP)-Arithmetic Interval Operations framework, Fig. 5 for the three alternatives are A1, A2, A3 stated below:

Alternatives:

A_1: Advance with the working system with insignificant changes or minimum modifications: continue with the proposed flawless system architecture

A_2: Advance with the working system with major revisions: continue with the proposed system architecture subjected to addressing major constraints
A_3: Advance with a customized package: Not to continue with the proposed system architecture.

4.1.4 BOCR-ANP: Control Network

In this section, we have carried out a comprehensive modeling for each of the benefits, opportunities, costs, and risks of BOCR–ANP control network. In each of these ANP based BOCR models, the factors or criteria are identified as follows: *management of software quality and assurance (b_1); reliability of system software prediction (b_2); financial aspects of I.T economics (b_3); organizational visibility (b_4).* Whereas the sub-factors or sub-criteria are distinct with respect to each of the BOCR models, defined in the respective modeling case. Most importantly, each of the BOCR–ANP are evaluated by three different experts.

Benefits Model

Figure 6 illustrates the ANP based benefits model with various factors and sub-factors with the alternatives. These identified factor elements reflect the benefits in realizing the objective of the case study *'evaluate prototype dependability assessment of software at prototype level'* The various factors under benefits model are as follows *management of software quality and assurance (b_1); reliability of system software prediction (b_2); financial aspects of I.T economics (b_3); organizational visibility (b_4).* The sub-factors under b_1 *are: superior early process enhancement (b_{11}); superior quality design with the application of various business approaches (b_{12}); effective change management process for improved project outcomes (b_{13}).* The sub-factors under b_2 are *probability of no. of defects predicted at various SDLC stages (b_{21}); Better coordination of client and end-user prerequisite (b_{22}); Forecasting the Mean Time Between Failures—MTBF (b_{23}).* The sub-factors under b_3 *are—prognostication of ROI (b_{31}); Ideal release time for a feasible working system (in respect to functional and non-functional) (b_{32}); Estimation of consequences with a lower budget and burden (b_{33}).* The sub-factors under b_4 as—*constructive road-map with an overall understanding of development process among the SDLC groups (b_{41}); Minimization of gap between software development team and end-users (b_{42}); Enhanced customer's genuine and actual feel of the product (b_{43}); Better employment credible user-friendly design tools and procedures (b_{44}).* The Fig. 6 ANP is accomplished via. non-linear architecture from various clusters relationships (i.e. network) that allows any relation among connecting influencing factors, sub-factors and alternatives. Further, shown various relations (influences) as the inner (self) and outer dependencies across the network. Table 3a, presents normalized local priory weights across the clusters of sub-factors and alternatives obtained from benefits model. Tabulated priory weights evaluated by three different experts.

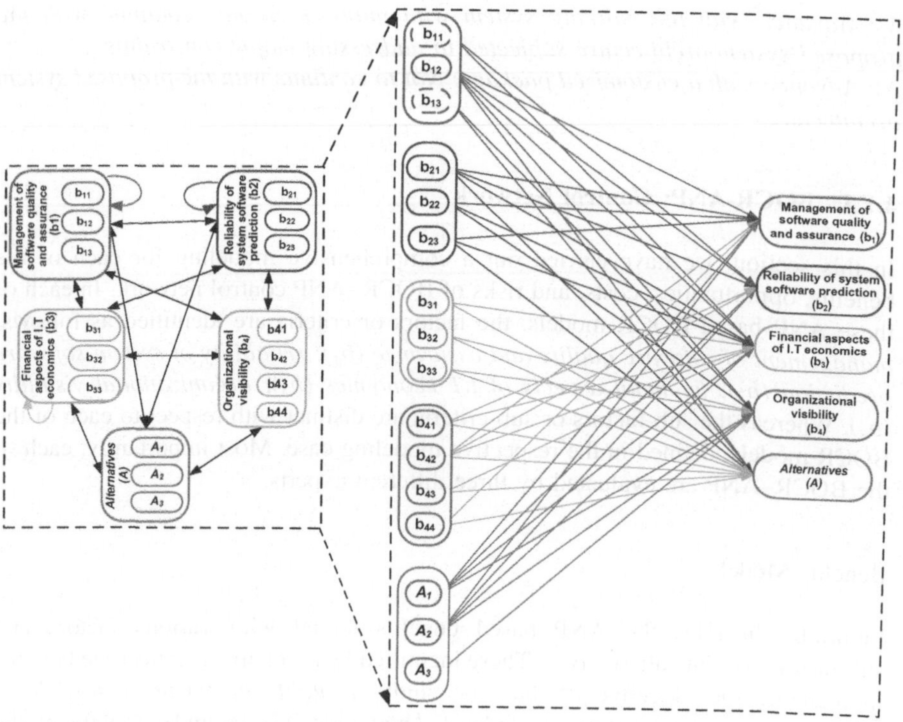

Fig. 6 ANP based benefits network model: inner (self) and outer dependencies

Opportunities Model

Figure 7 demonstrates ANP based opportunities model with various factors and sub-factors with the alternatives. These factors influence and responsible for opportunities to grow through several means via 'enhanced, additional, better and improved' resources. The identified factor elements reflect the opportunities in realizing the objective of the case study through the following sub-factors under each of the factors: Under *Management of software quality and assurance* (O_1), the sub-factors are: *Enhanced flexibility in deciding more analysts, programmers and designers from end-user perspective* (O_{11}); *Improved communications within the organization and effective traceability of undesirable state of affairs* (O_{12}); *Constructive deliberations and consultations* via. *planned meetings* (O_{13}); Next, under *Reliability of system software prediction* (O_2), the sub-factors are: *Simulation-based quantitative predictions to fix the bugs at an early stage* (O_{21}); *Early warning of missing functionality to address reliability issues* (O_{22}) *Faster user responses and quick feedback* (O_{23}); Subsequently, under *Financial aspects of I.T economics* (O_3), the sub-factors are: *Early detection of business misapprehensions and ambiguousness* (O_{31}); *Enhanced technical backing for a better success* (O_{32}); *Stronger statistical methods to predict performance at module or system*

Table 3 Normalized local priory weights by three different experts

(a): Benefits model

Benefits model merits	Criteria	Sub-criteria	Normalized by cluster		
			Expert 1	Expert 2	Expert 3
	B_1	b_{11}	0.4141	0.4094	0.4140
		b_{12}	0.3012	0.3047	0.2915
		b_{13}	0.2846	0.2858	0.2944
	B_2	b_{21}	0.2724	0.2684	0.2727
		b_{22}	0.5427	0.5581	0.5554
		b_{23}	0.1848	0.1734	0.1718
	B_3	b_{31}	0.3094	0.3159	0.3534
		b_{32}	0.3652	0.3794	0.3517
		b_{33}	0.3253	0.3071	0.2948
	B_4	b_{41}	0.2174	0.2200	0.2202
		b_{42}	0.4325	0.4398	0.4522
		b_{43}	0.1415	0.1412	0.1441
		b_{44}	0.2093	0.1989	0.1834
		$\mathbf{A_1}$	**0.5943**	**0.6188**	**0.6576**
		$\mathbf{A_2}$	**0.2626**	**0.2443**	**0.2356**
		$\mathbf{A_3}$	**0.1429**	**0.1367**	**0.1067**

(b): Opportunities model

Opportunities model merits	Criteria	Sub-criteria	Normalized by cluster		
			Expert 1	Expert 2	Expert 3
	O_1	O_{11}	0.5997	0.5990	0.6029
		O_{12}	0.1476	0.1475	0.1443
		O_{13}	0.2526	0.2535	0.2526
	O_2	O_{21}	0.3065	0.3020	0.3039
		O_{22}	0.3728	0.3785	0.3746
		O_{23}	0.3206	0.3194	0.3213
	O_3	O_{31}	0.4453	0.4535	0.4527
		O_{32}	0.3249	0.3166	0.3227
		O_{33}	0.2297	0.2298	0.2245
	O_4	O_{41}	0.4285	0.4220	0.4465
		O_{42}	0.3561	0.3581	0.3715
		O_{43}	0.2153	0.2198	0.1818
		$\mathbf{A_1}$	**0.6323**	**0.5841**	**0.6010**
		$\mathbf{A_2}$	**0.2551**	**0.3011**	**0.2732**
		$\mathbf{A_3}$	**0.1124**	**0.1147**	**0.1257**

(continued)

Table 3 (continued)

(c): Costs model

(c): Costs model

Costs model merits	Criteria	Sub-criteria	Normalized by cluster		
			Expert 1	Expert 2	Expert 3
	C_1	C_{11}	0.1271	0.1311	0.1275
		C_{12}	0.4059	0.4009	0.4087
		C_{13}	0.1827	0.1834	0.1829
		C_{14}	0.2842	0.2854	0.2807
	C_2	C_{21}	0.1687	0.1581	0.1681
		C_{22}	0.3578	0.4052	0.4057
		C_{23}	0.4737	0.4366	0.4260
	C_3	C_{31}	0.4314	0.4497	0.4407
		C_{32}	0.4071	0.3981	0.3924
		C_{33}	0.1614	0.1521	0.1668
	C_4	C_{41}	0.3748	0.3680	0.3895
		C_{42}	0.2619	0.2610	0.2429
		C_{43}	0.3631	0.3709	0.3674
		A_1	**0.6098**	**0.6292**	**0.6714**
		A_2	**0.2524**	**0.2531**	**0.2303**
		A_3	**0.1376**	**0.1176**	**0.0982**

(d): Risk model

Risks model merits	Criteria	Sub-criteria	Normalized by cluster		
			Expert 1	Expert 2	Expert 3
	R_1	R_{11}	0.5489	0.5591	0.5492
		R_{12}	0.2271	0.2348	0.2304
		R_{13}	0.2339	0.2159	0.2166
	R_2	R_{21}	0.3405	0.3534	0.3508
		R_{22}	0.3898	0.3911	0.3918
		R_{23}	0.2696	0.2553	0.2572
	R_3	R_{31}	0.3395	0.3261	0.3222
		R_{32}	0.4087	0.4376	0.4412
		R_{33}	0.2516	0.2361	0.2365
	R_4	R_{41}	0.3889	0.4356	0.4215
		R_{42}	0.2938	0.2812	0.2892
		R_{43}	0.3172	0.2830	0.2892
		A_1	**0.5812**	**0.6331**	**0.6725**
		A_2	**0.2763**	**0.2470**	**0.2060**
		A_3	**0.1424**	**0.1198**	**0.1213**

Bold signifies computed normalized priory values for each of the alternatives in the respective B-O-C-R network models

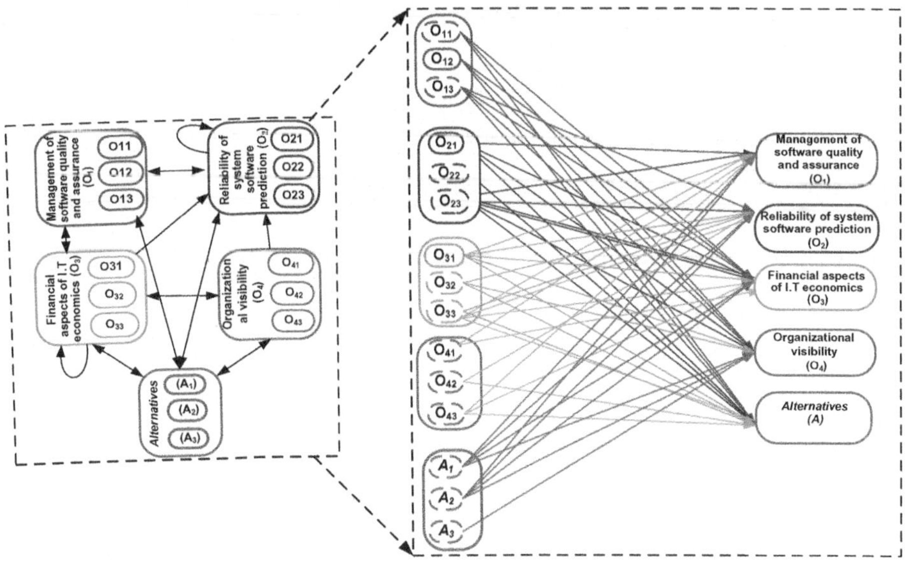

Fig. 7 ANP based opportunities network model: inner (self) and outer dependencies

level (O_{33}); Finally, under the Organizational *visibility* (O_4), the sub-factors are: *Early warning and warranty predictions for a product design decision* (O_{41}); *Better-quality organizational skills* via. *holistic approaches* (O_{42}); *Prompt upgradation of company's management strategies* (O_{43}). Table 3b, presents normalized local priory weights across the clusters of sub-factors and alternatives obtained from opportunities model. Tabulated priory weights evaluated by three different experts.

Costs Model

Costs Model: Fig. 8, shows ANP based costs model with various factors and sub-factors along with the alternatives. These factors focus on the cost incurred or spent during various stages of SDLC of project implementation. The identified factor elements responsible for the costs expenditure in realizing the objective of the case study through the following sub-factors under each of the factors: Under *Management of software quality and assurance* (C_1) factor, the sub-factors are: *Early discovery and fixing defect preventive cost* (C_{11}), *Cost due to Infrastructure set-up* (C_{12}), *Cost incurred from appraisal and improvement cost* (C_{13}), *Cost due to external and internal failures and other anomalies* (C_{14}); Next under Reliability of system software prediction (C_2) factor, the sub-factors are: *Cost of system maintenance and optimization* (C_{21}), *Cost owing allocating resources to enhance reliability and productivity* (C_{22}), *Cost to upkeep technology drop* (C_{23}), Subsequently under the factor *Financial aspects of I.T economics* (C_3), the sub-factors are: *Expenditure spent on resources—development team and infrastructure* (C_{31}),

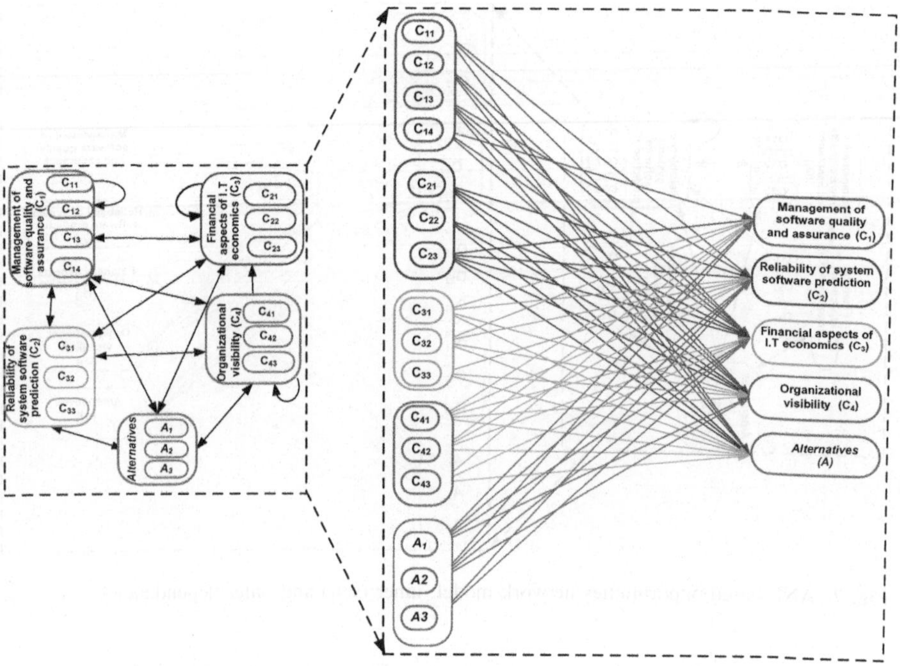

Fig. 8 ANP based costs network model: inner (self) and outer dependencies

Budget allocations for research innovation, implementation of new skills, quality regulatory and quality checks (C_{32}), Conducting in-house workshop sessions (C_{33}), Finally under the factor *Organizational visibility (O_4)*, the sub-factors are: *Expenses for prototype construction for a desired PoC (C_{41}), Spent over Innovative run-throughs (C_{42}), Overheads on un-anticipated modifications (C_{43})*. Table 3c, presents normalized local priory weights across the clusters of sub-factors and alternatives obtained from costs model. Tabulated priory weights evaluated by three different experts.

Risks Model

Figure 9, exemplifies ANP based risks model with various factors and sub-factors with the alternatives. Here, the factors deal with uncertainties associated with improper practices, lack of strategies and etc. These identified factor elements emulate related risks in realizing the objective of the case study. Under the factor *Management of software quality and assurance (R_1)*—the sub-factors are: *Conflicts due to perception issues among various stakeholders (R_{11}), Lack of end users maturity levels (R_{12}), Communication gaps and management traceability (R_{13})*; Next under *Reliability of system software prediction (R_2)* factor—the sub-factors are: *Precisions to apply appropriate reliability models (R_{21}), Selection of proper*

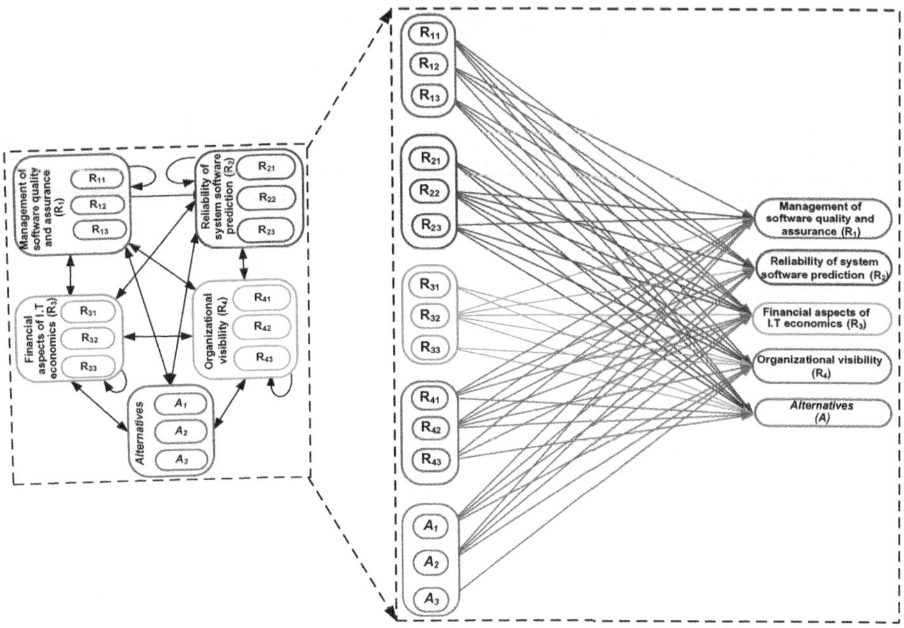

Fig. 9 ANP based risks network model: inner (self) and outer dependencies

data distribution for simulation-based quantitative predictions (R_{22}), Client's or end user's perspective of real knowledge/understanding of reliability techniques (R_{23}); Subsequently, under *Financial aspects of model economics (R_3)* factor—the sub-factors are: *Delayed exposure of errors (R_{31}), Incorrect perspective of operational applicative model (R_{32}), Application of inappropriate validation procedures (R_{33}),* Next, under *Organizational visibility (R_4)* factor—the sub-factors are: *Application of inadequate organizational project and skill driven strategies (R_{41}), Deficient organizational maturity levels (R_{42}), Inconsistencies in the practices of project management (R_{43}).* Table 3d, presents normalized local priory weights across the clusters of sub-factors and alternatives obtained from risks model. Presented priory weights evaluated by three different experts.

4.1.5 Strategic Network BOCR-AHP

Figure 10, demonstrates the AHP model to evaluate the importance of benefits, opportunities, costs, and risks in the form of strategic hierarchy model by applying BOCR–AHP. The underlying concept for the AHP strategic model is based on the external aspects governed by political, technological, economic and socialite features. The importance of BOCR (b, o, c, r) is evaluated via computed pair-wise strategic matrix based on the statistical data analysis (log of defect consolidation Table 4) and fuzzy slope evaluation (ref. Eq. 18). Table 5 tabulates the final evaluated priority values acquired from three experts as per BOCR-AHP model.

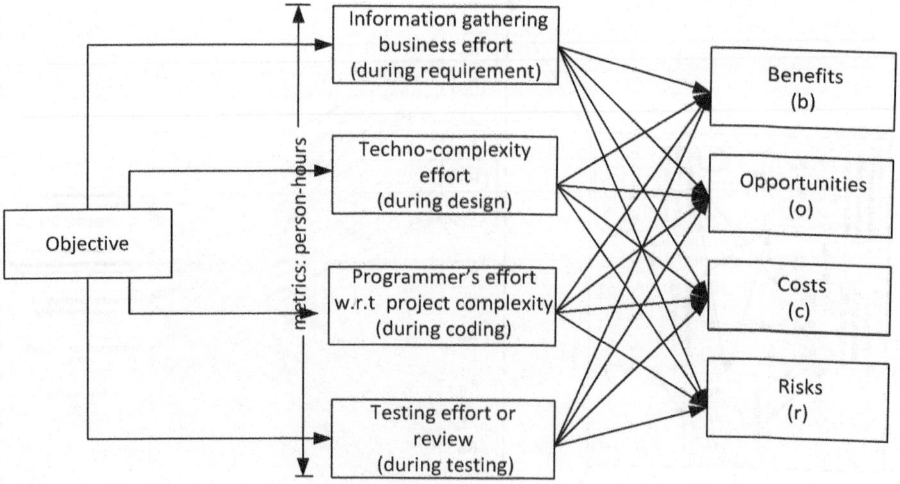

Fig. 10 AHP based BOCR model: strategic hierarchy

Table 4 Effort: log of defect consolidation (rounded to the nearest integer) [14]

Information gathering effort (w.r.t business topology)				
Effort in person-hours	Mean	S.D	Max	Min
Requirement (low)	32	27	60	0
Requirement (medium)	58	27	105	30
Requirement (high)	92	14	150	75
Technology-complexity effort (w.r.t application complexity)				
Effort in person-hours	Mean	S.D	Max	Min
Design (low)	32	27	60	0
Design (medium)	68	37	105	30
Design (high)	104	13.54	150	90
Programmers' level (w.r.t product familiarity and exp.)				
Effort in person-hours	Mean	S.D	Max	Min
Coding (low)	31	13	45	0
Coding (medium)	45	14	105	30
Coding (high)	104	14	150	90
Testing effort (w.r.t review during testing)				
Effort in person-hours	Mean	S.D	Max	Min
Unit testing (low)	52	46	100	0
Unit testing (medium)	98	45	175	50
Unit testing (high)	162	46	250	125

Table 5 BOCR-AHP model: priority values evaluations from three experts

	Expert 1	Expert 2	Expert 3
Benefits (b)	0.2892	0.2817	0.4051
Opportunities (o)	0.1109	0.0776	0.1116
Costs (c)	0.1740	0.1412	0.1118
Risks (r)	0.4257	0.4993	0.3713

4.1.6 Results Analysis

Table 6 presents pools-up priority vector values of the evaluated alternatives (A_1, A_2, A_3) under each BOCR-ANP control network models with min (L) and max (H) —assessed by respective three experts i.e. B_L, B_H, O_L, O_H, C_L, C_H, R_L, R_H. Table 7 pulls together the priority vector values—here, in this case, the values represent for presents the importance of benefits (b), opportunities(o), costs (c) and risks (r) along with min (L) and max (H) priority values acquired from different three experts via BOCR-AHP models (strategic hierarchy) i.e. b_L, b_H, o_L, o_H, c_L, c_H, r_L, r_H.

4.1.7 Final Results Evaluation: BOCR—Interval Arithmetic Operation Analysis

Tables 8 and 9, shows the final evaluated result outcomes of lower and upper bounds or limits of options and possible outcomes—i.e. alternatives (A_1, A_2, A_3) of

Table 6 Results min and max values of BOCR–ANP (B_L, B_H, O_L, O_H, C_L, C_H, R_L, R_H)

Alternatives	Expert 1	Expert 2	Expert 3	Min (L)	Max (H)
Benefits (B_L, B_H)					
A1	0.5943	0.6188	0.7576	0.5943	0.6576
A2	0.2626	0.2443	0.2356	0.1356	0.2626
A3	0.1429	0.1367	0.1067	0.1067	0.1429
Opportunities (O_L, O_H)					
A1	0.6323	0.5841	0.6010	0.5841	0.6323
A2	0.2551	0.3011	0.2732	0.2551	0.3011
A3	0.1124	0.1147	0.1257	0.1124	0.1257
Costs (C_L, C_H)					
A1	0.6098	0.6292	0.6714	0.6098	0.6714
A2	0.2524	0.2531	0.2303	0.2303	0.2531
A3	0.1376	0.1176	0.0982	0.0982	0.1376
Risks (R_L, R_H)					
A1	0.5812	0.6331	0.6725	0.5812	0.6725
A2	0.2763	0.2470	0.2060	0.2060	0.2763
A3	0.1424	0.1198	0.1213	0.1198	0.1424

Table 7 Results min and max values of BOCR–AHP: (b_L, b_H); (o_L, o_H); (c_L, c_H), (r_L, r_H)

	Expert 1	Expert 2	Expert 3	Min (L)	Max (H)
Benefits (b)	0.2892	0.2817	0.4051	0.2817	0.4051
Opportunities (o)	0.1109	0.0776	0.1116	0.0776	0.1109
Costs (c)	0.1740	0.1412	0.1118	0.1118	0.1740
Risks (r)	0.4257	0.4993	0.3713	0.3713	0.4993

Table 8 Final evaluation: BOCR (ANP) and interval arithmetic analysis—modified multiplicative (Eq. 17 in Sect. 3)

Alternatives	BO (min)	BO (max)	CR (min)	CR (max)	Lower	Upper
A1	0.3471	0.4158	0.3544	0.4515	0.7688	1.3510
A2	0.0601	0.0790	0.0474	0.0699	0.8594	1.6666
A3	0.0119	0.0179	0.0117	0.0195	0.6120	1.5268

our considered case study through proposed/modified multiplicative formula (via Eqs. 16 and 17 of interval arithmetic operations) and modified subtractive formula (via Eqs. 13 and 15 of interval arithmetic operations) respectively.

4.1.8 Results Interpretation/Discussion

Based on the results from the proposed methodology BOCR—interval arithmetic analysis, we have captured the outcome in a spectrum of minimum and maximum values against the traditional approach of having a single quantified value. In the case of modified multiplicative formula, Table 8 with the following outcomes: A_1 with a spectrum of 0.768–1.3510; A_2: from 0.8594 to 1.6666; A_3: from 0.6120 to 1.5268 with lower and upper limits. Figure 11, evidently there is a complete overlap between the three alternatives, illustrating no differences in the assessments provided by the experts. The victory is not so clean in this case.

Whereas, on the other hand, with modified Subtractive formula as: A_1 with the values −0.2398–0.0529; A_2 with a range of −0.0958–0.0377; A_3 with the values −0.0562–0.0164. Illustrated graphically in Fig. 12. When relate with the former case, Here, we observe a there is a significant overlap among the A_2 and A_3. However, with A_1, there is not much overlap with A_2 and A_3, which "stands" alone with a greater range of interval. Therefore, alternative A_1 appears to be the best alternative.

As a result of above outcome analysis showed, we conclude A_1 *(Advance with the working system with insignificant changes or minimum modifications: continue with the proposed flawless system architecture)* as the best alternative—evaluate prototype dependability assessment of software at the prototype level.

Table 9 Final evaluation: BOCR (ANP/AHP) and interval arithmetic analysis: modified subtractive (Eq. 16 in Sect. 3)

Alternatives	Bb (min)	Bb (max)	Oo (min)	Oo (max)	Cc (min)	Cc (max)	Rr (min)	Rr (max)	Lower	Upper
A1	0.1674	0.2663	0.0453	0.0705	0.0681	0.1168	0.2157	0.3357	−0.2398	0.0529
A2	0.0663	0.1063	0.0197	0.0336	0.0257	0.0440	0.0764	0.1379	−0.0958	0.0377
A3	0.0300	0.0578	0.0087	0.0140	0.0109	0.0239	0.0444	0.0711	−0.0562	0.0164

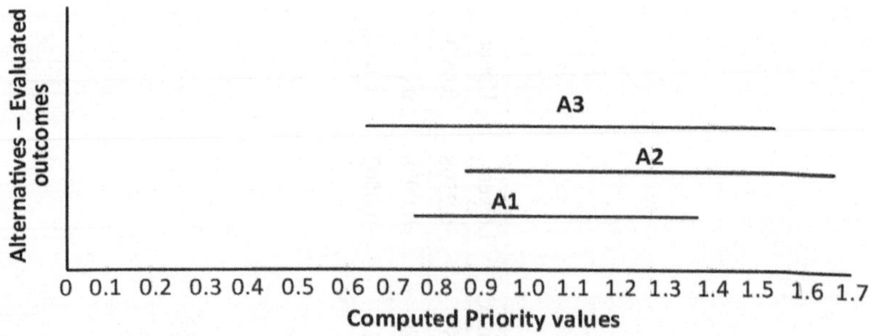

Fig. 11 Graphical representation of alternatives: modified multiplication

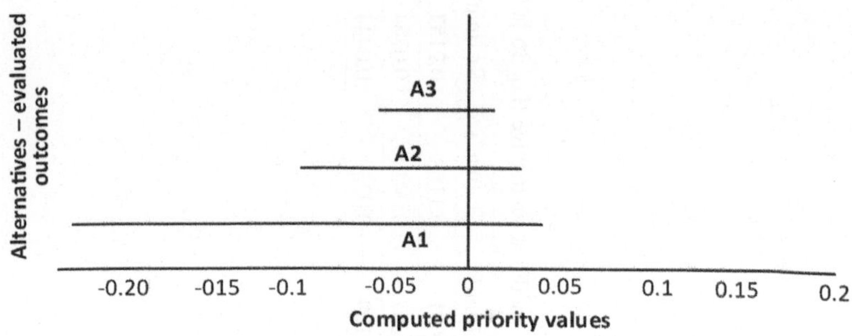

Fig. 12 Graphical representation of alternatives: modified subtractive

4.2 *Case Study 2*

4.2.1 Introduction

This Sect. 4.2, presents a second case study that illustrates a model validation with regards to the quality of services (QoS) in the SOA environment. Here, we have considered a different framework—a new validation based approach that has been suggested for SOA founded applications which is a contrast from the first case study demonstrated in the above Sect. 4.1. Additionally, employed a different aggregation technique—a fuzzy based methodology where the aggregation is applied before the modeling and assessment process. In other words, the fuzzy aggregation is performed on the input data obtained from three different experts. The aggregated input is applied to the BOCR (FANP) and BOCR (FAHP) models and hence decision output produced is the fuzzy-based outcome. Therefore, decision making process emphases to reduce the vagueness and improve customer satisfaction in endorsing real-world SOA applications. The algorithmic procedure portrays underlying FANP, FAHP and statistical evaluations for the BOCR models. The case study paves a way to carry out—the validation assessment to evaluate performance quantitatively through a presented fuzzy based analytical approach.

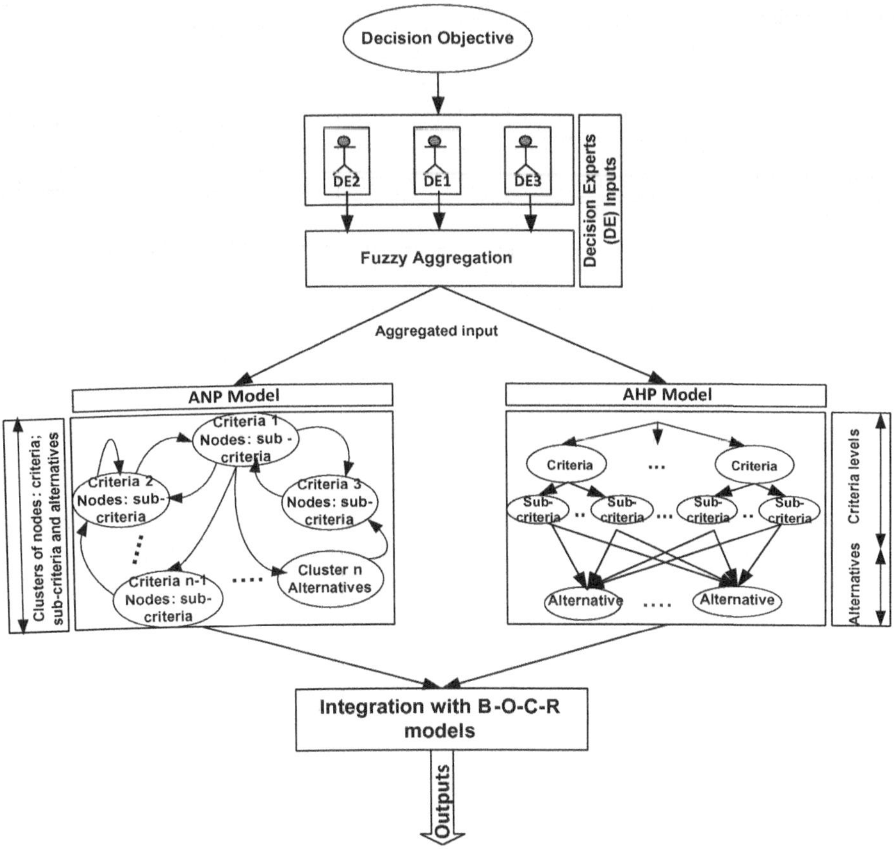

Fig. 13 Fuzzy based control network: BOCR–FANP model

Consequently, a systematic procedure has been employed to accommodate the execution of wider validation estimates of SOA applications in the software industry. The results presented in the case study can be referenced for future application problems. Here, in the considered case study, model inputs are obtained from various sources of availability based on 'in-field' and 'on-project' expertise provided by multiple experts. Figure 13 shows a decision model process flow that has been applied to the considered problem.

4.2.2 Proposed Steps

In this section, proposed procedural steps merging FANP, FAHP, and B-O-C-R is used for the assessment of model validation of the SOA applications. The steps as follows:

(1) Establish a precise problem statement, select suitable subject experts to provide input data to the decision model. Here in our case, we have selected three experts.
(2) Formulate fuzzy based pair-wise matrices for control network (FANP) and strategic hierarchy (FAHP) and evaluate priority vectors. Based on the multiple expert inputs (in our case three experts), fuzzy pair-wise matrices are accomplished via in-field' and 'on-project' expertise and fuzzy linguistic variable (refer Table 1, Sect. 3). Aggregate experts' input using the geometric mean and fuzzy aggregated matrices can be formed. Finally, defuzzified comparison matrices are prepared by the center of gravity (COG) method. Mathematically, the defuzzified comparison matrix is prepared through Eqs. 19–20

$$\widetilde{X_{ij}} = \left(\widetilde{x_{ij2}}\theta\widetilde{x_{ij2}}\ldots\widetilde{x_{ij(n-1)}}\ldots\theta\widetilde{x_{ijn}}\right)^{1/n} \tag{19}$$

X_{ij} represents fuzzy aggregated matrices

(3) Calculate priority vectors via defuzzfied pair-wise matrices with consistency check [42]. The computed priority vector mathematically represented as

$$\mathrm{XW} = \lambda_{max}\mathrm{W}; \tag{20}$$

where

X_{ij} defuzzified aggregated pair-wise matrix
W eigenvector;
λ_{max} argest eigenvalue of X [42];

(4) Construct FANP—control network and FAHP—strategic hierarchy for the BOCR analysis. In the FANP formulate various pair-wise matrices across the network with inner and outer relation dependencies with priority vectors evaluation, followed by supermatrix (unweighted and weighted) and finally limiting matrix.
(5) In FAHP, determine priority values of b.o.c.r merits via. formulation of pair-wise hierarchy based matrix with its priority vector calculation and estimate statistical based 5-steps scale for every strategic criterion.
(6) Finally, rank the alternatives using five BOCR formulas via. (a) additive; (b) probabilistic additive; (c) subtractive; (d) multiplicative priority powers; and (e) multiplicative (refer to Eqs. 11–15) via. the results obtained in step 4 and 5 i.e., through obtained priority vectors of FANP and FAHP.

4.2.3 Problem Statement

The objective of the second case study is to authenticate a 'novel model validation for the assessment of QoS in SOA environment' using FANP and FAHP. In order to endorse the presented objective, apply FANP and FAHP with BOCR from the obtained inputs from multiple experts (in our case 3 experts). Based on fuzzy aggregated inputs evaluate control network (FANP) and strategic hierarchy (FAHP) individually. Finally, evaluate A_1, A_2, and A_3 through FANP and FAHP with BOCR framework that offers more benefits and higher opportunities; at the same time with lesser costs and lowest risks. Eventually, the purpose is to endorse the stated objective: 'novel model validation for the assessment of QoS in SOA environment' using FANP and FAHP respect to the following alternatives A_1, A_2, and A_3.

A1: *Not to endorse the proposed model*
A2: *Intuition based approach*
A3: *Indorse the proposed model.*

4.2.4 BOCR-FANP Control Network

In this section, based on the problem statement, established a FANP based control network. After a thorough investigation, we identified various performance criteria attributes (factors and sub-factors) under control network (FANP) representing internal aspects of the application problem (i.e. functional attributes) forming BOCR-FANP network. Further, the input (pair-wise comparison matrix) to the FANP control network is provided by different experts based on their 'in-field' and 'on-project' expertise. There are three control criteria namely—*'response time (RT)'; 'throughput as (T)'; and 'CPU utilization (CU)'*. Under each of these criteria, identified sub-criteria classified with respect to BOCR models depicted in Fig. 14. Under benefits, the sub-criteria are: *Enhanced client flexibility and business agility (b_1); Better business workflows and improved end-user communication (b_2); Increased reuse and market portability (b_3) Customer contentment and extended application life span (b_4)*. Next, under the opportunities, the sub-criteria as follows: *Better Location transparency in the viable market place (O_1); Customer's reusability to facilitate opportunities for heterogeneous applications (O_2); Improved redundancy for enhanced scalability and high availability (O_3); Provision of parallel development to achieve independent based services orchestrate (O_4);* Under costs, the sub-criteria as—*Cost for data, process and service complexities (C_1); Expenses spent on technology strategies and security frameworks (C_2); Overheads on state-of-the-art practices (C_3);* Finally, under risks, the sub-criteria as—*Security liabilities and insights conflicts (R_1); Lack of knowledge of architectural of workable model details between various stakeholders (R_2); User acceptance and improper validation run-throughs impacting cumulative solution expenses (R_3)*. These criteria and sub-criteria are evaluated for the alternatives—*A_1: Not to endorse proposed model; A_2: Intuition based approach; A_3: Indorse the proposed model.*

Figure 14, demonstrates integrated B–O–C–R models with a comprehensive inter and outer relationships in the form of various dependencies associations between various factor elements and alternatives across the network. Here, we have

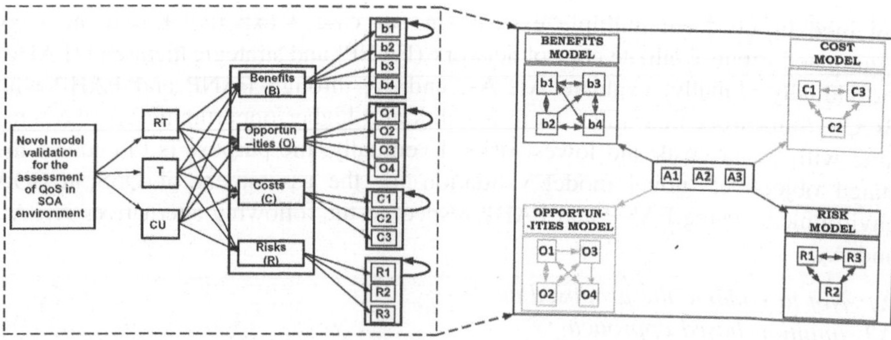

Fig. 14 Fuzzy ANP control model: with interrelationships across the network

considered fuzzy aggregated pair-wise matrices among factor and sub-factors and with **each alternative**. The aggregation is performed on the input data obtained from three different experts.

For example, **in case of** benefits model, we formulate and evaluate priority vector following fuzzy pair-wise comparisons between: (1) b_2, b_3, b_4 w.r.t 'b_1' (2) b_1, b_3, b_4 w.r.t 'b_2' (3) b_1, b_4 w.r.t 'b_3' (4) b_2, b_1, b_3 w.r.t 'b_4' (5) A_1, A_2, A_3 w.r.t 'b_1' (6) A_1, A_2, A_3 w.r.t 'b_2' (7) A_1, A_2, A_3 w.r.t 'b_3' (8) A_1, A_2, A_3 w.r.t 'b_4' (9) b_1, b_2, b_3, b_4 w.r.t 'A_1' (10) b_1, b_2, b_3, b_4 w.r.t 'A_2' and (11) b_1, b_2, b_3, b_4 w.r.t 'A_3'. Similarly, we evaluate fuzzy pair-wise for opportunities, cost, and risk models. For example, demonstrated a fuzzy pair-wise formulation and its priority vector evaluation in the Table 10 through fuzzy aggregated pair-wise matrix, defuzzified matrix and priory vector evaluation.

Now, founded on the algorithm proposed in Sect. 4.2.2, normalized priory vector weights in respect to benefits, opportunities, costs, and risks models are shown in Table 10. These priority values shown are normalized across each of the clusters, in other words, values are local w.r.t respective cluster. In case of benefits model, the alternatives priority values are calculated as A_1, A_2, A_3 as **0.1314, 0.3012, 0.5645** for the defined objective in our considered case of *novel model validation for the assessment of QoS in SOA environment* using FANP model. These alternatives values are used in the final evaluation of the BOCR model. Also, we can note that criteria factor element—*customer contentment and extended application life span (b_4)* with highest priority 0.3596 compared to b_1, b_2, b_3

Table 10 Example illustration of fuzzy pair-wise formulation and evaluation

	Fuzzy aggregated pairwise comparison				Simplified matrix					Priority vector – defuzzified aggregated pairwise matrix
	F1	F2	F3		F1	F2	F3			
F1	(1,1,1)	(0.32, 0.56, 1.00)	(0.19, 0.24, 0.58)		1	0.626	0.336		F1	0.1815
F2	(1.00, 1/0.56, 1/0.32)	(1,1,1)	(0.32, 0.56, 1.00)	→	1.597	1	0.626	→	F2	0.3052
F3	(1/0.58, 1/0.24, 1/0.19)	(1.00, 1/0.56, 1/0.32)	(1,1,1)		2.970	1.597	1		F3	0.5132

Table 11 Fuzzy BOCR-FANP control network model: priority evaluations

Benefits	Local priorities
Enhanced client flexibility and business agility (b_1)	0.2333
Better business workflows and improved end-user communication (b_2)	0.1520
Increased reuse and market portability (b_3)	0.2467
Customer contentment and extended application life span (b_4)	**0.3596**
$\mathbf{A_1}$	**0.1314**
$\mathbf{A_2}$	**0.3012**
$\mathbf{A_3}$	**0.5645**
Opportunities	
Better Location transparency in the viable market place (O_1)	0.1503
Customer's reusability to facilitate opportunities for heterogeneous applications (O_2)	**0.3789**
Improved redundancy for enhanced scalability and high availability (O_3)	0.2670
Provision of parallel development to achieve independent based services orchestrate (O_4)	0.2675
$\mathbf{A_1}$	**0.1170**
$\mathbf{A_2}$	**0.2387**
$\mathbf{A_3}$	**0.6441**
Cost	
Cost for data, process and service complexities (C_1)	0.3566
Expenses spent on technology strategies and security frameworks (C_2)	**03994**
Overheads on state-of-the-art practices (C_3)	0.2466
$\mathbf{A_1}$	**0.2041**
$\mathbf{A_2}$	**0.2967**
$\mathbf{A_3}$	**0.4990**
Risks	
Security liabilities and insights conflicts (R_1)	0.2024
Lack of knowledge of architectural of workable model details between various stakeholders (R_2)	**0.5071**
User acceptance and improper validation run-throughs impacting cumulative solution expenses (R_3)	0.2904
$\mathbf{A_1}$	**0.2149**
$\mathbf{A_2}$	**0.3487**
$\mathbf{A_3}$	**0.4363**

Bold signifies computed normalized priory values for each of the alternatives and as well highest normalized priority value (criteria factor element) in the respective B-O-C-R network models

towards having more benefits—these values are limited to benefits model itself. However, these are not used in the final evaluation of BOCR analysis. Similarly, the priority values are evaluated among other models i.e. opportunities, costs, and risks models, shown in Table 11.

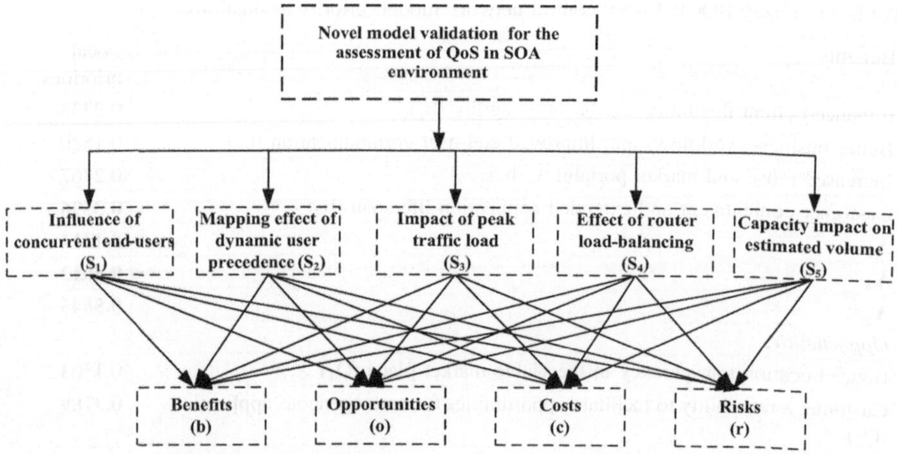

Fig. 15 BOCR–FAHP: strategic hierarchy

4.2.5 BOCR-FAHP: Strategic Hierarchy

Here, illustrated hierarchy strategic model (FAHP) that focus on sub-objective or sub-goal for model validation for the assessment of QoS in the SOA environment. Here, in this case study, Fig. 15, we have identified the various strategic criteria as *(a) Influence of concurrent end-users (S_1); (b) Mapping effect of dynamic user precedence (S_2); (c) Impact of peak traffic load (S_3); (d) Effect of router load-balancing (S_4); (e) Capacity impact on estimated volume (S_5)*. The purpose of the strategic model is to access the 'importance of benefits, opportunities, costs and risks' through *(i) BOCR-FAHP strategic hierarchy and (ii) through strategic 5-step scale values evaluation*

- *BOCR-FAHP strategic hierarchy*: the fuzzy pair-wise matrix is formulated between S_1, S_2, S_3, S_4, S_5 via three experts using a scale of fuzzy linguistic variable shown in Table 12. By the approach of geometric mean, the fuzzy aggregated matrix is obtained, the priority values are calculated shown in the Table 12.

Table 12 Priority values evaluation—defuzzified pairwise matrix (FAHP)

	S_1	S_2	S_3	S_4	S_5	priority values—defuzzified matrix
S_1	1	3.76	1.4	1.3	3.12	**0.3295**
S_2	0.265	1	0.440	0.752	1.76	**0.1242**
S_3	0.714	2.271	1	2	1.40	**0.2496**
S_4	0.769	1.329		1	2.76	**0.1955**
S_5	0.320	0.568	0.714	0.362	1	**0.1011**

Bold signifies computed normalized priory values in the strategic hierarchy (BOCR-FAHP) values

Table 13 Defuzzified based bocr priorities—fuzzy (b, o, c, r)

Criteria	Influence of concurrent end-users (S_1)	Mapping effect of dynamic user precedence (S_2)	Impact of peak traffic load (S_3)	Effect of router load-balancing (S_4)	Capacity impact on estimated volume (S_5)	Normalized Priorities
	0.3295	**0.1242**	**0.2496**	**0.1955**	**0.1011**	
Benefits (b)	very high (0.42)	Medium (0.16)	very high (0.42)	High (0.26)	Medium (0.16)	**0.4357**
Opportunities (o)	High (0.26)	Medium (0.16)	Medium (0.16)	Medium (0.16)	Low (0.1)	**0.2067**
Costs (c)	medium 0.16	low 0.1	High 0.26	Medium 0.16	Low 0.1	**0.2112**
Risks (r)	High 0.26	very low 0.06	High 0.26	High 0.26	very low 0.06	**0.1506**

Bold value signifies computed normalized priory values

- *Strategic 5—step scale values*: evaluated based on proposed by the authors [14], via normalized mean and standard deviation from the data sets collected (analyzed by the Eq. 18, Fig. 4) from three banking organizations. These values obtained as: very low—0.06, low—0.1, medium—0.16, high—0.26, very high 0.42.

The calculated results are shown in Table 13, the overall importance of the four merits is evaluated. For example, shown for benefits merit is as follows:

$$0.3295 \times 0.42 + 0.1242 \times 0.16 + 0.2496 \times 0.42 + 0.1955 \times 0.26 + 0.1011 \times 0.16 = 0.4357.$$

Similarly, calculated for the other three merits. Table 13 shows the calculated priorities for each value for BOCR merits as 0.4357, 0.2067, 0.2112, 0.1506.

4.2.6 Final Analysis BOCR (FANP and FAHP)

Eventually, in this section, illustrated a detail evaluation steps of BOCR with numerical steps. Table 14, intermediate steps to simplify BOCR evaluation of mathematical equations. Tables 15, 16, 17, 18, and 19 shows the evaluation of *additive; probabilistic additive; subtractive; multiplicative priority powers, multiplicative* formulas.

Finally, we have calculated the ranks of the final outcome—alternatives (A_1, A_2, A_3) based on the evaluations performed through Tables 15, 16, 17, 18, and 19, i.e. from BOCR-FANP (Sect. 4.2.4) and BOCR-FAHP (Sect. 4.2.5). The evaluated model validation results of alternatives alone with ranking has been presented in Table 20.

Table 14 Priority values obtained from FANP (A1, A2, A3) and FAHP (b, o, c, r)

Alternatives	Benefits	Opportunities	Costs			Risks		
	0.4357 (b)	0.2067 (o)	0.2112 (c)			0.1506 (r)		
	Normalized	Normalized	Normalized	Reciprocal	Normalized reciprocal	Normalized	Reciprocal	Normalized reciprocal
A_1	0.1314	0.1170	0.2041	4.8995	0.4768	0.2149	4.6533	0.4741
A_2	0.3012	0.2387	0.2967	3.3704	0.3280	0.3487	2.8677	0.2922
A_3	0.5645	0.6441	0.4990	2.0040	0.1950	0.4363	2.2920	0.2335

Table 15 BOCR (FANP/FAHP): additive mathematical evaluation

Additive formula: $P_i = bB_i + oO_i + c[1/C_i]_{normalized} + r[1/R_i]_{normalized}$
A_1: 0.4357 * 0.1314 + 0.2067 * 0.1170 + 0.2112 * 0.4768 + 0.1506* 0.4741 = 0.2535
A_2: 0.4357 * 0.3012 + 0.2067 * 0.2387 + 0.2112 * 0.3280 + 0.1506* 0.2922 = 0.2938
A_3: 0.4357 * 0.5645 + 0.2067 * 0.6441 + 0.2112 * 0.1950 + 0.1506* 0.2355 = 0.4557

Table 16 BOCR (FANP/FAHP): probabilistic additive mathematical evaluation

Probabilistic additive $P_i = bB_i + oO_i + c(1 - C_i)] + r(1 - R_i)$
A_1: 0.4357 * 0.1314 + 0.2067 * 0.1170 + 0.2122 (1 − 0.2041) + 0.1506 (1 − 0.2149) = 0.3685
A_2: 0.4357 * 0.3012 + 0.2067 * 0.2387 + 0.2122 (1 − 0.2967) + 0.1506 (1 − 0.3487) = 0.4278
A_3: 0.4357 * 0.5645 + 0.2067 * 0.6441 + 0.2122 (1 − 0.4990) + 0.1506 (1 − 0.4363) = 0.5702

Table 17 BOCR (FANP/FAHP): subtractive evaluation

Subtractive $P_i = bBi + oO_i - cC_i - rR_i$
A_1: 0.4357 * 0.1314 + 0.2067 * 0.1170 − 0.2112 * 0.2041 − 0.1506 * 0.2149 = 0.0059
A_2: 0.4357 * 0.3012 + 0.2067 * 0.2387 − 0.2112 * 0.2967 − 0.1506 * 0.3487 = 0.0653
A_3: 0.4357 * 0.5645 + 0.2067 * 0.6441 − 0.2112 * 0.4990 − 0.1506 * 0.4363 = 0.2079

Table 18 BOCR (FANP/FAHP): multiplicative priority powers evaluation

Multiplicative priority powers $P_i = B_i^b O_i^o [(1/C_i)_{normalized}]^c [(1/R_i)_{normalized}]^r$
A_1: $0.1314^{0.4357} * 0.1170^{0.2067} * 0.4768^{0.2112} * 0.4741^{0.1506} = 2.8036$
A_2: $0.3012^{0.4357} * 0.2387^{0.2067} * 0.3280^{0.2112} * 0.2922^{0.1506} = 2.9576$
A_3: $0.5645^{0.4357} * 0.6441^{0.2067} * 0.1950^{0.2112} * 0.2335^{0.1506} = 3.2045$

Table 19 BOCR (FANP): multiplicative evaluation

Multiplicative $P_i = B_iO_i/C_iR_i$
A_1: (0.1314 * 0.1170) /(0.2041 * 0.2149) = 0.3505
A_2: (0.3012 * 0.2387) /(0.2967 * 0.3487) = 0.6949
A_3: (0.5645 * 0.6441) /(0.4990 * 0.4363) = 1.6700

4.2.7 Results Interpretation/Discussion

Based on the final model validation results and ranking of alternatives presented in Table 20 of BOCR analysis from FANP (control network) and FAHP (strategic hierarchy) (BOCR), the following observations are noted. Here, in this case study, we observe a clear consensus among the evaluated alternatives. It is evident from

Table 20 BOCR(FANP/FAHP): final evaluation of ranking of alternative

	Additive Eq. 11/Table 15		Probabilistic additive Eq. 12/Table 16		Subtractive Eq. 13/Table 17		Multiplicative priority powers Eq. 14/Table 18		Multiplicative Eq. 15/Table 19	
Alternatives	Performance	Rank	Performance	Rank	Performance	Rank	Performance	Rank	Performance	Rank
A_1	0.2535	3	0.3685	3	0.0059	3	2.8036	3	0.3505	3
A_2	0.2938	2	0.4278	2	0.0653	2	2.9576	2	0.6949	2
A_3	0.4557	1	0.5702	1	0.2079	1	3.2045	1	1.6700	1

the rankings, alternative A_3 emerged as a victory having highest quantified value in respect to the five BOCR formulas, with ranking order of $A_3 > A_2 > A_1$. Therefore, based result analysis from BOCR(FANP/FAHP), we conclude *A_3 (Indorse the proposed model)* as the best alternative for the case study objective—*novel model validation for the assessment of QoS in SOA environment.*

5 Conclusions

In the book chapter, we have accomplished an integrated approach of BOCR modeling framework for decision tool evaluation. The analysis, modeling, and assessment are accomplished through a systematic approach comprising quantitative analytical models in conjunction with various aggregation techniques. The framework identifies key influencing factors at different levels of abstraction governing functional and non-functional aspects of the application. Importantly, the framework employs appropriate data aggregation techniques to eliminate input data discrepancies arising from multiple experts and data sets. In order to illustrate the proposed framework of the decision model, implemented with two diverse case studies. The first case study demonstrates a new approach of expressing the quantitative decision output through a spectrum of min and max values of BOCR models using arithmetic interval operations to access the project feasibility options at the prototype level. Here, the aggregation is performed on the multiple outputs obtained from BOCR models. Whereas, in the second case study, the aggregation is conducted on the input data using fuzzy logic via BOCR (FANP/FAHP) to ascertain the model validation for QoS consideration in SOA environment. Therefore, we have illustrated a road-map of how aggregation techniques can be applied "before" and "after" model evaluation through the considered case studies. The proposed framework can be used as a reference to provide future applications in the software industry and banking sector. Eventually, expressed the quantitative decision output with the highest level of accuracy pragmatically by applied appropriate aggregation techniques and statistical analysis. Finally, we conclude in the background of group decision-making in the BOCR framework process by considering two diverse aggregation techniques and their applicability.

6 Future Work

The authors of the book chapter at *Banking Labs, Toronto, Canada,*—are working towards integrating the proposed decision model frame-work to next level by the application of various machine learning (ML) models. The purpose of ML techniques is due to its high capability and clarity in representing the information. Further, ML can handle high volumes of data by splitting into training and testing within the range into the data sets. The application of ML through the knowledge-based decision tree translated into a hierarchy structure or node based non-linear structure. Therefore, this exploratory data analysis realizes to choose the appropriate machine learning algorithm for training, testing and cross-validation on the data sets chosen of the considered application study. Finally, ML easy adaptable, even by non-experts with maximum accuracy while minimizing handling capabilities for high-level architectural models.

Acknowledgements Part of the research work was carried out (funded and supported) at (i) University of Alberta (U of A), Edmonton, Canada and (ii) Banking Labs Inc, Toronto, a Canadian Architecture and Strategy consulting firm. Results from the labs are curated to measure and revise the framework.

References

1. Lee AHI (2009) A fuzzy AHP evaluation model for buyer–supplier relationships with the consideration of benefits, opportunities, costs and risks. Int J Prod Res 47(15):4255–4280
2. Saaty TL (2006) Fundamentals of the analytic network process—multiple networks with benefits, costs, opportunities and risks. J Syst Sci Syst Eng 13(3):348–379
3. Gedela RK, Mohan KK, Prasad VK (2018) Application of BOCR models in Service Oriented Architecture (SOA)—study on model validation through quantification for QoS considerations. Int J Syst Assur Eng Manag 9(6):1346–1354
4. Wijnmalen DJD (2007) Analysis of benefits, opportunities, costs, and risks (BOCR) with the AHP–ANP: a critical validation. Math Comput Model 46(7–8):892–905
5. Keeney RL, Kirkwood CW (1975) Group decision making using cardinal social welfare functions. Manag Sci 22(4):430–437
6. Keeney R, Raiffa H (1976) Decision with multiple objectives: preferences and value tradeoffs. Wiley, New York
7. Dyer RF, Forman EH (1992) Group decision support with the analytic hierarchy process. Decis Support Syst 8(2):99–124
8. Basak I, Saaty T (1993) Group decision making using the analytic hierarchy process. Math Comput Model 17(4–5):101–109
9. Ilbahar E, Cebi S, Kahraman C (2019) A state-of-the-art review on multi-attribute renewable energy decision making. Energy Strat Rev 25:18–33
10. Kaya İ, Çolak M, Terzi F (2019) A comprehensive review of fuzzy multi criteria decision making methodologies for energy policy making. Energy Strat Rev 24:207–228
11. Krishna Mohan K, Verma AK, Srividya A (2011) An effective early software reliability prediction procedure for process oriented development at prototype level employing artificial neural networks. Int J Reliab Qual Saf Eng 18(3):237–250
12. Krishna Mohan K, Verma AK, Srividya A, Papic L (2010) Integration of black-box and white-box modeling approaches for software reliability estimation. Int J Reliab Qual Saf Eng 18(3):261–273, 2010. (World scientific publication)
13. Krishna Mohan K, Verma AK, Srividya A (2010) Software reliability estimation through black box and white box testing at prototype level. In: Proceedings of the 2nd IEEE international conference on reliability, safety and hazard risk based technologies and physics of failure methods. Bhabha Atomic Research Center (BARC), Mumbai, India, pp. 517–522
14. Krishna Mohan K, Verma AK, Srividya A (2009) Early qualitative software reliability prediction and risk management in process centric development through a soft computing technique. Int J Reliab Qual Saf Eng 16(6):521–532

15. Krishna Mohan K, Verma AK, Srividya A, Gedela RK, Varaprasada Rao G (2008) Early quantitative software reliability prediction using petrinets. In: Proceedings of the IEEE colloquium and 3rd IEEE international conference on industrial and information systems. Indian Institute of Technology Kharagpur (IITK), India
16. Lantz K (1986) The prototyping methodology. Prentice-Hall
17. Patton B (1983) Prototyping a nomenclature problem. ACM SIGSOFT Softw Eng Notes 8 (2):14–16
18. Carey JM, Currey JD (1989) The prototyping conurdrum. Datamation, pp. 29–33
19. Lipp ME (1986) Analysis. In: Lipp ME (cd) Prototyping—state of the art report, vol 14, no 4. Pergamon Infotech Ltd, Maidenhead, pp 129–184
20. Klingler DE (1988) The ten commandments of prototyping. J Inf Syst Manag 5(5):66–72
21. Floyd C (1984) A systematic look at prototyping. In: Approaches to prototyping. Springer, pp 105–122
22. Budde R, Kautz K, Kuhlenkamp K, Zullighoven H (1992) Prototyping—an approach to evolutionary system development. Springer
23. Lichter H, Schneider M, Zullighoven H (1994) Prototyping in industrial software projects—bridging the gap between theory and practice. IEEE Trans Softw Eng 20(11)
24. Huang Y-S, Liao J-T, Lin Z-L (2009) A study on aggregation of group decisions. Syst Res Behav Sci 26(4)
25. Ossadnik W, Schinke S, Kaspar RH (2016) Group aggregation techniques for analytic hierarchy process and analytic network process: a comparative analysis. Group Decis Negot 25(2):421–457
26. Ho W (2008) Integrated analytic hierarchy process and its applications—a literature review. Eur J Oper Res 186(1):211–228
27. Krishna Mohan K, Srividya A, Gedela RK (2008) Quality of service prediction using fuzzy logic and RUP implementation for process oriented development. Int J Reliab Qual Saf Eng 15(2):143–157
28. Ly PTM, Lai W-H, Hsu C-W, Shih F-Y (2018) Fuzzy AHP analysis of Internet of Things (IoT) in enterprises. Technol Forecast Soc Chang 36:1–13
29. Ilbahar E, Karaşan A, Cebi S, Kahraman C (2018) A novel approach to risk assessment for occupational health and safety using Pythagorean fuzzy AHP & fuzzy inference system. Saf Sci 103:124–136
30. Wang Y-M, Chin K-S (2011) Fuzzy analytic hierarchy process: a logarithmic fuzzy preference programming methodology. Int J Approximate Reasoning 52:541–553
31. Calabrese A, Costa R, Menichini T (2013) Using Fuzzy AHP to manage Intellectual Capital assets: an application to the ICT service industry. Expert Syst Appl 40(9):3747–3755
32. Hickey T, Ju Q, Emden MHV (2001) Interval arithmetic: from principles to implementation. J ACM (JACM) 48(5):1038–1068
33. Benhamou F, Older WJ (1997) Applying interval arithmetic to real, integer, and Boolean constraints. J Log Program 32:1–24
34. Alefeld G, Herzberger J (1983) Introduction to interval computations. Academic Press
35. Kearfott RB, Kreinovich V (1996) Application of interval computations. Kluwer Academic Publishers, Dordrecht
36. Zadeh LA (1965) Fuzzy sets. Inf Control 8:338–353
37. Saaty TL (1980) The analytic hierarchy process. McGraw-Hill, New York
38. Krishna Mohan K, Prashanthi K, Hull R, Montemagno CD (2018) Risk assessment of a multiplexed carbon nanotube network biosensor. IEEE Sens J 18(11)
39. Acze J, Saaty TL (1983) Procedures for synthesizing ratio judgements. J Math Psychol 27 (1):93–102
40. Saaty TL, Vargas LG (2001) Models, methods, concepts und applications of the analytic hierarchy process. Kluwer Academic, Boston
41. Saaty TL (2001) Decision making for leaders—the analytic hierarchy process for decisions in a complex world, 3rd edn. RWS Publications, Pittsburgh, p 2001

42. Saaty TL (2003) Decision-making with the AHP: why is the principal eigenvector necessary. Eur J Oper Res 145:85–91
43. Saaty TL (1996) The network hierarchy process. RWS Publications, Pittsburgh
44. Saaty TL, Vargar LG (2006) Decision making with analytic network process. Springer's international series
45. Lee AHI, Hsing-Jen C, Chun-Yu L (2009) An evaluation model of buyer–supplier relationships in high-tech industry—the case of an electronic components manufacturer in Taiwan. Comput Ind Eng 57(4):1417–1430
46. Saaty TL (2001a) Decision making with dependence and feedback: the analytic network process. RWS Publications, Pittsburgh
47. Krishna Mohan K, Srividya A, Verma AK (2010) ANP based software reliability prediction using PoCs and subsequent employment of orthogonal defect classification measurements for risk mitigation during prototype studies. Int J Syst Assur Eng Manag 1(1):8–13
48. Krishna Mohan K, Srividya A, Ravi H, Gedela RK (2008) Improvement of Quality of Service (QoS) in process centric software development and application of Analytical Network Process (ANP) for the estimation of critical phases. J Oper Res (OPSEARCH) 45(4):369–380

Krishna Mohan Kovur received his Ph.D. from Indian Institute of Technology (IIT) Bombay. Krishna is presently working as a Principal Consultant and Chief Research Innovative Officer (CRIO) at Banking Labs Inc. Toronto, Canada. He is specialized in applied research for IT firms in Quantitative Analysis for Heterogeneous Complex Applications in different domain areas. His research interests are Quantitative Risk Analysis, Software Reliability, Predictive Analytics for Decision Modeling, Artificial Intelligence & Machine Learning Techniques, and Tool Development like Graphical User Interface (GUI). Prior to working at Banking Labs, worked as a researcher at IIT Bombay, India; University of Alberta (U of A), Canada; and as well a senior consultant for software firms like Envision Labs, India. He has I.T expertise in the various domain areas like ERP applications, Nano Biosensors, Agriculture Livestock, and Banking. He is on the editorial board on some international journals like Springer etc. Also, acted as a reviewer/referee in the international journals and conferences. He has addressed as a speaker at various international symposium and conferences around the world.

Ravi Kumar Gedela is a highly collaborative strategic change agent with a proven track record of leading and delivering advanced analytics solutions for Banking, specializing in digital innovation and data transformation. He has over two decades of professional IT experience in and has worked with leading North American banks and IT consulting firms. His main research interests are in Large Scale Software Architecture, Quality Attributes of Software Architecture, Advanced Analytics, and Deep Learning. He is a thought leader and frequent speaker in various global forums, seminars and conferences, and a renowned evangelist of Artificial Intelligence and Machine Learning. Ravi is currently the Chief Strategy Officer and Chief Architect at Banking Labs, a leading strategy and advisory consulting firm for banking, based of Toronto, Canada.

DevOps for IT Service Reliability and Availability

Mahesh Kuruba

1 Introduction

Today, business is being driven by Information Technology. Increasingly, customers are being serviced through digital applications like the web and mobile. Companies like Ola, Flipkart, RedBus and several others have demonstrated how business models can be shaped by IT. Indeed, technology has become indispensable for the survival of businesses. The technologies for Social Networking, Mobile Applications, Analytics, Cloud (SMAC) and Artificial Intelligence (AI) are redefining the business and the services that these businesses provide. Their widespread usage is changing the business landscape, increasing reliability and availability to levels that were unimaginable even a few years ago.

IT is the backbone of modern business in the digital world. Due to the wide range of business and industry sectors that IT serves, as well as their levels of sophistication, disruptive technologies coexist with legacy applications, resulting in a rather complex environment. Moreover, the following factors tend to impacting reliability and availability of IT solutions:

- operating models to deliver IT-based solutions or services,
- maintenance and support of IT from global locations,
- lack of (or) limited knowledge of appropriate software applications, products and infrastructure,
- onboarding applications and products without enough understanding of its impact on architecture and the maintainability of IT services,
- vendors supporting an organization's IT services in various ways,

M. Kuruba (✉)
Tata Consultancy Services (TCS), Tokyo, Japan
e-mail: mahesh.kuruba@gmail.com

D. R. Karanki et al. (eds.), *Advances in RAMS Engineering*,
Springer Series in Reliability Engineering,
https://doi.org/10.1007/978-3-030-36518-9_6

- lack of interdisciplinary knowledge and skills in application and infrastructure,
- dynamic customer requirements, including last-minute (or short-notice) changes and additions.

Agility has become a critical success factor for businesses to sustain and grow in their markets. The failure of IT in meeting business needs impacts business performance. However, on the other hand, frequent technology changes (and upgrades) can have negative effect on reliability of software and consequently, the services provided by the company.

1.1 Understanding Software Reliability and Availability

Let us first understand reliability and availability, two words that people often confuse with each other. Availability of software is a measure as the percentage of time the software is up, while reliability is a measure of how long a user can use the software (or) perform activities with the software without any errors.

As an example, let us discuss an Internet banking application which is available for a customer through a login facility. The customer should be able to login and able to view his account information. This is availability, meaning that he can access his account. But when he tries to perform a funds transfer, sometimes, the user cannot transact because of an error. This error may be due to a defect (or incident, as it is called in a production environment) in the application. Such errors point to the reliability of the software. In this example, the software application was available (because the user could log in), but not reliable (because he could not transact). Thus, obviously, from the user's perspective the application must be both available and reliable so that he (or she) can use it to perform the tasks the software is designed for.

1.2 Understanding the Meaning of Service and Software

To understand the meaning of service and software, let us discuss with the help of an example of a bank's customer who wants to access his or her 'account balance' in the Internet banking system (Fig. 1). In this example, Internet banking is the software application and 'Checking Account Balance' is a service. The activity of 'checking the account balance' may involve multiple applications. It may, for instance, depend on the bank's core banking software application. Thus, the software used to provide the user with information about his account balance—there may be multiple—may include the bank's internet banking application, core banking application and credit card application. Examples of other services in this context could include 'Money transfer to another person/party', 'Credit card payment', 'Checking credit card balance', etc.

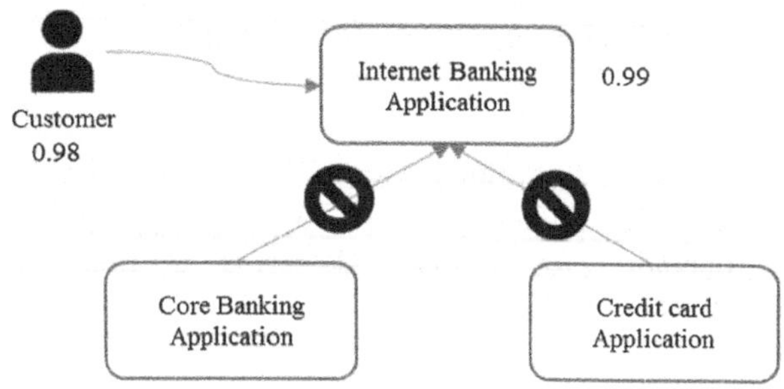

Fig. 1 Schematic for explaining Software availability

1.3 *What Does Service Availability Mean?*

Availability is the accessibility of a software to perform operations when required by a user. It is measured as the percentage of time the software is accessible for use when required. Availability can also be expressed as the probability that a software will not be brought down (or becomes non-functional) for maintenance purposes when it is needed by the user.

We will now try to understand software and service availability with the example of a user who is trying to access Internet banking (Fig. 1). The user wants to log into his bank account and view his account balance. Here, availability of the software will depend on the availability of Internet banking application. If the internet banking application is accessible even if the dependent applications are not available, the application will still be considered as available.

In arithmetic term availability is expressed as the ratio of uptime (of the software) to the total time that the software is in operation.

$$Availability = \frac{Total\ elapsed\ time - Down\ time}{\text{Total elapsed time}}$$

In this example of internet banking application, software availability is computed using the ratio of difference between total elapsed time of the application and downtime of the internet banking application to the Total elapsed time of the application.

However, if the user wants to access a service, such as checking account balance, it can sometimes happen that although the internet banking application is available, its underlying core banking application is not. For this reason, the service of 'checking account balance' is considered as not available. Thus, it needs to be understood that availability of software/application is different from availability of service.

Availability of the service ('Checking account balance' in our example) is computed as the ratio of the difference between total elapsed time and the sum of downtime of the internet banking application and downtime of the core banking application to the Total elapsed time.

1.4 *What Is Service Reliability?*

Software Reliability
Software reliability is defined as the probability of failure-free software operation for a specified period of time in a specified environment [IEEE-Std-729-1991].

Reliability is the ability of software to perform the operations for which it is designed. It is measured as the probability that an item will perform its intended function(s) for a specified time interval under stated conditions. There are two commonly used measures of reliability:

- Mean Time Between Failure (MTBF), which is computed as:

$$(\text{Total elapsed time} - \text{sum of downtime})/\text{Number of failures}$$

- Failure Rate (λ), which is computed from the relationship: number of failures (n)/Total time in service (T).

Let's understand 'service reliability' in today's context with an example of a user trying to access Internet banking. As shown in Fig. 2, the user wants to login into his bank account and view his bank balance and credit card transactions. The reliability of the service will now not only depend on the reliability of Internet banking application, but also depends on the reliability of the credit card application. If the reliability of each application is mentioned as 99% (or 0.99), it means that the reliability of the internet banking service, as experienced by the customer, is only 98% (0.98).

We will now try to understand the components of each application with the example shown in Fig. 3. The components of Internet banking application in Webserver, DB and OS will have different values of reliability. Let us now assume that the IT service provider has committed to a reliability of 0.99 for each component. Then, the reliability of the internet banking application will be 0.97

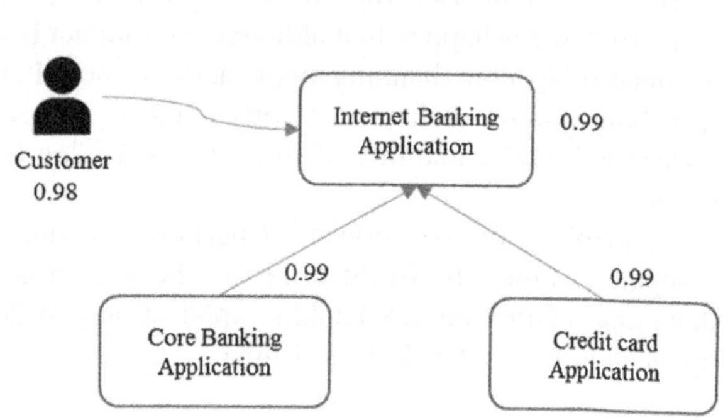

Fig. 2 Service reliability at software level

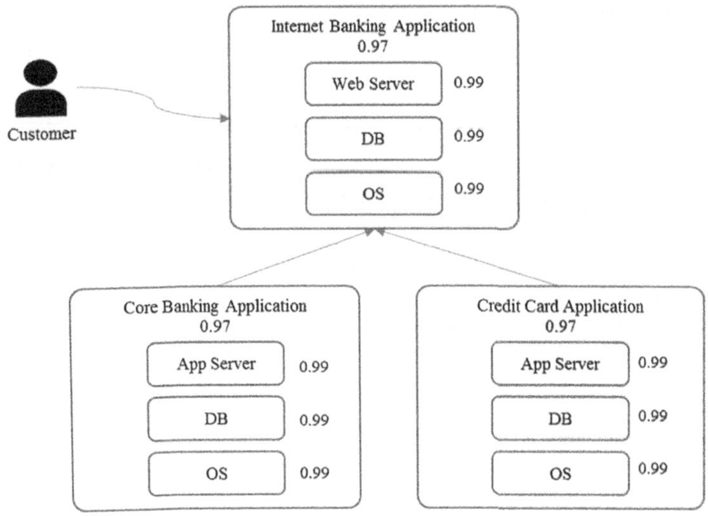

Fig. 3 An example of service reliability at component level

(0.99 × 0.99 × 0.99). But the customer will experience only 0.94 reliability. Thus, as we can see from the example, as the number of components increases in a service, the overall reliability of the service tends to reduce.

With multiple products and technologies in use, and their interplay, it is inevitable that the reliability of services provided will be impacted. For this reason, businesses (or the end-users) expect their IT service (or solutions) providers to be agile and release software increments at a much faster pace than they can at present. Because of this pressure to deliver, there is little time for development teams to test upgrades (or new releases) rigorously before their release. This increases the probability of failure in production conditions, thus affecting the reliability and availability of each software application.

1.5 Agile's Impact on Software Application Reliability and Availability

Adopting agile methodology calls for frequent releases. With frequent releases, the frequency of incidents in each application is also likely to increase. For example, an application with a release cycle of 4 weeks, the number of incidents will increase, resulting in unstable services (Fig. 4). This is not a situation that either the business or customers want. Irrespective of the changes that product teams are trying to achieve, reliability must improve, as shown in Fig. 5. However, in reality, that does not happen. The reason is that the operations team, which takes charge of the code from the development team to provide support during operations, often finds it

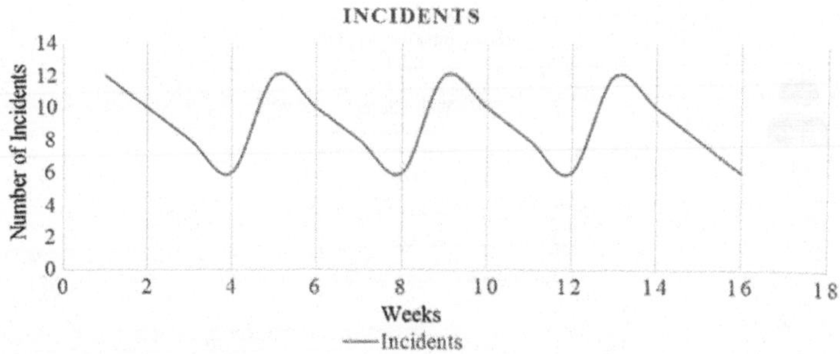

Fig. 4 A typical software reliability pattern

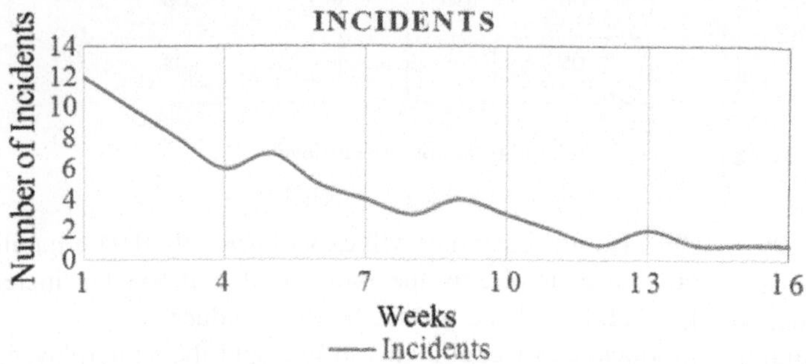

Fig. 5 Expected software reliability pattern

difficult to understand the changes made in such short cycle times. This difficulty usually results in increased incident resolution time.

Hence, along with agile ways of working, it is also essential to adopt DevOps practices for ensuring that the application remains stable despite several changes.

2 What Is DevOps?

Adoption of DevOps practices is gaining momentum due to their value in reducing the overall cycle time for delivery and increased reliability. There are several interpretations of DevOps, ranging from CI, CI/CD to the Dev and Ops teams working together. But organizations who want to adopt DevOps are often confused over how to go about it because there is no standard definition for DevOps.

How can an organization, which has adopted DevOps, know if it is benefitting from DevOps? There are two key indicators that can tell us if an organization is enjoying the DevOps advantage

1. Agility: The number of changes or the frequency at which releases are made in production
2. Reliability and Availability: The number of incidents raised after these changes or releases.

Under pressure, organizations release software quickly. However, these releases may carry several undetected defects. These defects could be functional or non-functional. For example, the programmer may have used memory for processing which has not been freed and which may not have been detected. Similarly, the developer may have written the code for connecting with a database, but he (or she) may not have released the connection, thus increasing the number of database connections in operations. With good controls, as part of Continuous Integration (CI), developers should be able to detect such defects. These defects result in incidents when the code is moved to production. The cost associated with fixing such defects at that stage is significantly higher, impacting the reliability and availability of service.

To explain DevOps better we will use a water filtration system as an analogy. As shown in Fig. 6, in a water filtration system, there is impure water coming in from one end and clean water coming out at the other. To check if the water has been cleaned (or purified) to the specified level, various parameters are measured: hardness, Total Dissolved Solids, Hardness, pH, etc. We expect the treatment process to be quick so that there is enough clean water to satisfy demand, usually measured as the rate of discharge (volume of water per unit time). Sensors are placed at the outlet for measuring the various parameters of water quality and provide feedback to the filtration system. In case of a problem, the system makes the necessary adjustments to fix the problem. We install filters in the system to produce clean water and then start monitoring its quality.

Similarly, the objective of the CI/CD in DevOps, is like the water filtration, to make sure that the code delivered is of good quality and is done at faster rate than usual. Any software quality issues that are detected in production need fixing as part of the CI process.

2.1 *How Does DevOps Differ from Other Methodologies?*

DevOps' footprint in the IT sector is growing (see Fig. 7 which shows how DevOps has evolved from existing methodologies). From traditional software development

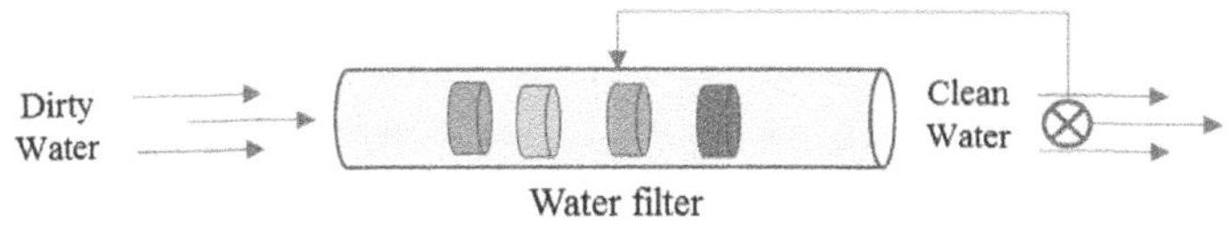

Fig. 6 A water filtration system

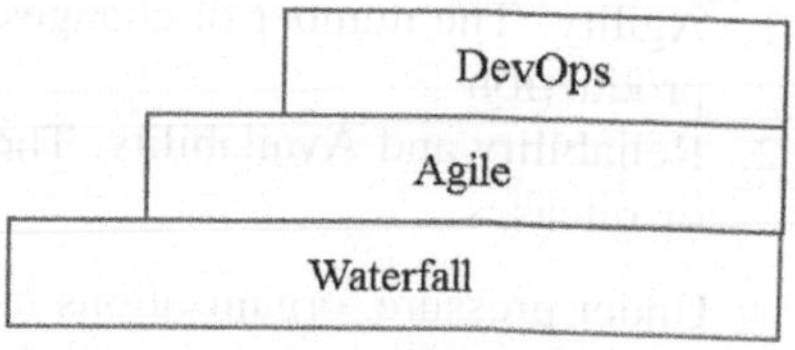

Fig. 7 Development Methodologies

methodologies like waterfall, iterative, agile and others, the industry is moving towards DevOps in a big way. What is the difference between Waterfall, Agile and DevOps methodologies?

In the waterfall methodology, feedback and the speed at which software being delivered is very slow. DevOps takes its cue from Agile. It establishes a platform from which the software is delivered at much faster pace. Software development is maturing towards a DevOps approach.

With conventional methods, the time taken for implementation is more and hence, is often the cause for dissatisfaction among users. Moreover, the development team does not allow too many changes to the software. Added to these constraints, the operations team follows a set of practices which includes receiving the documentation from the development team in a specific manner. This inflexibility has resulted in dissatisfaction with the waterfall methodology and hence, the shift to agile methods.

The agile methodology could deliver the software requirements at a much faster pace than was possible with waterfall. However, the challenge was in the speed with which the operations team receives the software. It was much faster because of which the operations team often found itself in a position where it could not cope with the changes and provide support. This led to an increase in the number of incidents and average incident resolution time in operations. What the user organization wanted is faster delivery of high-quality software with no or minimum incidents.

The limitations and challenges in existing methodologies for software development are summarized in Table 1:

> DevOps is a bringing together the cultures of Dev(elopement) and Op(eration)s Teams, the tools to automate activities, and practices to ensure that Software Development, Maintenance and Operations are managed seamlessly to increase the reliability of applications and services.

Table 1 Challenges for software development methodologies

Methodology	Limitations
Waterfall	Takes long time to deliver, resulting in increased time to market. Sometimes, by the time software is delivered, it is too late because the requirements have changed and thus, the initial requirements become irrelevant Consequently, business and end-users are not happy with the software delivered to them. Moreover, it is late to the market. But the Development and Operations teams are happy as they have completed their work within budget and according to schedule The risks with this method is high for business
Agile	Development teams are the winners here. They churn out software quickly. But, potentially, there could be several issues which challenge the operations teams who cannot make many changes. These can result in problems during operations leading to dissatisfied end-users of the application
DevOps	With DevOps, all teams—Business, Development and Operations teams—are happy. Business team is happy because it can take features to the market faster. The Development team is happy because it could deliver the product faster in small chunks. The operations team is happy as there are less issues in the product Thus, the key to 'happiness' is collaboration amongst Business, Development and Operations teams

2.2 *What Are the Practices of DevOps?*

DevOps consists of ten practices of which six are specific to the software life-cycle (Fig. 8) and four are common to all practices as shown in the reference architecture of DevOps in Fig. 8. The six in the lifecycle phases are continuous planning, Build and Integration, testing, Deployment, monitoring, feedback (which are outside the outermost circle in Fig. 8). These project specific practices are built around the

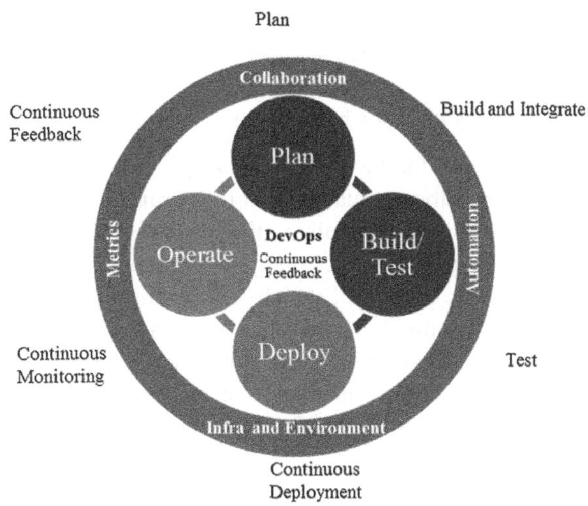

Fig. 8 DevOps reference architecture

project phases of Plan, Build and Test, Deploy and Operate. Common practices are foundational and common across the lifecycle. They are automation, collaboration, infrastructure and environment provisioning and the metrics.

Continuous planning includes breaking down a large piece of work into smaller parts. These parts become the individual work items which are prioritized according to need, built, integrated and tested.

Continuous build and integration are automated processes in which, that as soon as the code is developed and checked-in by the developer into source code version control system, the code is built and integrated with other modules or components.

Test automation is another key practice in DevOps. While the code is under development, automated test cases and scripts are prepared so that whenever there is a change in code, the developers can carry out the test automatically. Once the testing is complete, the code is moved to a QA environment for a regression test and then sent to the next stage. We can see that testing is a continuous process.

Continuous deployment is the deployment of code that has passed the QA environment to the next stage (system integration and testing environment). On successful completion of the test, the code is moved to Production. The conventional practice, which is followed by most organizations, is to move the code to production with the scripts, but with manual controls. In the continuous monitoring and feedback process, there are tools which constantly monitor the applications and the underlying infrastructure to identify any issues and then point to where the problem has occurred.

2.2.1 Continuous Planning

Planning is required for every release and sprint. The product owner needs to prioritize the backlog items which includes new feature requests, incidents/ problems and enhancements (Fig. 9). Depending on the volume of work, organizations can adopt different methods for this purpose. They can have

(1) Two separate teams—one for incident and one for new feature requests and enhancements, or
(2) One backlog for both teams.

Based on the available capacity and historical trends in incidents/problem reports, and the estimates from the scrum teams, the product owner prioritizes the backlog items and finalizes the sprint backlog. As part of the sprint, dedicated capacity is planned for managing incidents.

Feedback on usage, which will be discussed later, is considered for prioritization. For example, based on operational information, if certain features are not being used, the product owner may not want to spend too much money and effort on that feature. For features that are being widely used, the product owner may decide to allocate more effort and resources on them.

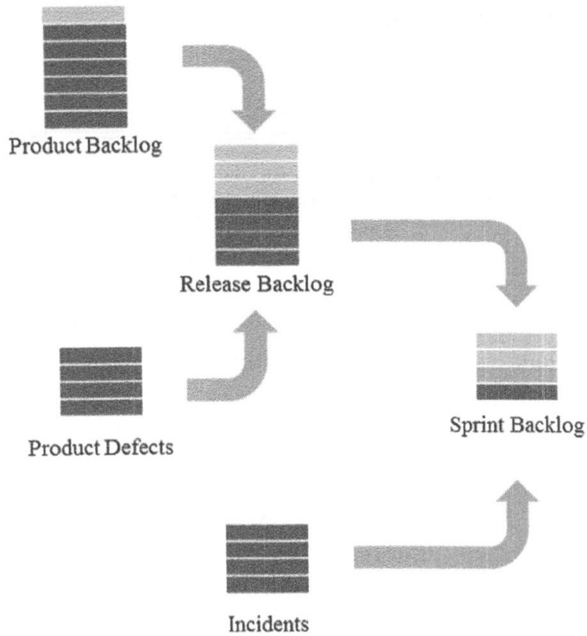

Fig. 9 Continuous planning of backlog

2.2.2 Continuous Build and Integration

Continuous Integration is the practice of incrementally committing the developed code and integrating it with the code base. As shown in Fig. 10, once the code is committed, it will be built and tested. If any issues are noticed in the build or during testing, the developer is notified immediately. Since the code is committed and integrated frequently, there is greater likelihood of problems like memory leak, functional defects, and other performance issues will be detected at once. In addition, the developer will not have to spend too much of time in analyzing the problems because the differential changes in code are minimal. Hence, as a practice, developers are expected to commit the changes they make as frequently as is necessary. There are various tools which help developers in integrating the code base— Jenkins, Bamboo, Team City, etc. Code analysis tools like Sonarqube are also integrated to ensure that developers adhere to the coding standards.

2.2.3 Continuous Testing

Testing is a crucial component of DevOps. After every build, developers need to test the changes they make and ensure that they do not result in other problems. The reliability of the software will depend on the extent (measured as percentage) of test automation and test coverage. Hence, the shift from Waterfall to DevOps will also

Fig. 10 Continuous integration

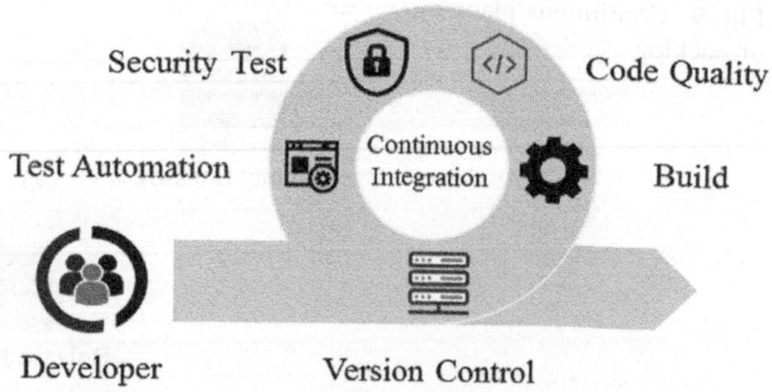

require a change in testing methods from manual to automate. While the developers are working on the code, the testers can write the scripts for testing it. Higher the percentage of test automation and test coverage, higher will be the reliability of software.

In addition to functional testing, performance and security tests are also necessary. These include Static Application Security Testing (SAST) and Dynamic Application Security Testing (DAST). It is vital to understand that security and performance issues will impact availability of the services for which the software is being developed. Any security issue will compromise the service and thus, impact its availability. Performance testing is usually performed in a separate environment.

2.2.4 Continuous Deployment

Continuous deployment of code which has been successfully tested is an automated process for moving the code to the next higher environment (see Fig. 11). It needs to be understood that manual deployment is a slow process and prone to errors due to which the release/deployment team use long windows for upgrades/maintenance activities. This has a negative impact on the overall availability of service. With automated deployment, the risk is reduced and turnaround is significantly faster as compared to manual processes of release management.

Continuous deployments will highlight inconsistencies in the configuration between environments. For this reason, deployments may fail in higher environments due to various causes: missing configuration, missing patches, environment mismatch, etc. Such failures are due primarily to dependency on the human factor. The deployment tools used in DevOps are intended to minimize or eliminate dependency on manual intervention.

After deployment, quick functional tests are needed to ensure that the deployed version has not disrupted other functionalities.

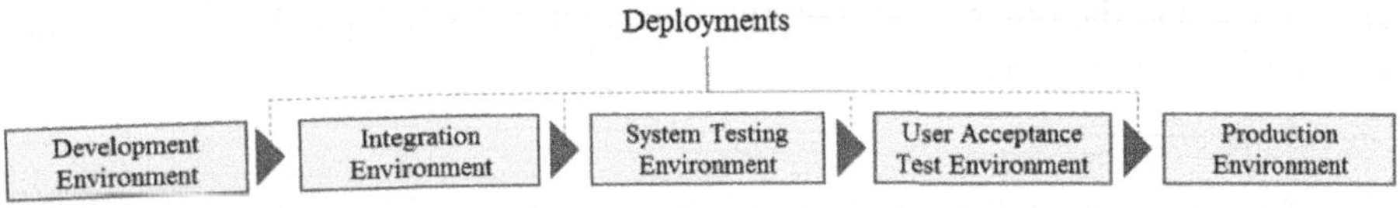

Fig. 11 Deployments

2.2.5 Continuous Monitoring

Monitoring is a critical practice in DevOps because the entire feedback loop depends on it. The data collected from the monitoring of various applications and the associated infrastructure is vital for making business and IT decisions.

As shown in the Fig. 12, the Operations and Support team follow these DevOps practices

- Monitor incidents and problems during production and perform causal analyses for taking corrective and preventive actions. Preventive actions may involve the Application Development teams, other Service providers or the infrastructure supplier.
- Monitor the pattern of end use and user behaviors using analytical tools for enabling businesses to make appropriate IT investment plans and decisions. Business decisions could also cover the company's marketing activities.
- Monitor application's various resource utilization parameters like CPU, Memory and traffic and provide feedback to Application Development teams. This helps the teams to implement the necessary controls as part of the Continuous Integration mechanism and prevent the incidents from recurring.

Data from different sources are monitored using various infrastructure monitoring and log management tools. The infrastructure monitoring tools constantly monitor the utilization of infrastructure resources like CPU, memory, disk, network

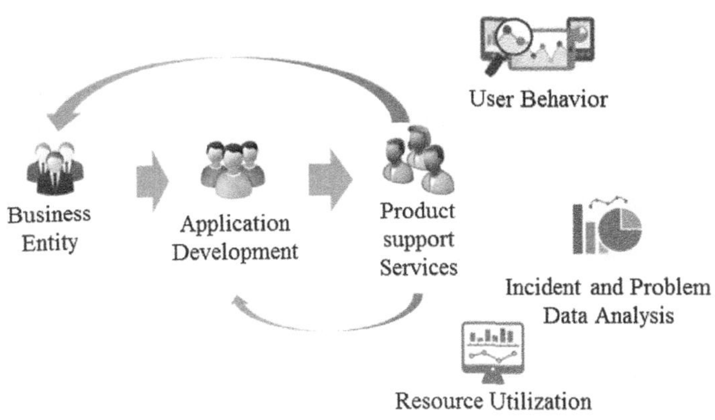

Fig. 12 Continuous monitoring

and storage. On the other hand, the log management tools typically parse through application logs to identify issues in an application.

Log Management

One of the several challenges faced by operations teams is errors in application logs. An effective log management system is necessary to enable an operations team to detect problems and highlight them for the production support team to fix them quickly even before the user reports it (see Fig. 13). For this purpose, log management tools like Elastic Search and Splunk are of great value to operations teams. With intelligent log management tools, a problem can be located quickly, thus reducing the MTBF and increasing reliability of the software and associated services.

2.2.6 Continuous Feedback

The logs generated while a software is executed, and monitoring these logs are the sources of 'real' data. Till the logs are generated and analysed, all claims and discussions about software design and design for reliability are only hypothetical. As seen in Fig. 12, user behavior provides useful insights to both business and IT. For example, based on the type of features being used, information can be gathered on user behavior which helps IT to prioritize the development of subsequent features or enhance the existing ones. Based on user interest, businesses can plan for targeted marketing. Based on the application's performance and associated programs, the engineering team can work on improving the performance of the application. Again, based on the features found useful by the end-users, as well as the manner of their use, testers can design their test cases accordingly.

2.2.7 Collaboration

In most medium and large IT organizations, the service desk receives calls from users whenever the service has failed or receive alerts from the monitoring systems. As shown in Fig. 14, the service desk logs these incidents and categorizes them according to their understanding of the issue—whether the problem is related with

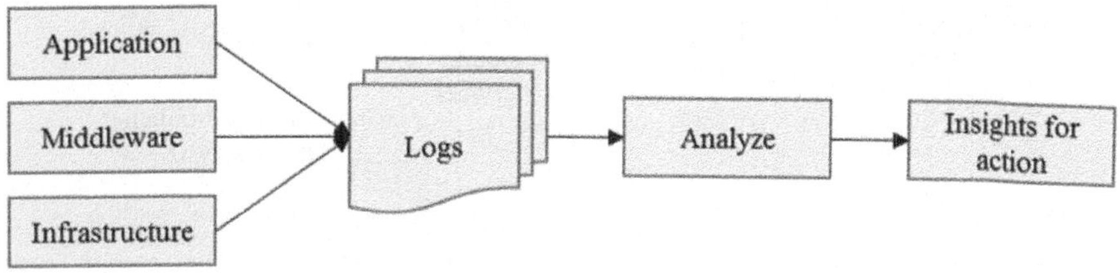

Fig. 13 Log management

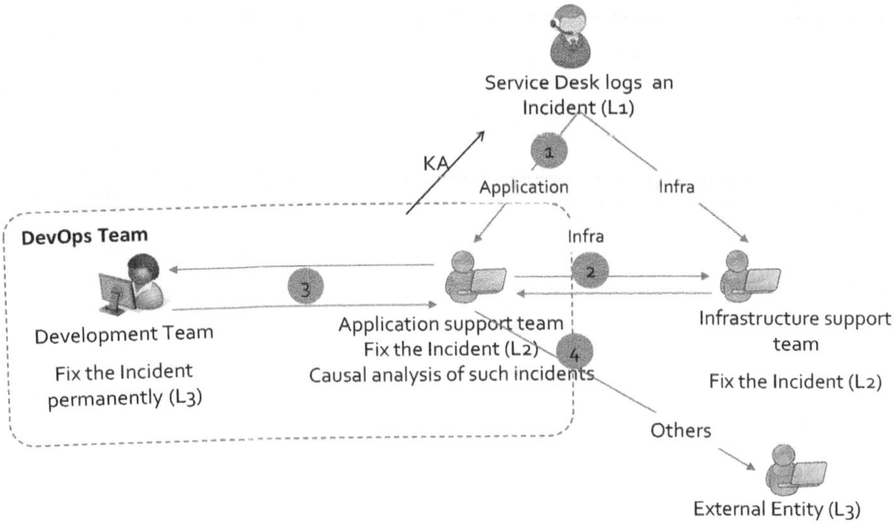

Fig. 14 Collaboration in a DevOps team

the application or infrastructure. If it is related to the Application, the incident is assigned to the Applications support team. Infrastructure-related incidents are assigned to the Infrastructure support team.

The Application support team will analyze the incident and aim for a quick turnaround while deciding on the fix and restoration of services. If the team is not able to restore the services, it takes help from Development team to fix the problem. If a code fix is required, the development team is involved for identifying the cause of the defect. It fixes the code, tests it and then releases it to production.

The objective of service management is to prepare the knowledge articles that can be used by the Service desk team so that the number of incidents assigned to L2 and L3 are minimized. Hence, L2's responsibility is to prepare/update the knowledge articles for sharing with the Service desk team.

This is the traditional approach. It poses several challenges to application support teams because, whenever there is a release, the Applications support team receives a last minute knowledge transfer and handover, which is inadequate for providing support services to the applications. In some organizations, due to warranty conditions, the development team provides support for a certain duration. In these situations, the support team does not have knowledge of the context for which the code was developed in a certain manner. This leads to an increase in Mean Time to Repair (MTTR) whenever a problem is reported. For this reason, it became imperative that the wall between Dev and Ops is broken down so that the two team can work closely together.

Organizations considering DevOps need to restructure for effective collaboration. The ideal way for bringing collaboration is by making the Application

Development and Application Support teams a part of a single team called "DevOps team". There are several organizations who keep the Development and Support teams separate for regulatory reasons. In such cases, the touch points for these teams need to be clearly defined so that the objectives of DevOps are achieved.

In organizations that have adopted DevOps, the collaboration model incorporates logical grouping of the applications and then having Dev and Ops team members working as part of one team. The organization structure of such organizations could look like one the following, considering that the collaboration between Dev and Ops team can happen in two ways

- The Ops team participates in release and sprint planning activities to provide NFRs
- Dev and Ops team members are rotated periodically.

Collaboration Practices:

Figure 15 shows the collaboration practices adopted by the DevOps teams. The Operations and Support teams need to be involved in Planning and Sprint activities during development, as mentioned here:

- Provide Non-functional requirements during release planning based on the previous experiences
- Participate in architecture review to identify Infrastructure requirements like capacity, server, etc. based on user needs
- Participate in Sprint planning activities to provide inputs on implementing and testing NFR's and technical debt items
- Participate in product backlog grooming sessions on issues/feedback from operations to the development team

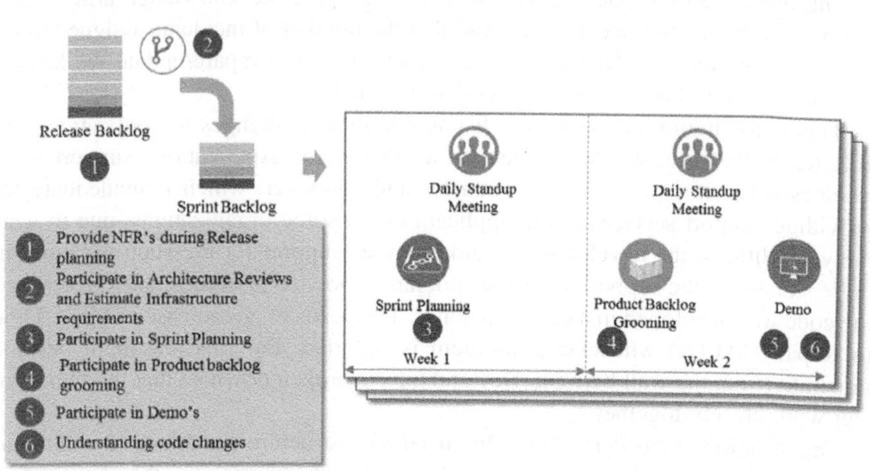

Fig. 15 Collaboration practices in DevOps teams

- Participate in Demos to understand the scenarios for which the application is tested and implemented
- Understand code changes that were implemented in the sprint for managing subsequent changes to the code.

2.2.8 Metrics

DevOps has changed the way software is developed, delivered and its performance monitored. Hence, there is a need for suitable metrics which are aligned with DevOps principles across all phases of the lifecycle. This practice includes capture, storage and analysis of performance data, and the appropriate reporting of results to the stakeholders with a view to improving performance and/or simplifying the IT landscape.

There are 6 key drivers of metrics and reporting:

- A culture of capturing and using measures throughout the organization;
- A repository at organizational level to store measurement data;
- A set of KPI with which to judge performance;
- A group to perform analysis on captured data;
- Standardized method for reporting of raw/analyzed metrics;
- A performance dashboard.

Some of the DevOps metrics across the lifecycle phases include

- Productivity metrics are the number of user stories (or features per person day effort),
- Successful build rate
- Code quality metrics (defects, security vulnerabilities, technical debt)
- Code coverage
- % of test automation
- Change frequency
- Number of rollbacks
- Number of releases per month/quarter
- Number of incidents post release
- Code Performance metrics.

2.2.9 Automation

In the present automation era, DevOps teams are constantly looking for opportunities to automate processes using tools and scripts. Automation is a key practice across all lifecycle phases. Automation eliminates dependency on individuals. A task once automated can be seamlessly integrated with the system to perform like an expert human.

Developers are familiar with automation; and hence, automation of operational tasks speeds up the DevOps cycle. For example, operations teams can automate activities like provisioning and configuring servers, installing and configuring databases, configuring applications and their dependencies, configuring networks, configuring firewall rules and monitoring the application and underlying infrastructure in a production environment.

Automation is invaluable to application, development and operations teams for:

- Eliminating manual errors
- Improving consistency
- Minimizing dependency on people
- Increasing the speed of detection of errors/issues.

Selection of appropriate tools across the lifecycle is essential for realizing the value of DevOps. There is a wide range of tools for automating several activities in the development lifecycle. Tools are available for activities like version control systems, DB automation, Continuous Integration, Test automation, Security, Cloud, Deployment, Containers, Monitoring, AIOps [1].

2.2.10 Infrastructure and Environment Provisioning

Before virtual and cloud computing, teams would wait for physical infrastructure to be made available. This wait varied from weeks to months. The time for procurement of infrastructure increase the overall duration of software development and thus increase in product-to-market time. Today, infrastructure provisioning has been accelerated with the help of virtual and cloud technologies and thus, virtual machines and associated computing power can be procured on demand.

However, it is important to understand that environment provisioning goes beyond infrastructure provisioning. It involves provision of the required operating system, database, middleware, networks and their configuration, web/application server and installing the application as well as its dependent applications. Automated tools like ansible help developers in provisioning the applications and the associated components.

3 Using DevOps Analytics to Improve Reliability and Availability

DevOps analytics helps Dev and Ops teams in improving the reliability and availability of software services by providing early insights about software quality during development and operations.

Traditionally, there is a wall between Dev and Ops teams because of a perception that the two must function separately. The result has been poor coordination

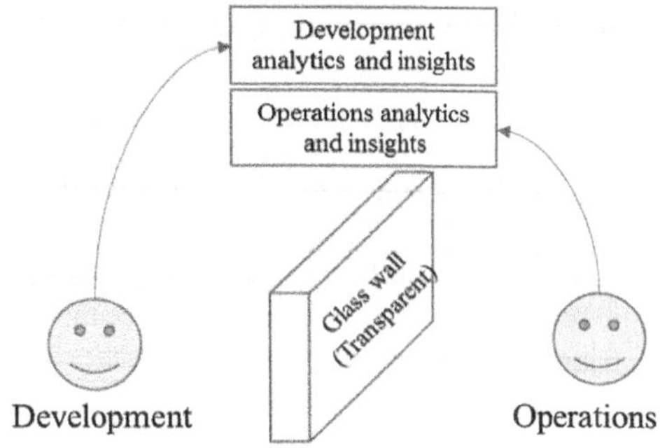

Fig. 16 Analytics for enabling transparency in DevOps teams

between the two as well as lack of trust in each other. This wall is difficult to break, even with DevOps, because of organizational structures and regulatory issues. But it is possible to improve transparency in their functioning by imagining the wall to be made of glass (as represented in Fig. 16). The wall helps the Ops teams to have 'visibility' on the quality and reliability of the software being developed. In turn Dev teams can 'see' their product in operation, the feedback on which will be useful for improving its quality.

> DevOps analytics is defined as: "…the integrated analytics on software data generated in runtime (production) and design time (test and development) to provide insights to the development (Dev) and operations (Ops) teams to improve software quality, productivity and to make informed decisions".

There is enough focus on analytics based on design data and there are tools in the market, for instance Snyder and Curtis [2] adopted analytics to monitor Agile-DevOps transformation program using CAST application intelligence platform [2]. The focus was on quality and productivity in the development phase. The quality was based on static code analysis tools. Augustine et al. [3] at ABB has deployed analytics based on the ALM tools data, like work item count and its status, cycle time of a work item, defect backlog count, and code ownership. The focus was on the progress and status. Gousios et al. [4] presented an approach for real-time software analytics, but do not mention any specific areas of application. Cito [5] proposed research focusing on data-driven decision making for software engineers during development based on the run-time information of method calls and execution times. Baysal et al. [6] proposed using qualitative information for developers, in addition to quantitative dashboards to improve code quality. Johnson [7] highlighted the need for automated and low barrier analytics tools.

The existing tools does not offer a blend of development and operations analytics. Hence, there is a need for DevOps analytics which provides insights to both Dev and Ops teams thus making the wall between Dev and Ops as a 'DevOps Glass wall' which is transparent between the teams and enable in building trust between teams.

DevOps analytics offers useful insights to

(a) the operations team, which are based on data across layers (application, middleware and infrastructure) for faster resolution of incidents;
(b) the application's owner and the operations team on the stability and quality of release, which is based on data obtained from development to operations processes;
(c) security operations team on the vulnerabilities in the application for suitable corrective actions that will result in savings in effort, resources and cost.

Thus, frequent delivery of software with minimal impact on the application's operations, together with operations team's ability to resolve incidents quickly, makes DevOps an effective tool. Considering the benefits that DevOps analytics offers by bringing in transparency between the Dev and Ops teams, and its ability to provide near real-time insights, there is a need to strengthen analytics with additional data sources and leverage algorithms to identify anomalies. Such an approach will further improve the quality of insights, thereby minimizing dependency on experts in the Dev and Ops teams.

3.1 Implementation of DevOps Analytics

This section is about implementing DevOps analytics for a product team. The product team in our example here has adopted agile and DevOps practices. To strengthen product quality, the product unit decided to leverage its data to make informed decisions on product quality and improving its performance. The initiative was expected to reduce incident resolution time of the product support team and ultimately, increase product reliability and availability.

The product development team built a CI pipeline with Jenkins as the orchestrator and maven for build. SAST and DAST are performed using a commercial product. Ansible is used for product deployment. The product support team used a cloud-based ticket management system for product support requests and incidents.

As shown in Fig. 17, data sources from development include features, defects (jira), code quality (sonarqube) and SAST/DAST tool. Data sources from operations include the ticket management tool, metrics and logs from servers, middleware and applications.

While DevOps analytics can potentially provide answers to several questions, the product team wanted to give more attention to the most frequently asked questions [8]. Thus, it identified the following focus areas and correlated them. The objective was to improve software quality and reduce incident resolution time for the operations team:

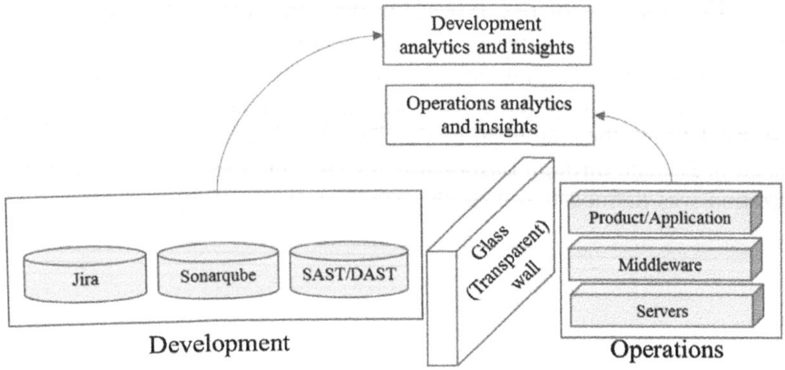

Fig. 17 Overview of DevOps analytics

- Application usage analytics
- Performance analytics
- Error analytics
- Code quality analytics
- Infrastructure analytics.

Application usage analytics helps the development and testing teams to focus on the most used features and the paths traversed paths by application users. The program files used for these features, together with their code quality metrics, will provide insights on the complexity and maintainability of these programs. Thus, teams can focus on these paths for improving their software quality and direct their efforts optimally and effectively to improve productivity.

In a similar manner, gaining an understanding of performance issues and frequent errors in program execution, which are observed in a production environment, will provide insights into programming issues which the development team can use fix them and improve software quality.

3.2 *Architecture*

The application was deployed on multiple servers and data generated in their respective servers. Development tools like SonarQube and SAST/DAST were deployed in different systems. Data had to be streamed from these data sources to feed analytics and gain insights.

The team used open source tools like File beat, Metric beat, Apache HBASE, Logstash, Solr and Banana for data aggregation, paring, transformation, storage, querying and visualization. The architecture for DevOps analytics and insights system is shown in Fig. 18.

Data from various sources in application, middleware and infrastructure (including httpd), application, database and syslogs, code quality, queue metrics were collected. The data was of two types: log data and metric data. Log data was in textual and metric data in numerical format. Real-time data was aggregated through file beat and metric beat from these data sources. File or metric beat was installed in each of these data source systems. Once aggregated, data was ingested and stored in HBASE by defining the canonical model. The data was then queried from Solr and visualized in Banana through dashboards for Dev and Ops teams in real-time. The visualization was defined based on the needs of Dev and Ops teams.

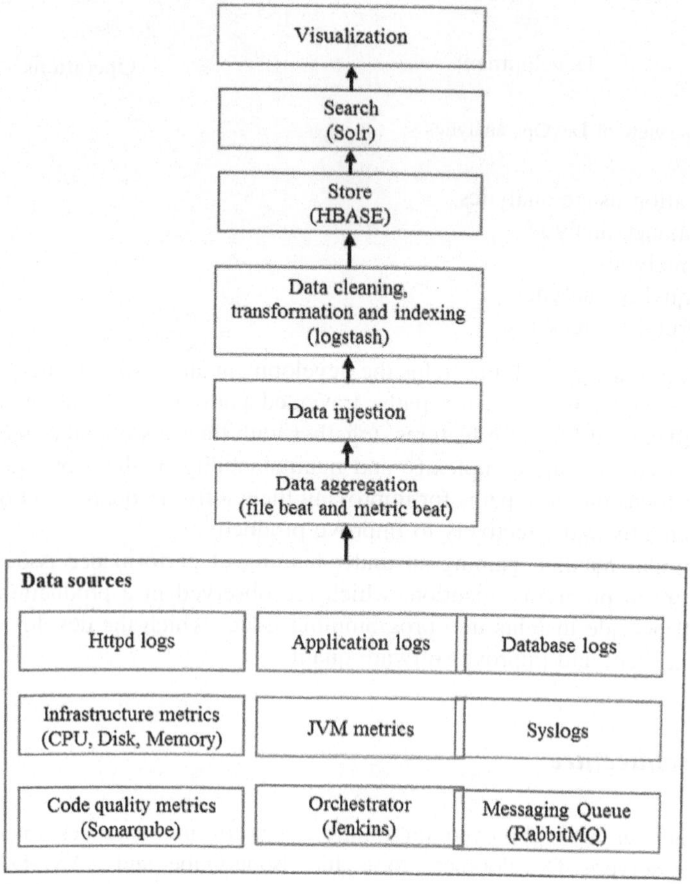

Fig. 18 Architecture representation of DevOps analytics system

3.3 *Descriptive Analytics*

The team in our DevOps example initiated visualization and descriptive analytics of various data like application usage, performance metrics, errors in logs, code quality metrics and infrastructure metrics. We will discuss each in some detail.

3.3.1 Application Usage Analytics

Application usage in runtime environment is a key parameter for providing several insights for decision-making by the application owner. Some of the measures used and their significance are given below.

- Number of active application users: This is usage in terms of number of users. It will assist the application owner in deciding if the investment in the application's features is justified and whether it should be continued with, increased or reduced.
- Daily Peak Usage. With this data the application owner or release manager can decide on the appropriate time for applying a patch. This activity is usually performed when there is no or minimal usage of the application. In most organizations the decision on when to apply a patch is based on intuition.
- Geographical location of the application users: The geographical location of application users can provide an indication of active or inactive users in that location. This triggers causal analysis for low usage in a geography which can help in understanding the reasons for low usage of the application in that location or whether the application owners need to improve communications for better promotion of the application.

Data about the number of active application users and their geographic location is of interest to the application owner. Data on peak application usage is of interest to the operations head because he (or she) is responsible for ensuring the high availability.

The usage pattern for the application in scope is shown in Fig. 19. Here, we can see that use of the application typically begins at 8:30 AM IST. It is also seen that there is no usage from midnight till 8:30 AM IST. The maintenance window for the application should, ideally, be during the lean usage period. Considering the usage and cost of operations, the application owner can decide on continuing with investment in this application. To illustrate this point, usage of the application for one day is represented in Fig. 19. By studying usage over a period—a week or month— the application's owner can determine the usage pattern. This can help in identifying anomalies and their reasons.

The use of the application in various geographies is shown in Fig. 20. This information shows the number of active users in a city and country. Large community of application users from Chennai, India corroborates with the fact of their presence in Chennai.

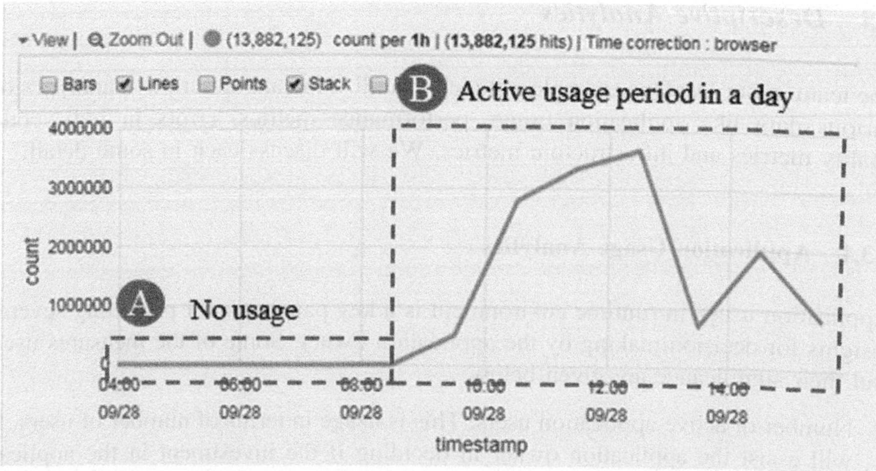

Fig. 19 Application usage analytics

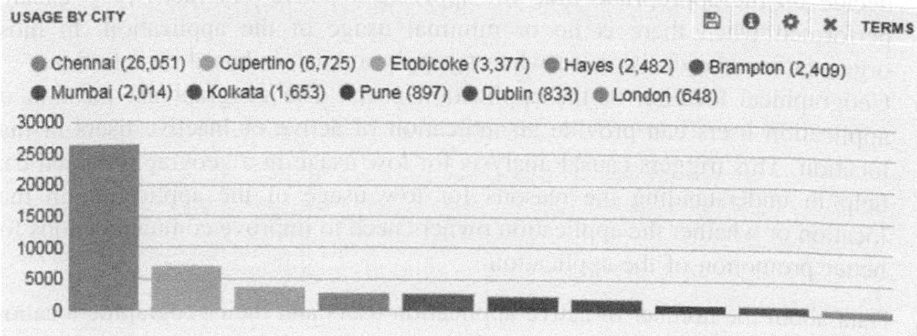

Fig. 20 Usage of the application in various locations

3.3.2 Performance Analytics

If the application development team in the example wants to improve performance, it is necessary to first understand the methods which are being called most frequently, as well as those that consume high processing times. This information from the runtime environment is useful to the programmers in their efforts to optimize their designs and code.

The Top-N modules and methods used in an application is shown in Fig. 21. The number of times a module is called is shown as the 'count' and the average processing time of the method is shown as the 'mean' (A). Thus, the aggregation of all methods in a module are known as 'count' and 'mean' at the module level (B).

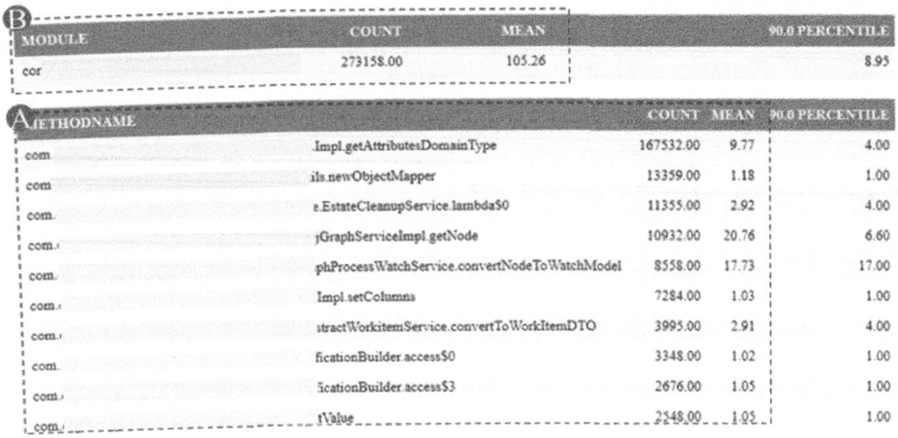

MODULE	COUNT	MEAN	90.0 PERCENTILE
cor	273158.00	105.26	8.95

METHODNAME		COUNT	MEAN	90.0 PERCENTILE
com	Impl.getAttributesDomainType	167532.00	9.77	4.00
com	ils.newObjectMapper	13359.00	1.18	1.00
com	e.EstateCleanupService.lambda$0	11355.00	2.92	4.00
com	jGraphServiceImpl.getNode	10932.00	20.76	6.60
com	phProcessWatchService.convertNodeToWatchModel	8558.00	17.73	17.00
com	Impl.setColumns	7284.00	1.03	1.00
com	stractWorkitemService.convertToWorkItemDTO	3995.00	2.91	4.00
com	ficationBuilder.access$0	3348.00	1.02	1.00
com	icationBuilder.access$3	2676.00	1.05	1.00
com	fValue	2548.00	1.05	1.00

Fig. 21 Top-N time consuming modules and methods

To begin with, the development team selected the top 'N' modules and methods for code optimization. After the performance was improved by optimizing the processing time, the next top 'N' time-consuming methods were called. This process was repeated till the team achieved the application performance levels required. If the development team had not succeeded in achieving the required performance level, deployment structure is modified by deploying the components on multiple servers for improving the availability of the application.

3.3.3 Error Analytics

There are various types of logs—application logs, error logs and diagnostic logs. As practice, operations teams browse through multiple log files at the time of incident to detect the location of the problem. This is a time-consuming process and needs expertise to visualize the problem and then detect its location. The problem gets more complex in a multi-vendor scenario in which there is no clear ownership of the incident. Error analytics will help the operations team to improve transparency and reduce resolution time, thereby improving reliability of the application.

To do this the application and error logs are ingested for the application in scope. To illustrate this, a dashboard showing the number of logs and the error logs by module is shown in Fig. 22. Based on the transaction id, the operations team can quickly diagnose the module and the method in which the error occurred, even before the user raises the incident. Then, the operations team can resolve the incident. It is possible for the operations team to drill down to the log statement in real-time in a single window instead of browsing through multiple logs to identify the location of the problem (Fig. 22).

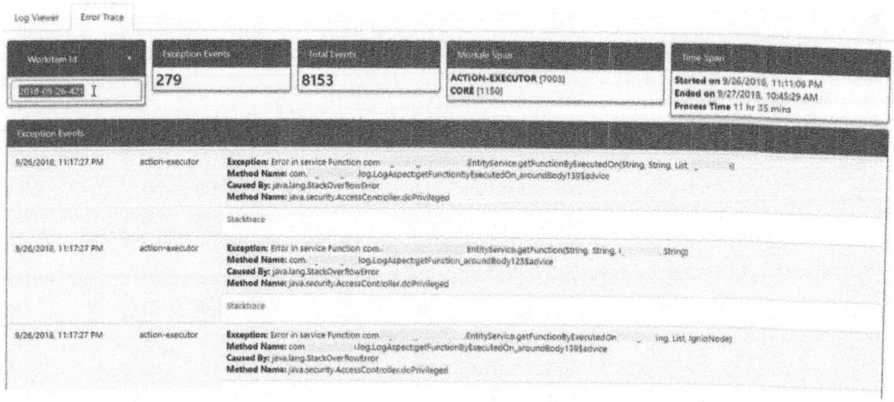

Fig. 22 Error analytics

This approach helps the development team to understand the most frequently-occurring error types and the methods in which the error is found. With this insight it is possible to provide a permanent fix and reduce the number of incidents that the operations team has to respond to.

3.3.4 Code Quality Analytics

As discussed earlier, it sometimes happens that the operations team does not have enough confidence in the build for sharing it at a rapid pace for deployment in production. Hence, data from build and code quality metrics like defects, vulnerabilities and complexity can provide insights and assurance about software quality and reliability to the operations teams.

The code complexity of an application is shown in Fig. 23, which was planned for released to production. As code complexity increases, its maintainability decreases. Code complexity, number of defects, number of build failures and vulnerabilities are indicators of the application's stability. Application incidents and quality parameters, provide insights to operations team on application stability.

3.3.5 Application Security Analytics

A key concern for the operations team is lack of confidence in the application releases receive by them. Sometimes they even test the application in their environment before it is moved to production. This consumes effort, resources and is a cost to the organization. In a scenario where the operations team had insights about the quality of development, some of these activities could have been eliminated. Similarly, the security operations team would also want to scan the application to make sure that there are no vulnerabilities or are not introduced in the system.

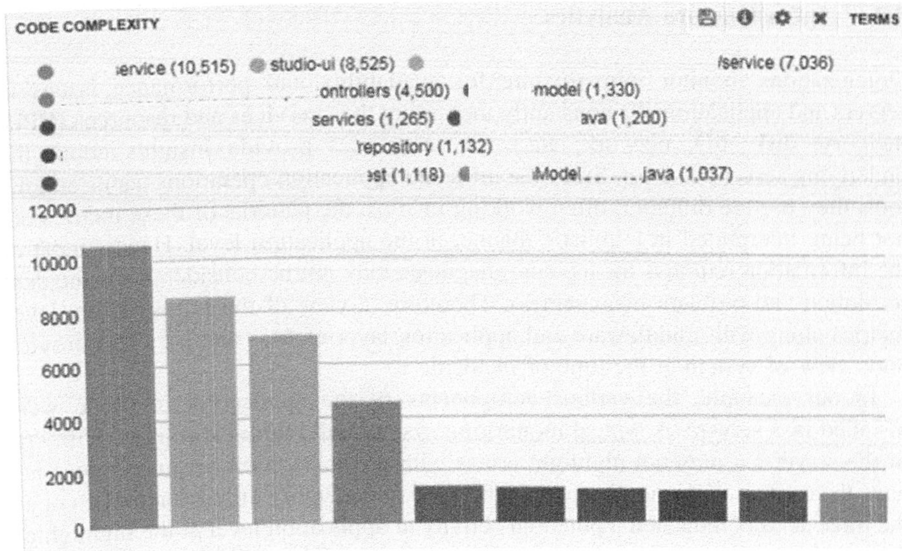

Fig. 23 Code complexity of the application

In our example, the team ingested application security testing tool results, from both SAST (Static application security testing) and DAST (Dynamic application security testing), which were run daily. As shown in Fig. 24, the trends provided valuable insights to the operations and security operations teams which they fixed following which, there was a drop in number of vulnerabilities.

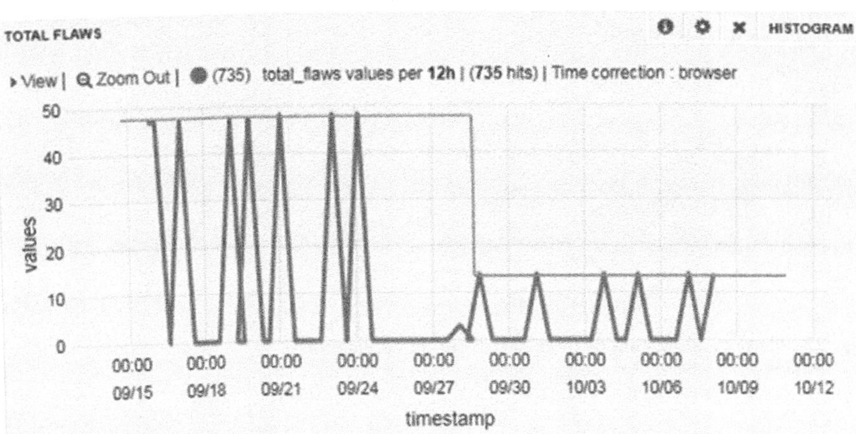

Fig. 24 Applications security vulnerabilities

3.3.6 Infrastructure Analytics

Organizations monitor infrastructure for availability and performance issues in servers and applications by constantly measuring their services and resources (CPU, memory, disk, I/O, network, etc.). These metrics provide insights about the underlying server behavior. Since the infra and application operations teams and the tools they use are different, often working in silos, the patterns of these metrics are not being interpreted in a holistic manner at the application level. Hence, most of the information required for incident diagnosis may not be considered for incident resolution and problem management. Therefore, a view of the infrastructure layer metrics along with middleware and application layer metrics and logs will provide 360^0 view of system at the time of incident.

In our example, the various components of the application in scope were installed in 3 servers. A typical monitoring system will highlight an issue with one of the servers; it may not highlight issues with the application because it does not have the context of the application. In this case, the memory utilization pattern of all the three servers indicated a potential activity at application level at the highlighted time (as shown in Fig. 25). On analysis, it was found that the application services went down during this period. We can see that analysis across layers will assist the operations team in quick detection of the cause of anomaly.

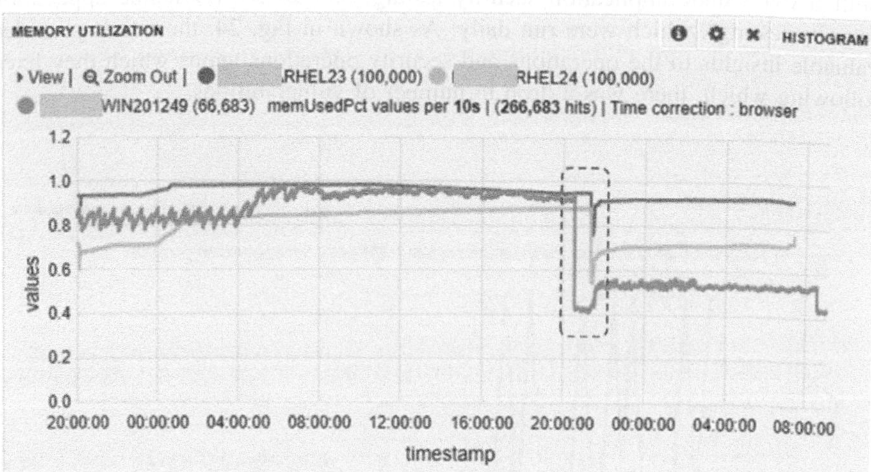

Fig. 25 Memory utilization of three servers of an application

3.4 *Correlation and Anomaly Detection*

Aggregation and visualization of logs and metrics alone are of little value to the Dev and Ops teams. Instead, insights which are based on correlation of multiple logs and metrics will be more useful. Hence, the Dev and Ops teams began correlating multiple metrics. Some of the correlations are discussed below.

3.4.1 Correlation of Deployment Structure and Resource Utilization Metrics

A large application was deployed on multiple servers. The operations team was facing challenges with unknown issues. To understand the cause of the problem, the team collected the resource metrics (CPU, disk and memory) and the deployment architecture (See Fig. 26).

The application was deployed on 3 servers namely, Server 1, Server 2 and Server 3. The 25 deployable components were deployed on these 3 servers. The database was deployed in Server 2 which had 32 databases. The CPU utilization of Servers 1 and 3 were found to be normal. Server 2 was experiencing frequent high CPU utilization. It became obvious that CPU utilization on Server 2 was consistently high due to which the operations team was facing issues. As a result of analysis, the operations team recommended modification of the deployment structure or increased resources for improved performance.

3.4.2 Detection of Anomalies in Application Usage

Analytics of the application in production environment showed a surge in its usage (as shown in Fig. 27). Considering the normal pattern of behavior, the spike was an anomaly. The operations team could see who was accessing the application from the IP addresses on the log and the country from where the users were logging in. With this information, the team could arrive at a quick diagnosis of the problem. In the absence of real-time analytics like this one, most events go unnoticed which is a risk to the application as well as the user. Analytics makes it possible for operations teams to relate all possible metrics like IP address, city, pages accessed by the users for obtaining insights.

3.4.3 Correlation of Errors with Methods and Programmer

Correlation of error-prone modules and methods are based on operations data of the application. This was useful for the development team to understand the characteristics of the errors. The methods which are called the most number of times and their corresponding error messages are shown in Fig. 28a, b respectively. This

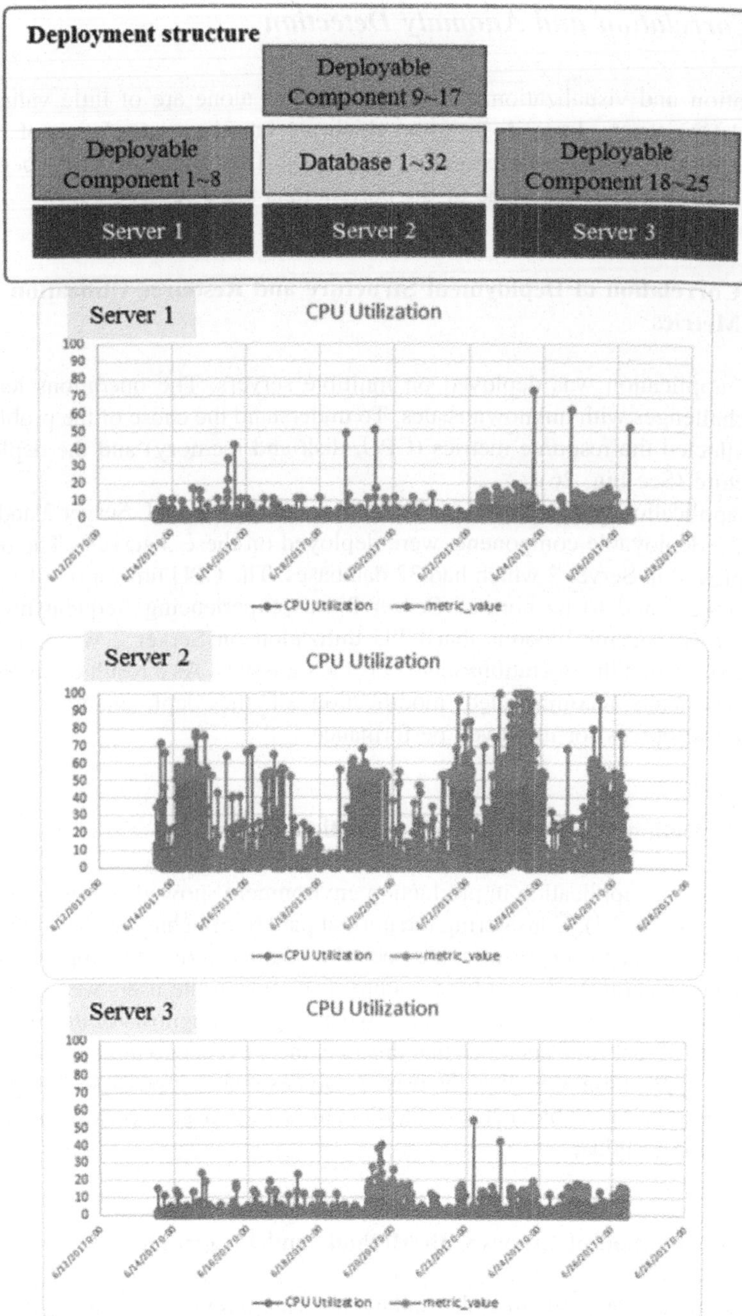

Fig. 26 Deployment architecture and resource utilization

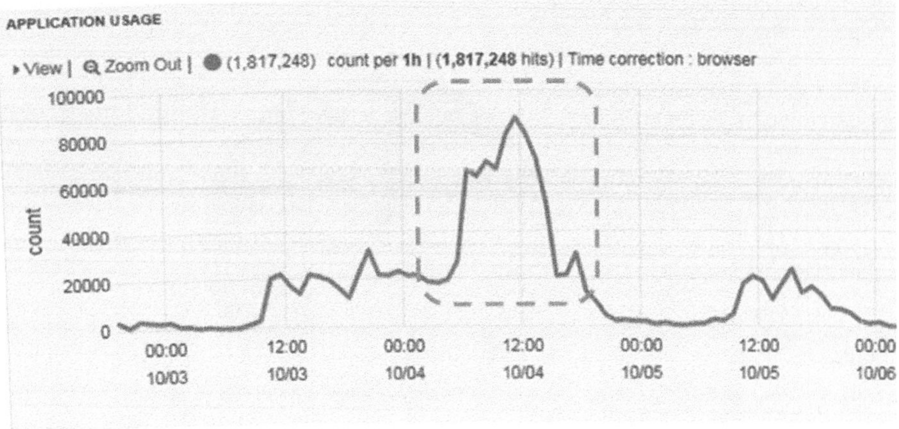

Fig. 27 Anomaly in application usage

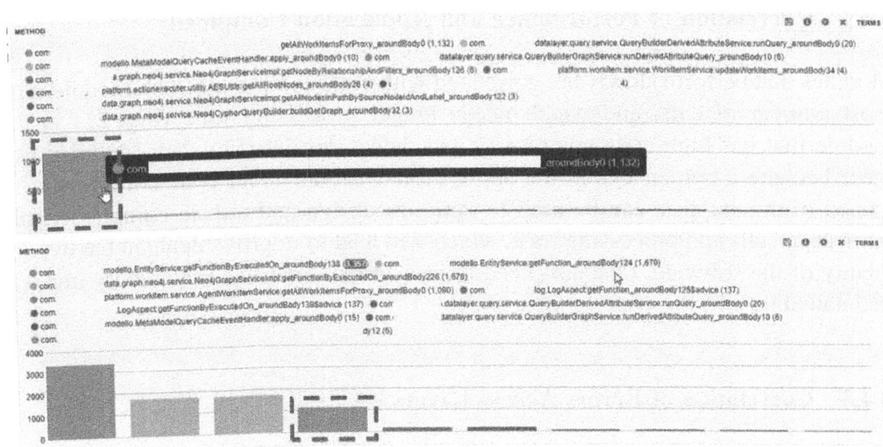

Fig. 28 **a** Errors in application logs by method. **b** Total number of logs by method

provided guidance and prioritization suggestions for the programmer regarding the methods that need refinement for improving quality.

In addition, the type of errors occurring most of the times will provide insights to development the team about the category of errors that need fixing. Correlating erroneous methods with the methods' programmers will provide insights about their programming abilities. This will help in making the necessary recommendations for appropriate training to avoid such errors. Where possible, code quality rules must be defined in static code analysis tools like sonarqube so that code quality issues are detected during the continuous integration stage.

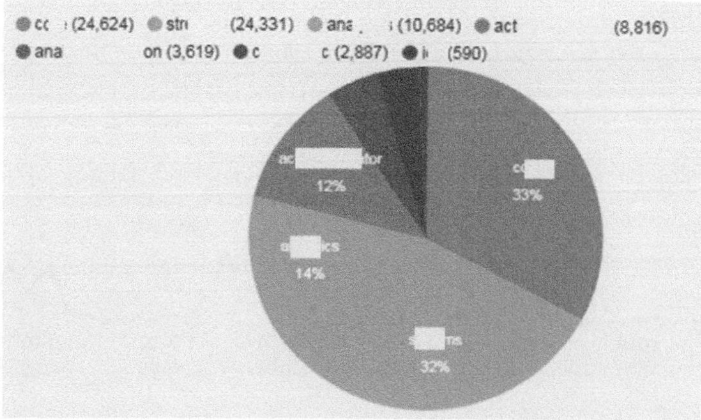

Fig. 29 Code complexity

3.4.4 Correlation of Performance and Application Complexity

Modules that perform slowly are correlated with code complexity. The module with most number of calls and which needs longer processing time (Fig. 21) is the module that has high code complexity and defect density (Fig. 29). Modules like these become a bottlenecks in the application. Insights about code complexity, as gained from analytics, can be used to optimize design and reduce complexity, the number of calls and processing time, which will lead to improvements in the overall quality of the software. One also need to bear in the mind that complexity impacts maintainability of the code by ops team.

3.4.5 Correlation of Errors Across Layers

Applications in production generate errors/exceptions in multiple logs. It is humanly impossible in real-time to determine where the problem is by observing the logs, and more so across multiple logs across layers. To assist in problem identification, a log aggregation and correlation is done across all layers as shown in Fig. 30. On clicking the specific type of error log (like application log or technology log), as shown in Figs. 31 and 32 respectively, the operations team can quickly diagnose the problem location.

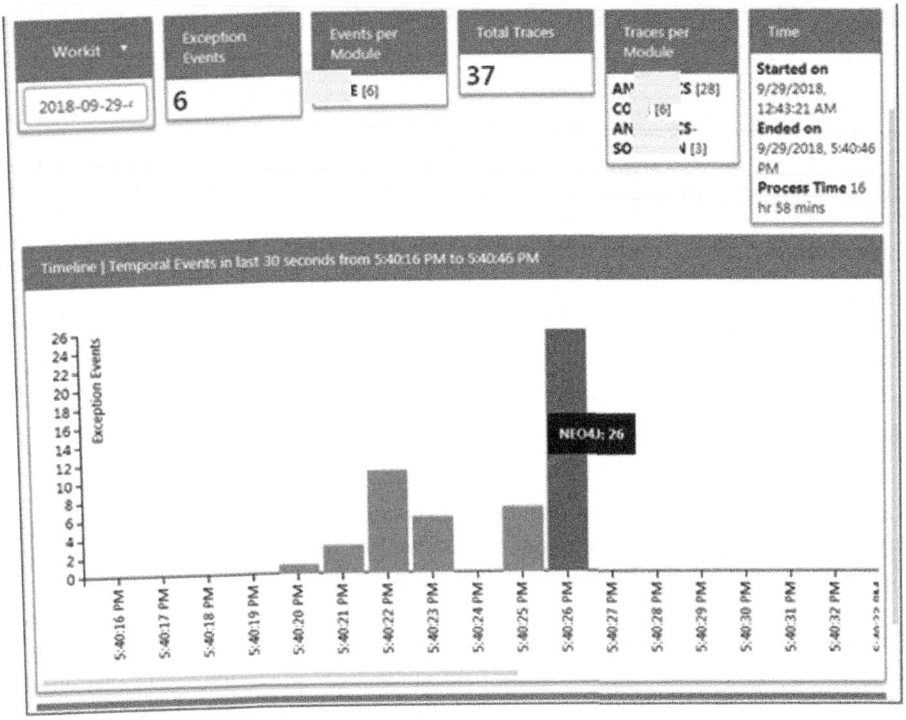

Fig. 30 Correlation of logs across layers

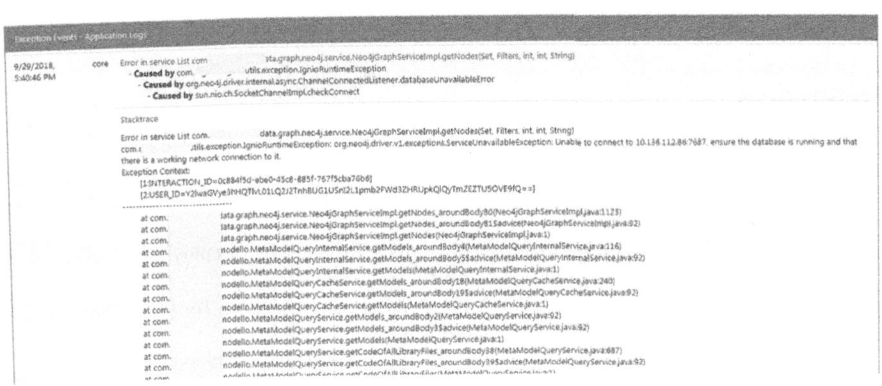

Fig. 31 Application log

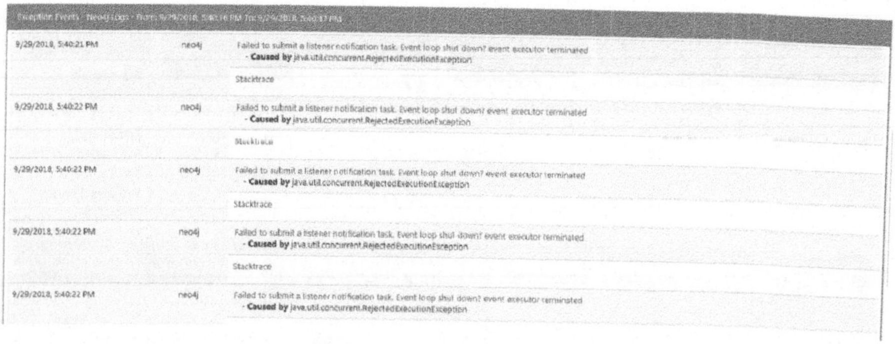

Fig. 32 Technology log

4 Summary

DevOps analytics offers useful insights to (a) operations team based on data across layers (application, middleware and infrastructure) to resolve the incidents faster (b) application owner and operations team on the stability and quality of the release based on the data from development to operations (c) security operations team on the application vulnerabilities, resulting into effort, resource and cost savings. Thus frequent delivery with minimal impact on application operations; and operations team ability to resolve incidents quickly, makes DevOps implementation effective. Considering the benefits that DevOps analytics offers by bringing transparency between the Dev and Ops teams, and its ability to provide insights near real-time, it improves service reliability and availability.

References

1. Periodic table. https://xebialabs.com/periodic-table-of-devops-tools/
2. Snyder B, Curtis B (2018) Using analytics to guide improvement during an agile-DevOps transformation. IEEE Softw 35(1):78–83
3. Augustine V, Hudepohl J, Marcinczak P, Snipes Will (2018) Deploying software team analytics in a multinational organization. IEEE Softw 35(1):72–76
4. Gousios G, Safaric D, Visser J (2016) Streaming software analytics. In: Proceedings of 2nd international workshop on BIG data software engineering, pp 10–11
5. Cito J (2016) Developer targeted analytics: supporting software development decisions with runtime information. In: Proceedings of 31st international conference on automated software engineering (ASE), IEEE/ACM, pp 892–895
6. Baysal O, Holmes R, Godfrey M (2013) Developer dashboards: the need for qualitative analytics. IEEE Softw 30(4):46–52
7. Johnson P (2013) Searching under the Streetlight for Useful Software analytics. IEEE Softw 30 (4):57–63
8. Menzis T, Zimmerman T (2013) Software analytics: so what. IEEE Softw 30(4):31–37

Dr. Mahesh Kuruba is an information technology (IT) professional with wide experience in research, consulting, as well as product and application development. He is author of "Role Competency Matrix" a quantitative competency management for effective workforce competency management practices. He is currently working as a Product Manager at Digitate, Tata Consultancy Services (TCS). He was Process Consulting Head for Japan and ASEAN countries until 2016. There, Mahesh managed transformation programs, achieving substantial quantitative benefits for various organizations. He also has wide experience in software development and management in both conventional and agile methodologies. Mahesh's doctoral research at the Indian Institute of Technology, Bombay, was on "Productivity loss reduction-based framework for software process improvement." He was also a Visiting Scientist at the National Institute of Information and Communications Technology (NICT), Japan, where he pursued research on network security. He has authored over 21 Papers in international conferences and journals and a speaker in several international conferences. He can be reached at Mahesh_kuruba@ieee.org.

The Unpopularity of the Software Tester Role Among Software Practitioners: A Case Study

Yadira Lizama, Daniel Varona, Pradeep Waychal and Luiz Fernando Capretz

Abstract As software systems are becoming more pervasive, they are also becoming more susceptible to failures, resulting in potentially lethal combinations. Software testing is critical to preventing software failures but is, arguably, the least understood part of the software life cycle and the toughest to perform correctly. Adequate research has been carried out in both the process and technology dimensions of testing, but not in the human dimensions. This work attempts to fill in the gap by exploring the human dimension, i.e., trying to understand the motivation/de-motivation of software practitioners to take up and sustain testing careers. One hundred and forty four software practitioners from several Cuban software institutes were surveyed. Individuals were asked the PROs (advantages or motivators) and CONs (disadvantages or de-motivators) of taking up a career in software testing and their chances of doing so. The results of this investigation identified 9 main PROs and 8 main CONs for taking up a testing career showing that the role of tester is perceived as a social role.

Keywords Testing career · Software testing · Software quality assurance

Y. Lizama · D. Varona
Culturеplex Laboratories, Western University, London, ON N6A5B9, Canada
e-mail: ylizama@uwo.ca

D. Varona
e-mail: dvaronac@uwo.ca

P. Waychal
CSIBER (Chhatrapati Shahu Institute of Business Education and Research), Kolhapur, India
e-mail: pradeep.waychal@gmail.com

L. F. Capretz (✉)
Department of Electrical & Computer Engineering, Western University, London, ON N6A5B9, Canada
e-mail: lcapretz@uwo.ca

Department of Computer Science, New York University, Abu Dhabi 129888, United Arab Emirates

D. R. Karanki et al. (eds.), *Advances in RAMS Engineering*, Springer Series in Reliability Engineering,
https://doi.org/10.1007/978-3-030-36518-9_7

1 Introduction

Researchers have been investigating practices to improve the work performance of individuals. Several theories were developed and utilized to enlarge the body of knowledge about this theme and to contribute to the improvement of practices. One of the key components which has an impact on the performance and productivity of individuals is the motivation to take up and sustain a job. Nevertheless, software engineering, particularly software testing, still lacks studies on motivation, especially motivation to take up testing careers. Therefore, it is important to focus on phases of the software process, since there are considerable differences in the mindset and skills needed to perform different software tasks.

The success of software projects depends on a precise balance among three pillars: person, process, and technology. Also, the commitment to quality from the team members and a well-organized quality assurance process are essential. In 2015, just a third of software projects were considered a success [1]. Software failures impact the economy [2, 3], cause several technological disasters [4, 5], and have negative social repercussions [6]. The accelerated spread of software in business processes, the increase of complexity and size of systems, and the dependence on third party components are factors that increase potential for failure that impact in software products.

Despite the attention that researchers and practitioners have paid to the process [7, 8] and technology dimensions [9–12], it is clear that more effort is needed to achieve better results in software testing. It has been pointed out that human and social aspects play a significant role in software testing practices [12–14]. In an academic setting, attention to human factors in software testing has been preached by Hazzan and Tomayko [15] and Capretz [16].

Aspects of the job that motivate software engineers include problem solving, working to benefit others, and technical challenges, though, the literature on motivation in software engineering appears to present a conflicting and partial picture. Furthermore, surveys of motivation are often aimed at how software engineers feel about the organization, rather than their profession. Although models of motivation in software engineering are reported, they do not account for the changing roles and environmental settings in which software engineers operate.

In a real-world environment, Shah and Harrold [17] and Santos et al. [18] found that software engineers with a positive attitude towards software testing can significantly influence those who have a negative attitude. However, there is no clear understanding of software engineers' jobs in general, what motivates them, how they are motivated, or the outcome and benefits of motivating software engineers.

For a long time, the term motivation was used as a synonym for job satisfaction and to describe several distinct behaviors of software engineers. This satisfaction/motivation disagreement among concepts represented a problem both for academic research and industrial practice, due to the need for the proper management of motivation in software companies, to achieve higher levels of productivity among professionals at work, and motivating software engineers continues to be a challenging task.

Motivational aspects have been studied in the field, including the need to identify with the task in hand, employee participation/involvement, good management, career path, sense of belonging, rewards and incentives, etc. Just like any profession in the world, software engineers also have their own de-motivators, such as the lack of feedback from supervisors, insufficient salary, lack of growth opportunities, etc.

The role of tester does not figure as a favored role among the population of software developers, according to previous study results [12, 19, 20]. Some studies point out the need for reversing people's perception regarding this role [21] by using career progression and other related mechanisms to reinforce the crucial dimension that a tester brings to the project. In the last decade, individual companies have defined competence profiles for the roles they assign to their projects, as stipulated by the methodology they selected. However, if human aspects are not taken into account in staffing software projects, an important piece of the puzzle for project staffing is overlooked.

We studied the chances of software practitioners taking up software testing careers and their reasons. To that end, we conducted a survey of several Cuban software practitioners; endeavoring to expose actual reactions to the role of tester in the software industry.

2 Methodology

In order to assess whether or not participants considered a software testing career, the authors conducted a survey of individuals from a variety of Cuban software institutes. Participants could either be software engineers, software testers, software developers, and/or have an interest in pursuing a career in software testing. The survey asked participants to share the advantages and disadvantages of pursuing a career as a software tester.

The survey asked four questions. The first two questions were open ended questions: (1) What are three PROs (in order of importance) of pursuing a career in software testing; and (2) What are three CONs (in order of importance) of pursuing a career in software testing. The third question asked participants to indicate their intentions of pursuing a career in software testing and were given the option to answer with either "certainly not," "no," "maybe," "yes," and "certainly yes." Participants were also invited to share their reasons behind their responses. Lastly, participants were asked to provide demographic information about themselves. The survey questions are shown in Appendix.

One hundred and forty-four software developers participated in the survey. 61% of participants were males and 39% were females. The average age of the participants was 31 years old. The average number of testing-related activities among participants was 9 years of experience.

Participants' performance evaluations and current professional roles were not required to participate in the survey. However, once participants consented to and participated in the survey, the authors approached their employers to get access to their performance evaluations as software professionals. These quarterly evaluations assessed professionals (i.e. those who participated in the survey) on a five-point scale and asked the participants' employers to evaluate participants as "Unacceptable," "Acceptable," "Good," "Very Good," and "Excellent"; 85% of participants were assessed as "Good" by the employers, whereas 5% were assessed as "Excellent". The remaining 10% of the assessments were spread across "Acceptable" and "Very Good".

3 Results

After a data analysis process, the main results are presented in Table 1. Common statements were combined during the refining of data. We found nine main PROs and eight main CONs in total.

It must be explained that the PROs and CONs were placed, in the table, and ordered reflecting the priority chosen by the respondents. The results related to the PROs and CONs individuals attributed to taking a testing career are presented in the two sections below.

3.1 PROs Related Results

Primarily, the items considered as PROs for taking up a testing career among the surveyed individuals are presented along with their frequencies. The first priority, as can be seen in Table 1, gathers four main trends. The most frequent, with a 36% of respondents, shows the perception that the role of the tester has more interactions with the project team members—better than by the role of analyst (PRO item 7). Followed by the belief of 29% of respondents pointing that testing tasks particularities make software tester focus on details—almost half of individuals in the survey ranked this reason as the second priority (PRO item 9). The remaining two reasons figuring in the first priority, with a 19% and 16% of recurrence respectively, were:

- Testing activities provides a full background of the project scope, modularization, and integration strategy in a short period of time (PRO item 2).
- There are test engines and other automated tools giving testers great technical support (PRO item 4).

In the second priority rank for PROs, the general position was mainly divided into two lines of thought: 49% of the subjects stated that a tester's activities are

Table 1 PROs and CONs descriptions and frequencies

No.	Frequency		Description
	Quantity	Perc. (%)	
PROs			
1	71	49	The activities related to testers are simple at first and then complexity is gradually increased; this helps fresh new testers to have a smooth curve of capacitation and specialization
2	27	18	Testing activities provide a full background of project scope, modularization, and integration strategy in a short period of time
3	18	12	The role of tester is a role in which related activities demand lots of creativity from the individual
4	23	15	Test engines and other automated tools give testers accurate support
5	5	3	Tester's responsibilities are spread along all project stages
6	71	49	Tester's activities are quite client oriented
7	52	36	After the analyst role, the tester has more interactions with the project team
8	55	38	A periodical rotation of project team members in the tester role will increase team commitment to product quality
9	42	29	Testing tasks particularities make the software tester focus on details
CONs			
1	59	40	Other project team members may be upset by a tester's findings when reviewing their releases
	40	27	
2	66	45	Other roles than tester enjoy more acceptance among software engineers
3	72	50	Too many detail-oriented skills are demanded from software testers
4	19	13	Gender-related issues from both parts (males (7) say it is a role for females to perform, and females (12) prefer more technical roles to show their abilities and skills to practitioners of the opposite gender
5	32	22	It is difficult to perform as a software engineer when a person is highly specialized in a particular role
6	39	27	Ability to handle abstraction is needed to have an adequate performance in role tester
7	48	33	The attention of testers has to be divided in two, between engineering artifacts and business process when other roles do not have this division
8	57	39	In some labor markets, the tester wages are less than the average wages of other roles

client-oriented (PRO item 6) while 47% stated that the particularities of testing tasks makes software tester focus on details (PRO item 9). These PROs were followed by a scarce 3% of individuals who noted that testers' responsibilities are spread within all project stages (PRO item 5).

The third priority contains PROs such as: 49% of responses state that testers start with simpler activities followed by gradual increase in complexity, which helps new testers to have a smooth learning curve (PRO item 1). In addition, 38% suggested that a periodical rotation of project team members in the tester role would increase team commitment to product quality (PRO item 8), and 12% noted the role demands lot of creativity (PRO item 3).

3.2 CONs Related Results

In contrast, the items tagged as CONs for taking up a testing career, respondents gave the most importance to the following reasons:

- Other roles enjoy more acceptance among software engineers (46%) (CON item 2).
- Other project team members may become upset after they receive the tester's results assessing their work (41%) (CON item 1).

There are gender-related issues in the choice of a testing career (13%) (CON item 4). Eight percent of male subjects in the sample stated that the tester role is a role for females to perform. On the other hand, 21% of female individuals expressed their preference for working in more technical roles such as: programmer or manager, in order to show their competence before a wide male-sexist opinion regarding women in software engineering.

The CONs listed in the second level of importance showed that 50% of the surveyed individual stated that the skills demanded of testers are too detail oriented (CON item 3). 28% of the respondents pointed out the perception that other project team members may become upset facing tester's findings at the reviewing process (CON item 1); similarly, 41% of individuals supported this same reason within their first priorities regarding cons. The remaining subjects (22%) noted that it is difficult to perform as a software tester when a person is highly specialized in another role, such as analyst or programmer (CON item 5).

Lastly, the third level of importance among CONs was found as follows: 40% of the subjects noted that in the labor market the role of tester is a role for which wages are lower than the average wages for other roles (CON item 8). Another 33% have the idea that the attention of the tester has to cover all engineering artifacts; while other roles only produce a specific type of artifacts (CON item 7). Furthermore, the remaining 27% of respondents believe that the ability to work with abstractions is required to perform adequately in the role of tester (CON item 6).

Table 2 Chances of taking up testing career among respondents

Third question distribution %		
Certainly not	24	17%
No	67	46%
Negative subtotal	*91*	63%
Maybe	22	15%
Yes	23	16%
Certainly yes	8	6%
Positive subtotal	*53*	37%
Total of answers	*144*	

Regarding the third question, Table 2 shows the responses to the actual chances of respondents taking up a testing career according to their personal preferences. The reason supporting each response for the items chosen are described below. The authors wish to note that similar responses were merged, and duplicates were eliminated to ensure a better understanding and further analysis. A total number of respondents who opted for the response option 'Certainly not' agreed they do not like the role of tester, as did the 25% who chose the response option 'No'. The remaining subjects, who picked the response option 'No', find the role of tester less attractive in comparison to other roles—such as analyst, programmer and designer, in that exact order. Nevertheless, some individuals with the same reason did not specify the role they found more attractive as compared to the tester role. Those individuals who selected the response option 'Maybe', are in total agreement by pointing out that they would perform as testers if no other job offer is available.

4 Discussions

According to the results, it can be pointed out that the most recurrent PRO item among respondents was that testing tasks make the tester focus on details. Furthermore, this statement is given most frequently as the second priority CON item. Consequently, the authors believe that focusing on details is a key competence of the role of tester; and at least one core item that software engineers take into account when considering a role.

As it can be seen in previous section, the most common response given as first and second level of importance for both PROs and CONs refers to the testers' interactions with project team members. Therefore, it is accurate to state that the role of tester involves a strong human interaction among software practitioners. The remaining PROs show that subjects identify the role as a way for better approaching a new project when they are newcomers; so, they see the opportunity to improve their soft skills and the demands of creativity as a positive item for their careers.

In addition, it was found some respondents perceive the following two aspects as constructive: the presence of the role of tester through all project stages, and the fact that testing activities provides full access to the project scope, modularization and integration strategy in a short period of time; authors believe that respondents may perceive the role of tester as a professional growth opportunity. Nevertheless, further empirical studies to investigate this aspect of the study needs to be conducted.

During the analysis of the cons, it was noted that the most frequently cited CON item was in regards to team members becoming upset with the tester due to the review of the team members' builds. This could be due to the fact that testers may be accustomed to auditing and criticizing the work of others. Also, the preference for roles other than tester due to their general acceptance constitutes a conclusive statement regarding the unpopularity of the role of tester among respondents.

Some other surprising, if not disturbing, reasons can be found when examining the cons, as it is the case of male-centered issues regarding who is better able to perform the role of tester; males or females. Consequently, a sexist stereotype surrounding individuals performing the tester role was noted; authors finds interesting that the career choices of female subjects are influenced by either male individuals' perceptions, or by their own sexist stereotyped perceptions. The first hypothesis could be the result of women's responses to men believing that the tester role is better performed by women. If added to the unpopularity of the role may re-enforce the devaluations of women; thus women become inclined to do more technical roles in order to prove their equality. On the other hand, the second hypothesis presents the possibility of some sort of feminism that incline women to seek general approval by taking up roles of assumed difficulty in the same way that an attitude of manliness inclines men to do so. Further studies into this area of gendered perceptions around software engineering roles are warranted.

Additionally, it must be highlighted in both PROs and CONs that there were individuals referring to abstraction, creativity and detail-oriented skills when considering to choose the tester role. When comparing percentages, it may be inferred that some of the individuals in the study offered those statements as both PROs and CONs for taking up a testing career in the software industry. The authors agree that to express those statements in either PROs or CONs is a deeply personal point of view, and it can be strongly influenced by the surroundings, the exact period of time when the decision is made, and even the individual's frame of mind. The present study demonstrates that the subjects perceive the role of tester demands: abstraction, creativity, and detail orientation. Further research is needed to clarify whether these adjunctions are exclusive to the role of tester or also conferred upon other roles within software development.

It is surprising that only 40% of subjects, not even a third of the number of participants in the study, refer to the lower wages of the tester role (40% is more than one third). As if the remuneration was not a factor in the career decision process.

When questioned about their chances of taking up a career in software testing, the percentage of negative responses nearly tripled the percentage of positive responses toward the role. An overwhelming quantity of respondents picked the option 'No' as a response. These results concur with prior studies [12, 19], which point to the tester role as one of the less popular roles among others such as project manager, analyst, designer, programmer, and maintenance.

The reasons given to support negative bias towards the tester role are mainly linked to personal preferences, followed by the perception of the role as less attractive than others, as mentioned earlier. On the other hand, supported reasons from those positives response taking up a testing career are related to: similar to personal preferences for most of the subjects in the study, the role's commitment to quality, and the opportunity the role offers to have an overall view of the project in a short time.

The reasons to consider or not a position as tester are directly related to the registered PROs and CONs for question number one in the questionnaire. Meanwhile, reasons supporting the 'Maybe' choice relates to the availability of better job offers, they also may reflect personal preferences and attraction to the role. Furthermore, it is the authors' belief that the tester is a role with more social connotations than technical inclinations, as reflected by the findings of the present investigation.

5 Limitations

This study has some intrinsic limitations. Although respondents represent a sample of currently active software practitioners throughout the country, their origin was not recorded. Such data could be used to determine if certain demographic zones are more inclined to certain kind of jobs, specifically testing tasks. Furthermore, the professionals were not asked to state their current role. By doing so, it would be possible to compare the correlation between people's current duties and their career related preferences; primarily those performing as testers at the time the questionnaire was conducted. Nevertheless, these limitations did not compromise the achievement of the study's main goals.

6 Conclusions

The top three PROs reasons for choosing a career as a tester were: (1) testers have more interactions with the project team, after the analyst role; (2) Particularities of testing tasks make the software tester focus on details; (3) Testing activities provide an overview of the project scope, modularization, and integration strategy in a short period of time. With item number 2 being the most frequent PRO item of all within the responses.

The top three CONs reasons for choosing a career as a tester were: (1) Too many detail oriented skills are demanded from software testers; (2) Roles other than tester enjoy more acceptance among software engineers; (3) Other project team members may be upset due to the tester's findings when their releases are reviewed. With item number 3 being the most frequent CON item of all within the responses.

Given the fact that pro and con more frequently reasons within the responses were: the particularities of testing tasks make software tester to focus on details and how upset a team member could become due to the testers' findings about his/her builds; these are the decisive items for software practitioners when making testing a career of choice.

Among the subjects of the study, the main reasons for taking or not taking up a testing career in the software industry were strongly related to individual preferences and the availability of a job offer involving a more attractive or a better-paid role. Some see the tester role as an opportunity to know quickly what the project entails, and see the benefits of several automated tools that support the tester's performance in their role. In addition, strong human interactions are attributed to the tester role, making it a role with more social connotations than technical implications.

The present study supports prior findings of the unpopularity of the role of tester, positioning the tester role among those less favored by software practitioners. In addition it prompts further research into: (1) why some software engineers considered that testing activities provide a full background of project scope, modularization and integration strategy in a short period of time and the presence of the role of tester through all project stages as a pro; (2) if and at what level, the required abstraction, creativity, and detail orientation are attributed only to the tester role or distributed among the other roles in software engineering; and (3) sexists issues surrounding software engineering roles.

Appendix: Survey Questions

We are very grateful to all participants for dedicating their time and attention to our study.

1. What are the three PROs (in the order of importance) for taking up testing career?
   (a)
   (b)
   (c)

2. What are three CONs (in the order of importance) for taking up testing career?
   (a)
   (b)
   (c)

3. What are chances of my taking up testing career?

Certainly Not No Maybe Yes Certainly Yes

Reasons:

4. Gender (optional):

5. GPA (optional):

References

1. The Standish Group (2015) Chaos Report 2015, Technical Report, Standish Group International
2. Summary of NIST Strategic Planning Results (2001) National Institute of Standards Technology Report
3. Zhivich M, Cunningham RK (2009) The real cost of software errors. IEEE Secur Priv 7 (2):87–90
4. Gleick J (1996) Little bug. Big bang. The New York Times
5. Geppert L (2004) Lost radio contact leaves pilots on their own. IEEE Spectr 41(11):16–17
6. Jee C (2016) Top software failures 2015/2016: Amazon, RBS, Starbucks—the worst software glitches this year, computer world UK, https://qesp.org/article/top-software-failures-20152016-amazon-rbs-starbucks-the-worst-software-glitches-this-year/. Accessed 21 Nov 2016
7. Burnstein I, Homyen A, Suwanassart T, Saxena G, Grob R (1999) A testing maturity model for software test process assessment and improvement. Softw Qual Prof 1(4):8–21
8. Koomen T, Pol M (1999) Test process improvement. Pearson Education, Essex
9. Kasurinen J, Taipale O, Smolander K (2010) Software test automation in practice: empirical observations. Adv Softw Eng 1:1–18
10. Graham D, Fewster M (2012) Experiences of test automation: case studies of software test automation. Addison-Wesley
11. Capretz LF, Varona D, Raza A (2015) Influence of personality types in software tasks choices. Comput Hum Behav 52:207–214
12. Astigarraga T, Dow EM, Lara C, Prewitt R, Ward MR (2010) The emerging role of software testing in curricula. In: IEEE transforming engineering education: creating interdisciplinary skills for complex global environments, pp 1–26
13. Rooksby J, Rouncefield M, Sommerville I (2009) Testing in the wild: the social and organizational dimensions of real world practice. Comput Support Coop Work 18(15–16):559–580
14. Beecham S, Baddoo N, Hall T, Robinson H, Sharp H (2008) Motivation in software engineering: a systematic literature review. Inf Softw Technol 50(9–10):860–878
15. Hazzan O, Tomayko J (2004) Human aspects of software engineering: the case of extreme programming. In: Extreme programming and agile processes in software engineering. LNCS, vol 3092. Springer, pp 303–311
16. Capretz LF (2014) Bringing the human factor to software engineering. IEEE Softw 31 (2):102–104
17. Shah H, Harrold MJ (2010) Studying human and social aspects of testing in a service-based software company: case study. In: Proceedings of the ICSE workshop on cooperative and human aspects of software engineering (CHASE 2010), pp 102–108

18. Santos RES, Magalhaes CVC, Correia-Neto JS, Silva FQB, Capretz LF, Souza REC (2017) Would you like to motivate software testers? Ask them how. In: ACM/IEEE 11th international symposium on empirical software engineering and measurement (ESEM industry track 2017), pp 95–104
19. Capretz LF, Varona D, Raza A (2015) Influence of personality types in software task choices. Comput Hum Behav 52:373–378
20. Capretz LF, Waychal P, Jia J, Varona D, Lizama Y (2009) Studies on the software testing profession. In: 41st ACM/IEEE conference on software engineering (ICSE–SEIP 2019), 2 pp
21. Weyuker EJ, Ostrand TJ, Brophy J, Prasad B (2000) Clearing a career path for software testers. IEEE Softw 17(2):76–78

Yadira Lizama is a Ph.D. student of Hispanic Studies at the Faculty of Arts and Humanities, Western University. She graduated in Software Engineering at the University of Informatics Sciences (UCI) in Havana, Cuba (2009). She has 10 years of experience working for the software industry as a tester, programmer, analyst, and lately, as a software leader in projects related to digital innovation, digitization, big data, and business process management. She also has taught Software Engineering and Software Management for 4 years at UCI. She is a co-founder of the Research Group of Computer Science and Law. She is Research Assistant and Project Manager at Cultureplex Lab (https://cultureplex.ca), where she develops several projects and partnerships with companies in the London area. She currently holds a Mitacs Accelerate Fellowship. Her research interests are focused primarily on software development, network analysis, data science, and artificial intelligence to solve practical issues on the Transitional Justice and Digital Humanities fields. She can be reached at yadiralizama@gmail.com.

Daniel Varona is a Ph.D. student working at Cultureplex, in the Faculty of Arts and Humanities at Western University. He graduated in Software Engineering at the University of Informatics Sciences (UCI) in Havana, Cuba (2008). He has 12 years of experience working for the software industry as a tester, analyst, and quality manager in projects related to digital innovation, eGovernment, big data, and business process management. He also has taught Software Engineering, Software Management, Research Methods, and Quality Assurance for 8 years at UCI. His current research interests are in human factors on software engineering, empirical studies of software engineering, and human–computer interaction, and more recently, he is working on Machine Learning's limitations avoiding automation of bias. He is member of the Ibero-American Network of Engineering Project and Co-founder of Research Group of Informatics and Laws in E-Government Software Development Institute at UCI. He can be reached at dvaronac@uwo.ca, and further information about him can be found at http://www.danielvarona.ca.

Pradeep Waychal is Director at CSIBER (www.siberindia.edu.in), India that has computing, management, social works, environment, commerce, and economics schools and chairs Guruji Education Foundation (www.gurujifoundation.in) that provides holistic education to the underprivileged students. His research interest are at the intersection of human sciences and software engineering, engineering education, and innovation management. He has done Ph.D. in developing innovation competencies for Information System Businesses from IIT Bombay, India. He is a senior member of IEEE, a member of ACM and ASEE, and a life member of CSI (Computer Society of India). He was associated with organizations such as ISO-JTC1 SC7, NASSCOM, and ISBSG. He has 23 years of industry and 13 years of academic experience. He can be reached at pradeep.waychal@gmail.com.

Dr. Luiz Fernando Capretz has vast experience in the software engineering field as practitioner, manager, and educator. Before joining the University of Western Ontario (Canada), he worked at both technical and managerial levels, and taught and did research on the engineering of software in Brazil, Argentina, England, Japan, and the United Arab Emirates since 1981. He is currently a Professor of Software Engineering and Assistant Dean (IT and e-Learning), and former Director of the Software Engineering Program at Western. He was the Director of Informatics and Coordinator of the computer science program in two universities in Brazil. He has published over 200 academic papers on software engineering in leading international journals and conference proceedings, and co-authored two books: *Object-Oriented Software: Design an Maintenance* published by World Scientific, and *Software Product Lines* published by VDM-Verlag. His current research interests are software engineering, human aspects of software engineering, software analytics, and software engineering education. Dr. Capretz received his Ph.D. from the University of Newcastle upon Tyne (U.K.), M.Sc. from the National Institute for Space Research (INPE-Brazil), and B.Sc. from UNICAMP (Brazil). He is a senior member of IEEE, a voting member of the ACM, a MBTI Certified Practitioner, and a Certified Professional Engineer in Canada (P.Eng.). Dr. Capretz is currently spending his sabbatical leave as Visiting Professor of Computer Science at New York University in Abu Dhabi, United Arab Emirates; and part-time Visiting Professor of Computer Science at Universiti Teknologi PETRONAS in Malaysia. He can be reached at lcapretz@uwo.ca, and further information about him can be found at http://www.eng.uwo.ca/people/lcapretz/.

A Study on Reliability of Rotors Using XLrotor

H. N. Suresh, N. Madhusudan, D. Saravana Bhavan and S. B. Murgayya

Abstract The work is focused on computation of rotating machines in XLrotor in order to predict the failure due to disk offset and to check reliability. Reliability of rotary machines is not dependent on static design stress but also dynamic forces generated during operating speeds. These rotating machines are subjected to various forces such as misalignment, bend shaft, looseness, imbalance and so on. Though all influencing parameters are diagnosed, still the problems in rotating machines are faced when operated on site (Ambience condition etc.). These problems are severe and cause failure in system within no time. The on-site technological tool helps the plant to know the health of rotating machine, but to avoid resonance and shifting/bypassing critical speeds the simulation tool XLrotor is benefited compared to FEM. To check reliability of rotating machines, the XLrotor computational tool is considered optimum in incorporating the effect of mass, stiffness, inertia and imbalance effectively. The simulation based methodology is used for modeling and analysis for the simple and complex rotors. The analysis in XLrotor due to disk offset on system model shows the vibration level (failure) of rotor and load acting on bearing due to imbalance. The vibration results determine the impact of disk offset on rotor model and its performance.

Keywords Reliability engineering · Rotor dynamics · Critical speeds · Stiffness and XLrotor

---

H. N. Suresh · S. B. Murgayya (✉)
Department of Automobile Engineering, Dayananda Sagar College of Engineering, Bengaluru, India
e-mail: murgayya-au@dayanandasagar.edu

N. Madhusudan
Diagnostic Engineers, Bengaluru, India

D. S. Bhavan
School of Mechanical Engineering, Dayananda Sagar University, Bengaluru, India

D. R. Karanki et al. (eds.), *Advances in RAMS Engineering*,
Springer Series in Reliability Engineering,
https://doi.org/10.1007/978-3-030-36518-9_8

1 Introduction

Vibration analysis is a technique that is used to look for irregularities and monitor significant changes from the vibration signature. The signature of any rotating machine is characterized by variations of frequency, amplitude and intensity. Although a number of sophisticated techniques are existing and used in modern era, but the two most common fundamental methods for presenting vibration data are time and frequency domain. Machine faults can also increase the vibrations and excite them. The vibration behavior of machine due to natural frequency and imbalance is one main important criteria in rotating machinery which should be thoroughly studied analyzed in design stage [1]. In modern engineering era, the design and development of rotating systems is at higher critical speeds due to its light weight. In rotor dynamic system, the accurate prediction is vitally important at design stage. This prediction is based on dynamic behavior of rotor system and rate of failure. The reliability sensitivity is analyzed with respect to stiffness of shaft and damping, stiffness of stator and damping, clearance in radial and stiffness of stator in radial direction [2]. The lateral and torsional rotordynamic analysis is carried out by the machinery industries in order to investigate and mitigate the risk of such critical speed problems [3]. In the condition of bending vibration, the natural frequencies are obtained and the functions of frequencies with variables are matched based on the method of the artificial neural network (ANN) technique. The frequency reliability sensitivity analysis method is provided with a criterion that the absolute value of the difference between the natural frequency and forcing frequency, the reliability mode and the system safety probability is defined [4]. Through the published literature it is well known that model reduction existing methods for large scale non rotating dynamics problems are very practical with symmetric matrices [5]. For rotors on bearing stiffness and damping, the rotor models involve relatively simply supported beam condition representation. These models do not need reduction the rotors systems analysis includes rotor critical speeds with damping, imbalance response and stability [6, 7]. Before studying and analyzing the complex rotors, it is important to know how Jeffcott rotors behave under different rpm and its critical speeds. The resonance occurs in rotors due to various parameters and this can be controlled by design modifications. The critical speeds can be shifted by altering inertia mass, disk configurations and stiffness and damping properties of rotors. The influencing parameter that causes failure in real time in rotordynamic is due to imbalance. The imbalance response on Jeffcott Rotor is analyzed using analytical and Dyrobes simulation method. Here, when rotor mass center M does not lie along the axis of rotation C of the rotor an unbalance occurs, as illustrated in Fig. 1a. By statically (static condition) the unbalance may be detected if the rotor is located on supports with appropriately low friction. Due to imbalance, this will cause the rotor to rotate away from axis of rotating. Figure 1b shows the unbalanced rotor in downward position [8, 9].

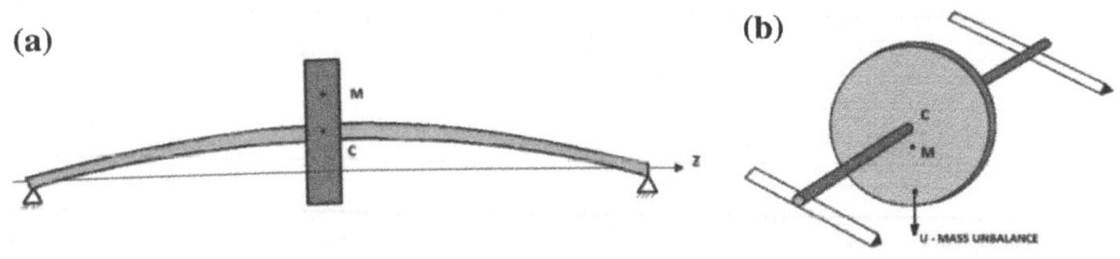

Fig. 1 **a** Jeffcott rotor. **b** Jeffcott rotor unbalance mass

Based on the assumption that the rotor disk does not affect the rigidity of the massless shaft, the lateral bending stiffness at the axial center of a simply supported beam is given by

$$K_s = \frac{48EI}{L^3} \tag{1}$$

where E is the beam elastic modulus, L length between the two bearings and I is area moment of inertia of shaft. For a cylindrical shaft of uniform cross section with diameter D, the area moment of inertia is given as,

$$I = \frac{\pi D^4}{64} \tag{2}$$

Additionally, we also assume that damping acting on the lateral motion of the disk at the rotor mid span which is relatively small and the damping constant is given by C_s this viscous damping is combination of the shaft structural damping, fluid damping and the effective damping added by the end bearings due to flow in turbo machines [8, 9]. Considering a case study of Jeffcott rotor for analysis to find the rotor unbalance eccentric city force.

Given a rotor mass has an unbalance of, U = 10 oz—in (7.064 N-mm) and weight of W = 500 lbf (2225 N) speed N = 8000 rpm

To find the unbalance eccentricity e_u, shaft angular velocity ω and rotating force on a rigid rotor F_u.

$$e_u = \frac{U}{W} = \frac{10\,oz - in}{16 \times 500\,oz}$$

$$e_u = 0.00125 \text{ in} = 1.25(0.0318\,\text{mm})$$

The rotor mass is

$$M = \frac{W}{g} = 1.295\frac{lbf - sec^2}{in}$$

The angular velocity is,

$$\omega = \frac{2\pi N}{60} = 8000\frac{rev}{min} \times 60\,\sec \times 6.28\frac{rad}{rev} = 837.7\frac{rad}{sec}$$

Force on a rigid rotor F_u,

$$F_u = me_u\omega^2 = 1.295 \times 0.00125 \times (837.7)^2$$

$$F_u = 1135\,\text{lbf}(5050\text{N})$$

It is easily seen that, the rotating unbalance force can be quite large for a Jeffcott rotor. This author describes an unbalance force acting on simple rotor and one can analyze the force acting on complex rotors with supporting conditions [8, 9]. The methods of reduction are discussed comprehensively applied to control systems and in control theory. In this paper, they give a brief summary for controls [10]. The author gives a summary of methods used in analysis of structural mode by Finite Element Method (FEM) [11]. The current period of reduction methods in rotor dynamics reduces the complexity of problem in large rotors [12]. The author suggests using advanced validation tools for in-depth analysis of large machines, design modifications at the design stage to achieve higher operating speed, torsional design to ensure the smooth operation in industry and finally machine cost efficiency [13]. The following Table 1 shows the use of simulation to solve the cases with design modifications of the systems in torsional mode of vibration. The below case studies [13] shows causes of failure in complex rotors and its behavior. These case studies demonstrate the power of simulation and experimental method.

The development of the rotor crack changes parameters significantly, which clearly describes the rotor behavior during the run-up period [14]. Reliability depends not only on the static design stress, but also on the dynamic forces under operational conditions [15]. For Jeffcott Rotor, the analysis is carried out using experimental and FEM method and confirms the moment of the forward and reverse whirl through the Campbell diagram [16]. In this Dynamics R4, the system functions, elements and algorithms are shown and advanced rotor dynamics analysis and simulation are carried out [17]. The focus is on capturing the data and analyzing the stress waves in rotating equipment by measuring impacts, fatigue and friction through theoretical and experimental approaches [18]. The work is on the Vibration Analysis Expert System (VAES), which is analyzed using Matlab simulation for diagnosis and prognosis [19]. On rotor bearing system models, XLrotor tool can perform absolutely for any type of rotordynamic analysis. XLRotor is suitable for the use of a wide range of rotating machinery in design, maintenance assessment and evaluation. XLrotor is a proprietary computing engine based on sophisticated algorithms for mathematical modeling. The XLrotor tool is classified as intelligent, fast and easy to use, adding an advantage to rotordynamics for modifications. This results in, changes in critical speeds and changes the resonance during the design stage their by reducing costs and downtime [20]. The sensitivity analysis of rotor is to determine how the independent variables (mass, stiffness, inertia, disk configuration, material properties and bearing stiffness and damping properties) impact

Table 1 Case studies to solve machinery vibration using rotordynamic analysis

S. no.	Cases	Method to analyze the vibrations	Corrected measures
1	Case 1—Fixing a critical speed interference problem	Finite element method and experimental method	Re-design the bearings from vibration response of the compressor
2	Case 2—Aero-derivative gas turbine instability	Finite element method and experimental method	A hybrid 3-pad, 2-pocket pressure dam bearing was conceived
3	Case 3—Solving a critical speed problem with a damper bearing	Experimental method	Bearings fabrication and installation of the turbine which is rebuilt
4	Case 4—Changing the excitation mechanism to avoid torsional vibration	Experimental method	New impeller installation with four vanes by slightly moving and shifting the pressure pulsations to avoid torsional resonance
5	Case 5—Changing the couplings to avoid torsional vibration	Experimental method and virtual simulation	By changing stiffness of coupling the easiest and quickest way to shift torsional resonances
6	Case 6—Designing the system to accommodate torsional excitation and stress	Experimental method and virtual simulation	Re-designed the system to sustain the known torsional resonance and fatigue modes
7	Case 7—Changing the system inertia to avoid torsional vibration	Experimental method and virtual simulation	To avoid torsional resonance and move away from the interfering operating speed

dependent variables (critical speeds). By several metric independent variables a single metric dependent variable is predicted. The main aim of the multiple linear regression traditional method is to use the completely independent parameter whose values determine the single dependent value. Each independent variable is measured by the logistic regression analysis procedure to achieve maximum prognostication from the parameter set. The XLrotor is ideal for design, reliability, maintenance and evaluation and also checks for a wide variety of failures in rotating machines. The benefits of XLrotor are as follows,

- Easy to learn and gets well interfaced between component tabs.
- Saves time in automated calculations and formulation.
- XLrotor is powerful and accurate (model scale reduction factor avoided, calculation speed and high precision results delivered).
- Easy to customize.

2 Applications of Rotordynamics Using Simulation Tools

The vibration significantly influences the physical functioning of rotating machines. The simulation based methodology plays vital role in design, development, investigation and evaluation. This literature shows importance of simulation based research using XLrotor and FEM for predicting failure and to enhance the performance.

From the Table 2, it is observed the Finite Element Modeling is exhaustively used for analysis compared to XLrotor. In our research work, for vibration analysis the XLrotor tool is used because of its benefits over FEM as highlighted in introduction. This work clearly emphasis on usage of XLrotor in carrying out the rotordynamic analysis for lateral and torsional vibration [40]. The cost of XLrotor tool is high compared to FEM tool. Due to its features and benefits, the XLrotor tool is used exhaustively in the industries for development of large scale rotary machines like steam turbines and gas turbines which is difficult and time consuming in FEM [43].

3 Simulation Case Studies

3.1 Working Methodology

For Rotordynamics analysis, the XLrotor on system model with bearings works well in rotating machines in determining the failure results compared to other simulation methods. The procedure of using XLrotor is to perform a complete rotordynamic analysis that is outlined in Fig. 2.

3.2 Analysis and Results of Case Study Using XLrotor

The rotor model with disk configuration is modeled in XLrotor for analysis as shown in process flowchart Fig. 2, Spectra Quest's Machinery Fault Simulator (MFS-MG) Magnum is used to introduce, simulate and investigate faults in rotating machine. The MFS-MG equipped with a resonance kit is the ideal instrument for gaining practical experience in rotating machine and learning techniques to mitigate resonance. Rotating components consist of rotor and bearings, beam coupling and spinning shaft supported with two bearings of rolling elements separated 28.5 inch apart. Aluminum disks at different places are mounted on the shaft and the imbalance is intentionally introduced on disks to determine the amplitude at critical speeds (1X, 2X etc.). For analysis [1], the shaft diameter of 0.625 and 0.5 inch with two aluminum disks is considered. The results determine variation in critical speeds and imbalance response with configuration of both disks close to and away from

Table 2 Applications of rotordynamics

S. no.	Case studies	Methods incorporated	Worked on
1	Rotordynamic analysis using XL rotor [1]	Experimental and XL rotor	Shift of critical speeds by changing disk configuration
2	Rotordynamics considerations in refurbishing the turbomachinery [22]	XL rotor	Reviewed to study between theoretically and virtual to know unbalance, instability, requires light weight couplings and balancing on the overhung to control the forces
3	Transient response of rotor on rolling element bearings with clearance [23]	Analytical system procedure in COBRA AHS	Rotor behavior due bearing clearance is studied and analyzed
4	Rotordynamic modeling and analysis [24]	Analytical system models	Rotordynamic analysis.
5	Rotordynamics [25]	Analytical models	Review on rotordynamics and suggestion virtual simulation
6	Rotor dynamic analysis of steam turbine rotor using ANSYS [26]	FEM technique	Unbalance response from steam turbine and safe working under bearings valves and rotor loads
7	3D solid finite element modeling and rotordynamics of large rotating machines: application to an industrial turbo engine [27]	FEM technique	Unbalance response
8	Dynamic analysis of rotor-bearing system for flexible bearing support condition [28]	FEM technique	Campbell diagram for FW and BW and modal analysis
9	Solid model rotor dynamics [29]	FEM technique	Campbell diagram with forward and backward Whirl
10	Static and dynamic analysis of lathe spindle using ANSYS [30]	FEM technique	Static and transient analysis, Campbell diagram for FW and BW
11	A comparison of the finite element modeling methods for the natural frequencies computation of stepped shaft [31]	FEM technique	Modal analysis
12	Rotor dynamic analysis of 3D-modeled gas turbine rotor in ANSYS [32]	FEM technique	Campbell diagram, modal analysis and ANSYS overview

(continued)

Table 2 (continued)

S. no.	Case studies	Methods incorporated	Worked on
13	A time-domain methodology for rotor dynamics: analysis and force identification [33]	FEM technique	The propfan engine developed in the scope of the European project DUPRIN time domain identification methodologies for simple rotating systems and aeronautics
14	Introduction of rotor dynamics using implicit method in LS DYNA [34]	FEM technique	For rotor dynamics analysis and force identification time-domain methodology and suggested a rotor dynamics time- domain methodology: analysis and strength identification
15	Analysis of dynamic characteristics of the main shaft system in a hydro-turbine based on ANSYS [35]	FEM technique	Modal analysis and calculates the critical speed of rotation and dynamic analysis and a foundation for the design or improvement
16	Unbalanced response and design optimization of rotor by ANSYS and design of experiments [36]	FEM technique	Critical frequency of rotor, unbalance response and DOE
17	Dynamic analysis of a spindle-bearing system based on finite element method [37]	FEM technique	Dynamic behaviors of spindle-bearing systems and unbalance response
18	Rotor dynamic analysis of RM12 Jet engine rotor using ANSYS [38]	FEM technique and DyRoBeS	Understanding, modeling, simulation and post processing techniques
19	Analysis of rotor dynamics acceptance criteria in large industrial rotors [39]	FEM technique	Campbell diagram with gyroscopic effects
20	A new dynamic model of rotor systems [41]	Experimental and FEM technique	Lateral and torsional vibration of rotor shaft with disk
21	Vibration analysis of functionally graded rotating shaft system [42]	Analytical and FEM technique	Comparison of stainless steel and Alumina Oxide material mode shapes and frequency with Campbell diagram

bearing housing. The author has not highlighted on disk offset which is considered as one of the factor that leads to change in critical speeds of rotor and its performance. The rotor is analyzed for following specifications as listed below [1],

- Motor: Marathon "Four In One" CAT No—D 391
- Shaft diameter 0.625 (inch)
- Rotor bearings span 28.5 (inch)

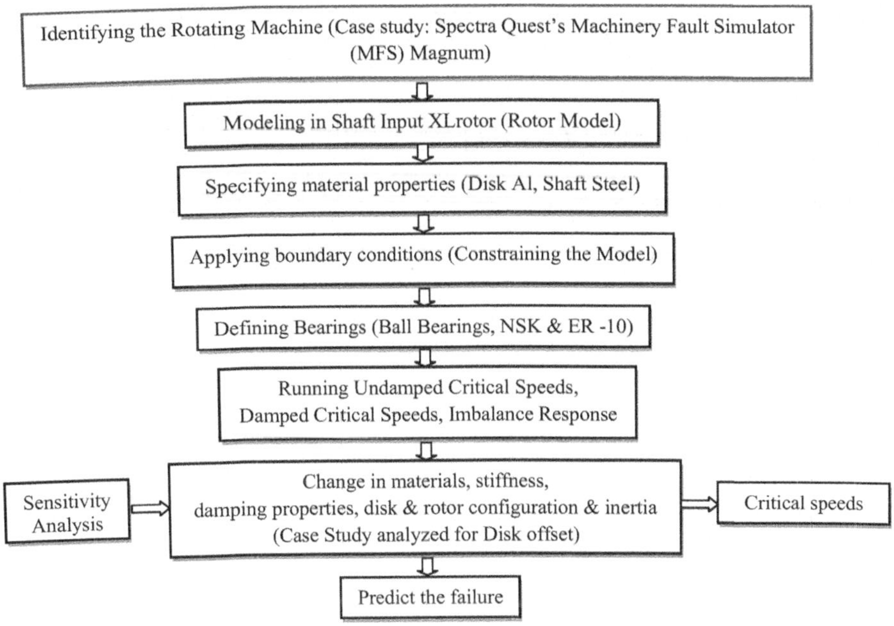

Fig. 2 Process flowchart for lateral analysis

- Disks diameter (aluminum): 6 (inch)
- Disks thickness 0.625 (inch)
- Rolling element bearings used for the 5/8″ shaft NSK 6203 (Left)
- Rolling element bearings used for the 5/8″ shaft ER-10 K (Right).

The first step in analysis is to input Eigen Values Analysis Speeds, i.e. the maximum and minimum rpm of rotating machines and undamped critical speeds stiffness values of bearings to constrain the model and to check whether the model is intentionally constrained. The boundary condition of rotor is free-free to fixed i.e. soft to rigid bearings to calculate and estimate force required in constraining the model. The shaft input sheet is to model the rotor with material properties, the equation below shows added mass, polar moment of inertia and transverse effect. In case model is loaded with extra added mass and inertia properties on disks or shaft. The speed factor should be 1 for all beams making up the first level and 1.4 for beams in the second level, i.e. when they are inter connected the secondary rotor spins about 40% faster than primary rotor. The spreadsheet of the shaft input station that summarizes all the properties of the beam used as an input to build the mass and stiffness matrix in each station based on Timoshenko's beam theory [20, 21] and in the model, the transverse shear effect is considered as a parameter (Ø) as defined in Eq. 4 and is given by,

$$m_i = \partial\pi \frac{(D_o - D_i)}{2} l_i \quad (1)$$

$$I_p = \frac{1}{2} m_i \left(\frac{(D_o - D_i)}{2}\right)^2 \quad (2)$$

$$I_t = m_i \left[\left(\frac{l_i}{12}\right) + \left(\frac{(D_o - D_i)}{16}\right)\right] \quad (3)$$

$$\emptyset = \frac{12EI}{kGAl^2} \quad (4)$$

where:

D_o Outer Diameter
D_i Inner Diameter
E Material Young's Modulus (Steel and Al)
G Material Shear Modulus
I Inertia Moments
A cross- section area
l length of beam
k is a shape factor equals to 0.883 for poisson ratio (υ) = 0.3.

The Geo plot in sheet shows the 2D view of rotor and location of bearings as shown in Fig. 3, the bearings located shows nothing on damping and stability.

Running the analysis (Undamped Critical Speed Analysis) of XLrotor i.e. (Considering bearing stiffness and neglecting bearing damping this is to determine the free—free to fixed—fixed condition of end bearings). The UCS analysis in

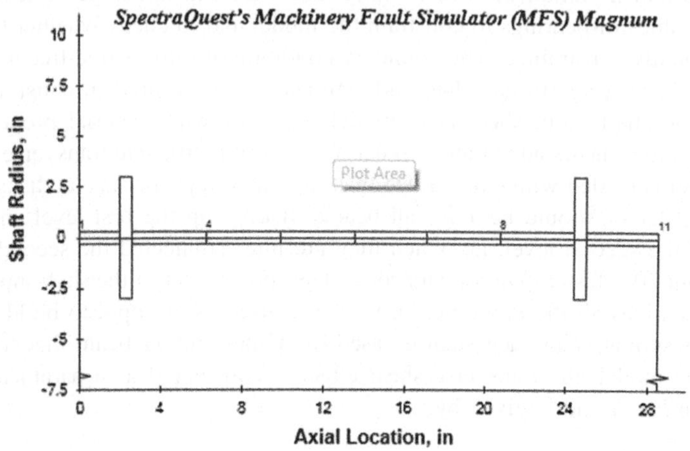

Fig. 3 Geo plot of rotor [1]

Table 3 Stiffness and critical speeds

Stiffness (lb/in)	Critical speeds in cycles per minute (cpm)			
	cpm 1	cpm 2	cpm 3	cpm 4
10000	3233.172	9294.767	17069.07	38837.41
31622.78	3313.341	10578.08	20743.59	42125.34
100000	3339.582	11092.6	22907.82	44668.65
316227.8	3347.971	11269.06	23813.28	46055.88
1000000	3350.633	11326.35	24130.43	46615.25
3162278	3351.475	11344.62	24234.18	46808.48
10000000	3351.742	11350.41	24267.34	46871.41

Table 3 also shows how stiff bearings are and what speed is required to achieve them.

The rotor model is analyzed for the condition from soft to rigid bearings. When the shaft is supported with stiffness of 10000 lb/in the critical speed is observed at 3233.172 cpm and for 10000000 lb/in the critical speed is at 3351.742 cpm. The results shows critical speed is directly proportional to stiffness as shown in Fig. 4. As stiffness of rotor increases the natural frequency of rotor also increases.

The critical speed factors depend on the steadiness of the shaft and its support, the total mass of the shaft and the attached parts, the weight imbalance in direct relation to the axis of rotation and the amount of damping in the system. The computed eigenvalues are automatically synchronous (i.e., whirl frequency = rotor speed). Eigenvalue frequencies normally vary with rotor speed because of gyroscopic effects. The stiffness and damping properties of bearings and type of bearings are required in bearing sheet to compute the Campbell diagram, which determines the values at zero spin speed and rotated till maximum rpm.

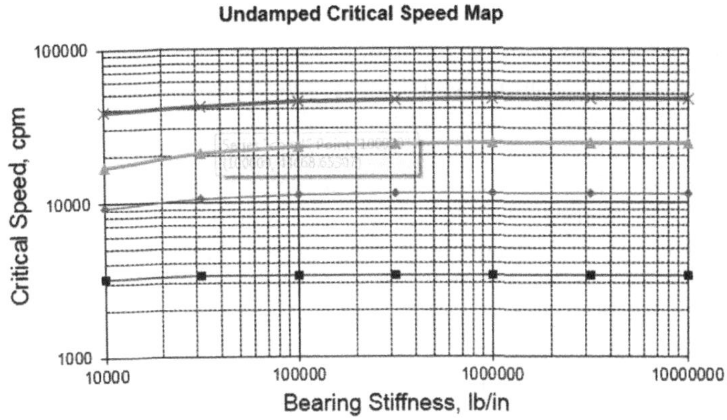

Fig. 4 Critical speeds versus bearing stiffness

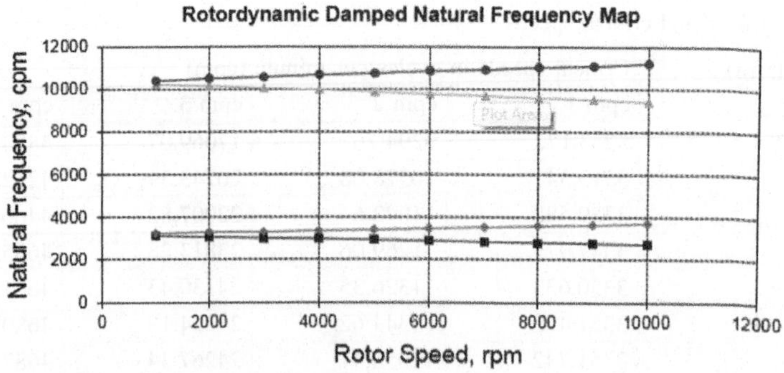

Fig. 5 Natural frequency versus rotor speed

The 1st damped critical speed is determined at 3400 cpm which shows good agreement with paper referred [1]. As the rotor speed increases the forward whirling increases and backward whirling decreases with natural frequency as shown in Fig. 5. The graph shows rotating frequencies with rotating speeds. The Campbell diagram shows the system's vibration frequencies in different operating RPM. A precise benefit of solid models is the presence of spin softening, stress stiffening and temperature effects in the rotor model analysis which are not considered in the conventional beam element modelling [29]. The effect of spin softening has a significant influence on the modes of backward whirl and the effect of stress stiffening on the modes of forward whirl [29]. The traditional Campbell diagram uses a rotor motion equation to express the external force as a periodic function caused by the rotational frequency. This function is mapped to a graph allowing you to analyze a system's vibration characteristics. Roots Damped worksheet displays the results of a damped analysis for lateral rotor models. The analysis results are a row of roots showing only the imaginary parts of the eigenvalues versus rotor speed since the real parts are always zero. To run the synchronous imbalance response analysis, we need to specify the intentional imbalance distribution and the speeds at which to generate the response and also the way of output. Imbalance is created by adding mass of 0.454 gm-in on the disk to determine response of imbalance and imbalance effects on bearing load. The graph shown in Fig. 6a,b reflects the imbalance response at 1st critical speed.

The rotor is connected to Marathon "Four In One" CAT No—D 391 [1] and rigidly supported which falls under Group 3 ($P \geq 15$ kW) which states displacement above 56 microns or 2.20 mils is considered as case producing severe vibrations [44]. A small amount of (0.454 gm-in) imbalance intentionally distributed on station 2 and station 9. By disk offset of 1.5 inch on the rotor, the amplitude is found to be increased twice on station 9 (disk 2) compared to station 2 (disk 1). The small amount disk offset of 1.5 inch majorly impacts on failure of rotor which is vibrating beyond the acceptable limits of vibration severity criteria [44] and also increase load on right end of bearing. It also determined form paper [1] the critical speed is found to be increased with increase in shaft diameter. Based

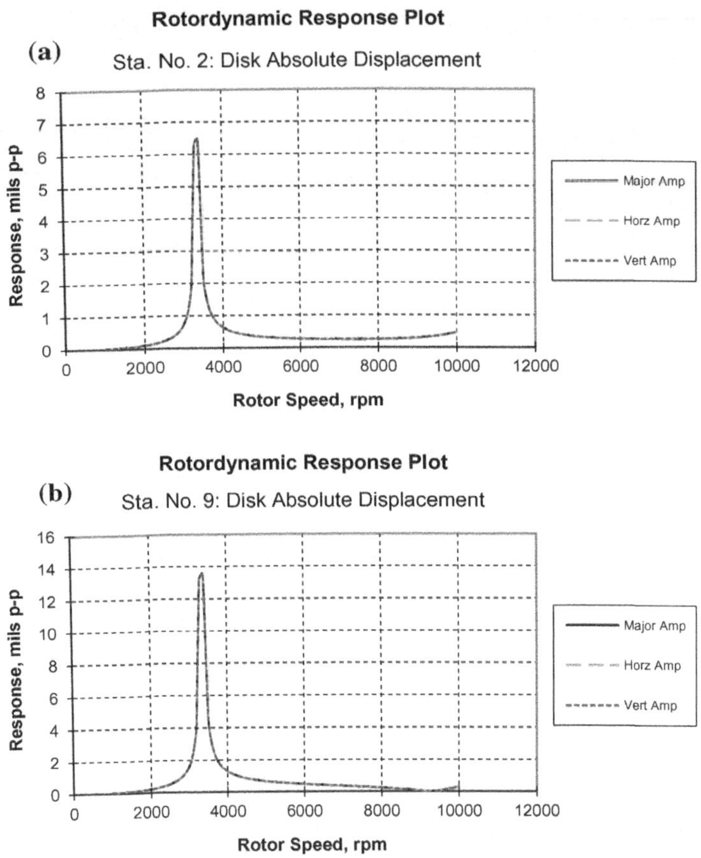

Fig. 6 **a** Imbalance response at station 2 (AL Disk) geo plot. **b** Imbalance response at station 9 (AL disk) geo plot

Table 4 Imbalance response analysis results

S. no	Stations	Displacements	
1	2	165.1 microns	6.5 mil
2	9	330 microns	13.5 mil

Table 5 Bearing load response analysis and results

S. no	Stations	Loads
1	1	561 bf
2	11	681 bf

on studies, the imbalance response and load on bearing vary due to offset of disk as shows response in Table 4 and load on bearing in Table 5.

The amplitude of imbalance response is doubled at station 9 due to offset of disk. Since the values are beyond acceptable limits, this is considered as severe and catastrophic failure of rotor can occur.

The bearings are designed for high load carrying capacity, i.e. static load of 730.62 lbf and dynamic load of 1432.03 lbf at station 1 and station 11. Table 5 shows results.

The load acting on bearing is increased at 1st critical speed at station 11 compared to station 1 as shown in Fig. 7a, b. This variation of load on bearings is due to disk offset. The load acting on right end bearing is more compared to left end bearing. Further, as discussed in papers [1, 4, 6, 8] imbalance on rotor due to various factors can be increased as they are more prone to imbalance and considered as one of the primary cause in failure of the rotors. Due to disk offset of 1.5 inch, the load distributed is more on station 11. As we can see the load has increased 18% at station 11 compared to station 1 and also with increase in the imbalance mass, the load on the bearing can also be increased which will cause

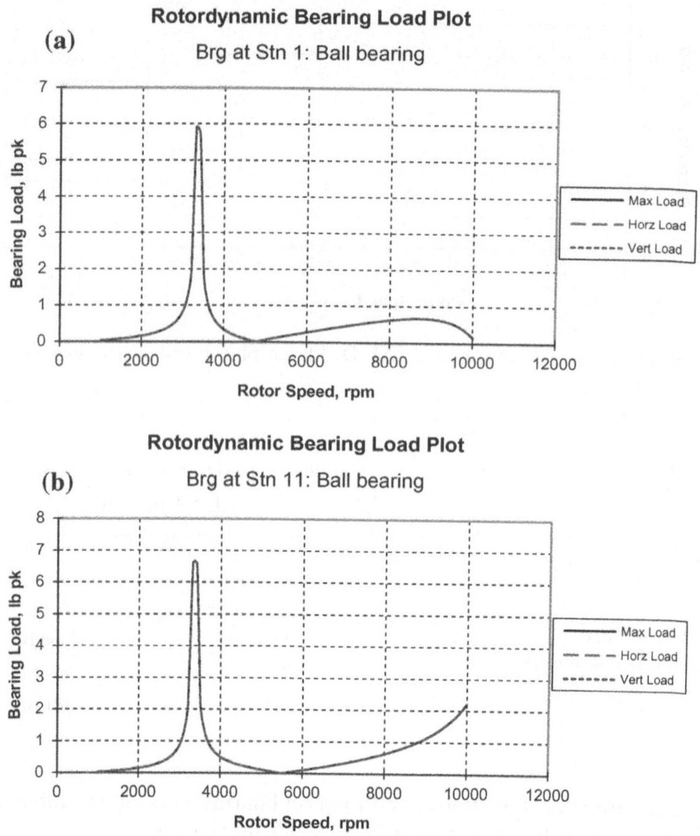

Fig. 7 **a** Bearing load response at station 1. **b** Bearing load response at station 11

Table 6 Comparison of results due to disk offset

Disk 1 offset from bearing 1	Disk 2 offset from bearing 2	Resonance (1st critical speed, cpm)	Amplitude at Disk 1 (mils)	Amplitude at Disk 2 (mils)	Load on bearing 1 (l bf)	Load on bearing 2 (l bf)
2 inch	2 inch	3500	6.3	8.22	61.6068	61.5087
3.5 inch	2 inch	3400	11.63	8.43	64.6479	56.4075
2 inch	3.5 inch	3400	6.46	13.5	56.1132	64.6479

failure to bearings and rotors. Though, the load acting on bearings is at 1st critical speed it is considered to be safe as it is vibrating in acceptable limits of load carrying capacity of bearings. Hence, in this analysis disk offset of 1.5 inch has impacted the rotor performance.

From the Table 6, due to disk offset the critical speeds (3400–3500 cpm) have not influenced much to the rotor model. When disks are equally located 2 inch away from both the bearings, vibration are close and load acting on bearings are equal. The disk offset of 1.5 inch has increased the vibration levels at offset location and vibrations are extreme as shown in vibration severity criteria chart [44] in Table 7.

This clearly highlights it does not affect the rotor critical speeds but extreme vibrations which are vibrating above acceptable limits can fail the rotor model. Though load acting on bearing is increased it is considered to be safe due to high load carrying capacity of bearings. For design modifications in rotor model, i.e. shifting/bypassing critical speeds and failure due to imbalance can effect rotor performance by disk offset parameter. One should careful in designing the model as small change in disk position can lead to failure of system.

4 Conclusion

In this work, the analysis of rotary model includes,

(1) Undamped critical speeds
(2) Damped critical speeds
(3) Imbalance response.
(4) Bearing load due to imbalance.

The failures are analyzed and controlled by XLrotor simulation technique. The applications listed above in Table 2, illustrates various types failures and reason for failure using XLrotor and FEM technique. The study and analysis due to disk offset on Spectra Quest's Machinery Fault Simulator (MFS) Magnum using XLrotor which shows results on critical speeds, effects of bearing stiffness and damping on critical speeds, imbalance response and load on bearings due to imbalance. Hence, in this lateral analysis the disk offset of 1.5 inch has impacted the rotor

Table 7 Vibration severity chart

Group	Machines	Foundation	Displacement in mils											
			0.43	0.71	0.87	1.1	1.42	1.77	2.2	2.8	3.54	4.45	5.51	
Group 4	Pumps>15kW	rigid												
		flexible												
Group 3		rigid												
		flexible												
Group 2	Medium Machines 15kW <P≤ 300kW	rigid		A			B			C		D		
		flexible												
Group 1	Large Machines 300kW< P<50mW	rigid												
		flexible												
Conditions			A - Good Condition			B - Allowable			C- Short Term Allowable			D- Causes Damage		

performance. Therefore it is determined the rotor vibrates severely at 1st critical speed due to disk offset which causes failure to it. As small change in disk position can lead to catastrophic failure. The real dynamics of rotors is difficult to model theoretically; hence the calculations are based on equations of algorithms used in XLrotor.

Acknowledgements The authors would like to thank XLrotor Rotordynamics Software for inputs and the authors of refereed papers for their valuable suggestions/comments to improve the quality of this paper.

References

1. Mohsen N, Suri G (2008) Rotordynamic analysis using XLrotor. Spectra Quest Inc., Richmond, VA
2. Zhang YM et al (2003) Reliability sensitivity for rotor-stator system with rubbing. 259 (5):1095–1107 (Elsevier)
3. Overview of Rotordynamic analysis of rotating machinery (1961) Toyo engineering corporation. https://www.toyo-eng.com/jp/en/products/analysis/rarm/. Accessed 22 Aug 2019
4. Lü CM et al (2013) Frequency reliability and sensitivity analysis on the rotor system. Appl Mech Mater 271–272:1296–1300 (Trans Tech Publications, Switzerland)
5. Antoulas AC, Sorensen DC, Gugercin S (2001) A survey of model reduction methods for large-scale systems. Contemp Math 280:193–219
6. Chaudhry J, Wagner M, He F et al (2010) A complete rotordynamic analysis of a rotor-disk system using finite element method. In: Proceedings of the 8th IFToMM international conference on Rotordynamics, KIST, Seoul, Korea
7. Nelson FC (2007) Rotor dynamics without equations. Int J Cond Monit Diagn Eng Manage 10(3):2–10
8. Gunter EJ (2004) Critical speed analysis of offset jeffcott rotor using English and metric units. Fellow ASME, RODYN Vibration Inc., Charlottesville, VA
9. Yoon SY et al (2013) Introduction to Rotordynamics control of surge in centrifugal compressors by active magnetic bearings, Advances in Industrial Control, © Springer, Berlin
10. Das AS, Dutt JK (2008) Reduced model of a rotor shaft system using modified SEREP. Mech Res Commun 35(6):398–407
11. Qu Z (2004) Model order reduction techniques. Springer, New York, NY, USA

12. Wagner MB et al (2010) Model reduction methods for rotor dynamic analysis: a survey and review. Int J Rotating Mach. 273716:17
13. Malcolm E, Leader PE (2015) Using rotordynamics to solve serious machinery vibration problems. Applied Machinery Dynamics Co., Dickinson, TX
14. Todorovic P et al (2008) Vibration analysis of cracked rotor during run-up. Tribol Indus 30:1–2
15. Hussain VMS (2013) Reliability modeling of rotary systems subjected to imbalance. Int J Perform Eng 9(4):423–432
16. Jahromi AF et al (2015) Forward and backward whirlingof a rotor with gyroscopic effect. In: Sinha JK (ed) Vibration engineering and technology of machinery, mechanisms and machine science, vol 23. https://doi.org/10.1007/978-3-319-09918-7_78
17. Alfa-Transit Co., Ltd. (2016) Power system for turbo machinery rotordynamics analysis. Dyn R4
18. Emerson (2017) High frequency vibration analysis. Machinery Health Analyzer, Emerson Reliability Solutions, Knoxville, TN
19. Bharath KL, Basappaji KM (2014) Development of vibration analysis expert system for condition monitoring. Master thesis, Department Of Mechanical Engineering Jawaharlal Nehru National College Of Engineering, Shimoga
20. Ganeriwala S et. al Numerical and experimental dynamic system identification for the development of operational modal analysis in physics—based on diagnostic/prognostic model. Spectra Quest Inc., Richmond, VA
21. Cowper GR (1996) The shear coefficient in Timoshenko's beam theory. ASME J Appl Mech 335–340
22. Murthy TN et al (2015) Rotordynamics considerations in refurbishing the turbomachinery. In: 7th international conference on latest trends in engineering and technology (ICLTET). Irene, Pretoria (South Africa), pp 26–27
23. Fleming DP et al (2006) Transient response of rotor on rolling element bearings with clearance. In: 7th IFToMM conference on rotor dynamics. Vienna, Austria, pp 25–28
24. Dyrobes (2012) Rotordynamics modeling and analysis analytical models and essential components. https://dyrobes.com%3Euploads%3E2012/12%3Ebook_pract_rdffbd_Chapter-6. Accessed 22 Aug 2019
25. Alsala J et al (2015) Rotordynamics. Department of Mechanical Engineering College of Engineering, University of Basrah, Basrah
26. Tenali N et al (2014) Rotor dynamic analysis of steam turbine rotor using ANSYS. Int J Mech Eng Rob Res
27. Ma WM et al (2012) 3D solid finite element modeling and rotordynamics of large rotating machines: application to an industrial turbo engine. Adv Mater Res 591–593:1879–1885 ISSN: 1662-8985
28. Tiwari S et al (2017) Dynamic analysis of rotor-bearing system for flexible bearing support condition. Int J Mech Eng Technol (IJMET) 8(7):1785–1792
29. Vastrad V (2012) Solid model rotor. Quest Global, White paper
30. Anand A, et al (2018) Static and dynamic analysis of lathe spindle using ANSYS. Int J Appl Eng Res 13(9):6994–7000 ISSN 0973-4562
31. Yue L et al (2013) A comparison of the finite element modeling methods for the natural frequencies computation of stepped shaft. Transactions, SMiRT-22, San Francisco, California, USA, pp 18–23
32. Samuelsson J (2009) Rotor dynamic analysis of 3D-modeled gas turbine rotor in ANSYS Supervisor: Thomas Domeij GRCRM, Siemens Industrial Turbomachinery AB Examiner: KjellSimonsson IEI. Linköping University, Finspång
33. Paulo PVDL (2011) A time-domain methodology for rotor dynamics: analysis and force identification. Master thesis, Dissertação para obtenção do Grau de Mestreem Engenharia Aeroespacial
34. Li L et al (2016) Introduction of rotor dynamics using implicit method in LS DYNA. In: 14th international conference LS-Dyna session simulationm, pp12–14

35. Bai B et al (2012) Analysis of dynamic characteristics of the main shaft system in a hydro-turbine based on ANSYS. In: Procedia engineering international conference on advances in computational modeling and simulation, vol 31, Elsevier, pp 654–658
36. Fegade R et al (2013) Unbalanced response and design optimization of rotor by ANSYS and design of experiments. Int J Sci Eng Res 4(7):1521 ISSN 2229-5518
37. Jauhari K (2017) Dynamic analysis of a spindle-bearing system based on finite element method. Int J Appl Eng Res 12(15):5110–5126 ISSN 0973-4562
38. Deepak S (2012) Rotor dynamic analysis of RM12 jet engine rotor using ANSYS. Master thesis, Department of Mechanical Engineering, Blekinge Institute of Technology, Sweden
39. Razi M (2013) Analysis of rotor dynamics acceptance criteria in large industrial rotors. Master of Applied Science (Mechanical Engineering) at Concordia University Montreal, Quebec, Canada
40. XLrotor tutorial (2007) XL Rotordynamics hand book, RMA, pp 55–55. www.xlrotor.com
41. Ma H et al (2015) A new dynamic model of rotor blade systems. J Sound Vib 357:168–194 (Elsevier)
42. Rao DK et al (2015) Vibration analysis of functionally graded rotating shaft system. In: 12th international conference on vibration problems, ICOVP, Elsevier
43. XLrotor Rotordynamics (2012) Manual Rotating Machinery Analysis, Inc., Austin, TX pp 918–9140. www.xlrotor.com
44. AMP Maintenance (2019). https://www.maintenance.org/topic/iso-standard-values-for-displacement. Accessed 22 Aug 2019

Dr. H. N. Suresh received his Ph.D. degree from IITB, Mumbai. Currently, he is working as Professor and Head of Automobile Engineering Department at Dayananda Sagar College of Engineering, Bengaluru. He has got 36 years of teaching experience and is a Six Sigma Green Belt. He has ISTE and Society for Reliability and Safety membership. He has a number of national and international publications and has executed research projects. He has been an active consultant in condition monitoring projects. His area of interest includes quality, reliability, and maintenance. He can be reached at hnsiitb@gmail.com.

Mr. S. B. Murgayya received his M.Tech. degree in Machine Design from Basaveshwara Engineering College, Bagalkot. Currently, he is working as Assistant Professor at Dayananda Sagar College of Engineering, Bengaluru. He has got 2 years of industrial experience and 5 years of teaching experience. He is currently pursuing Ph.D. under Dr. H. N. Suresh, in School of Mechanical Engineering from Dayananda Sagar University, Bengaluru. He is also "Associate Member of the Institution of Engineers". He has authored several national and international publications. His area of interest includes vibration monitoring and rotordynamics. He can be reached at murgayya-au@dayanandasagar.edu.

Time Variant Reliability Analysis of Passive Systems

Hari Prasad Muruva

1 Introduction

In recent years most of the advanced nuclear reactors implement passive systems, aimed at improved safety and availability, in order to reduce human error and active component malfunctions. The passive systems play an important role on various innovative reactor safety systems and can result in substantial economic benefits and simplicity in operation over the active systems. However, drawbacks of their use come from the larger difficulties in the thermohydraulic design (compared with active systems) and from reliability considerations specifically connected with the system operation. Hence, it is very important to ensure the availability of passive systems whenever the demand arises. With the help of the reliability analysis one can identify the weak links in the system and one can improve the design of the system so as to improve the reliability of the system. So far the reliability of the passive systems has been estimated only by means of static analysis i.e., it doesn't consider the time dependency of the reliability of the passive systems. But in actual practice reliability of the systems degrades with time due to ageing, random/stochastic loading or strength degradation and the reliability is just not a single value but it is a function of time.

Hence, it is very important to evaluate the passive system reliability by considering both static and time variant analysis. In the present chapter passive system reliability has been estimated by using both static and time variant analysis and the methodology is explained with a case study. In static analysis system reliability has been estimated by using Fuzzy approach and in time variant analysis random loading has been considered as a stochastic fatigue loading. A new model has been developed to treat stochastic fatigue loading with Poisson arrival and the reliability

H. P. Muruva (✉)
Bhabha Atomic Research Centre, Trombay, Mumbai 400085, India
e-mail: hari_m@barc.gov.in

D. R. Karanki et al. (eds.), *Advances in RAMS Engineering*,
Springer Series in Reliability Engineering,
https://doi.org/10.1007/978-3-030-36518-9_9

of the system has been estimated as a function of time. The salient feature of the study and the results obtained are discussed in this chapter.

2 Passive Systems

In general, a power plant (nuclear, thermal, chemical etc.) consists of operating and emergency safety systems. These systems vary from very complex to simple systems. A system normally consists of active components and passive components. The failure of any operating system will lead to a change in the state of the plant. The availability of the plant depends on the successful operation of the operating systems and the operation of the components in the systems. In order to ensure the availability of the plant reliability of the systems/components should be ensured.

Passive systems are being utilized extensively in advanced reactors in order to reduce human error and active component malfunctions mainly to improve the safety of the plant. As per IAEA definition [1], a passive component is a component, which does not need any external input to operate, and a passive system is either a system, which is composed entirely of passive components, or a system, which uses active ones in a very limited way to initiate subsequent passive operation. Consider a simple passive system as shown in Fig. 1. The purpose of the system is to reject the heat from the steam which is produced in a steam drum to the heat sink by condensing primary fluid in the heat exchanger tube bundle and the condensate will return to the steam drum. The process is entirely a natural circulation process. Passive systems are mainly depend on the natural laws (e.g. gravity, natural circulation) to accomplish their designated safety function. The deviations of natural forces or physical principles from the expected conditions can impair the performance of the system.

The reliability estimation of passive systems is somewhat different from active components. The reliability of passive systems can be defined as the probability of the system to carry out its safety function under the prevailing conditions for the given duration of time when required. In practice the reliability of the system has

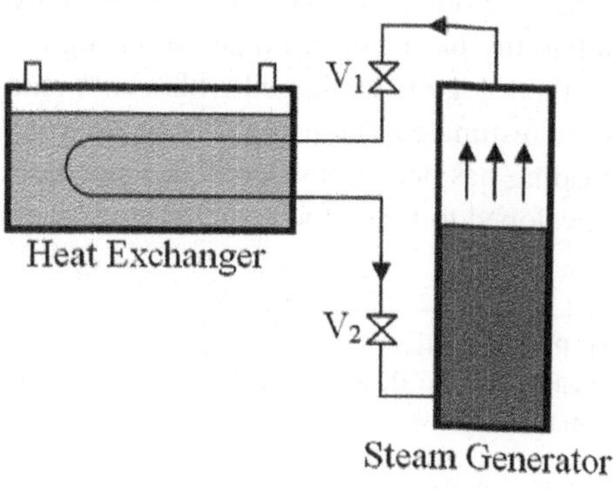

Fig. 1 Passive system depicting the natural circulation

contribution from static and time variant analysis. In static analysis the failure contribution comes from the malfunctioning of the active/passive components in the system including human failures. Whereas in time variant analysis the contribution comes from the time variant failure probability of the components due to ageing, stochastic loading or strength degradation.

There are well established techniques for estimating the reliability of active systems/components. Whereas for passive systems very few techniques are available and the techniques are not yet frozen. Many researchers, academicians and engineers in various fields are working extensively to develop methods for carrying out reliability analysis of passive systems. Few papers [2–4] described the methodology for performing the passive system reliability. In these papers only a point value has been estimated for reliability of passive systems i.e., reliability has been estimated through static analysis only and they do not consider the time dependency of reliability of the system. Also in these papers even though enough information is not available for the key parameters that affect the system performance, the corresponding distributions have been assumed based on the expert judgment. Hence, to take care of the sparseness in the data in the present paper fuzzy approach has been introduced for reliability estimation as a part of static reliability analysis. As a part of time variant analysis stochastic loading has been treated as a fatigue loading and the failure probability of the system has been estimated as a function of time. A case study on reliability analysis of passive decay heat removal system (PDHRS) of large sized PHWRs has been discussed in the paper.

3 Passive System Reliability

The reliability of passive systems refers to the ability of the system to carry out its safety function under the prevailing conditions for the given duration of time when required. Reliability estimation of components has two aspects. The first one is deterministic, which deals with finding various failure modes and causes of failure, while the second one is probabilistic where probability of failure and uncertainty in parameters is estimated. These two aspects together make reliability prediction more accurate and complete. There are several components which contribute to the failure of the passive systems. The failure contribution can come from both static and time variant analysis. In the static analysis the following component failures cause the failure of the system

1. Virtual component
    a. Design and process parameters
    b. Partial component failures

2. Active component
    a. Valves
    b. Piping
    c. Other Components

In the time variant analysis the failure probability contribution come from the following factors

1. Stochastic loading
2. Strength degradation

The different aspects of static and time variant reliability analysis have been explained with appropriate methodologies with a case study in the following sections.

4 Static Reliability Analysis

The passive system reliability is evaluated from the evaluation of the failure probability of the system to carry out the desired function. In principle, in a natural circulation system, the operational mechanism of buoyancy driven natural circulation should never fail as long as there is a heat source and sink with an elevation difference between them. However, even though the mechanism does not fail, it may not be able to drive the required flow rate whenever called in, if there is any fluctuation or deviation in the operating parameters even though the system geometry remains intact. In the case of a motor driven pump, the head versus flow characteristics are not so much susceptible to a slight change or fluctuation in operating parameter to cause the failure of the system unless there is any mechanical failure of the pump itself. Hence, its performance characteristics are well known and can be simulated accurately while assessing the overall safety of the plant. On the other hand, the characteristics of buoyancy driven natural circulation cannot be accurately predicted under all operational conditions or transients due to the inherent complex phenomena associated with natural convection systems as discussed before. Hence, the methodology to treat the passive system reliability differs from that of active systems. Figure 2 shows the structured methodology for the evaluation of reliability of passive systems. The methodology is explained in the following steps with a case study on passive decay heat removal system (PDHRS).

4.1 *A Case Study on PDHRS*

The procedure described in the previous section has been demonstrated with a case study on PDHRS of a typical nuclear power plant.

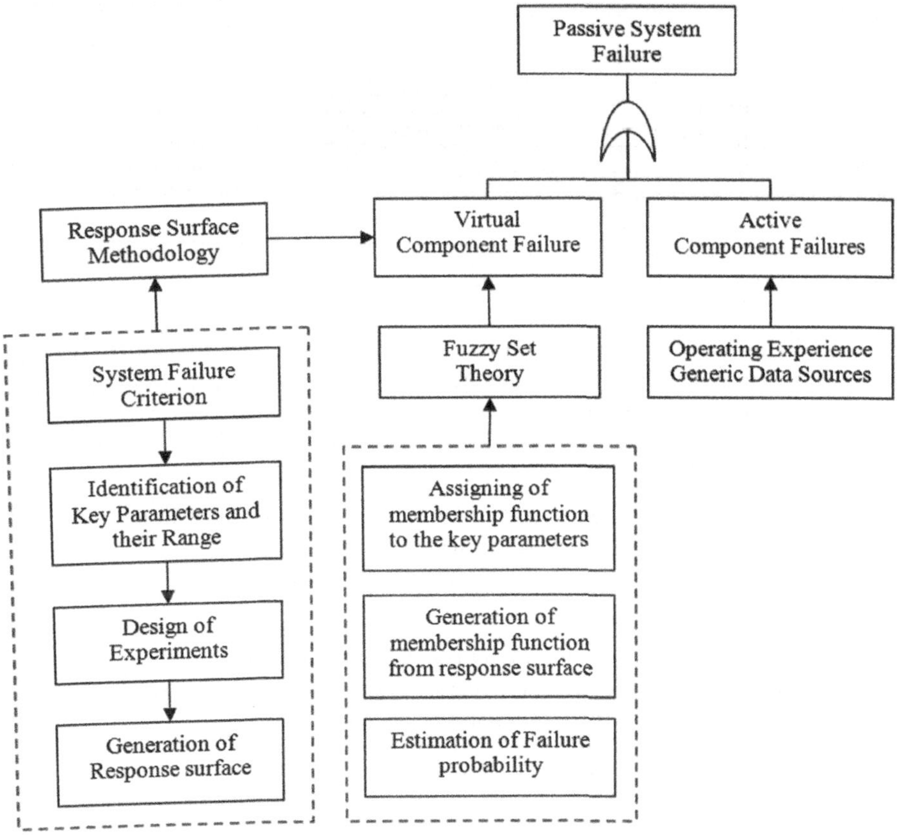

Fig. 2 Procedure for estimating the passive system reliability

System Description

The primary function of the PDHRS is to ensure continued availability and recirculation of the inventory on the secondary side of the Steam Generators (SG) in the event of station black out (SBO) and in the process ensure continued removal of decay heat from the core. The system comprises of four tanks, each containing horizontal V-tube condenser bundles connected to their respective steam generators through take-off lines from main steam line and condensate line joining the steam generator as shown in Fig. 1. During normal shutdown, initially the decay heat is removed from primary circuit in steam generators. The steam produced on the secondary side of the SGs is dumped to condenser or to atmosphere through atmospheric steam discharge valves. Feed water make up to steam generators is by main or auxiliary boiler feed water pumps. Further cool down of primary circuit to room temperature is by shutdown cooling system. SBO is characterized by simultaneous failure of class IV and class III power supplies. The normal heat removal through the SGs initially, and by shutdown cooling system later are

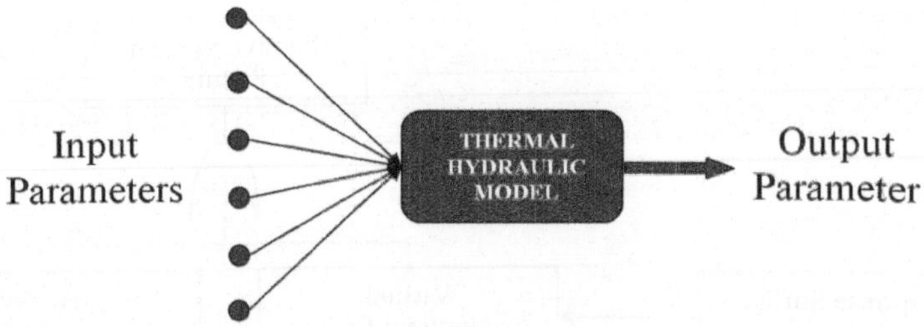

Fig. 3 Input and output interface of the thermal hydraulic model

affected during SBO condition. In case of station blackout, the available heat sink is the PDHRS for recirculating the steam generator side inventory through the V-tube condenser inside PDHRS tanks.

Reliability Analysis

In carrying out the reliability analysis the methodology illustrated in the Fig. 3 has been utilized and is explained step by step as follows:

Step 1: Passive system to be analyzed

As a first and foremost step the system to be analyzed for reliability estimation should be identified. In this case study passive decay heat removal system has been considered for the passive system reliability analysis.

Step 2: System Operating Mechanism

The objective of providing PDHRS is to enhance reactor safety during the station blackout event (Class IV and Class III power supply failure) by ensuring continuance of thermosyphon cooling of the reactor through steam generators in the absence of makeup to the Steam Generator (SG) drums. This can be accomplished by recirculating the steam through the PDHRS condensers provided along with each of the SGs. In this process heat is transferred to the inventory outside the tubes contained in the PDHRS tanks.

Step 3: Setting up of Failure Criterion

Since failure of natural circulation in secondary system affects the natural circulation in Primary Heat Transport (PHT) system during station blackout condition, the failure of natural circulation in PHT system has been considered as the failure criterion. This is characterized by increase in clad surface temperature above 400 °C.

Step 4: Key Parameters that affect the system operation

More than 20 parameters (PHT system inventory, SG inventory, PDHR inventory, decay power, non-condensable in PHT, PHT system pressure, SG pressure, primary coolant flow rate, thermal conductivities of various material involved, thermodynamic properties, reactor power, fouling on heat transport surfaces, SG feed water flow and temperature, non-condensable in SGs, etc.) that affect the system operation and in effect cause the failure of the system have been identified. Out of these parameters few key parameters have been identified based on the sensitivity studies which will have more impact on the failure of the system and are listed below.

1. PHT inventory
2. SG inventory
3. PDHR Tank Inventory
4. Decay heat
5. Non condensable gases in PHT
6. Non condensable gases in PDHR tube

Step 5: Identification of key parameter ranges and Membership functions

In finding out the membership functions for the key parameters the range of the parameters should be identified as a first step. The ranges for these parameters have been assigned based on the expected deviations in the values of these parameters in normal operating conditions. The ranges of the key parameters that are identified are shown in the Table 1. Once the ranges for key parameters are identified next step is to assign a proper membership functions. Each of these key parameter deviation has been analyzed for identifying the membership functions. The different key parameters ranges and their membership functions are shown in the Table 1.

Step 6: Generation of Response Surface

Thermal hydraulic analysis has been carried out to generate the response surface (from various ranges of identified key parameter values). The input and output interface of the thermal hydraulic model is shown in Fig. 3.

In the present analysis we have six key parameters that affect the clad surface temperature. According to the 2K factorial design method one can get 64 combinations from these six key parameters. For all the 64 combination of the key parameters, thermal hydraulics (process dynamics) simulation model runs were taken for the case of case of station blackout scenario.

Table 1 Key parameters range and their membership functions

S. no.	Key parameter	Range	Membership function
1	PHT Inventory (tons)	0–15	Trapezoidal
2	Decay heat (%)	0–20	Triangle
3	PDHR tank Inventory (tons)	0–20	Trapezoidal
4	SG Level (m)	10.4–15.4	Trapezoidal
5	Non condensable gases in PHT (%)	0–7.5	Triangle
6	Non condensable gases in SG secondary side (%)	0–5	Triangle

Table 2 Regressor Coefficients of the Polynomial

Coeff.	Value	Coeff.	Value	Coeff.	Value
a_1	732.838	b_1	−1.8289	c_1	−10.191
a_2	2666.33	b_2	−45.296	c_2	2.5778
a_3	62.5492	b_3	0.6364	c_3	−2.3831
a_4	−156.79	b_4	1.9211	c_4	−1.3553
a_5	105.511	b_5	12.357	c_5	−20.623
a_6	42.4151	b_6	5.4610	a_0	−71639

Based on the results obtained from thermal hydraulic analysis for all the cases using, response surface has been generated (from various combinations of identified key parameter values as input). From the analysis different response points have been obtained and response surface has been developed by using non linear regression analysis. Polynomial used for fitting the data is given in Eq. (1). The regression coefficients obtained from the analysis is shown in the Table 2.

$$\begin{aligned} T = {} & a_0 + a_1x_1 + a_2x_2 + a_3x_3 + a_4x_4 + a_5x_5 + a_6x_6 \\ & + b_1x_1^2 + b_2x_2^2 + b_3x_3^2 + b_4x_4^2 + b_5x_5^2 + b_6x_6^2 \\ & + c_1x_1x_2 + c_2x_2x_3 + c_3x_3x_4 + c_4x_4x_5 + c_5x_5x_6 \end{aligned} \tag{1}$$

Step 7: Generation of membership function

Once the response surface is available, next step is the generation of membership function for the clad surface temperature. In this approach uncertain parameters are considered as fuzzy variables and the corresponding membership functions have been derived from available information (which is shown in the Table 1) and are shown in Fig. 4. By using fuzzy arithmetic membership function [5, 6] for the clad surface temperature has been obtained which is shown in Fig. 5.

Step 8: Failure probability estimation

Once the membership function for the clad surface temperature is available, then next step is to convert the membership function into probability density function (PDF) or cumulative distribution function (CDF). For this purpose one can use the possibility to probability transformations [7, 8]. The obtained PDF and CDF are shown in Figs. 6 and 7 respectively.

From the distribution the failure probability of the system has been estimated by considering the failure criteria as clad surface temperature exceeding a critical value i.e., 400 °C. The failure probability can be obtained as the area under the PDF curve above 400 °C. For the present analysis the estimated failure probability obtained is 4.203×10^{-4}. This is only the contribution from process and design parameters deviation. The failure probability contribution from all the components (Virtual component and active component failures) is found to be 7.513×10^{-4}.

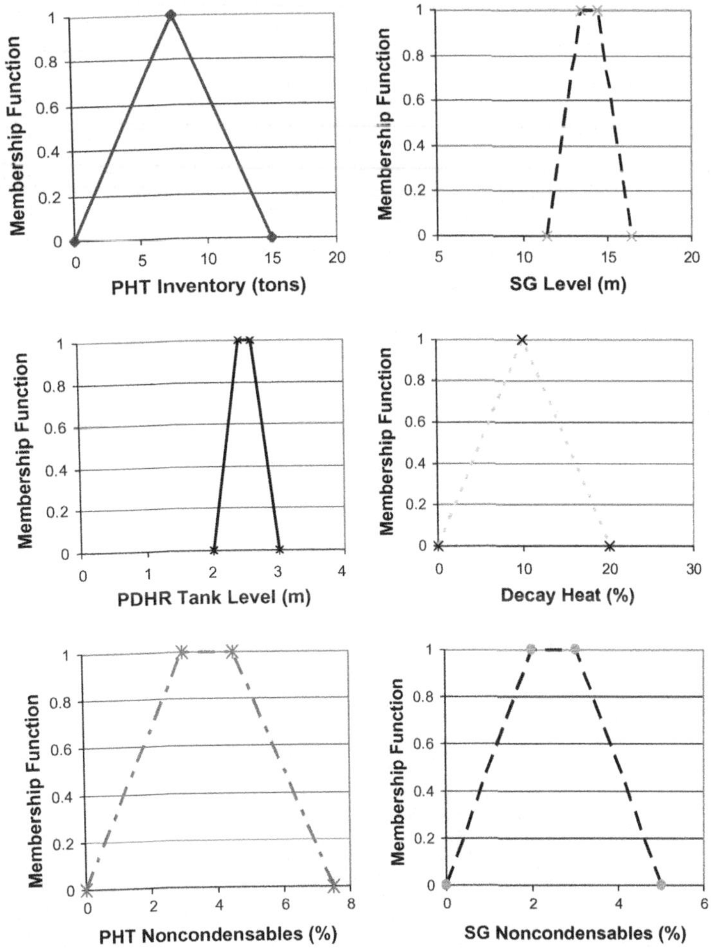

Fig. 4 Membership functions of different key parameters

5 Time Variant Reliability Analysis

In real life situations the components/structures are in general not only subjected to the static loads but also random loads. This means that the structural reliability analysis not only considers random variables only but also considers random variables which are functions of time, referred to as stochastic processes. As a result the failure probability is no longer just a single number, but also a function of the time. In general one should always keep in mind that just mentioning a value for the failure probability does not make any sense without specifying the period of time for which it was derived.

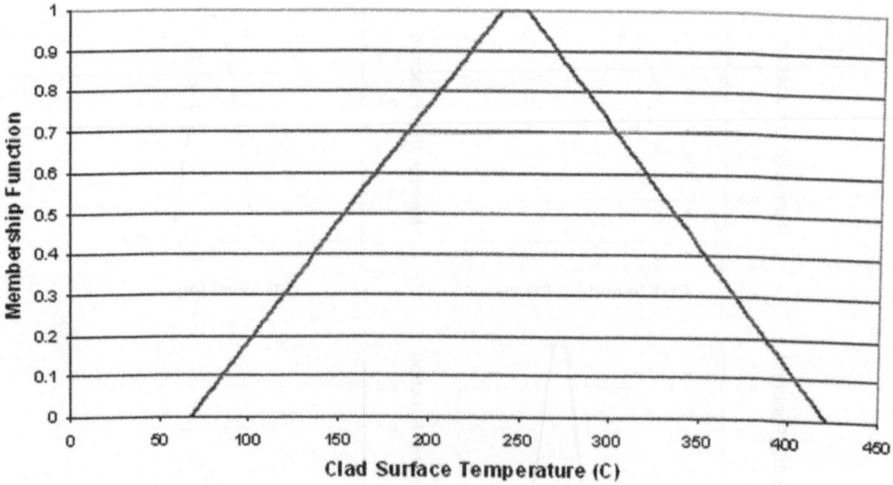

Fig. 5 Membership function of clad surface temperature

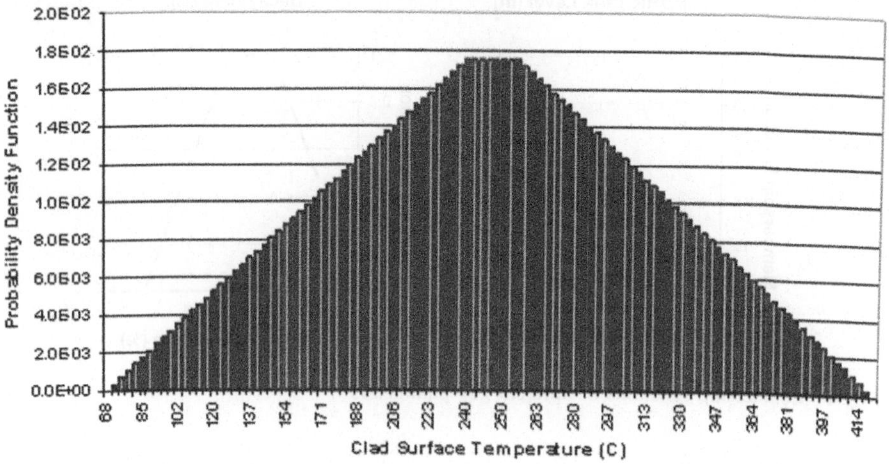

Fig. 6 Probability density function of clad surface temperature

Stochastic Process

A stochastic process can be viewed as a family of random variables. A collection of time functions for a stochastic process is typically called an ensemble as shown in Fig. 8. Hence, Ensemble of all the possible time functions (time histories) that might result from the experiment is known as a random process or stochastic process. The two main classes of random processes are the stationary and non stationary processes. The basic feature of stationary processes is that their statistical

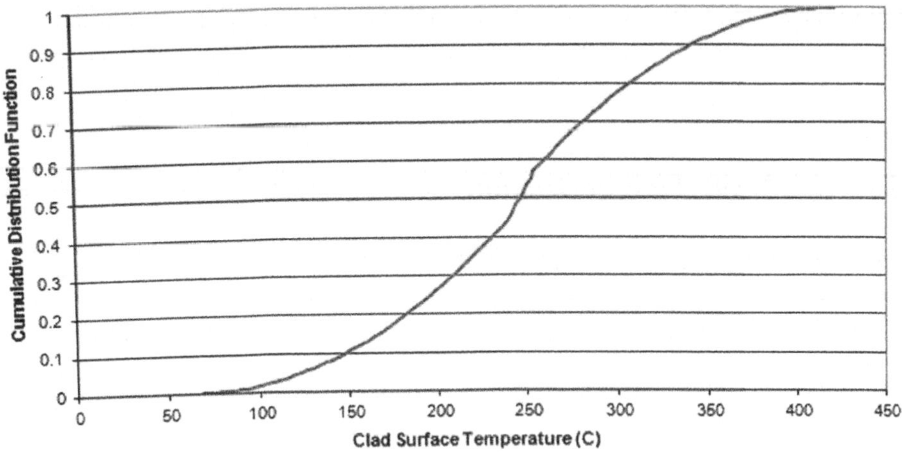

Fig. 7 Cumulative distribution function of clad surface temperature

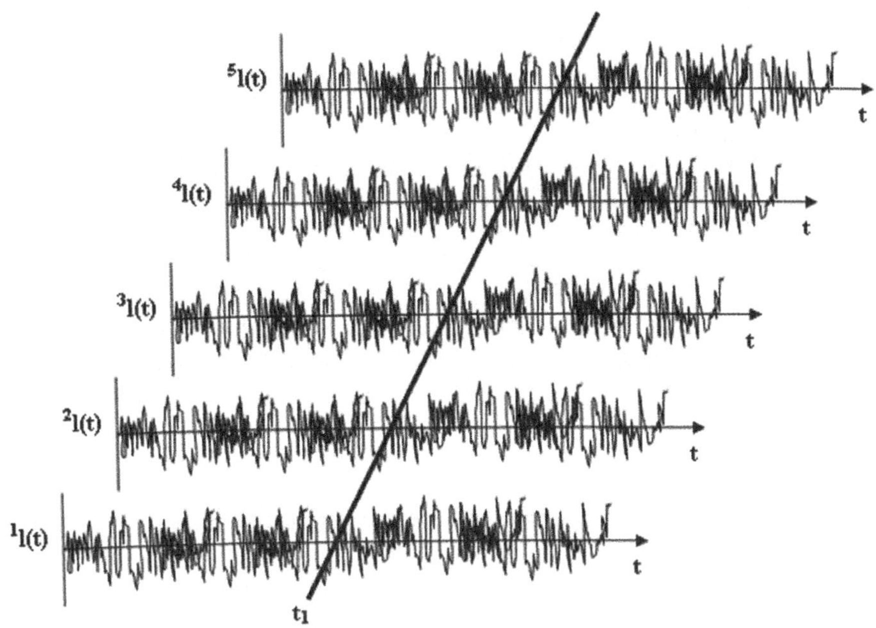

Fig. 8 Ensemble of time histories

properties (mean, standard deviation etc.) do not change with time. Whereas in the case of non stationary processes these properties changes with time.

Conventional reliability models seldom considered the times of load action. The reliability calculated by these models is actually the reliability when random load acts for specified times. For general components and systems, these models can't reflect the effect of times of load action on reliability explicitly. In the present work the time variant reliability analysis has been explained for stochastic fatigue loading with a case study.

5.1 Stochastic Fatigue Loading

When a metal is subjected to repeated cycles of stress or strain, it causes its structure to break down, ultimately leading to fracture, this behavior is called fatigue. When a body (specimen or structural component) is subjected to cyclic loads, the process of rupture starts with damage nucleation (voids, slip lines, micro cracks, etc.) at stress concentrators of the body. At some time later, a crack appears that propagates until fracture of the body. Generally the maximum stress values are less than the ultimate tensile strength limit, and may be below the yield strength limit of the material.

The nature of this failure apparently results from the fact that there are microscopic regions, usually on the surface of the member, here the localized stress becomes much greater than the average stress acting over the cross section. As this higher stress is cycled, it leads to the formation of minute cracks. Occurrence of these cracks causes a further increase of stress at their tips or boundaries, which in turn causes a further extension of the cracks into the material as the stress continues to be cycled. Eventually cross-sectional area of the member is reduced to the point where the load can no longer be sustained, and as a result sudden fracture occurs. The material, even though known to be ductile, behaves as if it were brittle.

Fatigue occurs when a material is subjected to repeated loading and unloading. If the loads are above a certain threshold, microscopic cracks will begin to form. Eventually a crack will reach a critical size, and the structure will suddenly fracture. The process of fatigue failure can be divided into the following successive stages:

- Formation of a micro crack nucleus
- Propagation (growth) of the fatigue crack and
- Final rupture

In the present case study, a model has been developed for reliability calculation of the system under stochastic cyclic loading with Poisson arrival rate. In this model stress amplitude, number of cycles to failure for a given stress amplitude and the number of cycles over a given period of time have been considered as uncertain parameters. In the traditional fatigue analysis the fatigue data is represented as a S-N diagram [9] as shown in Fig. 9. In this approach a series of specimens are each subjected to a specified stress and finds the number of cycles to failure. The results are plotted as a graph representing the stress S as the ordinate and the number of cycles to failure N as the abscissa. Most often the values of N are plotted on a logarithmic scale since they are generally quite large. However, if a large number of specimens are tested for a given stress amplitude there will be large scatter in the number of cycles to failure for a particular value of the stress amplitude.

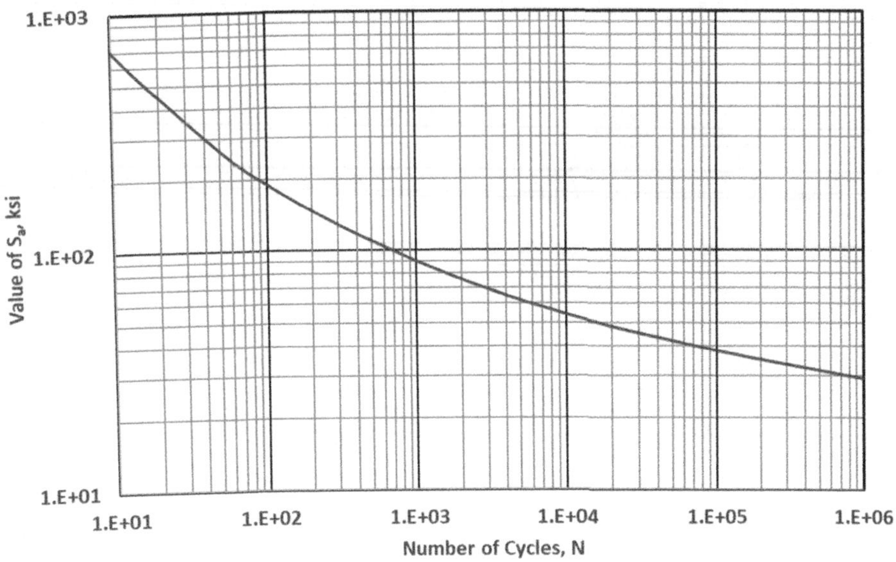

Fig. 9 S-N diagram for a typical metal [9]

Hence, one gets a distribution for the number of cycles to failure for a given stress amplitude. The uncertainty in the number of cycles to failure can be attributed to variation in material properties. This type of analysis leads to the introduction of probability of failure P with other quantities S and N and the S-N diagrams will be represented with P-S-N diagrams [10].

Now consider uncertainty in the stress amplitude. One can represent the variation in the parameter by a probability density function $f_S(s)$, it means, the stress amplitude can take the values of s_i ($i = 1, 2,\ldots,$ m) with the respective probabilities of p_i ($i = 1, 2,\ldots, m$) depending on the probability density function. By the total probability principle, under the loading condition that the stress amplitude takes on the m different values of $y_1, y_2,\ldots, y_m$ with the respective probability of $p_1, p_2,\ldots, p_m$, respectively. In this case the failure probability can be given as (see Fig. 10):

$$P_F(N, s_i) = \int_0^N f_N(n, s_i)dn \tag{2}$$

In other words, the cyclic stress amplitude follows a certain statistical distribution, while the individual cyclic stress samples are constant amplitude cyclic stresses with deterministic values. What is highlighted here is the methodology of calculating fatigue reliability according to stress amplitude distribution and fatigue life distribution directly. In the fatigue analysis it is important to find the stresses that are acting on the system. In the present work the fatigue analysis has been

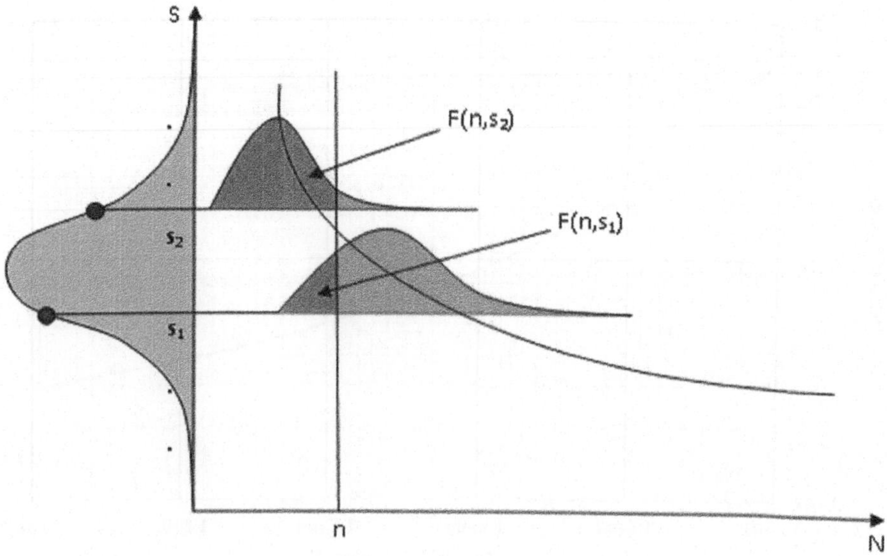

Fig. 10 Random stress amplitude and corresponding failure probabilities [9]

carried out for a piping system. The following stresses are acting on the piping system viz., stresses due to dead weight, pressure stress due to internal pressure and thermal gradient stress due to temperature difference across the thickness of the piping. The stress calculation is described in the following sub sections.

5.2 A Case Study on PDHRS Piping System

To carry out the fatigue analysis a case study on piping system corresponding to passive decay heat removal system (PDHRS) has been considered. The Passive Decay Heat Removal System comprises four PDHRS tanks of 200 m^3 capacity each containing horizontal V-tube condenser bundles connected to their respective steam generators through take-off lines from main steam line and condensate line joining the steam generator shell just above the tube sheet as shown. The diameter and length of the tank is 5 m and 12 m respectively. The tank consists of 50 V-tubes and length of each tube is 16 m. The piping system has been divided into three parts viz., a piping line which carries steam from steam generators to the PDHRS tank, V-tubes inside the tank which condense the steam and a piping line which carries the condensate to the steam generator. The details of the piping system are given in Table 3. Throughout the plant life time the system will be subjected to different loading conditions (Fig. 11). Whenever the reactor goes from operating state to other states like hot shut down, cold shut down and full power operation the pressure and the temperatures in the system change accordingly.

Table 3 Details of the piping system

Piping	Parameters	Value (mm)
Inlet line	Outer diameter Inner diameter Thickness	150.0 122.3 30.00
V-Tube	Outer diameter Inner diameter Thickness	56.40 50.80 2.800
Condensate line	Outer diameter Inner diameter Thickness	80.00 73.20 6.800

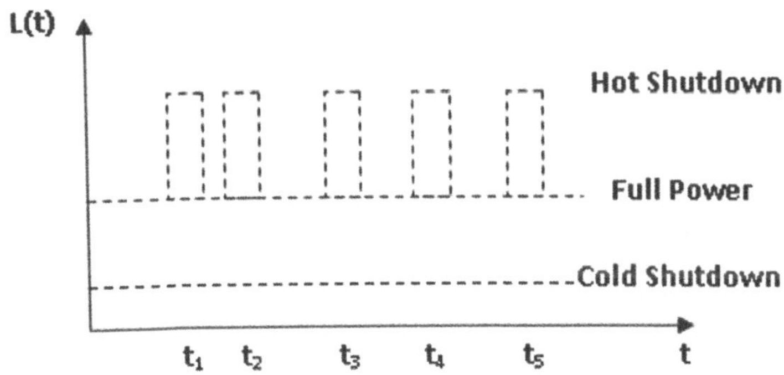

Fig. 11 Loading on the piping system

Table 4 Different loading conditions

Case	Type of loading	Parameters	Value
1	Full power	Pressure Temperature	40×10^5 N/m^2 250 °C
2	Hot shutdown	Pressure Temperature	50×10^5 N/m^2 260 °C

Eventually this leads to the change in the stresses in the system. In the present analysis one such type of loading cycle considered is reactor changing it state from full power operation to hot shut down condition and coming back to full power operation. The temperature and pressure in the system is given in Table 4.

Stress Calculations

The stress calculations have been carried out for the V-tube piping which is inside the PDHRS tank. The total number of tubes is 50 and is symmetric in nature. The material properties used in the analysis is shown in the Table 5. The loadings conditions are considered as mentioned in Table 4. The stress amplitude is calculated as 213×10^6 N/m^2. Since the material properties are not constant and they are random in nature the stresses acting on the system also will be random in nature.

Table 5 Material properties

S. no.	Property	Value
1	Material	SS-316L
2	Density (ρ)	8×10^3 (kg/m^3)
3	Young's Modulus (E)	1.93×10^{11}(N/m^2)
4	Ultimate Tensile Strength (UTS)	560×10^6 (N/m^2)
5	Yield Strength (YS)	290×10^6 (N/m^2)
6	Coefficient of thermal expansion(α)	17.3×10^{-6} (/°C)
7	Thermal conductivity (K)	16.2 (W/mK)
8	Poisson's ratio (ν)	0.3

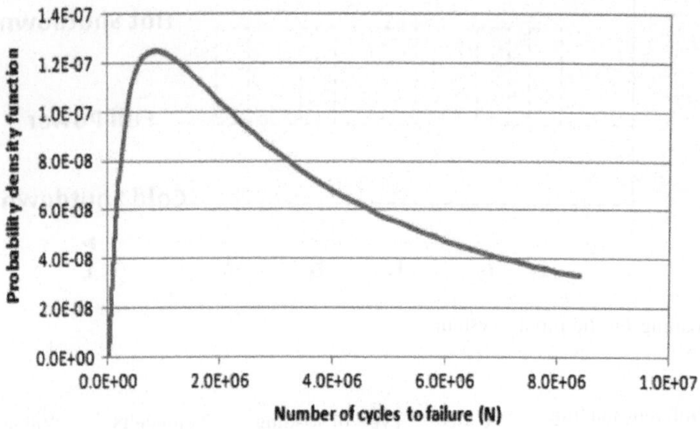

Fig. 12 Lognormal distribution for Number of cycles to failure

Hence, the stress amplitude has been considered as a random variable with the mean value as given above and the coefficient of variation is considered as 0.02 and the distribution is considered as truncated normal distribution.

The number of cycles to failure for the corresponding stress amplitude is obtained from the S-N curves and is given as 2×10^5 cycles. But this number of cycles to failure is not constant and there is always variation involved and can be treated as a random variable. In this analysis it is assumed that the number of cycles to failure follows lognormal distribution with median value as 2×10^5 cycles and error factor as 10 (large uncertainty in the fatigue analysis). The corresponding distribution is shown in the Fig. 12. From the distribution failure probability for different cycles (which represents time) has been evaluated and the same is plotted against time as shown in the Fig. 13. The total failure probability of the system will be the contribution from the static analysis and the time variant analysis. The same is shown in the Fig. 14.

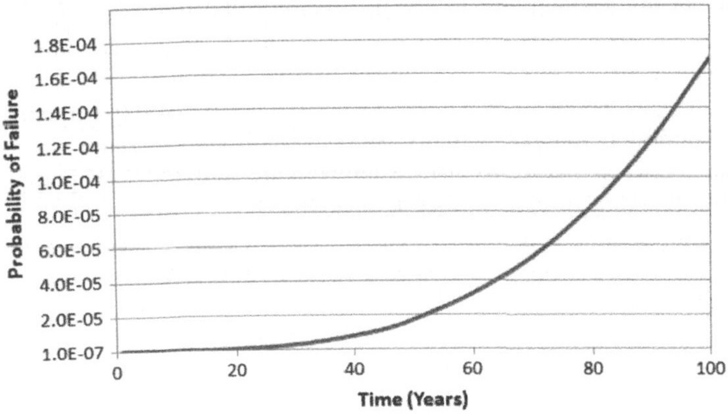

Fig. 13 Failure probability variation with time

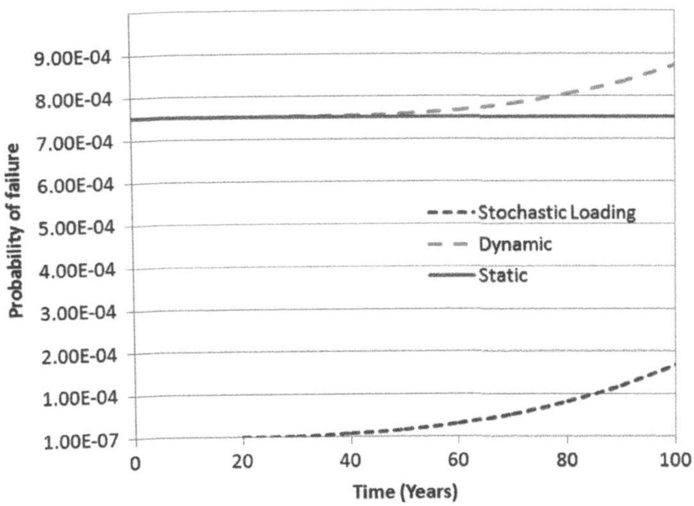

Fig. 14 Failure probability variation with time

6 Summary

In this chapter methodology for performing time dependent reliability analysis of passive system has been discussed. A case study on PDHRS has been presented. The failure contribution from both virtual and active components has been estimated. The failure of virtual component has been modeled based on the uncertainty present in the design and process parameters. The key parameters that affect the system performance and their ranges have been identified based on the available information. The concept of fuzzy set theory has been utilized in modeling the uncertainty in the key parameters. In this analysis the uncertain parameters have been treated as fuzzy numbers and the variability is characterized by the membership function which has been obtained from the available information. Thermal hydraulic analysis has been performed as per the design of experiments. By using the linear regression analysis response surface has been generated which gives the relationship between the clad surface temperature and the key parameters. The membership function of clad surface temperature has been generated by using the key parameter membership function and response surface. Later this membership function has been transformed to probability density function by fuzzy probability transformation methods. Finally the failure probability has been estimated based on the failure criteria. This analysis has been presented as static reliability analysis. In order to account for reliability variation with respect to time, time dependent reliability analysis has been carried out for the case of stochastic fatigue loading. The analysis has been carried out for the same system and the failure probability is represented as a function of time.

References

1. IAEA-TECDOC-626 (1991) Safety related terms for advanced nuclear plants. International Atomic Energy Agency, Vienna
2. Bianchi F, Burgazzi L (2001) Evaluation of the reliability of a passive system. In: International conference nuclear energy in Central Europe 2001, September 10–13, 2001
3. Jafari Jalil, Auria Francesco D (2003) Reliability evaluation of a natural circulation system. Nucl Eng Des 224:79–104
4. Marques M, Pignatel JF (2005) Methodology for the reliability evaluation of a passive system and its integration into a Probabilistic Safety Assessment. Nucl Eng Design 235:2612–2631
5. Zadeh LA (1965) Fuzzy sets. Inf Control 8(3):338–353
6. Zadeh (1973) Outline of a new approach to the analysis of complex systems and decision processes. IEEE Trans Syst Man Cybernet 3(1)
7. Hari Prasad M, Gaikwad AJ, Srividya A, Verma AK (2011) Failure probability evaluation of passive system by using fuzzy montecarlo simulation. Nucl Eng Des 241:1864–1872
8. Rao KD, Gopika V, Kushwaha HS, Verma AK, Srividya A (2004) Uncertainty analysis of safety systems of nuclear power plant by fuzzy set theory. In: International conference on Probabilistic Safety Assessment and Management, Berlin, Germany, 14th–19th June 2004, pp 2283–2288
9. Verma AK, Ajit S, Muruva HP (2015) Risk management of non-renewable energy systems, Springer Series in Reliability Engineering, Springer International Publishing, Switzerland
10. Provan JW (1987) Probabilistic fracture mechanics and reliability, 1st edn. Martinus Nijhoff Publishers, The Netherlands

Hari Prasad Muruva is a Scientific Officer at Bhabha Atomic Research Centre, Mumbai, India. Since year 2000, he has been working in the areas of probabilistic safety assessment (PSA), seismic PSA, risk-based inspection of power plants, dynamic PSA, dynamic reliability-based risk analysis of passive safety systems, and risk monitor tool development. He has got a credit of two books and several research publications. He can be reached at hari_m78@yahoo.co.in.

Reliability Considerations in Analysis of Tunnels in Squeezing Rock

G. Venkatachalam, Mallika Sreedharan and A. Srividya

Underground openings and excavations are increasingly being used for civilian and strategic purposes all over the world. Pressure on tunnel support depends on deformation and stress release and tunnels can be classified [9] into: (a) conforming; (b) non-conforming. Evaluation of safety of a tunnel, especially a non-conforming one, in soft ground or poor rock mass needs an interactive analysis. Interaction is an issue of great relevance in the case of tunnels in lower Himalayas. Many valuable field observations on actual Himalayan tunnels have been documented by [2, 7, 8]. Based on these, several empirical correlations have been developed, which are very useful in evaluating, particularly, the rock pressure on such tunnels. Although the correlations account for a wide variety of ground conditions, there is, inevitably, considerable empiricism. Jimenez and Recio [3] have attempted a probabilistic modelling of the variable ground conditions and evolved classifiers to distinguish between squeezing and non-squeezing conditions. However, evaluating the response of tunnels considering the material uncertainties has not been included in their scope. The present study is an attempt to evaluate tunnel safety and stability through reliability analysis under squeezing conditions.

G. Venkatachalam
Mukesh Patel School of Technology Management and Engineering, Mumbai, India

M. Sreedharan (✉)
Dar, Pune, India
e-mail: Mallika.Sreedharan@dar.com

A. Srividya
Luleå University of Technology, Luleå, Sweden

Western Norway University of Applied Sciences, Haugesund, Norway

D. R. Karanki et al. (eds.), *Advances in RAMS Engineering*,
Springer Series in Reliability Engineering,
https://doi.org/10.1007/978-3-030-36518-9_10

1 Methodology

Response Surface Method (RSM) is a collection of statistical and mathematical techniques used for developing, improving and optimizing processes [5]. Based on the observed data from the system or process, an empirical model, which is a polynomial function of the random variables is built using regression analysis of the selected deterministic analysis results. This polynomial, called the response surface, can serve as a basis for further simulations and hence, in a better estimation of the probability of failure. Once the response surface model is obtained, suitable reliability analysis can be adopted to compute the Reliability Index of the system. Thus, RSM reduces the number of analyses for a process that involves large number of random variables and helps in arriving at a more reliable estimate of the response of a system.

The twin factors primarily affecting squeezing are the rock quality expressed in terms of Q Index and the overburden depth. Accordingly, the study is carried out in stages as follows:

1. The squeezing potential of an opening in rock, in particular, circular, is studied through RSM based reliability analysis
2. The squeezing potential of an opening in rock, in particular, circular, is studied through RSM based reliability analysis
3. In addition, for given squeezing potential and overburden combination, Reliability Index is evaluated for different probabilities of squeezing and effect of overburden depth is examined
4. Reliability analysis is carried out for several cases of tunnels (13 in all) covering different combinations of Q Index, overburden depth and probability of squeezing with a view to relate the Reliability Index with the factors governing squeezing.

A case study of a tunnel located in lower Himalaya midlands zone and which is known to have been subjected to squeezing related problems has been chosen. The case of Khimti tunnel in Himalayas is considered for all analyses. The typical rock material properties around the tunnel with 4 m diameter are selected based on those reported in Shrestha [8] as listed in Table 1.

Table 1 Properties of the rock medium

Elastic modulus E (MPa)	375
Cohesion c (MPa)	0.134
Angle of internal friction φ (deg)	21.2
Overburden depth (m)	112
Q	0.06
Bulk unit weight (kN/m^3)	26

Table 2 Random variables chosen for reliability analysis

Ground material property	Mean	COV (%)
Elastic Modulus E (MPa)	375	13
Cohesion c (MPa)	0.134	24
Angle of internal friction φ (deg)	21.2	6

2 Reliability Analysis of an Opening in Squeezing Ground

Based on the criterion set by Singh et al. [6], the above stated tunnel is categorized as non-squeezing. However, using the criteria suggested by Goel et al. [2], the tunnel is predicted to be subjected to mild squeezing conditions. This issue of categorization is revisited here through a reliability based analysis. To perform the reliability analysis, the material properties of the ground are considered as normally distributed random variables with the Means and COV chosen by studying the case studies of Himalayan tunnels [8] and other literature on reliability analysis [4], as shown in Table 2.

3 Application of RSM

There are many approaches for classifying and analyzing tunnels in squeezing ground conditions. However, attempts to predict the squeezing phenomenon considering the variability in the material parameters of the surrounding ground medium and evaluating the influence on stability of an underground opening are not evident. RSM combined with numerical analysis is suitable for this purpose. RSM is therefore used here to analyse the effect of variability in the ground material properties on tunnels subjected to squeezing conditions. The numerical analysis is performed using PLAXIS FE package. The FE mesh model used is shown in Fig. 1.

In view of the availability of well-documented basic data on actual Himalayan tunnels, the present study uses data of actual tunnels in order to make it more relevant. However, information such as the construction sequence or stress-strain properties were not reported. Considering that a reliability analysis accounts for uncertainties, in order to generalize the results, specific issues of tunneling such as three dimensional or stage excavation and non-linearity are excluded. A plane strain model with six hundred and ninety two elements is used to model the tunnel and the surrounding medium. Fifteen-noded triangular elements are used to model the surrounding medium whereas, the tunnel is modeled as a beam element. A total of 5726 nodes are used to model both the tunnel and the surrounding medium. Considering the Elastic modulus E, cohesion c and angle of internal friction φ of the rock as the random variables (Table 2), the coded variables and the results of the FE analysis are shown in Table 3.

Considering the strain development of 1% as the indicator of onset of squeezing condition, the performance criterion G(x) is chosen such that the ratio of the tunnel

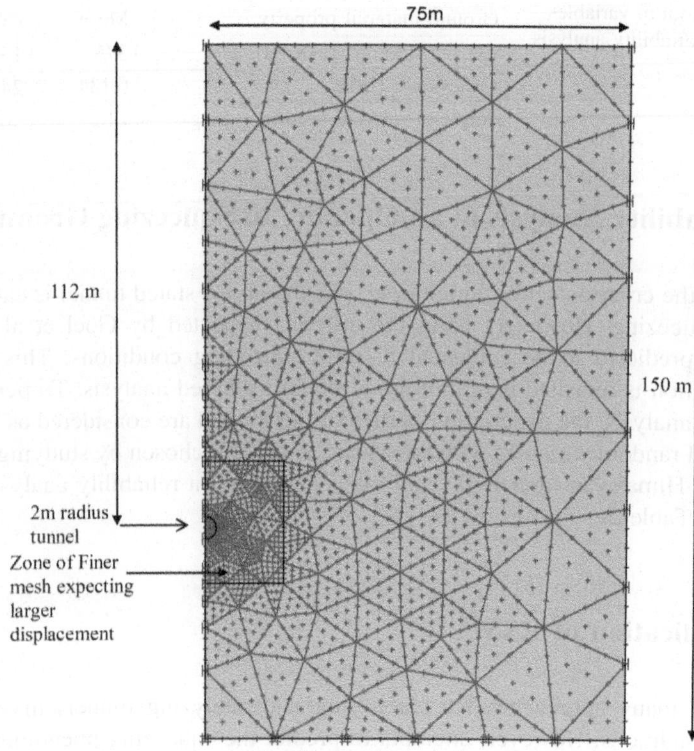

Fig. 1 FE discretization of the tunnel and the medium

closure (u) to the radius of the tunnel (r) does not exceed 1% as shown in the equation below:

$$G(x) = 0.01 - (u/r) \geq 0 \tag{1}$$

Based on these values, using second order regression, the response surface is built using Eq. (1) as:

$$u = \alpha_1 + \alpha_2 E + \alpha_3 c + \alpha_4 \varphi + \alpha_5 Ec + \alpha_6 E\varphi + \alpha_7 c\varphi + \alpha_8 E^2 + \alpha_9 c^2 + \alpha_{10}\varphi^2 \tag{2}$$

The coefficient matrix $[\alpha]$ consisting of the terms $\alpha_1, \alpha_2\ldots, \alpha_{10}\ldots$, is calculated by

$$[\alpha] = \left[X^T X\right]^{-1}\left[X^T\right][Y] \tag{3}$$

where the X matrix consisting of coded variables x_1, x_2, $x_1 x_{2\ldots}$ is as follows:

Table 3 Coded variables and corresponding FEM results

Coded variables	E (MPa)	c (MPa)	φ (deg)	Tunnel closure (m) u (m)
$a_{min}b_{min}c_{min}$	326.65	0.1005	19.986	0.214
$a_{min}b_{min}c_{avg}$	326.65	0.1005	21.2	0.155
$a_{min}b_{min}c_{max}$	326.65	0.1005	22.414	0.163
$a_{min}b_{avg}c_{min}$	326.65	0.134	19.986	0.186
$a_{min}b_{avg}c_{avg}$	326.65	0.134	21.2	0.155
$a_{min}b_{avg}c_{max}$	326.65	0.134	22.414	0.137
$a_{min}b_{max}c_{min}$	326.65	0.1675	19.986	0.131
$a_{min}b_{max}c_{avg}$	326.65	0.1675	21.2	0.115
$a_{min}b_{max}c_{max}$	326.65	0.1675	22.414	0.101
$a_{avg}b_{min}c_{min}$	375.4	0.1005	19.986	0.177
$a_{avg}b_{min}c_{avg}$	375.4	0.1005	21.2	0.176
$a_{avg}b_{min}c_{max}$	375.4	0.1005	22.414	0.131
$a_{avg}b_{avg}c_{min}$	375.4	0.134	19.986	0.162
$a_{avg}b_{avg}c_{avg}$	375.4	0.134	21.2	0.139
$a_{avg}b_{avg}c_{max}$	375.4	0.134	22.414	0.119
$a_{avg}b_{max}c_{min}$	375.4	0.1675	19.986	0.114
$a_{avg}b_{max}c_{avg}$	375.4	0.1675	21.2	0.100
$a_{avg}b_{max}c_{max}$	375.4	0.1675	22.414	0.088
$a_{max}b_{min}c_{min}$	424.15	0.1005	19.986	0.146
$a_{max}b_{min}c_{avg}$	424.15	0.1005	21.2	0.156
$a_{max}b_{min}c_{max}$	424.15	0.1005	22.414	0.126
$a_{max}b_{avg}c_{min}$	424.15	0.134	19.986	0.142
$a_{max}b_{avg}c_{avg}$	424.15	0.134	21.2	0.120
$a_{max}b_{avg}c_{max}$	424.15	0.134	22.414	0.105
$a_{max}b_{max}c_{min}$	424.15	0.1675	19.986	0.101
$a_{max}b_{max}c_{avg}$	424.15	0.1675	21.2	0.089
$a_{max}b_{max}c_{max}$	424.15	0.1675	22.414	0.078

[Y] stands for the response of the tunnel which is deformation in this case.

1	326.6	0.101	19.99	32.83	6528	2.009	106697	0.0101		399
1	326.6	0.101	21.20	32.83	6925	2.131	106697	0.0101		449
1	326.6	0.101	22.41	32.83	7322	2.253	106697	0.0101		502
1	326.6	0.134	19.99	43.77	6528	2.678	106697	0.018		399
1	326.6	0.134	21.20	43.77	6925	2.841	106697	0.018		449
1	326.6	0.134	22.41	43.77	7322	3.004	106697	0.018		502
1	326.6	0.168	19.99	54.71	6528	3.348	106697	0.0281		399
1	326.6	0.168	21.20	54.71	6925	3.551	106697	0.0281		449
1	326.6	0.168	22.41	54.71	7322	3.754	106697	0.0281		502
1	375.4	0.101	19.99	37.73	7503	2.009	140925	0.0101		399
1	375.4	0.101	21.20	37.73	7958	2.131	140925	0.0101		449
1	375.4	0.101	22.41	37.73	8414	2.253	140925	0.0101		502
1	375.4	0.134	19.99	50.3	7503	2.678	140925	0.018		399
1	375.4	0.134	21.20	50.3	7958	2.841	140925	0.018		449
1	375.4	0.134	22.41	50.3	8414	3.004	140925	0.018		502
1	375.4	0.168	19.99	62.88	7503	3.348	140925	0.0281		399
1	375.4	0.168	21.20	62.88	7958	3.551	140925	0.0281		449
1	375.4	0.168	22.41	62.88	8414	3.754	140925	0.0281		502
1	424.2	0.101	19.99	42.63	8477	2.009	179907	0.0101		399
1	424.2	0.101	21.20	42.63	8992	2.131	179907	0.0101		449
1	424.2	0.101	22.41	42.63	9507	2.253	179907	0.0101		502
1	424.2	0.134	19.99	56.84	8477	2.678	179907	0.018		399
1	424.2	0.134	21.20	56.84	8992	2.841	179907	0.018		449
1	424.2	0.134	22.41	56.84	9507	3.004	179907	0.018		502
1	424.2	0.168	19.99	71.05	8477	3.348	179907	0.0281		399
1	424.2	0.168	21.20	71.05	8992	3.551	179907	0.0281		449
1	424.2	0.168	22.41	71.05	9507	3.754	179907	0.0281		502

Once the initial response surface is built, the reliability analysis is performed using the Hasofer Lind reliability estimation method. This procedure is repeated until subsequent steps lead to convergence. The response surface equation corresponding to this minimum reliability is also obtained. For the squeezing criteria, the response surface equation is:

$$\begin{aligned} u = & -4.7684 + 0.003E + 16.94c + 0.23\varphi - 0.0048Ec - 0.00006E\varphi \\ & - 0.3993c\varphi - 0.000001E^2 - 17.981c^2 - 000293\varphi^2 \end{aligned} \tag{4}$$

The minimum Reliability Index and the corresponding limit state values of the random variables E, c and φ are:

$$\begin{aligned} \beta &= 1.28 \\ E_c &= 552.4\,\text{MPa} \\ c_c &= 0.326\,\text{MPa} \\ \varphi_c &= 24.7 \\ p_f &= 0.10 \end{aligned}$$

This indicates that there is a 10% probability of the tunnel failing under squeezing condition. Hence adequate tunnel support system to resist the squeezing effect needs to be designed. This shows that the characterization of the material as per Goel et al. [2] is corroborated. This also shows that (i) Reliability Index or the probability of failure can be related to the squeezing condition and (ii) the just-squeezing condition can be represented in an alternative form by the response surface equation (Eq. 5) in terms of the Modulus and strength properties of the material as well.

4 Reliability Analysis of Tunnel Support System Under Squeezing Conditions

For the above mentioned tunnel subject to squeezing action, a typical support system is now chosen consisting of steel sets and shotcrete as shown in Table 4. The shotcrete lining is 0.3 m thick with steel sets embedded at 0.6 m spacing.

A reliability based evaluation of this support system is conducted considering the lining-ground interaction as well as the variability in the ground material properties. The FE model is chosen representing the composite tunnel support system having equivalent stiffness as that of shotcrete and steel sets as shown in Fig. 2. The displacement, thrust and moment distributions in the lining are given in Fig. 3 and the capacity curves (M-N interaction diagram) for the chosen support system obtained through deterministic FEM computations are shown in Figs. 4 and 5.

Table 4 Details of chosen tunnel support system

Steel sets	
Steel set spacing (m)	0.60
Steel set height (m)	0.162
Area of steel set (m2)	0.00475
Moment of Inertia (m4)	2.23E-05
Modulus of steel (MPa)	200000
Poisson's ratio	0.25
Compressive strength (MPa)	500
Tensile strength (MPa)	−500
Shotcrete lining	
Shotcrete thickness (m)	0.3
Modulus of shotcrete (MPa)	36050
Poisson's ratio	0.15
Compressive strength (MPa)	40
Tensile strength (MPa)	−4.4
Area of shotcrete (m2)	0.18
Moment of Inertia (m4)	0.0014

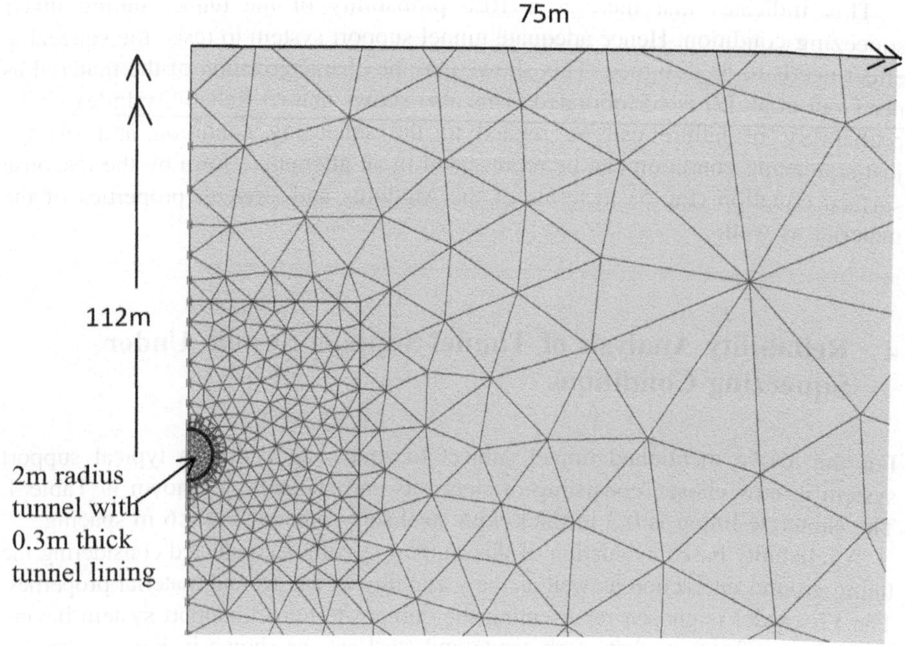

Fig. 2 Close view of the FE mesh around tunnel and support

From the M-N interaction diagrams it is evident that, the deterministic analysis finds the steel support system safe for the tunnel loading conditions while it is just safe for the shotcrete lining. Hence, a reliability based evaluation is carried out for the chosen support system of the tunnel in squeezing rock, considering the variabilities in the rock properties already mentioned in Table 2. The coded variables are given in Table 5.

For the performance of the tunnel lining, the Serviceability limit state (SLS) is used, considering the maximum permissible displacement in the lining as 5 mm. The response surface is built for the deformation, moment and thrust occurring in the tunnel lining. The performance functions for the Thrust, moment and deformation responses are as follows:

$$\text{ULS of Thrust: } N_c - N = 0 \tag{5}$$

$$\text{ULS of Moment: } M_c - M = 0 \tag{6}$$

$$\text{SLS of Deformation: } y_c - y = 0 \tag{7}$$

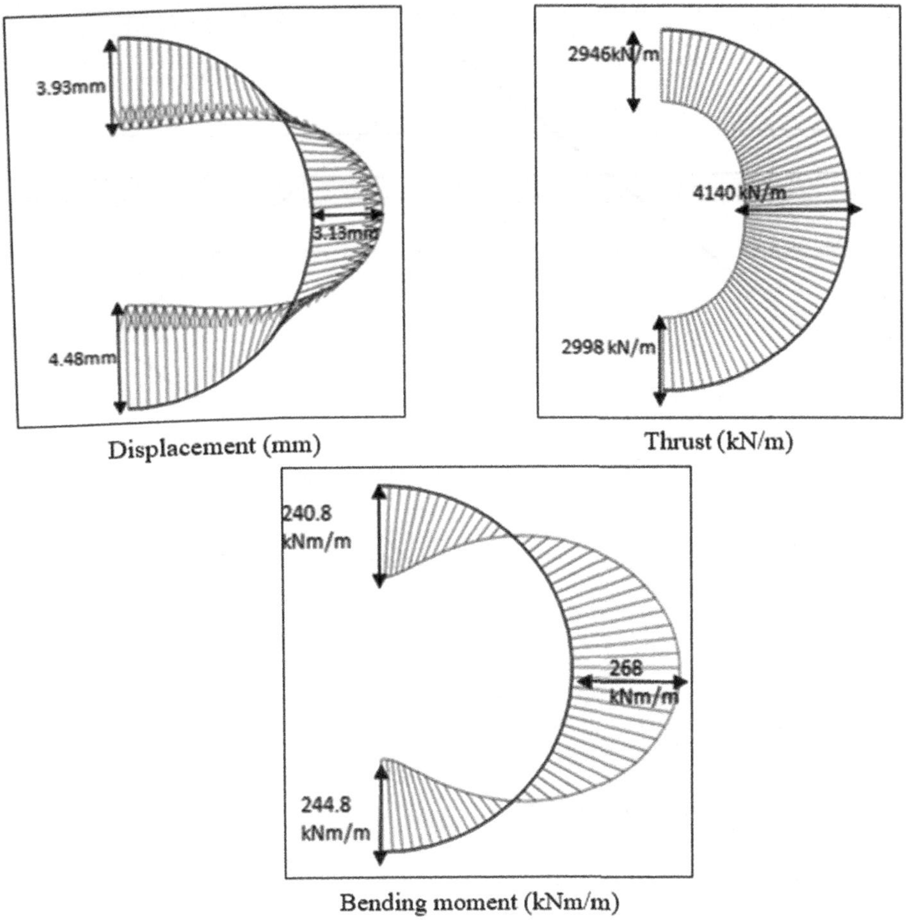

Fig. 3 Straining actions in the tunnel lining (displacement, thrust and bending moment)

Once the initial response surface is built, the reliability analysis is performed using the Hasofer Lind reliability estimation method. This procedure is repeated until subsequent steps lead to convergence. The response surface equation corresponding to this minimum reliability is also obtained.

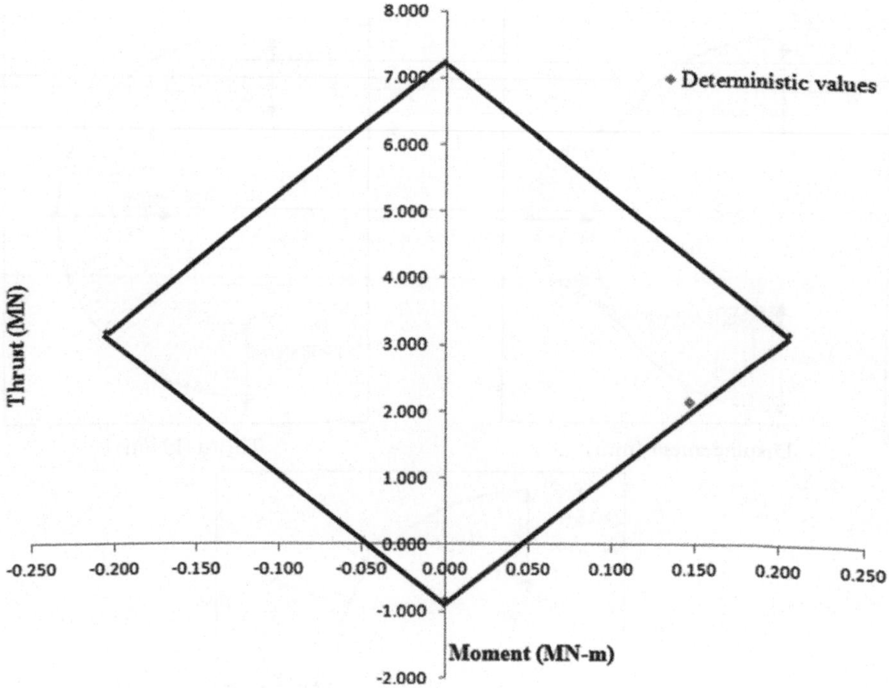

Fig. 4 M-N interaction diagram of the shotcrete lining

5 Target Reliability Indices

Table 6 lists the target reliability index for linings falling under class RC2 as per [1].

6 Results of Reliability Analysis of Lining

The minimum Reliability Index β and the corresponding limit state values of the random variables E, c and φ for the three responses (thrust, moment and deformation) of the lining support system are given in Table 7.

It is found that the reliability is very low in comparison to the target reliability, especially when the serviceability limit state is considered. However, the probability of failure (p_f), for the moment response is less compared to that of thrust and displacement. The critical values of the Elastic modulus, E at which the failure occurs is found to be relatively high for the thrust response. Therefore, the shotcrete

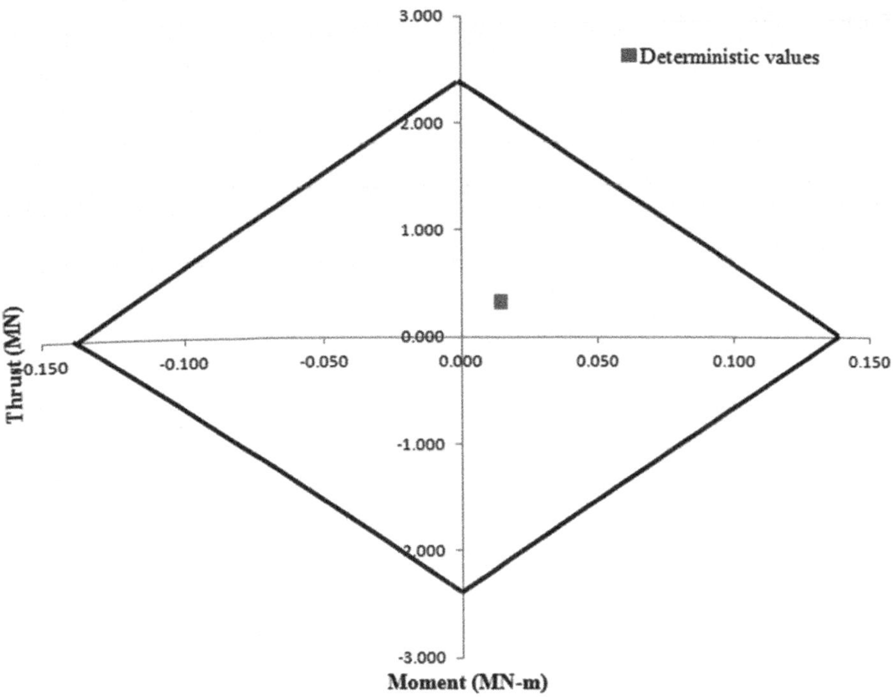

Fig. 5 M-N interaction diagram of the steel sets

lining moment capacity needs to be improved so as to lessen the probability of failure. Figure 6 represents graphically the variation of the thrust and moment response for the tunnel lining studied.

7 Effect of Tunnel Radius

In order to study the influence of radius of the tunnel on its reliability, the above RSM based reliability analysis is performed for tunnels of larger radius keeping all other properties unaltered. Table 8 lists the Reliability Index values and the corresponding critical values at which the limit state is reached for the moment and thrust response, considering tunnels of 2 m, 3 m and 4 m radius, respectively.

It can be seen that the Reliability Index values have increased for both moment and thrust response conditions of the tunnel. Thus, with all the conditions similar except for the tunnel diameter, a smaller diameter tunnel has lesser Reliability Index. It could be because a smaller diameter tunnel is more rigid compared to the larger tunnel and hence is subject to comparatively higher moment and thrusts which exceeds the design capacity of the lining resulting in higher probability of

Table 5 Coded variables used for RSM

Coded variables	E (MPa)	c (MPa)	φ (deg)
$a_{min}b_{min}c_{min}$	326.65	0.1005	19.986
$a_{min}b_{min}c_{avg}$	326.65	0.1005	21.2
$a_{min}b_{min}c_{max}$	326.65	0.1005	22.414
$a_{min}b_{avg}c_{min}$	326.65	0.134	19.986
$a_{min}b_{avg}c_{avg}$	326.65	0.134	21.2
$a_{min}b_{avg}c_{max}$	326.65	0.134	22.414
$a_{min}b_{max}c_{min}$	326.65	0.1675	19.986
$a_{min}b_{max}c_{avg}$	326.65	0.1675	21.2
$a_{min}b_{max}c_{max}$	326.65	0.1675	22.414
$a_{avg}b_{min}c_{min}$	375.4	0.1005	19.986
$a_{avg}b_{min}c_{avg}$	375.4	0.1005	21.2
$a_{avg}b_{min}c_{max}$	375.4	0.1005	22.414
$a_{avg}b_{avg}c_{min}$	375.4	0.134	19.986
$a_{avg}b_{avg}c_{avg}$	375.4	0.134	21.2
$a_{avg}b_{avg}c_{max}$	375.4	0.134	22.414
$a_{avg}b_{max}c_{min}$	375.4	0.1675	19.986
$a_{avg}b_{max}c_{avg}$	375.4	0.1675	21.2
$a_{avg}b_{max}c_{max}$	375.4	0.1675	22.414
$a_{max}b_{min}c_{min}$	424.15	0.1005	19.986
$a_{max}b_{min}c_{avg}$	424.15	0.1005	21.2
$a_{max}b_{min}c_{max}$	424.15	0.1005	22.414
$a_{max}b_{avg}c_{min}$	424.15	0.134	19.986
$a_{max}b_{avg}c_{avg}$	424.15	0.134	21.2
$a_{max}b_{avg}c_{max}$	424.15	0.134	22.414
$a_{max}b_{max}c_{min}$	424.15	0.1675	19.986
$a_{max}b_{max}c_{avg}$	424.15	0.1675	21.2
$a_{max}b_{max}c_{max}$	424.15	0.1675	22.414

Table 6 Target reliability indices [1]

Limit state	Target reliability index	
	1 year	50 years
Ultimate	4.7	3.8
Fatigue	–	1.5–3.8
Serviceability (irreversible)	2.9	1.5

Table 7 Minimum reliability index and corresponding limit state parameters

	Deformation	Moment	Thrust
β	0.046	0.477	0.094
E_c (MPa)	323.94	243.40	427.68
c_c (MPa)	0.13	0.13	0.13
φ_c (deg)	21.77	22.13	21.67
p_f (%)	48.17	31.67	46.26

failure. In short, with the same random variation in the material properties, a smaller diameter tunnel is more susceptible to failure (Fig. 7). This can be explained with the help of two important findings [6], namely (i) support pressure is independent of tunnel size in the present range and (ii) rock mass quality Q estimated from a larger tunnel would be smaller than that estimated for a smaller tunnel in a similar rock mass. In the present study, value of Q is kept constant at 0.06 in order to study the effect of overburden depth.

The deformation encountered in the tunnel-medium FE model is compared with the deterministic case and the critical case at which minimum reliability is obtained. Following are displacement distributions (Figs. 8, 9 and 10) obtained for a given tunnel radius of 2 m, 3 m and 4 m and corresponding to deterministic case and critical case, respectively. It is clearly observed that at the point of minimum reliability, the maximum displacements around the tunnel have increased compared to the deterministic case.

The results of reliability analysis considering deformation as response variable are given in Table 9.

8 Conclusion

The reliability analysis of a tunnel susceptible to squeezing action has been studied. A typical unlined tunnel is chosen and its susceptibility to squeezing action is first analyzed. The conventional approaches of squeezing classification treat the lining as 'non-squeezing to mild squeezing' for a given deterministic set of material properties. Thereafter, using the RSM based reliability analysis, the performance of the tunnel is checked against squeezing criteria. It is found that the Reliability Index is low and corresponds to a probability of failure of 10% under squeezing action.

The same tunnel is further analyzed after assuming a lining support system. It is found that the Reliability Index of the tunnel lining under serviceability conditions of moment, thrust and deformation is far from the target reliability. The maximum probability of failure is found to correspond to the deformation criteria. Also, the probability of failure is higher for the thrust response compared to moment response.

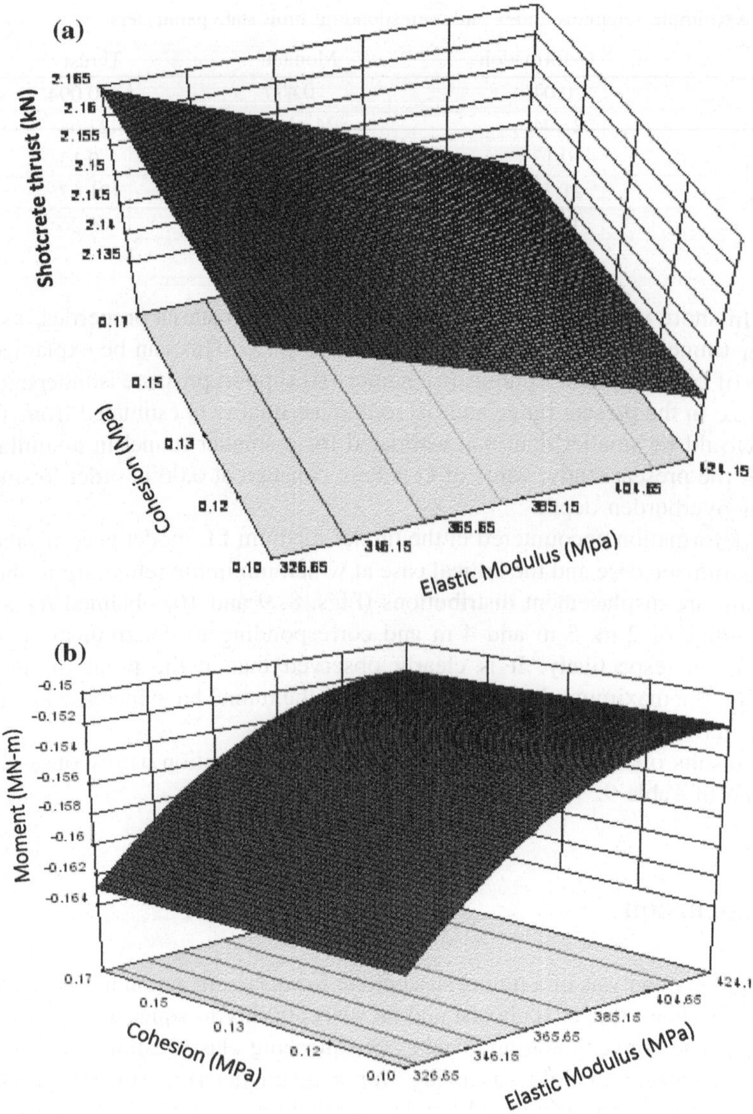

Fig. 6 Three dimensional representation of **a** Thrust and **b** Moment response of 4 m diameter tunnel lining

Table 8 Results of reliability analysis for varying tunnel radius

Radius (m)	2.0		3.0		4.0	
Response	Moment	Thrust	Moment	Thrust	Moment	Thrust
β	0.47	0.09	0.98	0.92	1.59	1.39
E_c (MPa)	243.40	427.68	167.30	2644.96	244.21	3103.53
c_c (MPa)	0.13	0.13	0.13	0.13	0.13	0.13
φ_c (deg)	22.13	21.67	22.76	30.45	17.84	38.33
p_f (%)	31.67	46.26	16.39	17.88	5.59	8.23

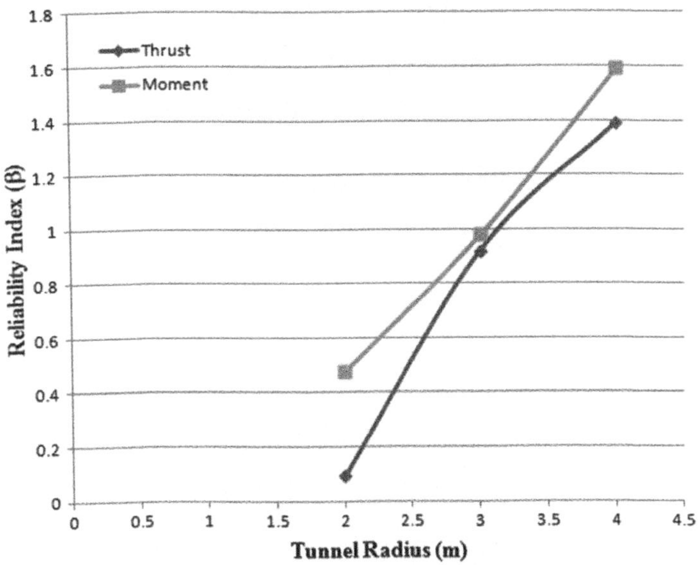

Fig. 7 Variation of reliability index with respect to tunnel radius

All the conditions remaining the same, the tunnel lining reliability is computed for varying tunnel radii. It is found that the reliability index values have increased for both moment and thrust response conditions of the tunnel. It also shows that the Reliability Index with respect to thrust and moment is lesser for a smaller diameter tunnel. This indicates that a smaller diameter tunnel is more rigid compared to the larger tunnel and hence is subject to comparatively higher moment and thrusts which exceeds the design capacity of the lining resulting in higher probability of failure. However, considering the deflection as the performance criteria, it is observed that the reliability index reduces with increase in the tunnel radius.

(a)

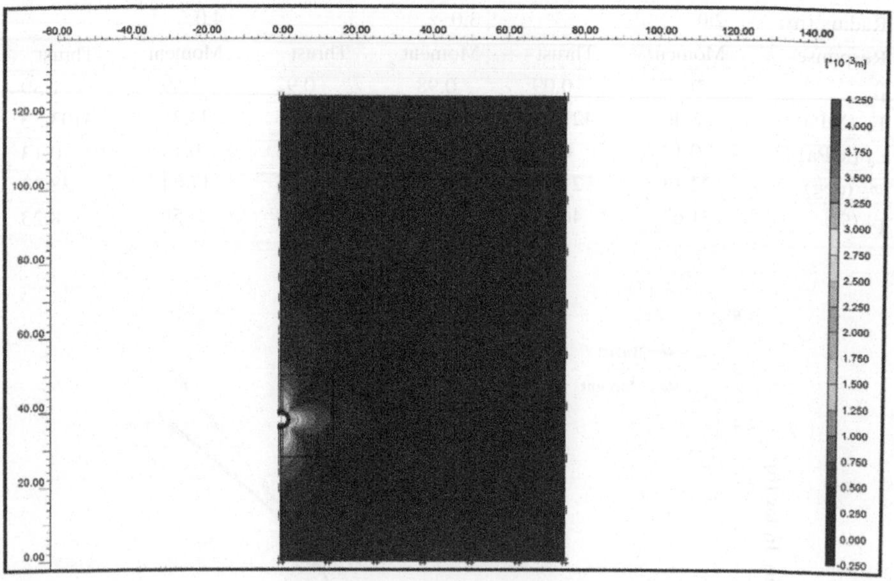

(b)

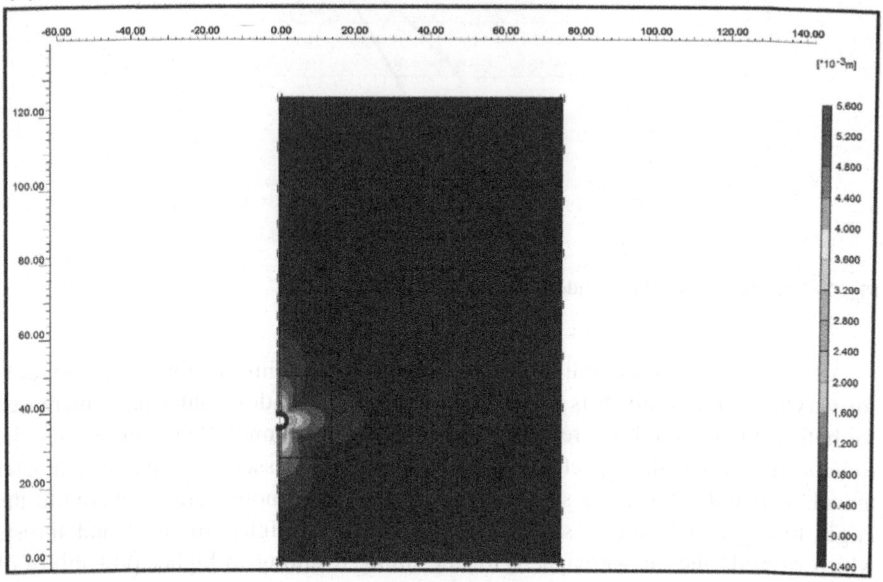

Fig. 8 Displacement contours in tunnel of 2 m radius, **a** deterministic case, **b** at minimum reliability index

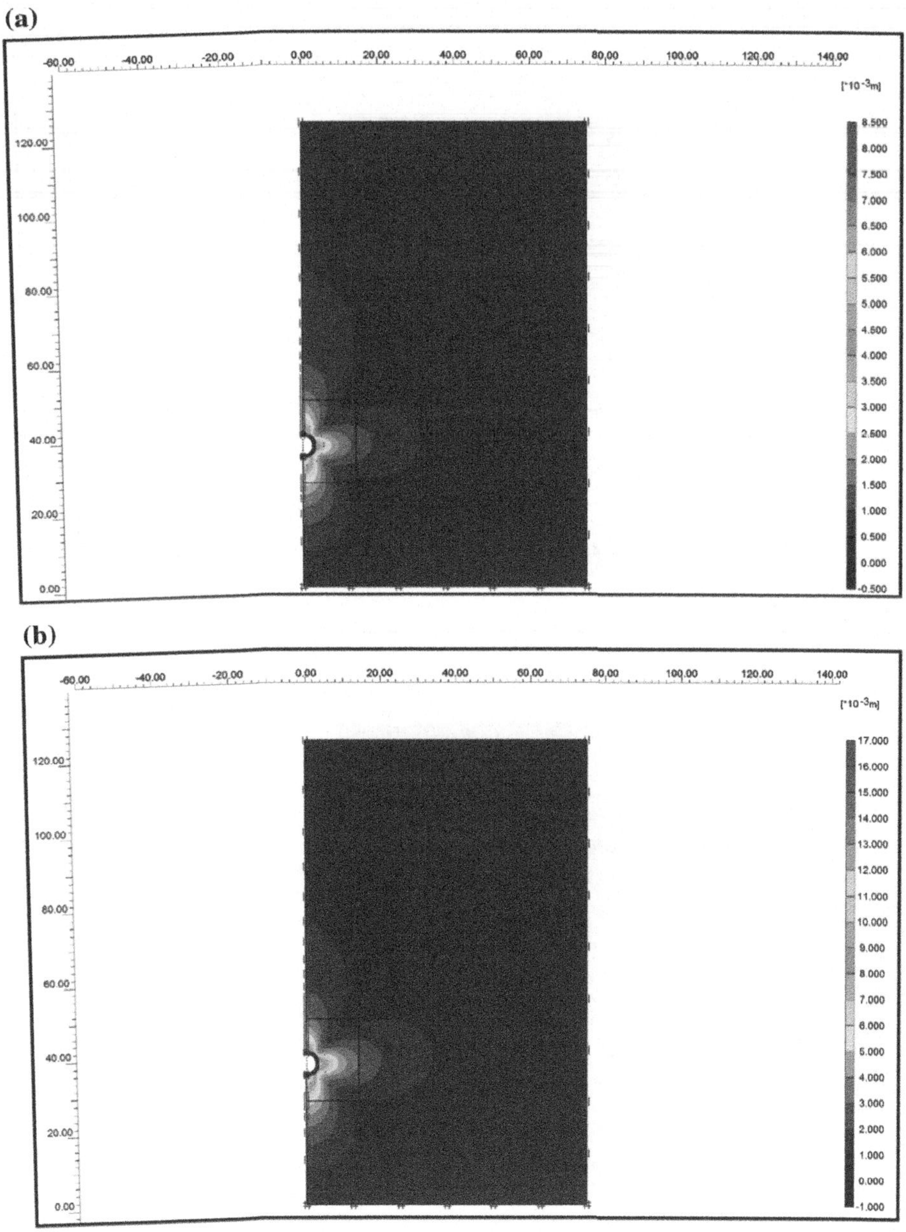

Fig. 9 Displacement contours in tunnel of 3 m radius, **a** deterministic case, **b** at minimum reliability index

(a)
(b)

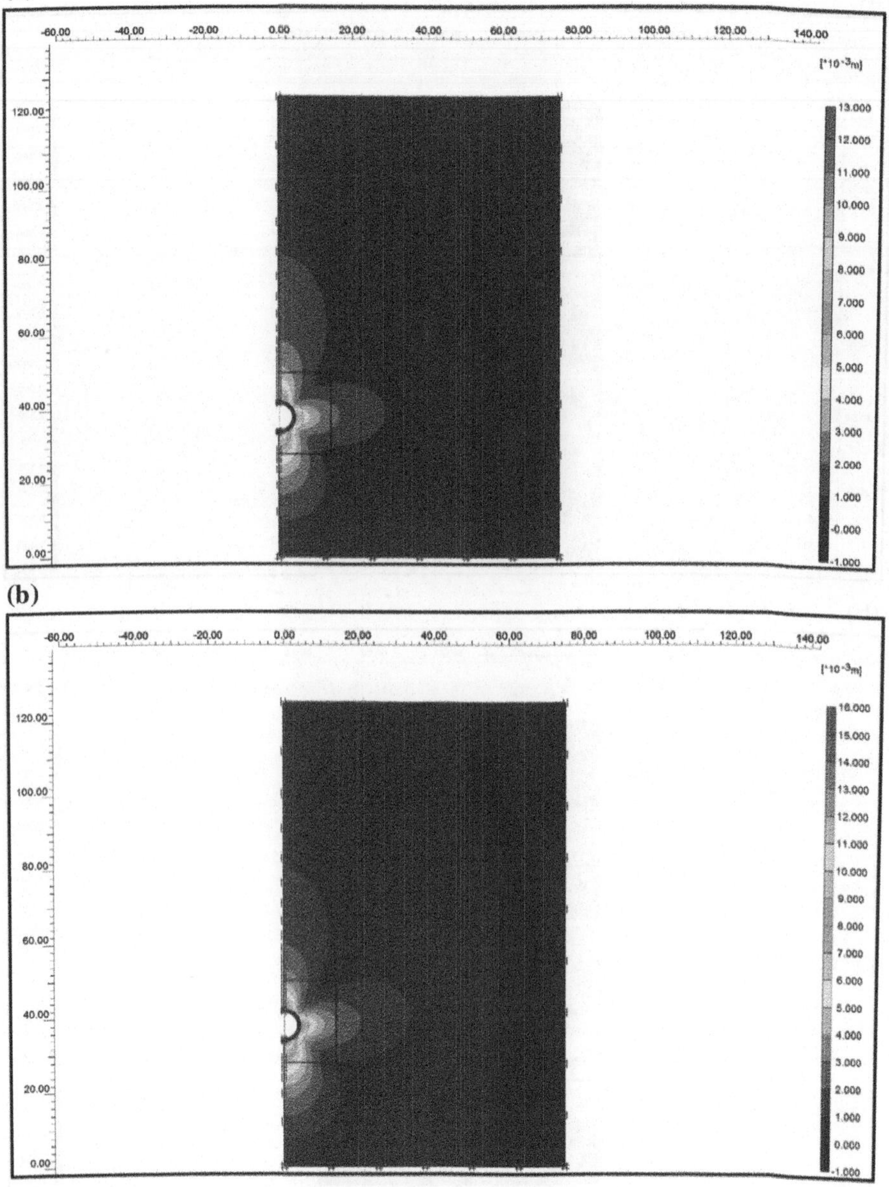

Fig. 10 Displacement contours in tunnel of 4 m radius, **a** deterministic case, **b** at minimum reliability index

Table 9 Results of reliability analysis for varying tunnel radius considering deformation as the response

Radius (m)	2.0	3.0	4.0
Response	Deformation	Deformation	Deformation
β	0.374	0.295	0.255
E_c (MPa)	243.40	167.30	244.21
c_c (MPa)	0.134	0.134	0.134
φ_c (deg)	22.13	22.76	17.84
p_f (%)	35.42	38.40	39.94

References

1. Eurocode (1990) Basis of structural design BS EN1990:2002
2. Goel RK, Jethwa JL, Paithankar AG (1995) Indian experiences with Q and RMR systems. Tunn Undergr Space Technol 10(1):97–109
3. Jimenez R, Recio D (2011) A linear classifier for probabilistic prediction of squeezing conditions in Himalayan tunnels. Eng Geol 121(2011):101–109
4. Mollon G, Dias D, Soubra AH (2009) Probabilistic analysis of circular tunnels in homogeneous soil using response surface methodology. J Geotech Geoenvironmental Eng ASCE 135 (9):1314–1325
5. Myers RH, Montgomery DC (1995) Response surface methodology process and product optimization using design experiments. Wiley, New York
6. Singh Bhawani, Jethwa JL, Dube AK, Singh B (1992) Correlation between observed support pressure and rock mass quality. J Tunn Undergr Space Technol 7(1):59–74
7. Singh Bhawani, Goel RK, Jethwa JL, Dube AK (1997) "Support pressure measurement in arched underground openings through poor rock masses. Eng Geol 48:59–81
8. Shrestha GL (2005) Stress induced problems in Himalayan Tunnels with special reference to squeezing. Ph.D. Thesis. Department of Geology and Mineral Resources Engineering. Norwegian University of Science and Technology, Trondheim
9. St. John CM, Zahrah TF (1984) Seismic design considerations for underground stuctures. J Adv Tunn Technol Subsurf Use 4(3):105–112

G. Venkatachalam is currently Professor Emeritus (Adjunct) at Mukesh Patel School of Technology Management & Engineering of NMIMS University at Mumbai. Prior to this, he was with the Civil Engineering Department of IIT Bombay from 1972 to 2015 and served in various capacities. He holds a B.E. degree from BITS, Pilani, India (1964), M.Tech. in Soil Engineering from IIT Bombay (1967), and Ph.D. from Leningrad Polytechnic Institute (1972). His research interests include slope stability, reliability analysis of slopes and tunnels, GIS and GPR. He has supervised 16 Ph.D. and 72 M.Tech. theses and authored more than 150 papers. He was a contributor to the establishment, at IIT Bombay, of Centre of Studies in Resources Engg. (CSRE), funded by DST and National Geotechnical Centrifuge Facility (NGCF), funded by DST, DRDO, and MHRD. He delivered the IGS Annual lecture in 2005. He was the National Coordinator for Quality Improvement Program of AICTE (1998–2000) and coordinated the preparation of the Model Curriculum for Engineering Education. He coordinated and contributed to a book on Soil

Mechanics for IGNOU and to Video Lectures on Soil Mechanics for NPTEL (2002–5). He was Convener, Bombay High Court-appointed Committee for evaluation of slope stability at Twin-tower site, M.P. Mill Compound, Cumballa Hill, Tardeo (2007), Member of International Society for Soil Mechanics and Foundation Engineering Technical Committee (TC–102) on In-Situ Testing (2010–13). He is also a consultant and academician with wide experience. He is an experienced Team Chair for accreditation of Engineering Colleges on behalf of National Board of Accreditation (NBA).

Dr. Mallika Sreedharan has 16 years of experience in engineering design and joined Dar Al-Handasah Consultants (Shair and Partners) in 2007 as Geotechnical Engineer at the Firm's Cairo office, Egypt. She is currently Senior Geotechnical Engineer in Pune office, India. She completed her master's degree in Geotechnical Engineering from IIT Bombay in 2005 following which she pursued Ph.D. in Reliability Considerations in Analysis of Underground Circular Openings Subjected to Squeezing and Seismic Effects. Presently at DAR, she is involved in various design, design review, and coordination activities for major projects in Africa, the Middle East, and the Arab Gulf region.

A. Srividya is currently a Guest Professor at Lulea Technical University, Sweden since August 2016 and has been an Adjunct Professor in the Maritime Studies in the autumn 2019 at the Western Norway University of Applied Sciences. She was an Adjunct Faculty at UiS Stavanger, Norway in the spring semesters of 2014–2016. She was Professor II at Stord/Haugesund University College from August 2014–July 2015 and a Guest Professor from August 2012–July 2014 at Stord/Haugesund University College, Haugesund, Norway. She joined IIT Bombay as a faculty in 1988 where she was a Professor in Civil Engineering for over 8 years with research focus in the areas of reliability and safety engineering & quality management. She has supervised/co-supervised 28 Ph.D. and 59 master's theses in the area of structural reliability, reliability-based optimisation, simulation studies for reliability estimation, quality benchmarking studies for service industries, quality systems, accelerated life testing, and condition-based maintenance. She has executed various research projects to promote industry interface and has been course coordinator for industry CEPs like Reliability Engineering, Quality Management, and Six Sigma for IT industries. She has jointly edited books on "Reliability Engineering and Quality Management" and "Quality, Reliability and Information Technology". She is a joint author in books titled (1) Fuzzy Reliability Engineering Concepts and Applications (Narosa), (2) Optimal Maintenance of Large Engineering System-Practical Strategies for Effective Decision Making (Narosa), (3) Reliability and Safety Engineering (Springer),

(4) Dependability of Networked Computer Based Systems (Springer), and (5) Risk Management of Non-Renewable Energy Systems (Springer). She has over 200 publications in various national/international journals and conferences. She is an Editor of International Journal of Systems Assurance Engineering and Management (IJSAEM) published by Springer and also as a member in the Communications in DQM editorial board.

Maintainability Engineering

Integrated RAMS, LCC and Risk Assessment for Maintenance Planning for Railways

Adithya Thaduri and Uday Kumar

Abstract As of today, about 70% of the transportation infrastructure has already built for the needs of customers, business and society, where Railways is the major infrastructure. Due to huge investment for renewal and overhaul, there is emergent need to maintain the infrastructure with high availability with minimum cost and risk, being, transportation is the backbone of the economy. These infrastructures normally lead to degradation due to operational loads, environmental factors and frequent interventions. Hence, planning and optimization of the maintenance actions with the constrained resources is implemented properly for the efficient operation. Due to the hierarchical nature of Railways, there is necessary for railway infrastructure managers to design a generic framework for the decision-making process when planning maintenance and interventions, which is an important functional block of asset management in railway infrastructures. This chapter proposes an integrated methodology to perform maintenance decision making using definitive "building blocks" namely Reliability, Availability, Maintainability and Safety (RAMS), Life Cycle Costing (LCC) and Risk assessment. It has to incorporates the "building blocks" at different planning levels in asset hierarchy; namely network, route, line and component and planning hierarchy; namely Strategic Asset Management Plan (SAMP), Route Asset Plan (RAP), Route Delivery Plan (RDP) and Implementation of Asset Maintenance Plan (IAMP) as proposed in IN2SMART which was renamed from ISO 55000 Asset Management Framework.

Keywords RAMS · LCC · Risk assessment · Maintenance planning · Railways

A. Thaduri (✉) · U. Kumar
Division of Operation and Maintenance Engineering, Luleå University of Technology, Luleå, Sweden
e-mail: adithya.thaduri@ltu.se

U. Kumar
e-mail: Uday.Kumar@ltu.se

D. R. Karanki et al. (eds.), *Advances in RAMS Engineering*,
Springer Series in Reliability Engineering,
https://doi.org/10.1007/978-3-030-36518-9_11

1 Introduction

Of all the transportation systems, railways are one of the most prioritized infrastructures due to its increase in reliance of customers and cargo over a period. Due to frequent loading of trains, environmental conditions, changing environment, inefficient operation, material degradation, the systems and components in the Railway infrastructure deteriorate over a period. To maintain the demands of the society and improve the performance and utilization of trains, there is a necessity of maintenance actions and process that needs to be performed on corrective or preventive basis. If the infrastructure degrades in such a way that any maintenance action doesn't make enough improvement, then infrastructure managers (IM) proposes to renewal or overhaul. In addition to these two main maintenance strategies, there are a combination of several maintenance strategies that are implemented to increase the availability and safety of the railway infrastructure. These collection of failures and maintenance actions that have effect on availability and safety constitute Reliability, Availability, Maintainability and Safety (RAMS). Each of these maintenance actions incur direct and indirect costs and the total cost shouldn't be more the prescribed budget. Hence, there is necessary to conduct life cycle costing (or analysis) (LCC) of the railway infrastructure during various phases of life cycle such as design, manufacturing, installation, operation, maintenance and disposal [50]. Although some studies considered RAMS and LCC separately (see e.g. [62, 67, 71]), still there is a need for an integrated approach for RAMS and LCC for railway industry [44]. If the maintenance actions are performed beyond a certain number of failures or reaching threshold for degrading components, then these actions will have higher consequences that could lead to higher risk.

To calculate RAMS, LCC and risk parameters, IMs are required to store different kinds of databases for monitoring and measurement of several infrastructure elements. Due to the centralized or decentralized system of different IMs in the World, these databases are stored in different locations with different partners which are quite disparate in nature [20]. For some assets, RAMS calculation is a straightforward calculation based on frequency of failure and maintenance actions. But due to the advancements of technologies such as condition-based maintenance [69], predictive maintenance and Maintenance 4.0 [31], IMs are interested more in degradation behaviour of the critical assets. In that case, RAMS need to be calculated considering the degradation behaviour, prediction of future condition and estimating the remaining useful life with different threshold limits. Traditionally, there are few methods are used to estimate RAMS parameters in this context, such as Markov modelling [7, 9], Petri-net approach [56], Particle Filter [46] and Bayesian approaches [37]. Due to the existence of different database storages, there will be a lot of challenges to calculate RAMS parameters such as delay in processing of data, localization, contractual limitations, data cleaning etc. Hence, IMs need to adopt a comprehensive architecture to store or structure the existing data to improve the monitor RAMS performance of railways.

Railway infrastructure constitute of different systems (of systems), subsystems, components and items that exit in the hierarchical nature with dependability on its operational functions. There is a variation of degree of functional and degradation of the components that will have an influence on the quality and performance of the infrastructure. To maintain the quality of the infrastructure, there is a need for measurement (or opportunistic inspection) of these components at specific times [21]. The quality will be quantified by several indicators, or key performance indicators (KPIs) such as service reliability, availability, cost-effectiveness, infrastructure safety, etc. [13]. As an IM, he/she is more interested to maintain these indicators within the prescribed or allocated budget constraints as recorded in their capital expenditure (CAPEX) and operational expenditure (OPEX). They are also constrained with number of resources available at their expense, such as human and machines for measurement and maintenance. In addition, as a responsibility of infrastructure, IMs tends to maintain the infrastructure in such a way that could lead to a smaller number of accidents, disruptions, CO_2 emissions according to government guidelines. Hence, there is a necessity for optimization of the maintenance activities in terms of cost effectiveness, RAMS and risk assessment. Although some studies and EU projects proposed RAMS, LCC and Risk assessment in different "building blocks" [30, 39, 42], there are only few studies that implement an integrated approach for all these building blocks for improving the performance of Railways.

To optimize RAMS, LCC and Risk, there is necessary to implement maintenance planning and scheduling of maintenance actions to obtain preferred KPIs [35, 65]. Traditionally, planning can be implemented at three main stages; strategic, tactical and operational planning. Intelligent Innovative Smart Maintenance of Assets by integRated Technologies (IN2SMART), an EU project, proposes four main maintenance planning schemas, mainly, strategic asset management plan (SAMP), route asset plan (RAP), route delivery plan (RDP) and implementation of asset maintenance plan (IAMP) based on ISO 55000 framework [27]. Maintenance planning and decision making also need to be performed based on context with detection and prediction of assets for estimation of RAMS [64, 63]. These planning schemas differ in time horizon and selection of asset in hierarchical structure of infrastructure [36]. There is still a lot that needs to be done in the context of planning hierarchy and asset hierarchy within the asset management framework on how to plan for the assets and how all assets interdependent on the planning and asset hierarchy. In addition, there is also a lot of scope on how RAMS will be defined at each planning schema. To implement maintenance planning, several combinations of strategies to be selected based on extraction of RAMS parameters and the best alternative strategies will be selected based on optimization algorithms. Within this context, this book chapter proposes an integrated RAMS, LCC and Risk assessment methodology for the maintenance planning of Railway infrastructure within the context of asset and planning hierarchy.

This chapter starts by introducing the basics of RAMS, LCC and Risk assessment building blocks in Sect. 2. Section 3 briefs about different maintenance planning levels incorporated in Railways in accordance with IN2SMART

(EU) project. Next Sect. 4 describes different types of databases exist in Railways. Section 5 provides a summary of the existing methodologies of RAMS and LCC in other European Union (EU) projects based on RAMS, LCC and Risk Assessment and further highlighted the importance of selection of maintenance planning at asset level or planning level. Section 6 proposed a novel methodology for an integrated RAMS, LCC and Risk assessment for maintenance and intervention planning to optimize the performance.

2 Introduction to RAMS, LCC and Risk Assessment

RAMS, LCC and Risk assessment are often used as optimization targets in the strategic maintenance planning for entire networks or subnets. At this level, they are one of the key decision-criteria for investments and maintenance strategies.

While RAMS describe the technical performance of an asset (system, module, component), LCC deals with the economic performance and Risk assessment deals with the consequences of actions. There are strong relationships and dependencies between RAMS, LCC and Risk assessment.

2.1 Definition of RAMS

RAMS of a system can be characterised as a qualitative and quantitative indicator of the degree that the system, or the sub-systems and components comprising that system, can be relied upon to function as specified and to be both available and safe [6].

RAMS according to EN 50126 is an abbreviation describing a combination of Reliability (R), Availability (A), Maintainability (M) and Safety (S):

Reliability is defined as *"the probability that an item can perform a required function under given conditions for a given time interval".*

Availability: "the availability of an object being in a condition in order to fulfil a required function under given terms and given period or during an alleged span of time provided that the required auxiliary materials/external tools are available".

Maintainability: "the feasibility that a certain maintenance measure could be executed for a component under existing boundary conditions within a defined span of time, if the maintenance will be made under defined conditions and defined process and auxiliary materials will be used".

Safety: *"the non-existence of an unacceptable damage risk. With respect to safety the common safety method has to be taken into account".*

Typical key parameters and risk related analyses are summarized in Fig. 1.

Key parameters and analyses

Reliability - *R*

- λ (t) - Failure Rate
- **MTTF - Mean Time To Failure**
- **MTBF - Mean Time Between Failures**
- MTTFF - Mean Time To First Failure
- MDBF - Mean Distance Between Failures
- MCBF - Mean Cycles Between Failures
- MFDT - Mean Failure Detection Time
- RBD - Reliability Block Diagram

Availability - *A*

- Theoretical or interior availability considers corrective maintenance
- Technical or engrained availability considers corrective and preventive maintenance

$$A = \frac{MTTF}{MTTF + MTTR}$$

Maintainability - *M*

- MMH - Mean Maintenance Hours
- MDT - Mean Down Time
- MCDT - Mean Corrective Downtime
- MPDT - Mean Preventive Downtime
- **MTTR - Mean Time to Restore (Repair)**

Safety - *S*

- HR - **Hazard Rate**
- THR - Tolerable Hazard Rate
- FME(C)A - Failure Mode, Effects (and Criticality) Analysis
- FTA - Fault Tree Analysis
- HA - Hazard Analysis
- ETA - Event Tree Analysis

Fig. 1 Key parameters and analysis for RAMS

2.2 *Definition of LCC*

The following explanations are based on the work done in INNOTRACK. A more detailed description can be found in the "Guideline for LCC and RAMS analysis" [26].

A life cycle cost analysis calculates the cost of a system, asset or component over its entire life span. The method is one of the most recommended for investment projects, assessment of different solutions over the whole life cycle and comparison of various strategies.

LCC phases are:

- Concept and definition,
- design and development,
- production,
- installation,
- operation and maintenance and
- disposal.

For a LCCA a system description and a cost structure are necessary shows the life cycle phases according EN 60300 and the related "headers" in the cost matrix shown in Fig. 2. Life cycle cost analysis (LCCA) is a layered method to assess all incurred costs within the technical life cycle of the system. In general, the main phases of a life cycle of the system are design and development, manufacturing, installation, operation and maintenance and disposal phases.

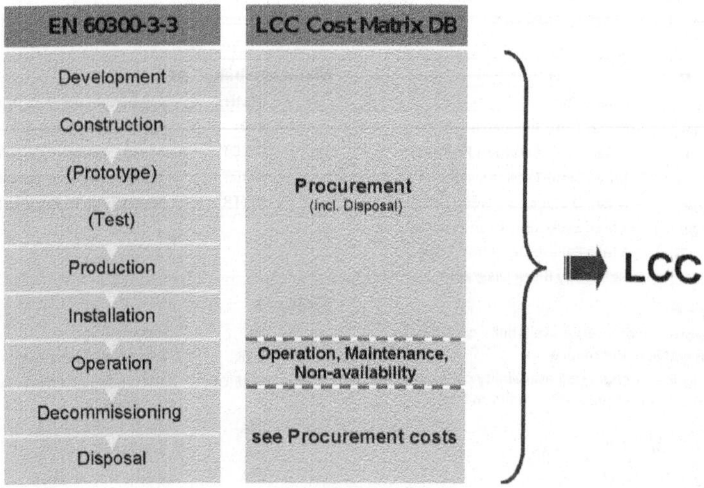

Fig. 2 Life cycle phases according (EN 60300-3-3 2017)

2.3 *Definition of Risk Assessment*

Risk is usually described by certain expected or unexpected events with subsequent consequences [54]. The outcome of the risk is either positive or negative that can be measured by its associate likelihood or occurrence of those events with respective consequences. Risk can be a quantitatively defined as product of frequency of events (occurrence) and respective consequences. Risk can be defined from different perspectives; financial, business, environmental, social, etc.

In Railways, the risk needs to be contained based on condition of the asset and respective type of maintenance action has to be taken. Hence, risk management aims to maintaining the control over uncertainty with reduced negative outcomes with proper treatment on the assets, for example maintenance actions. The three main steps in the risk management process are identification, analysis and evaluation of risks. For the informed decision making, one must consider the several possible events with their respective causes, consequences and probabilities of risk to take necessary action to reduce the overall risk of the asset [45] (see Fig. 3).

The first step in the risk management is the Risk identification that aims to generate a thorough list of risks based on events that will impact the condition of the assets. For each risk event, there is need to collect possible causes and sources that can potentially impact the performance of the system [54]. In the context of the conditional data, several failure events and its possible combination of precursors, drivers, failure modes and failure effects are identified.

The second step is the risk analysis that aims to develop either qualitative or quantitative approach to analyse the impact of the identified risks that might affect the performance of the asset [47]. This step involves the calculation of several risk

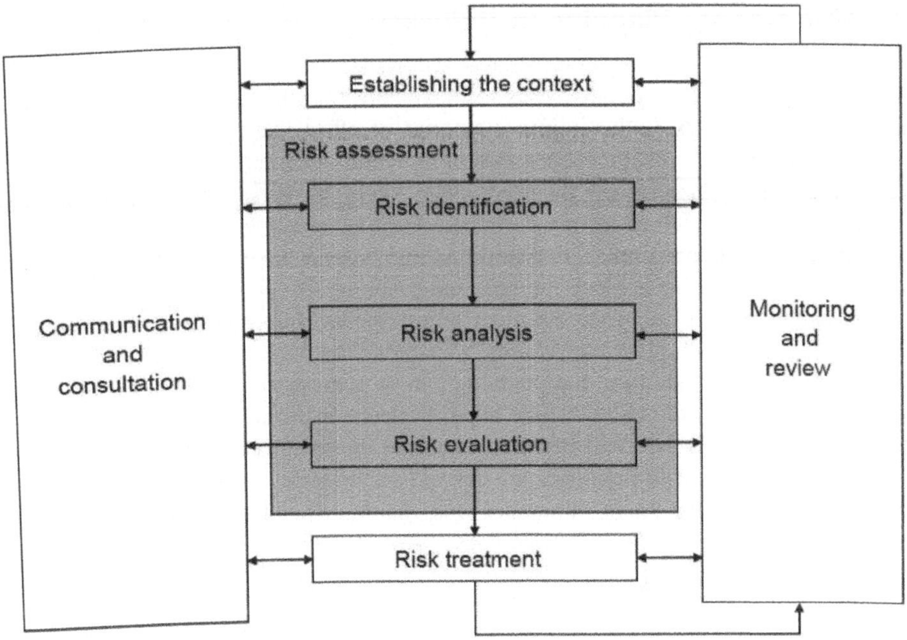

Fig. 3 Risk management framework

events with their respective frequency (probability of occurrence) and its level of consequences on the system. In some cases, the probability of detection of specific failure modes are also considered.

The next step in the risk management framework is the risk evaluation. This step is used to calculate the level of significance of each type of risk. If the overall risk is evaluated against the risk criteria, then this process is called risk assessment. The risk criteria will be determined by the legal and regulatory authorities. This step also signifies in establishing the main criteria for decision making w.r.t to their treatment and prioritization of tasks [28].

Risk assessment and management is a crucial task for many parties and agents like asset owners, maintainer or operators [48]. Although the terms safety and risk sometimes seem to be interchangeably, they refer to two separate aspects of system conditions. However, the task of Safety Management and Risk Management are not independent. Where Safety refers to a current condition Risk refers to the likelihood of maltreatment occurring in the future. Risk could be expressed by the severity of harm multiplied with the probability of the occurrence of harm.

3 Maintenance Planning in Railways

The main goals of IM of railways is to make sure that the assets are operating at maximum availability with minimum amount of effort. The effort here, signifies about budget, human resources, possession time and machine resources. Due to the several actions taking place on the assets, such as operation, inspection and maintenance, there is a necessary to implement maintenance planning. The maintenance process is a complex decision-making process in the asset management framework that able to achieve the required goals of IM. In practice, not a single model cannot able to achieve the asset requirements. Instead, several models in conjuncture with each other are required to achieve the prescribed key performance indicators (KPI). In addition, the planning can be segregated into several planning levels that are well defined with time horizon, decision variables, constraints, asset level and restrictions that will be imposed based on objectives.

In general, these planning levels are defined as; strategic, tactical and operational planning. Mostly, the main parameter that distinguishes among these planning levels is time horizon. In addition, these time horizons also varies with each stakeholder, infrastructure manager and application that depends on objective. The definitions of these planning levels are as follows [36].

- ***Strategic planning*** is defined at the decisions that mainly influence the maintenance management at the top-level with long-term horizon. The scope of the strategic planning mainly focuses at the systematic level and planning of long-term policies or methods that need to be implemented against KPIs. The planning process involves the selection of maintenance strategies for maintenance possession, allocation of the budget, improvement of the capacity and defining contractual agreements [2].
- ***Tactical planning*** is defined at the network level that involves more concrete specifications of the maintenance activities to be performed with predefined specific objectives within short-term horizon [3]. A simple example of tactical planning is the planning of maintenance actions within traffic management system with specific possession times on the field.
- ***Operational*** (or ***dynamic***) ***planning*** is defined at the on-field which has more concrete description of maintenance optimization with maintenance activities, maintenance scheduling of man and human resources. This planning level focuses on day-to-day planning of the maintenance activities. Online information and processing of the existing maintenance schedules with new maintenance optimization of the schedules is crucial [55].

The following subsections provide Intelligent Asset Management framework introduced by D2.1 of IN2SMART [11] as shown in Fig. 4.

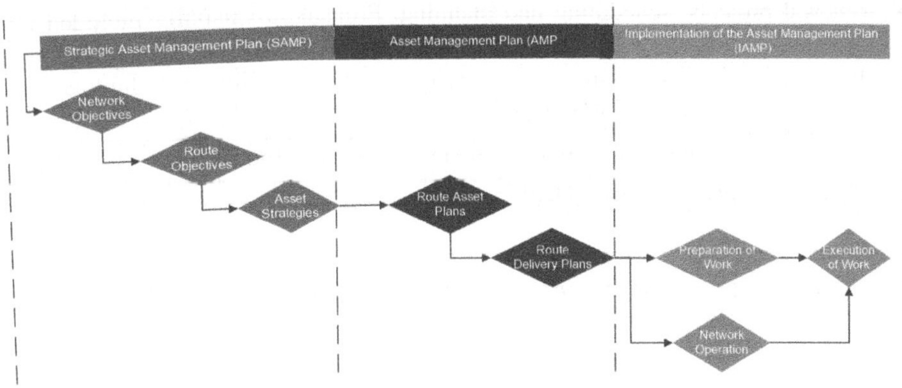

Fig. 4 Intelligent asset management system from IN2SMART D2.1

3.1 *Strategic Asset Management Plan (SAMP)*

Existing optimization articles are aimed at supporting:

- Design choices, e.g. unballasted versus ballasted track, wood/concrete S&C. The main objective is to ensure that maintenance issues and constraints are considered during the design phase [17].
- Significant maintenance policies and contracts, such as grinding and tamping [16]. These optimization studies are aimed at:
  - Testing and/or identifying different maintenance organisations in order to select the most efficient.
  - Supporting technical decision making, e.g. operational parameters, planning constraints and time tolerances.
- Renewal policies regarding specific components, such as signalling components and track equipment.

These articles mainly focused on track assets and maintenance processes and only a few discussed about other technical domains. Some multi-objective analyses have also been performed in order to support Asset Management decision making, based on both performance and costs objective functions.

3.2 *Tactical and Route Asset Management Plan (RAMP)*

The tactical Asset Management mainly focuses on the following points.

- Supporting investment decision making and, more precisely, the choice of the best renewal configuration (complete/partial/no track renewal). In this case, OPEX-CAPEX studies are used to analyse the profitability of the options considered.

- Renewal projects, scheduling and planning. Projects are usually included into programs and/or CAPEX trajectories, which are decided and dimensioned at the strategic level. The main issue here is to support allocating budgets and resources over time and assets groups.

In summary, the main objective at this level relates to the prioritisation of asset renewal, subject to strategic decisions.

3.3 Implementation of the Asset Management Plan (IAMP)

The last level of optimization studies focuses on operational planning and resources management. Providing optimization methods for this specific scope is particularly challenging, as it is necessary to integrate realistic operating conditions while there is a huge increase in the number of factors to be integrated, as this level is closer to the specifics of each elementary component.

There are two main topics discussed in literature:

- Maintenance planning and specific operation scheduling. Based on concept of condition-based maintenance operation
- Resources scheduling such as track possessions, budget or maintenance teams. Optimising inspections agenda based on journey times between assets, supporting asset prioritisation based on risks of failure and other indicators.

3.4 Interactions Among Planning Levels

Figure 5 also provides a summary of the inputs that are repeated in the various levels, along with the possible models applied at each stage. From the Figure, the definition of RAMS parameters varies with the planning schema and they are interrelated from one planning level to another planning level.

However, not much has been specified regarding the feedback among different stages of planning levels, i.e. about the information going from the lower levels of planning to the higher levels of planning. These connections will be briefly presented in this section. The information coming from the Route Asset Plan stage to the Strategic Asset Management Plan stage can be summarised in these two points.

- First, changes will be triggered in the overall strategy and policies. The aim will be to satisfy the needs at a lower level, for instance, at route level.
- Hence, there will be an adaptation of restrictions such as budget depending on these needs.

The two stages in the Asset Management Plan also have a strong connection, as the planning becomes closer to the asset and the particularities of the different areas of the infrastructure.

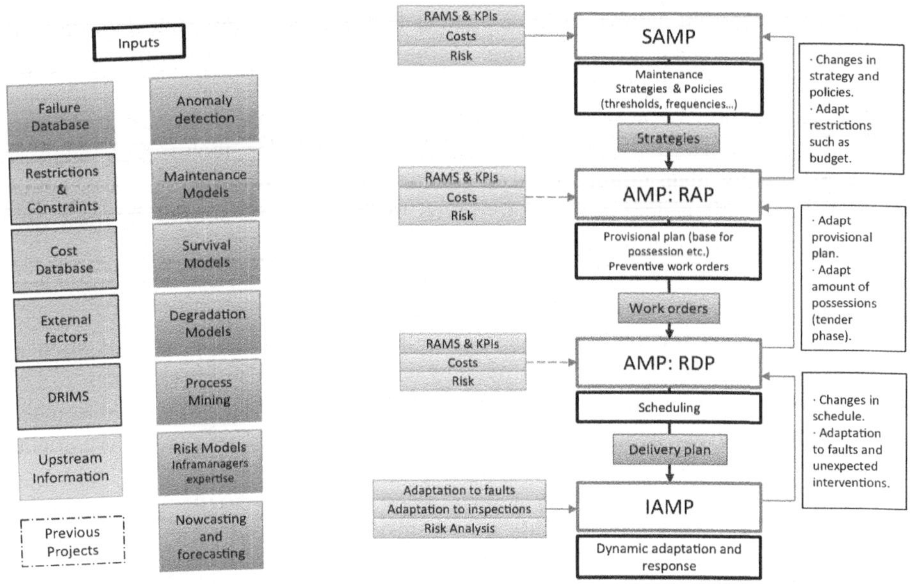

Fig. 5 Interaction among planning levels

- The RDP must adapt the provisional plan to satisfy external factors and further demands.
- Therefore, the amount of possessions might vary to adapt to these changes, which is a key issue in the tender phase.

Finally, there will be a strong feedback interaction from the Implementation of the Asset Management Plan to the Route Delivery Plan.

- Unscheduled events would trigger changes in the execution scheme coming from the RDP, which would also affect the preparation of work and the operation of the network.
- The adaptation to faults and unexpected events could also induce changes in the budget and resources required, which might encourage a change in the strategy to be followed from that point onwards.

4 Architecture of Failure and Cost Database

From the experiences gained from IN2SMART Research and Innovation (R&I) activities, a generic architecture of failure and cost database must be constructed. The following are the challenges faced to extract RAMS parameters from databases of each IMs are as follows:

- There are different database source sources that will store the asset data depends on the type of data such as failure, maintenance, inspection, traffic, measurement, registry, weather database etc.
- The information of a single asset needs to be extracted from all the above database sources that will need to be conducted for techniques such as pre-processing, data alignment, data cleaning and data mining
- This data is often accessed and retrieved by not only personnel in their own organization but also from other multiple stakeholders
- Data is stored in different formats and there is a need to standardize and harmonize the data to conduct RAMS and LCC analysis
- There are some direct RAMS parameters from these database sources such as time to failure, time to carry out maintenance, number of failures, etc.
- There is need for indirect way to extract RAMS parameters such as time to reach maintenance limit, availability, safety indicators, number of preventive maintenance actions etc.
- Collection of cost database is a huge issue that needs to be sorted out by individual IMs. Though some IMs are organized the cost database, there is still a standard type of method is necessary to collect and store the data
- The indicators of safety and risk are often qualitative or quantitative depends on the objectives of the use case.

A generic architecture of failure and cost database is developed from the lessons learned from R&I activities is shown in Fig. 6. The description of the architecture is as follows.

Each of IMs are several basic databases sources as listed in blue boxes such as Failure data, maintenance data, measurement data, condition data, inspection data and traffic data. There are two types of parameters as discussed above; direct and indirect parameters. These databases incur respective costs as shown in green boxes. Respectively, these databases incur respective costs such as cost of failures,

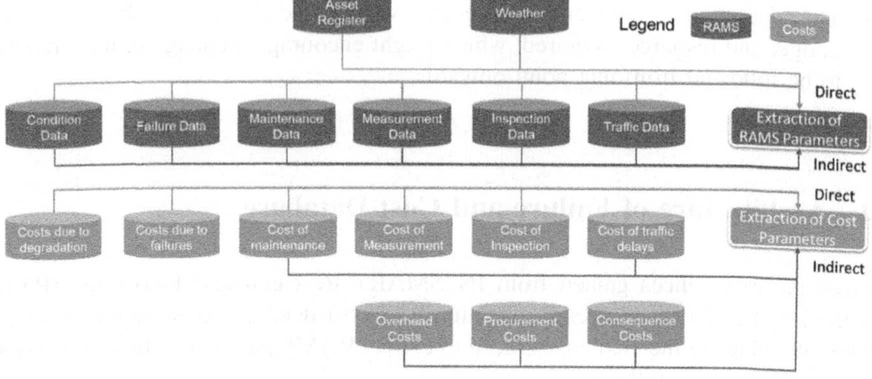

Fig. 6 Generic Architecture of Failure and Cost Database

cost of maintenance, cost of measurement, cost of inspection and cost of traffic delays. These extractions of costs are divided into two types; direct and indirect costs. The indirect costs are mainly due to unexpected costs in addition to the above costs such as overhead costs, procurement costs and consequence costs. These costs may vary depending on organization to organization and country to country. In addition, Asset register and weather database are generic database sources that does not incur costs. This vast amount of database requires data standardization and harmonization to get better benefits from the integrated RAMS, LCC and Risk methodology.

4.1 Guidelines for Future Data Standardization

There is a need for harmonization of the data for the railway to accomplish the main objective of UIC, International Union of Railways acting upon the SSO (Standards Setting Organisation) with leaflets [66]. It comprises of several railways that aims to achieve common rules for ensuring data for design, construction, operation, safety, security and maintenance of the railway system. Several deliverables consist of the activities that will facilitate uniform data standardization.

In that aspect, to provide attractive modes of transport for customer, there is necessary to make a technical harmonization and compatibility in terms of [58]:

- Compatibility among infrastructure, rolling stock and different stakeholders
- Compatibility among traction units
- Compatibility among different units of railways; traffic management systems, traffic signalling, operation and maintenance and management.

Hence, it is necessary to follow standardized rules based on technical harmonization of the data to ensure safe and reliable transport. The harmonization must be followed not only within a single organization but also among several organizations.

With the above background, it will be beneficial for carrying out RAMS analysis with the implementation of standard practices, compatibility and harmonization of the data. The business objectives can be achieved including the individual goals of the units into consideration to apply RAMS, LCC and Risk assessment with obtained uniformed data [68].

The accuracy of the implementation of integrated methodology depends on the quality of the collected data. In addition, there is also a lot of need for pre-processing of the data and this might influence the prediction accuracy of the methodology. In order to reduce these impacts, the standardization of failure, maintenance and conditional databases is required to perform an efficient analysis [59]. The advantage of the data harmonization is that more simpler models can be performed with better accuracy being harmonization will reduce the uncertainty issues in the database. Complex models can have significant impact on the maintenance planning but the development of these models will induce additional cost and time for IMs to implement in the demonstrators [59, 66].

In addition to asset failures, there are secondary failures induces into the system that also needs to be collected such as [59, 66]:

- Failures induced by software
- Failures induced due to human factors
- Failures induced due to uncontrollable factors such as environmental
- Failures induced due to common cause failures
- Data corrupted due to security issues.

The existing reliability and maintenance data often do not consider the data from the above factors and assume this data is a random failure that could reflect the uncertainty of the data there by predictions. Hence, there is a necessary to incorporate standardized data encapsulation and harmonization methods to acquire data and store in an organized way. This will further helpful in improving the Architecture of failure and cost database.

Several factors that will influence RAMS as shown in Fig. 7, which are common and can be considered on various industrial applications. Precisely, not every lack of information poses a serious issue to perform RAMS analysis and care should be taken to prioritize those parameters according to the specific objectives of the business, i.e., planning level, asset level and organization level [4]. Markeset and Kumar [43] defined some of the factors influencing the management of RAMS data.

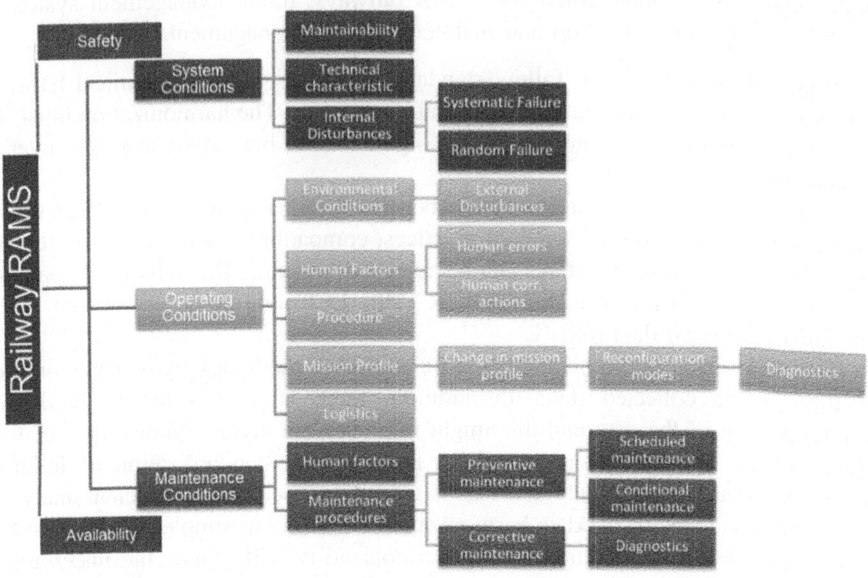

Fig. 7 Factors influencing railway RAMS

5 Methodology for RAMS and LCC

The main objective of this section is to provide an integrated methodology for RAMS and LCC for the maintenance planning.

5.1 *Railway Asset Hierarchy*

The hierarchical representation of railway infrastructure is shown in Fig. 5. Railway system comprises of complex system that are interconnected, interrelated and integrated together to perform the goals of the organization. Each block either system/subsystem/component/item has specific RAMS parameters or indicators. If RAMS analysis is performed from top-down approach, the output of RAMS analysis conducted at system level are considered as RAMS indicators at the subsystem level. Similarly, if RAMS analysis is performed from bottom-up approach, the RAMS analysis conducted at component level has indicators from subsystem level. The interesting aspect in this hierarchical nature of railway infrastructure is that the railway system is so complex such that predefined RAMS parameters/indicators cannot be extracted until there is a clear definition of objectives, planning level and requirements is known. Traditionally, railway IMs adopt hierarchical planning process according to the network; strategic, tactical, operational and/or short-term planning [22, 38]. It is also important to note that these RAMS parameters also change depends on the chosen asset. There exists a complex relation to even RAMS parameters among different hierarchical levels of the railway infrastructure [65].

The other way of looking into asset hierarchy will be based on whole railway network perspective that is shown in Fig. 8. Using a conventional top-down approach, the whole railway network can be broken down into operational routes where operation routes represents different areas of network. Each route further decomposes into several lines conventionally these lines are named between two important stations. Furthermore, these lines are further divided into track segments. Conventionally, each track segment is of 200 m or type of asset. As mentioned above, the RAMS targets/indicators are performed at network level and these parameters further expanded to the bottom levels.

Based on a hierarchical representation of the railway network, a framework for modelling railway infrastructure is presented by Rama and Andrews [56]. This framework is applicable for conducting RAMS analysis at various levels from individual item to large multi-asset networks. The hierarchical models are used as building blocks for infrastructure models with varying degrees of complexity. The data and information gathered on asset degradation, infrastructure utilisation levels, and intervention strategies are used to determine RAMS parameters.

Figure 9 shows the modified conceptual structure of the model [19]. Several degradation models/maintenance models/LCC models are developed using existing

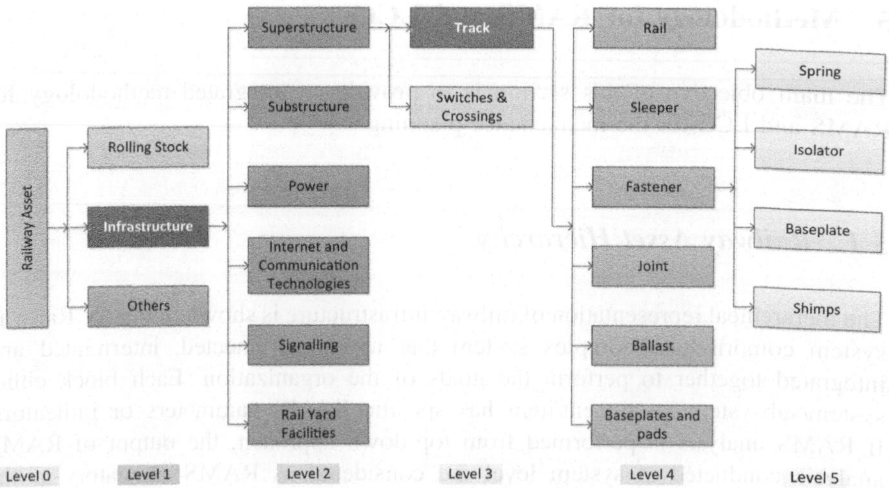

Fig. 8 Hierarchical representation of railway infrastructure [25]

Fig. 9 Hierarchical representation of railway network

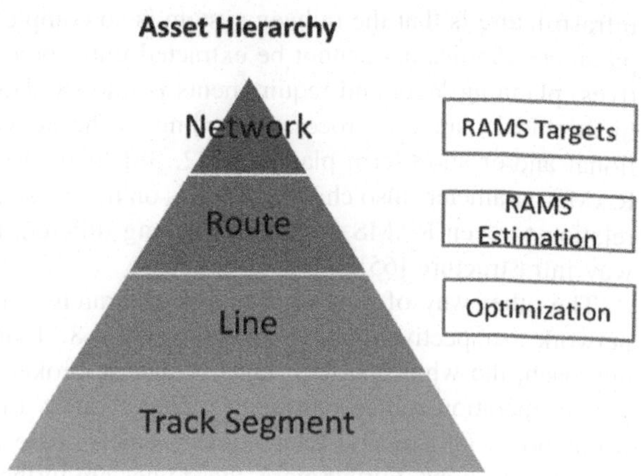

failure records, maintenance records or any other measurements data with associated costs databases. These modelling will be performed at the abstract level with comprehensive network and system details to obtain a system state model. The system state models can be any systems whether electrical system or track system. The system performance will be predicted based on KPIs defined in the asset management strategy [35].

Macchi et al. [41] implemented family-based approach (clusters which has common features) and emphasized that there must be a relation between RAMS parameters and the train service. The two-step methodology (building families of similar items and building a railway model) incorporates the reliability targets that

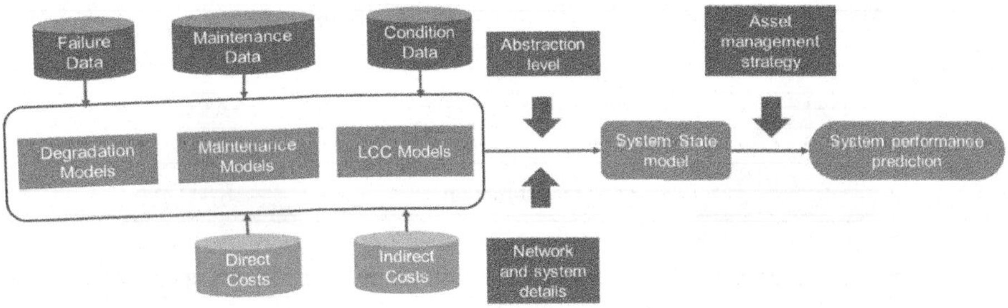

Fig. 10 A framework for modelling railway infrastructure

can be homogenous by infrastructure managers, as seen in Fig. 10. Using system RAMS analysis, the second step defines failure consequences on service levels expected from the train's circulation.

5.2 *Asset Hierarchy Versus Planning Hierarchy*

One of the main requirements for asset management perspective is to develop a generic methodology for RAMS, LCC and Risk assessment. However, the definition of the RAMS and cost parameters differs w.r.t asset hierarchy or planning hierarchy. For instance, the maintenance planning hierarchy is associated with asset hierarchy as shown in the Fig. 11. There is always a connection with maintenance optimization between railway assets and planning levels as shown in Fig. 12. The most presumptuous approximation for defining RAMS parameters are

- Setting RAMS targets for network at SAMP planning level
- Implementing RAMS optimization at RAP planning level
- Conducting optimization of resources at RDP and IAMP planning level.

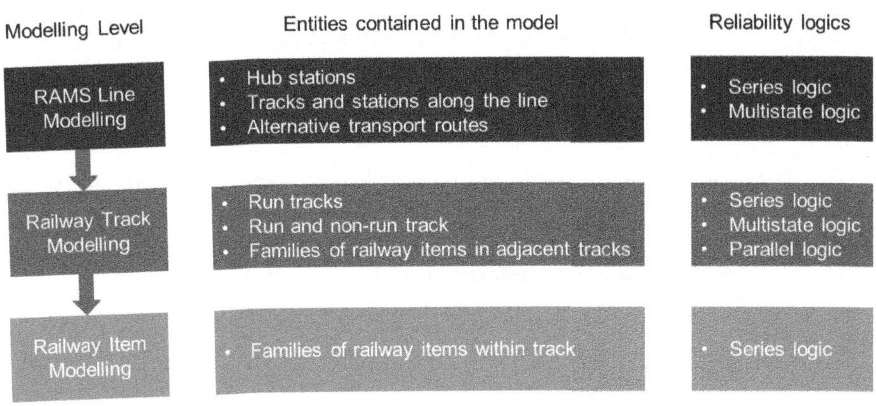

Fig. 11 Modelling at asset level

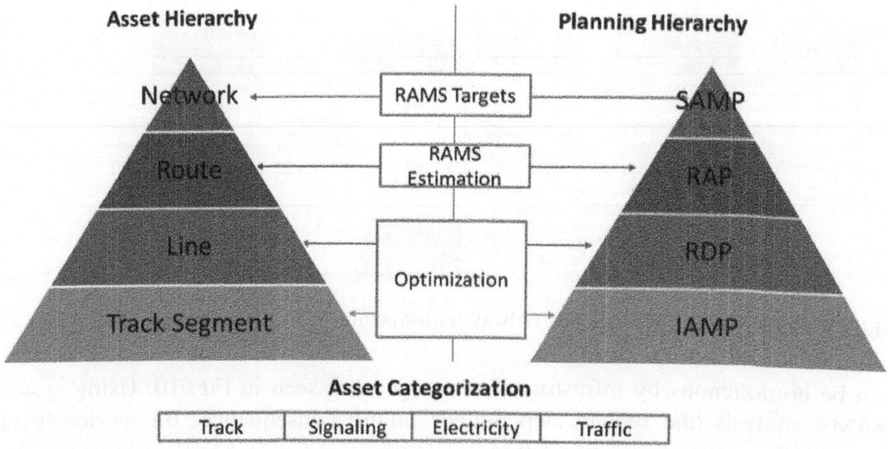

Fig. 12 Asset hierarchy versus planning hierarchy

However, this will not always be right in most cases. Because the main issues that, hinder to this reasoning are:

- often that the selection of asset and planning level depends on the business objectives which are sometimes at crossways
- the asset management framework developed in the railway organization is not always makes important to the technical details at the IAMP level
- the definition of asset varies with organization to organization because of their priority/criticality of the assets
- segregation of defining planning levels from different divisions within the railway organization
- due to involvement of multiple stakeholders and it is difficult to access resources and assets
- understanding of RAMS parameters at different levels. Each asset level defines RAMS parameters differently and often they won't interact together. For instance, at the SAMP level, most of the IMs are interested in availability of the asset but much focus on reliability. At the RAP/RDP level, it is a combination of different RAMS parameters. At the IAMP level, it is mostly about supportability and logistics of the resources.

Due to the above issues, it will not be practical per perceive that the RAMS & LCC methodology can be too strict to the definition of asset hierarchy and planning hierarchy. Hence, care should be considered in defining the assets and planning levels to understand and extract relevant RAMS parameters.

5.3 Methodologies from Previous Projects

This subsection briefly describes the aims and objectives of past and present, EU framework projects and the relevant assessment methodologies used within their respective projects. The main importance of this section is to provide an extended list of efforts to put the concept of RAMS, LCC and sometimes Risk on previous projects and their applications in Railway sector.

5.3.1 INNOTRACK

The main objective of INNOTRACK [15] was to reduce maintenance and infrastructure costs by 30%, by the application and implementation of existing innovative infrastructure for optimising maintenance and inspection [26]. Each parameter in RAMS incurs respective costs and technical specifications reflects economic specifications. It was highlighted that main driver for RAMS analysis is to conduct LCC analysis and cost/benefit analysis for optimising the maintenance. The project also provided several guidelines to conduct RAMS approach and include LCC analysis by using various methods such as present value, sensitivity analysis and Monte Carlo probabilistic approaches.

5.3.2 D-Rail

The main objective of D-Rail was to identify the root causes of freight traffic derailments [10]. The project identified the relation among reliability, availability and supportability parameters and maintainability for several scenarios for setting up RAMS targets as. Each of these scenarios incur several costs that was attributed to it such as preventive maintenance (PM) cost, corrective maintenance (CM) cost, logistics costs, ownership costs, energy costs, etc.

5.3.3 SUSTRAIL

In SUSTRAIL, the dependability parameters were considered to perform RAMS and LCC analysis for a wagon [57]. Within the dependability of the systems, the availability was a combination of reliability, maintainability and maintenance support. Each of these indicators were extracted from parameters like failures, repairs, preventative maintenance and logistical support respectively. In this project, the costs were clearly segregated based on specific parameters. For example, the logistical support was part of the investment cost; preventative maintenance and material cost for preventive maintenance cost; failure and repairs for corrective maintenance cost and other indirect costs for consequences cost.

This methodology will be further revisited in the next subsections because it has laid a significant contribution that relates to the asset hierarchy and planning hierarchy. In a way, the top down approach highlighted can be replicated to the methodology. Means that the methodology developed can be naively perceived as

- availability can be considered at the network level on SAMP planning
- reliability, maintainability and maintenance support at the route on RAP planning
- failure, repairs and other maintenance actions at the line on RDP planning
- and corrective and consequences at the track segment on IAMP planning.

5.3.4 Automain

Automain provided a relation in planning hierarchy where the objectives come from the network to the implementation of the work [1]. RAMS (Reliability, Availability, Maintainability and Safety) and LCC (Life Cycle Cost) methodology were acknowledged as the two main methods for supporting the optimization process. It was highlighted that the Key Performance Indicators (KPIs) and categories of performance killers/drivers for RAMS and LCC need to be identified, developed and transformed. These KPIs and audits were part of the reviewing mechanisms to ensure the railway system operates w.r.t required performance.

5.3.5 MAINLINE

The main aim of MAINLINE project was to develop methods and tools that contribute to cost efficient and effective improvement of European railway infrastructure based on whole life cycle [40]. The Life Cycle Assessment Tool LCAT that was being developed in MAINLINE covers main assets such as bridges, tunnels, track, cuttings and retaining walls. The LCAT discourses on the asset condition and respective planning of future interventions by considering

(a) maintenance and replacement strategies of the assets,
(b) lifecycle performance of the assets and
(c) conditional indicators from measurements.

The LCA results (related to CO_2 and waste) was given as an input to LCC analysis with the financial objectives.

5.3.6 ON-TIME

The main aim of the ON-TIME project is 'to improve railway customer satisfaction through increased capacity and decreased delays for passengers and freight'

(ON-TIME). Infrastructure planning, timetabling and operations, a general high-level measure of 'Quality of Service' (QoS) was incorporated in this project. The top factors show the key performance indicators (KPI) for QoS and the associated quantitative key measures and the bottom factors shows the static and dynamic factors that influence the QoS.

From the experiences from the previous project, it was clearly understood there is strong relation among RAMS, LCC and Risk parameters of the assets defined at asset hierarchy. The projects Innotrack and Sustrail highlighted the mainly on maintenance decisions based on RAMS and LCC. However, the generic methodology is not developed for implementing the maintenance planning at different planning levels. It is further discussed within R&I activities such that a generic methodology cannot be implemented for all planning levels at a once as it will too complex to perceive RAMS parameters at individual level and relation among them. The similar projects and standards for reliability and maintainability parameters with LCC can be found in [5].

6 Maintenance Decisions Based on RAMS, LCC and Risk

The previous section discusses on RAMS & LCC methodology in several existing projects. In order to improve the availability of the railway system with reduced overall cost and risk, there is a need to implement maintenance optimization process at several planning levels. The maintenance decisions can make use of building blocks of RAMS & LCC analysis including additional building block of risk analysis. The modified maintenance decisions can be made starting with business requirements as shown in Fig. 13 [50]. Business requirements define RAMS targets to achieve the goal of overall aim and vision of the railway organization. For optimizing the existing resources, there is necessary to include several strategies that can facilitate the better maintenance decision making. This further helps to define their maintenance budgets and contracts in short term and long term. RAMS, LCC and Risk methodology will facilitate these maintenance decisions. A single maintenance strategy consists of different decision variables or combination of several maintenance actions based on specific objectives and requirements. The above requirements can be defined in several maintenance strategies as M1, M2 and M3. From the previous projects, it is clearly highlighted LCC calculations plays a major role in defining the maintenance decision making. Hence, each maintenance incurs LCC costs as LCC1, LCC2 and LCC3. By including risk factors into the decision-making process, these strategies have individual risks as R1, R2 and R3. Now, a specific maintenance strategy(strategies) are to be selected based on higher performance with lower LCC cost and lower Risk.

Cost-effectiveness analysis produces quantitative results to assist the decision support system with risk analysis. Table 1 shows the calculation of cost-effectiveness from the LCC values of different maintenance strategies. When taking an optimized decision on maintenance strategy, it is necessary to measure the

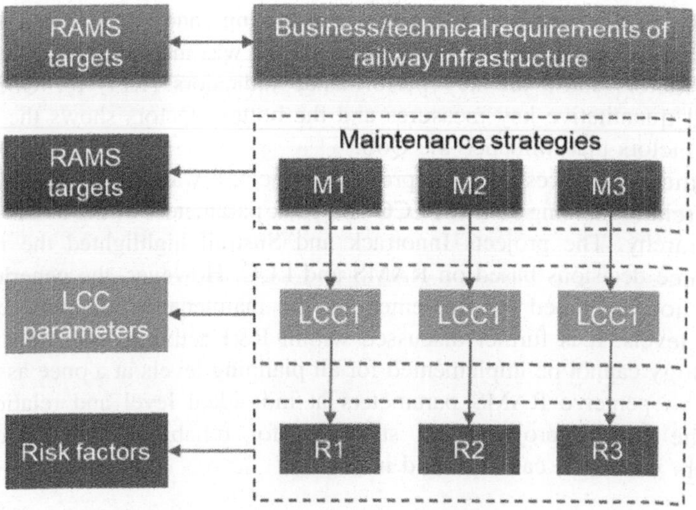

Fig. 13 Maintenance decisions based on RAMS, LCC and risk

Table 1 Cost-effectiveness of maintenance strategy

Maintenance strategy	System effectiveness	Life cycle cost	Risk effectiveness	Total effectiveness
M1	SE1	LCC1	RE1	SE1/(LCC1 * RE1)
M2	SE2	LCC2	RE2	SE2/(LCC2 * RE2)
M3	SE3	LCC3	RE3	SE3/(LCC3 * RE3)

cost-effectiveness of different maintenance strategies. The best maintenance strategy can be selected which highest cost-effectiveness.

Figure 14 illustrates the modified relationship among maintenance management, asset performance and asset maintenance [39]. The asset management of the railway system consists of an integrated RAMS, LCC and Risk management that derives the asset performance. This performance depends on the optimized combination of maintenance strategies consists of asset maintenance. Asset maintenance management affects activities varying from minimal maintenance actions to the higher maintenance and also ranges from short-term to long-term.

6.1 Integrated RAMS, LCC and Risk Assessment

The following conclusions are drawn from experience from previous EU projects:

- Most of the developed methodologies consistently dictates that there is a cohesive relationship between RAMS parameters and cost parameters

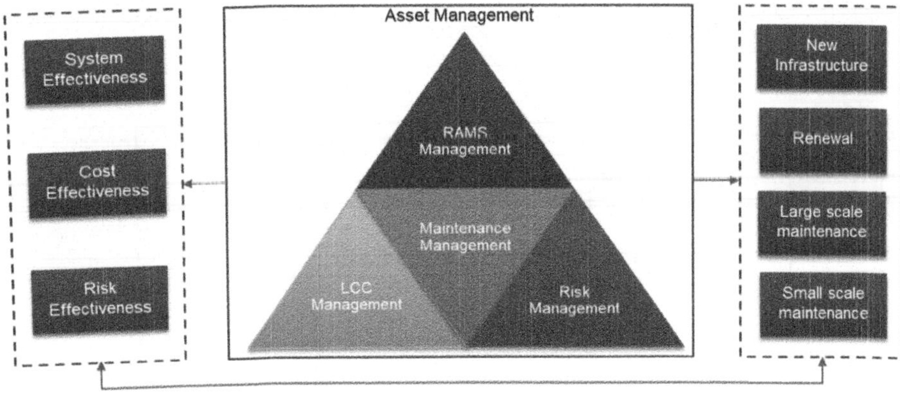

Fig. 14 Factors influencing maintenance management

- The methodologies were not applied to the maintenance planning
- Only few methodologies directly/indirectly identify risk as a defining parameter, but it was explicitly mentioned in their methodologies
- In addition, specific maintenance scenarios were not discussed though optimization is pointed out
- The bridge between RAMS and LCC and the hierarchical nature of railway assets and different maintenance planning levels were not clearly defined in the methodologies
- It was identified that LCC is an important building block for implementation of maintenance planning
- It is discussed such that these building blocks are more important for the strategic and tactical planning levels. But at the implementation level, these parameters can be considered indirectly because in day-to-day or week-to-week operations, mainly supportability and reducing risk are the highest priority
- It is also discussed that RAMS acts as leading indicators for strategic level and as lagging indicators for tactical planning
- The definition of RAMS and LCC parameters itself changes on asset and planning hierarchy.

Hence, to fill the gaps, there is a need to develop an integrated RAMS, LCC and Risk Assessment methodology is developed involving individual building blocks.

6.2 *Generic Framework*

This section provides a bridge between the developed generic framework and the building blocks. As part of IAMS asset management and decision support system, a generic framework is developed from maintenance and intervention planning in [24], which is shown in Fig. 15. The main building blocks in this framework are:

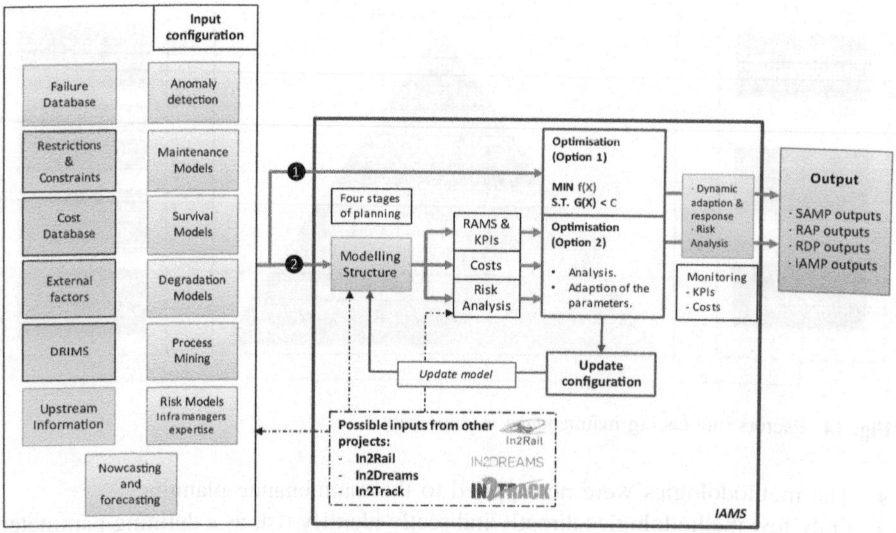

Fig. 15 Generic framework for maintenance planning

- Architecture of failure database, i.e., data required to define RAMS
- Architecture of cost database
- Modelling structure to extract RAMS KPIs
- RAMS, Costs and Risk Analysis blocks
- Optimization of the parameters to obtain optimum maintenance and intervention plan.

Each of the building blocks are connected to develop an integrated RAMS, LCC and Risk methodology.

6.3 *Proposed Methodology*

From the experience from the previous projects and IN2SMART R&I activities, there are some major gaps on facilitating the RAMS and LCC methodology for maintenance planning. Therefore, the main novel contributions of this Deliverable are highlighted in the following points;

1. Development of a generalized architecture of failure and cost database with further requirement of data harmonization from additional sources
2. Development of an integrated RAMS, LCC and Risk assessment methodology
3. Implementation of the developed methodology for maintenance planning
4. Emphasised the importance of maintenance planning between asset hierarchy and planning hierarchy.

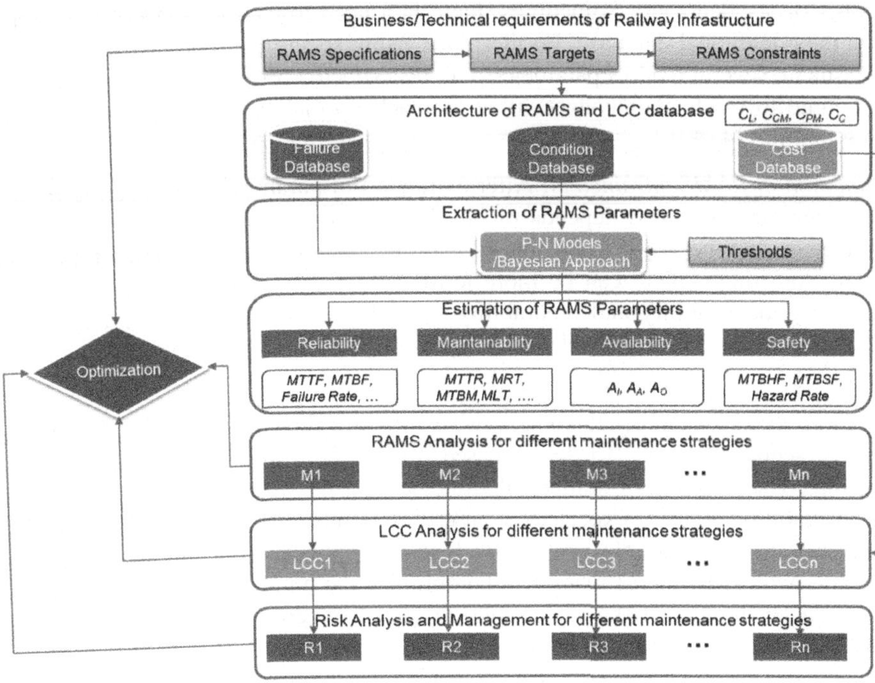

Fig. 16 Methodology for integrated RAMS, LCC and RISK assessment

Based on the discussions with partners and infrastructure managers in specific, Fig. 16 illustrates the generic integrated RAMS, LCC and Risk assessment methodology for maintenance planning. There are eight building blocks to implement the maintenance planning and those building blocks are explained below.

Business/technical requirements of railway infrastructure: For every railway organization, to operate different assets as per the operational requirements, there is necessary to define the RAMS targets. These targets are defined from both RAMS specifications and RAMS constraints as per the existing asset infrastructure. These specifications will include required asset availability, maintenance capabilities and safety regulations down to asset level [33]. The constraints will include budgetary requirements, available resources and limitations of the risk. RAMS targets often reflect the Key Performance Indicators (KPIs) of the organization [52].

Architecture of RAMS and LCC database: Each IM has its own architecture for failure and cost database. The main objective is to collect failures, inspections, interventions, measurements and associated costs from different databases of their own IM. The International Union of Railways (UIC) also developed universal infrastructure data exchange format called RailTopo Model and RailML [8].

Extraction of RAMS parameters: There are specific parameters that exactly reflects the RAMS parameters. These parameters include time to failure, time to repair, number of accidents etc. Sometimes, there is a need to extract RAMS parameters from the conditional database. Furthermore, using specific thresholds on condition assessment, several RAMS parameters can be extracted. For instance, time to reach alert, time to reach maintenance, time to reach normal failure and time to reach acute failure [34, 42].

Estimation of RAMS parameters: After extraction of parameters, respective RAMS parameters are estimated depends on the RAMS targets defined in step 1. These parameters can be segregated into reliability, maintainability, availability and safety. Sometimes, there can be more terms associated with RAMS such as supportability, detectability and reparability [32]. The definition of these parameters will change according to the selection of planning schema at asset level or planning level [36].

RAMS analysis for different maintenance strategies: Several strategies (or alternatives or scenarios) needs to define based on decision variables for maintenance planning. The selection of number of strategies can be lower or higher depends on the availability of data, expected accuracy of optimization, level of constraints, number of decision variables and its levels [50]. It is better to include strategies that covers a wide spectrum of planning. To optimize the number of strategies, several tools can be implemented for example design of experiments to select optimum runs [53]. RAMS parameters are to be extracted either from deterministic or probabilistic approach for each specific maintenance strategy [14].

LCC analysis for different maintenance strategies: Each maintenance strategy needs to be assessed by implementing LCC analysis [72]. Sometimes, depends on the availability of cost database, there is necessary to implement either total LCC analysis or cost-benefit analysis [61]. Each maintenance action incurs specific costs and these individual costs needs to be summed during their entire lifecycle [12].

Risk Analysis and management for different maintenance strategies: Each maintenance alternative need to be assessed by implementing appropriate Risk Analysis or risk assessment or risk management based on specific targets defined in Step1 [18, 51]. There can be qualitative or quantitative methods. There are several risk analysis methods available in literature such as Failure Mode Effect Analysis (FMEA), Fault Tree Analysis (FTA), Monte Carlo Methods, etc. to analyse the risk associated with each specific strategy [45].

Optimization of maintenance plans: The optimization of outputs extracted from RAMS analysis, LCC analysis and Risk analysis will be carried out against constraints and objective function of maintenance plan. To be general, a specific or set of maintenance alternatives to be selected based on maximum RAMS, minimum cost and minimum risk. Further, these alternatives can be further assessed individually to consider the other direct or indirect parameters such as availability of resources, long term goals of organization, business risk, and short- or long-term

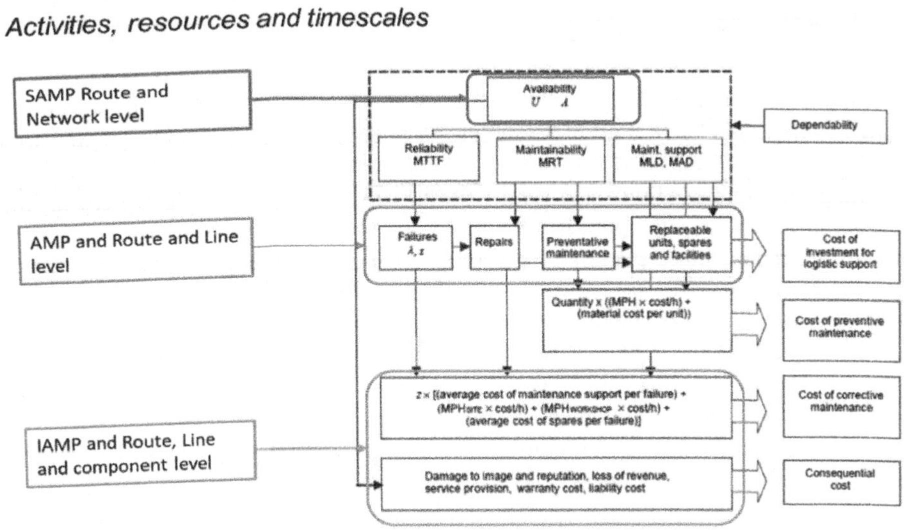

Fig. 17 Different planning hierarchy associated with RAMS parameters (modification from [57])

strategies. Several algorithms can be used to conduct optimization of maintenance plans such as cost optimization [60], linear programming [29], multi-criteria optimization [23], stochastic programming, online optimization [70], etc. and others mentioned in [4].

As several times discussed, it is difficult to conceptualize a generic RAMS and LCC methodology for different planning levels. Hence, the combined idea of the infrastructure managers is to define these parameters at each asset hierarchy level and emphasize to be laid on interaction between these levels. It is also necessary to convert these parameters from bottom-up approach or top-down approach. This will be further useful in defining the asset management framework of each Infrastructure Manager. The integrated methodology can be further illustrated in the simplest manner as shown in Fig. 17. As discussed in [57], it is categorized based on activities, resources and timescales as:

- SAMP route or network level can consists of availability as a basis for defining maintenance strategies.
- RAP/RDP or line level can consist of failures, repairs and preventive maintenance actions as defining maintenance strategies.
- IAMP or route, line or component level can consist of corrective actions and consequential costs as basis for maintenance strategies.

7 Conclusion

Railway infrastructure is a complex system which has different systems and components in hierarchical nature. In order to manage its performance with high availability and track performance, it is necessary to implement maintenance planning for short, medium- and long-term horizons. In order to implement it, this chapter proposed an integrated RAMS, LCC and Risk assessment methodology for supporting maintenance planning decision process. To increase the performance of Railways, there is still a lot of work to be done in formalizing the combinations of data formalization and standardization, planning hierarchy, asset hierarchy, estimation of RAMS parameters, complexities in degradation behaviour and uncertainties, etc. Due to advancements in technologies from other infrastructure, there will be a scope of improvement on above aforementioned areas to make the Railways sustainable due to competition from other transport infrastructure, degrading infrastructure, budget restrictions and increase in demands from the society.

References

1. AUTOMAIN (2013) D4.2 Optimised maintenance activities like, grinding, tamping and other maintenance processes. https://automain.uic.org/IMG/pdf/wp4_11-09-09-2013-f-d4_2-optimised_maintenance_activities_like_grinding_tamping_and_other_maintenance_processes.pdf
2. Al-Turki U (2011) A framework for strategic planning in maintenance. J Qual Maint Eng 17 (2):150–162
3. Baldi MM, Heinicke F, Simroth A, Tadei R (2016) New heuristics for the stochastic tactical railway maintenance problem. Omega 63:94–102
4. Blischke WR, Murthy DP (2011) Reliability: modeling, prediction, and optimization. Wiley
5. CAPACITY4RAIL (2014) D5.2.1 Compendium and evaluation of RAMS, LCC and migration tools, and methods and sources of data. http://www.capacity4rail.eu/IMG/pdf/d5.2.1_-_compendium_and_evaluation_of_rams_lcc_and_migration_tools_and_methods_and_sources_of_data.pdf
6. CENELEC E (1999) 50126: railway applications—the specification and demonstration of Reliability. Availability, Maintainability and Safety (RAMS)
7. COMMISSION IE (2006) IEC 61165: application of Markov techniques
8. Ciszewski T, Nowakowski W, Chrzan M (2017) RailTopoModel and RailML–data exchange standards in railway sector. Arch Transp Syst Telemat 10
9. Consilvio A, Di Febbraro A, Sacco N (2016) Stochastic scheduling approach for predictive risk-based railway maintenance. In: 2016 IEEE international conference on intelligent rail transportation (ICIRT) 2016. IEEE, pp 197–203
10. D-RAIL (2014) D7.2 RAMS analysis and recommendation. http://d-rail-project.eu/IMG/pdf/DR-D7_2-F2-RAMS_analysis_and_recommendation-20140926.pdf
11. Dambra C, Papa F, Mazzino N, Crovetto C (2017) IN2SMART D2.1 system requirements and functional, form fit and interfaces specifications. EU
12. D'Ariano A, Meng L, Centulio G, Corman F (2017) Integrated stochastic optimization approaches for tactical scheduling of trains and railway infrastructure maintenance. Comput Ind Eng

13. EN 15341 (2007) Maintenance key performance indicators. European Committee for Standardization (CEN), Brussels, Belgium
14. Ebeling CE (2004) An introduction to reliability and maintainability engineering. Tata McGraw-Hill Education
15. Ekberg A, Paulsson B (2010) INNOTRACK: concluding technical report. International Union of Railways (UIC)
16. Famurewa SM, Asplund M, Galar D, Kumar U (2013) Implementation of performance based maintenance contracting in railway industries. Int J Syst Assur Eng Manag 4(3):231–240
17. Famurewa SM, Xin T, Rantatalo M, Kumar U (2015) Optimisation of maintenance track possession time: a tamping case study. Proc Inst Mech Eng Part F J Rail Rapid Transit 229 (1):12–22
18. Fecarotti C, Andrews J (2016) Producing effective maintenance strategies to control railway risk
19. Fourie CJ, Zhuwaki NT (2017) A modelling framework for railway infrastructure reliability analysis. S Afr J Ind Eng 28(4):150–160
20. Galar D, Thaduri A, Catelani M, Ciani L (2015) Context awareness for maintenance decision making: A diagnosis and prognosis approach. Meas J Int Meas Confed 67:137–150
21. Garambaki A, Thaduri A, Seneviratne A, Kumar U (2016) Opportunistic inspection planning for railway emaintenance. IFAC-PapersOnLine 49(28):197–202
22. Gent T (2001) Tactical tools for integrated rail service planning. In: Proceeding of the world congress on railway research 2001
23. van Horenbeek A, Pintelon L, Muchiri P (2010) Maintenance optimization models and criteria. Int J Syst Assur Eng Manag 1(3):189–200
24. IN2SMART D9.2 (2019) Generic framework for decision support system in maintenance and interventions planning [Homepage of IN2SMART], [Online]
25. INFRALERT (2017) D3.2 Application of physical, data-driven and symbolic models from component to system level. http://infralert.eu/wp-content/multiverso-files/4_56128ae071b45/INFRALERT-D3.2.pdf. Accessed 15 Sept 2019
26. INNOTRACK (2006) INNOTRACK Guideline for LCC and RAMS Analysis. http://www.charmec.chalmers.se/innotrack/deliverables/sp6/sp6.html. Accessed 16 Feb 2019
27. ISO 55000 (2014) Asset management
28. ISO I (2009) IEC 31010: 2009. Risk management-risk assessment techniques
29. Jayakumar A, Asgarpoor S (2006) Maintenance optimization of equipment by linear programming. Probab Eng Inf Sci 20(1):183–193
30. Jiménez-Redondo N, Calle-Cordón A, Kandler U, Simroth A, Morales FJ, Reyes A, Odelius J, Thaduri A, Morgado J, Duarte E (2017) Improving linear transport infrastructure efficiency by automated learning and optimised predictive maintenance techniques (INFRALERT). In: IOP conference series: materials science and engineering 2017. IOP Publishing, pp 012105
31. Kans M, Galar D, Thaduri A (2016) Maintenance 4.0 in railway transportation industry. In: Proceedings of the 10th world congress on engineering asset management (WCEAM 2015) 2016. Springer, Cham, pp 317–331
32. Karim R, Jagare V, Lindberg L, Kumar U (2015) Robust and Reliable Transport System (RTS)—a strategic agenda for resource pooling, collaboration, innovation and research. https://www.ltu.se/cms_fs/1.129828!/file/RTS_2015_webb.pdf. Accessed 10 Feb 2019
33. Kumar S, Espling U, Kumar U (2008) Holistic procedure for rail maintenance in Sweden. Proc Inst Mech Eng Part F J Rail Rapid Transit 222(4):331–344
34. Kumar U, Ahmadi A, Verma AK, Varde P (2015) Current trends in reliability, availability, maintainability and safety: an industry perspective. Springer, Berlin
35. Kumar U, Galar D, Parida A, Stenström C, Berges L (2013) Maintenance performance metrics: a state-of-the-art review. J Qual Maint Eng 19(3):233–277
36. Lidén T (2015) Railway infrastructure maintenance-a survey of planning problems and conducted research. Transp Res Procedia 10:574–583

37. Lin J, Asplund M, Parida A (2013) Bayesian parametric analysis for reliability study of locomotive wheels. In: 2013 proceedings annual reliability and maintainability symposium (RAMS) 2013. IEEE, pp 1–6
38. Lusby RM, Larsen J, Ehrgott M, Ryan D (2011) Railway track allocation: models and methods. OR Spectrum 33(4):843–883
39. Lyngby N, Hokstad P, Vatn J (2008) RAMS management of railway tracks. Handbook of performability engineering. Springer, Berlin, pp 1123–1145
40. MAINLINE (2013) Deliverable 5.4: Proposed methodology for a Life Cycle Assessment Tool (LCAT). http://www.mainline-project.eu/IMG/pdf/ml-d5.4-f-proposed_lcat_methodology.pdf
41. Macchi M, Garetti M, Centrone D, Fumagalli L, Pavirani GP (2012) Maintenance management of railway infrastructures based on reliability analysis. Reliab Eng Syst Saf 104:71–83
42. Mahboob Q, Zio E (2018) Handbook of RAMS in railway systems: theory and practice. CRC Press
43. Markeset T, Kumar U (2003) Design and development of product support and maintenance concepts for industrial systems. J Qual Maint Eng 9(4):376–392
44. Martorell S, Villanueva JF, Carlos S, Nebot Y, Sánchez A, Pitarch JL, Serradell V (2005) RAMS C informed decision-making with application to multi-objective optimization of technical specifications and maintenance using genetic algorithms. Reliab Eng Syst Saf 87 (1):65–75
45. Meyer T, Reniers G (2016) Engineering risk management. Walter de Gruyter GmbH & Co KG
46. Mishra M, Odelius J, Thaduri A, Nissen A, Rantatalo M (2017) Particle filter-based prognostic approach for railway track geometry. Mech Syst Signal Process 96:226–238
47. Modarres M (2016) Risk analysis in engineering: techniques, tools, and trends. CRC Press
48. Mokkapati C, Tse T, Rao A (2009) A practical risk assessment methodology for safety-critical train control systems
49. ON-TIME, D1.2—A framework for developing an objective function for evaluating work package solutions (Cost function). http://kajt.org/onewebmedia/ONT-WP01-D-NRI-030-02%20-%20D1.2%20-%20A%20framework%20for%20developing%20an%20objective%20function%20for%20evaluating%20work%20package%20solutions%20.pdf
50. Patra AP (2009) Maintenance decision support models for railway infrastructure using RAMS & LCC analyses
51. Podofillini L, Zio E, Vatn J (2006) Risk-informed optimisation of railway tracks inspection and maintenance procedures. Reliab Eng Syst Saf 91(1):20–35
52. Praticò FG, Giunta M (2017) Proposal of a key performance indicator for railway track based on LCC and RAMS analyses. J Constr Eng Manag 144(2):04017104
53. Prescott D, Andrews J (2013) Modelling maintenance in railway infrastructure management. In: 2013 proceedings annual reliability and maintainability symposium (RAMS) 2013. IEEE, pp 1–6
54. Purdy G (2010) ISO 31000: 2009—setting a new standard for risk management. Risk Anal: Int J 30(6):881–886
55. Quiroga LM, Schnieder E (2010) A heuristic approach to railway track maintenance scheduling. WIT Trans Built Environ 114:687–699
56. Rama D, Andrews JD (2015) A holistic approach to railway infrastructure asset management. Int J Perform Eng 11(2):107–120
57. SUSTRAIL (2015) D5.1 RAMS and LCC analysis. http://www.sustrail.eu/IMG/pdf/sustrail_d5_1_13042015.pdf
58. Simões GMP (2008) RAMS analysis of railway track infrastructure
59. Smith DJ (2005) Reliability, maintainability and risk: practical methods for engineers; 1993
60. Soleimanmeigouni I, Ahmadi A, Letot C, Nissen A, Kumar U (2016) Cost-based optimization of track geometry inspection. In: World congress of railway research: 29/05/2016–02/06/2016
61. Stenström C, Norrbin P, Parida A, Kumar U (2016) Preventive and corrective maintenance–cost comparison and cost–benefit analysis. Struct Infrastruct Eng 12(5):603–617

62. Swier J (2004) Exposing the real infrastructure cost drivers. Railw Gaz Int 160(1)
63. Thaduri A, Kumar U, Verma AK (2014) Computational intelligence framework for context-aware decision making. Int J Syst Assur Eng Manag 5(1):1–12
64. Thaduri A, Galar D, Kumar U (2015) Railway assets: a potential domain for big data analytics, 2015. Elsevier, pp 457–467
65. Tzanakakis K (2013) Maintenance planning using LCC KAI RAMS. In: The railway track and its long term behaviour. Springer, Berlin, pp 293–296
66. UIC SSO (2019) UIC—International union of railways—Standards Setting Organisation. https://uic.org/spip.php?page=leaflets-irs-table. Accessed 16 Feb 2019
67. Vatn J (2002) A life cycle cost model for prioritisation of track maintenance and renewal. Innov Cost Eff Railw Track 2:20–25
69. Villemeur A (1992) Reliability, availability, maintainability and safety assessment, assessment, hardware, software and human factors. Wiley
68. Wenzel R (2011) Condition-based maintenance: tools to prevent equipment failures
70. Wu Q, Cole C, McSweeney T (2016) Applications of particle swarm optimization in the railway domain. Int J Rail Transp 4(3):167–190
71. Zoeteman A (2006) Asset maintenance management: state of the art in the European railways. Int J Crit Infrastruct 2(2–3):171–186
72. Zoeteman A (2001) Life cycle cost analysis for managing rail infrastructure. Eur J Transp Infrastruct Res 1(4)

Adithya Thaduri He has been working as Associate Senior Lecturer in the Division of Operation and Maintenance Engineering at Luleå University of Technology. He has experience in coordination of four European projects (IN2RAIL, INFRALERT, IN2SMART, and FR8RAIL) and three national projects (InfraSweden, Mindi, and SKF) in the area of railways and has worked in collaboration in other seven projects. I recently got funding for one European project for railways (IN2SMART2) and two national projects: one from Vinnova to railway and other from Coal India Limited to mining. He is part of over 35 deliverables/reports within above mentioned projects. He has authored over 40 research publications (28 after Ph.D.) in journals, book chapters, and conference proceedings. He has been teaching Maintenance Engineering course for master's program for 2 years. His areas of research are machine learning and context-aware maintenance decision-making within the framework of Maintenance 4.0 in railways, asset maintenance analytics, prognostics and degradation modeling of railway infrastructure, reliability predictions, maintenance planning and optimization, RAMS, LCC and risk assessment, predictive analytics of mining machines, and cybersecurity.

Uday Kumar is the Chair Professor of Operation and Maintenance Engineering at Luleå University of Technology (LTU), Sweden. He is also Director of Research and Innovation (Sustainable Transport) at LTU and Director of Luleå Railway Research Center at LTU. Dr. Kumar is also a Visiting Guest Professor at National Science Foundation sponsored Center of Excellence for Intelligent Maintenance Systems at University of Cincinnati, Ohio, USA, since April 2011. He is also an external examiner at University of Manchester, UK for the graduate program in the area of Asset and Reliability Engineering. Earlier, he has been a Visiting Faculty at Imperial College London, Helsinki University of Technology, Helsinki and Stavanger University in Norway, Tromsö University, among others. He has published more than 250 papers in international journals and conference proceedings and has co-authored two books on maintenance. He is an Honorary Professor at Beijing Jiao Tong University, Beijing, China. He is an elected member of the Royal Swedish Academy of Engineering Sciences.

Implementation of Predictive Maintenance Systems in Remotely Located Process Plants under Industry 4.0 Scenario

P. G. Ramesh, Saurav Jyoti Dutta, Subhas Sarma Neog, Prandip Baishya and Indrani Bezbaruah

Abstract Rapid developments in technologies such as Robotics, Digital Automation, Internet of Things and AI have heralded the Fourth Industrial Revolution, commonly referred to as Industry 4.0 (i4.0). Industrial operations and products have since become more competitive and hence more demanding. Systems have also become more complex and inter-disciplinary in nature. Diligent surveillance of operating conditions of such systems and initiation of appropriate actions based on monitored conditions have become indispensable for sustainability of businesses. Significant amount of research is being undertaken world over to meet this requirement of the day. In line with the ongoing research, this paper highlights the need for identifying the needs of condition monitoring preparedness of process plants located in remote places, especially in a logistic sense. Issues related to assessment of the need for the new paradigm in condition monitoring, challenges faced by such plants in the transition from legacy systems to a new system and customisation and optimisation of Predictive Maintenance under Industry 4.0 (PdM 4.0) have been discussed. A Case Study pertaining to remote monitoring of a gas compressor system of a petroleum refinery in North Eastern

P. G. Ramesh (✉)
Mechanical Engineering Department, The Assam Kaziranga University, 785006 Jorhat, Assam, India
e-mail: pgramesh@kazirangauniversity.in

S. J. Dutta
Mechanical Maintenance Department, Numaligarh Refinery Ltd., Numaligarh, Assam, India

S. S. Neog
Predictive Maintenance Department, Numaligarh Refinery Limited, Numaligarh, Assam, India

P. Baishya
Mechanical Maintenance Department and Area-in-Charge of Diesel Hydro-Treater Unit of Numaligarh Refinery Limited, Numaligarh, Assam, India

I. Bezbaruah
Mechanical Engineering Department, Kaziranga University, Jorhat, Assam, India
e-mail: indrani.b@kazirangauniversity.in

D. R. Karanki et al. (eds.), *Advances in RAMS Engineering*, Springer Series in Reliability Engineering,
https://doi.org/10.1007/978-3-030-36518-9_12

India and a Case Discussion on Basic Technical Requirements for the implementation of Industrial internet of Things (IIOT) based predictive maintenance system are presented to highlight the benefits and issues associated with the radical shift in paradigm from legacy systems to Industry 4.0 based predictive maintenance (PdM 4.0) system. Frameworks for PdM 4.0 system decision making and development are also suggested for supporting future work in this area.

Keywords Industry 4.0 · PdM 4.0 · Condition based maintenance · Predictive maintenance · Remote health monitoring

1 Introduction

Technologies such as Internet of Things, Artificial Intelligence, Cloud Computing, Robotic Automation and Big Data/Data Analytics have brought about a completely new paradigm of conducting business so much so, that this change is being rightly termed in the industrial circles as the fourth Industrial Revolution or in short, "Industry 4.0". In order to maximise the benefits of Industry 4.0, business operations need to transition to the new paradigm in totality which means that maintenance of plant machinery and systems also need to follow at the same pace and mode.

There are multiple challenges in the implementation of Industry 4.0. Today's plant floors need access to a variety of information, but are often burdened with issues around large data volumes, integration of multiple systems, and security. As a result, there is a greater need to share expertise across facilities to optimise safety, production and recovery. Subject matter experts are becoming increasingly difficult to locate and companies need to find ways to use them more efficiently [1]. Remote management allows authorized individuals (specialists, experts) to monitor the automation systems, help diagnose problems, tune loops, optimize processes, and generally improve production [2]. Industrial process plants can be monitored remotely by system architecture having General Packet Radio Service (GPRS) and wireless Internet connection in conjunction with Distributed Control System (DCS), Supervisory Control And Data Acquisition (SCADA) with consequent improvements in reliability, response time, etc. under adverse environmental conditions of process plants for maximizing plant operational conditions [3].

In this chapter, the importance of adoption of Industry 4.0 technologies for Predictive Maintenance of machinery and systems of remotely located engineering plants and aspects related to its implementation are discussed. In Sect. 2 the distinguishing features of remotely located process and manufacturing plants is discussed highlighting the indispensability of their transition to predictive maintenance under Industry 4.0. Frameworks for such a system for large engineering plants that are located remotely are presented in Sect. 3 which will prove useful for further work in this field. Two case studies are presented at Sects. 6 and 7. The first case study is aimed at highlighting the importance of making use of global expertise and resources on an Industry 4.0 platform for process critical machinery based on a case of implementation of such a system for a heavy duty process critical gas compressor of a petroleum refinery in North Eastern India. The second case study brings out the complexity of predictive maintenance systems under Industry 4.0 and highlights the need for comprehensive but optimum specifications for such a system.

2 Remotely Located Process Plants

A number of process and manufacturing plants are situated at locations which pose serious challenges related to supportability of the plant machinery and systems due to logistic constraints. The challenge on account of remoteness of such plants is aggravated by poor connectivity for transportation, geographically scattered locations, inadequate local expertise or resources and, in some cases, poor communication links. Such remotely located plants include oil & gas drilling sites, both land based and off shore, refineries, coal and mining locations, power plants, solar and wind power farms, cement, sugar, chemicals and other process industries. In addition to these stationary plants there are mobile platforms such as ships and aircraft which are required to operate over remote regions which are not easily reachable for logistic support. These platforms too, therefore, can be included in the category of remotely located process plants.

Most of these plants are also large and complex in their equipment or system configurations and operations, thereby posing additional demands on support infrastructure. Comprehensive maintenance strategies are required to be planned, optimised and executed for these plants. Optimal availability of hardware, software and support from remote expert centres as well as deployment of adequate maintenance personnel are some of the essential requirements to ensure requisite standards of reliability and availability [4].

Maintenance of such plants has constantly been evolving driven by the needs of industry and increasingly available technologies and skills. Efforts are being made to make available, details such as, equipment information, failure data, maintenance resources and material availability more readily and develop more and more optimised maintenance decisions. Remotely located plants have benefitted from such developments.

However, in the recent past, there has been radical shift in the paradigm of operation of these plants and such changes have also influenced maintenance philosophies and strategies.

2.1 Industry 4.0 and Changing Operational Philosophies of Remote Process Plants

Technologies such as Industrial Internet of Things (IIOT), Artificial Intelligence (AI), Cloud Computing (CC), Autonomous Robotics, Information and Communication Technologies (ICT) and big data analytics have revolutionised the way Industries can function and have been collectively termed as the fourth Industrial Revolution or Industry 4.0. Industries are leveraging these technologies for their business operations for being more competitive and profitable. The impact of Industry 4.0 technology on business models has been disruptive over the entire value chain.

Under this scenario of increased competitiveness, asset maintenance strategies are also required to be aligned on like terms so that equipment and systems are maintained in requisite states of availability or readiness to match with the overall business expectations.

Engineering assets and operations under Industry 4.0 will have following major components [5]:

(a) Cyber Physical Systems (CPS)
(b) Internet of Things (IoT)
(c) Internet of Services (IoS)

Industries are increasingly configuring their machinery and systems as Cyber Physical Systems (CPS) by employing the above technologies. CPS involves control and surveillance of physical systems through computational and supervisory prowess of computers including web based software resources and expertise. Such web based resources are available as CC applications. Internet of Things (IoT) refers to even the smallest of components or sub-systems that are capable of sending and receiving signals from the internet. Internet of Services (IoS) refers to services that can be provided over the internet and that which can be availed by industries aligned with the concept of Industry 4.0.

Factories that are in line with the above configuration can be called Smart Factories. The Internet of Services and Internet of Things are two basic concepts that should be implemented in factories as a precondition for the smart factory of the future (Fig. 1) [6].

These changing operational scenarios of industries are, in fact, more relevant to remotely located plants since the benefits so derived can offset many of the constraints related to being remotely located.

Industry 4.0 is seen to generally spread from developed countries to developing ones through a rather slow process of diffusion and adoption of constituent technologies [8, 9]. Thus usually different behaviour patterns are seen and perceptions regarding the usefulness of different technologies differ widely from one region or sector to another. Several industries, especially those in the developing economies of the world have commenced application of Industry 4.0 technologies. However, the scope of application is rather non-comprehensive thereby denying the

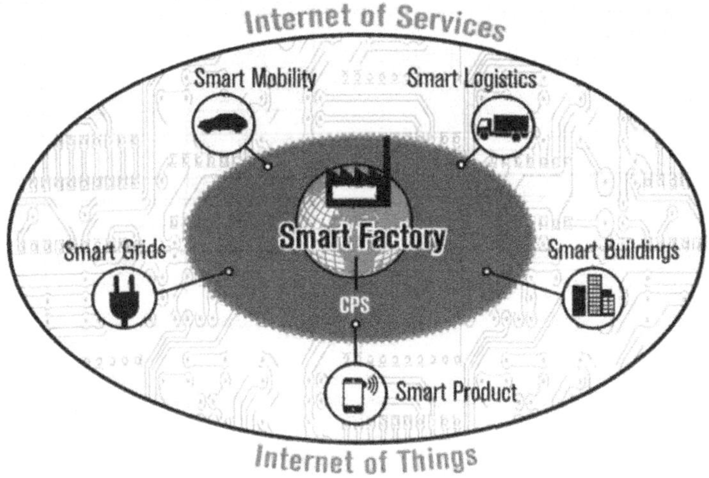

Fig. 1 Schematic representation of a smart factory [7]

possibilities of truly revolutionary changes. For instance, it has been shown by Dalenogare et al. [10] as to how industry in a developing country such as Brazil has not yet taken advantage from some promising technologies such as product big data analysis and cloud services for manufacturing. In fact, benefits to be accrued from Industry 4.0 are largely derived from studies in developed economies. As regards developing countries, more comprehensive studies into how the benefits could be leveraged are required to be undertaken.

Studies thus show that there is significant dispersion between countries in the readiness of their industries in terms of ability to adopt Industry 4.0. The reasons that determine the differences among countries in the ability to adapt to Industry 4.0 require further research: the structure of the industrial sector, its role within each country's economy and differences in business models or management styles even within the same sectors, should be elements of study when looking for the Industry 4.0 drivers [11].

Going forward, enterprise level implementation of technologies of Industry 4.0 encompassing all elements of supply/production/service chain will provide the best results from such an enterprise. The level of application can range from individual machines, individual processes or complete factory in the increasing order of automation or complexity. Also, the level of intelligence embedded in the machines/process/entire factory can range increasing order from merely having an automated control, integration of automated controls of several functions or embedding higher level intelligence in the machines/process/entire factory for autonomous and intelligent functions. This aspect has been well presented by Qina et al., as nine intelligence applications going from low-intelligence and simple automation to high-intelligence and complicated-automation as shown in Fig. 2. From applications I–IX, the production system becomes increasingly automated, flexible, and intelligent [12].

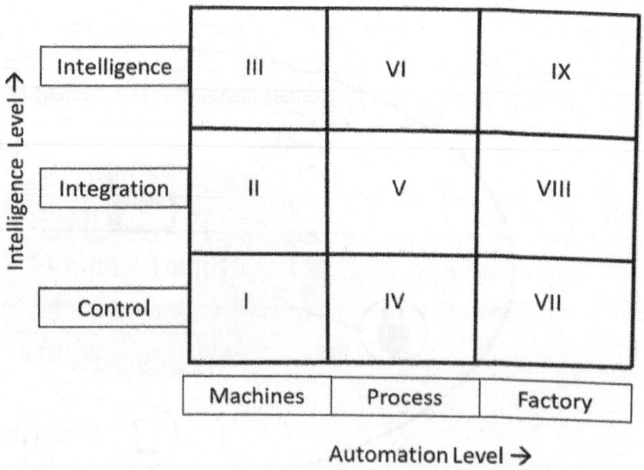

Fig. 2 A categorical framework of manufacturing for Industry 4.0 and beyond [12]

Experts such as Moubray [13] believe that not more than 20% to 35% failures can be prevented by Predictive maintenance depending on whether only condition monitoring inputs or other additional sensing of failure mechanisms are employed. Under this condition, in order to render predictive maintenance under Industry 4.0 to be an applicable and significantly effective strategy for total maintenance, there is a need to combine technology advancement with basic maintenance. Cross-functional working to handle large amount of data, leverage development in technology along with basic maintenance practices will be the key to better maintenance under Industry 4.0.

Augmented Reality (AR) is an important technology of Industry 4.0 and its application for maintenance was known for quite some time. However practical limitations such as lack of adequate knowledge on its application made it difficult for realising its benefits. Now that some of the technological limitations have been overcome and AR seems to be ready to become a tool for industry, it is believed that the scientific community can focus on trying to solve the real industrial problems [14]. This includes maintenance. Technologies like AR and Additive Manufacturing (AM) can provide better way to carry out maintenance operations with respect to a traditional approach as shown in the case of aircraft maintenance [15]. AR can support the operators with user-friendly manuals where virtual models and instructions are mixed with real world, while AM can be useful to avoid large warehouses and cut the logistic chain.

2.2 *Condition Monitoring of Remote Process Plants Under Industry 4.0*

Industry 4.0 assumes limited participation of machine operators in monitoring and diagnosis of manufacturing and technological processes [16]. This is the case not only about machinery and system controls but also about their condition monitoring. Industrial platform for condition monitoring [17] consists of the three major modules: monitoring and feature extraction (MFE), real-time anomaly detection (RTAD) and fault diagnosis (FD). These three modules are also true of conventional or legacy systems. However, the scope and reach of these modules under Industry 4.0 is much extended. MFE module can consist of IIOT based sensing equipment, collection of a large number of sensed and computed data from which feature extraction becomes more accurate and providing better coverage to the maintained system. The MFE information is passed on RTAD which can leverage Artificial Intelligence (AI) and Machine Learning (ML) tools among others to detect anomalies. Likewise FD module makes use of AI, ML at plant locations as well as remote expertise and analytical tools by means of internet communication and cloud resources.

3 Maintenance Frame Work Under Industry 4.0 for Remote Process Plants

Maintenance framework under Industry 4.0 for Remote Process Plants is very much an extension of e-maintenance framework shown in Fig. 3 [4].

The local O&M platform has machinery and systems to be monitored. Monitored condition date are processed in the local platform as far as feasible for detection and prognosis of impending defect conditions. The capacities of local units being limited, remote centres of expertise are configured into the system as

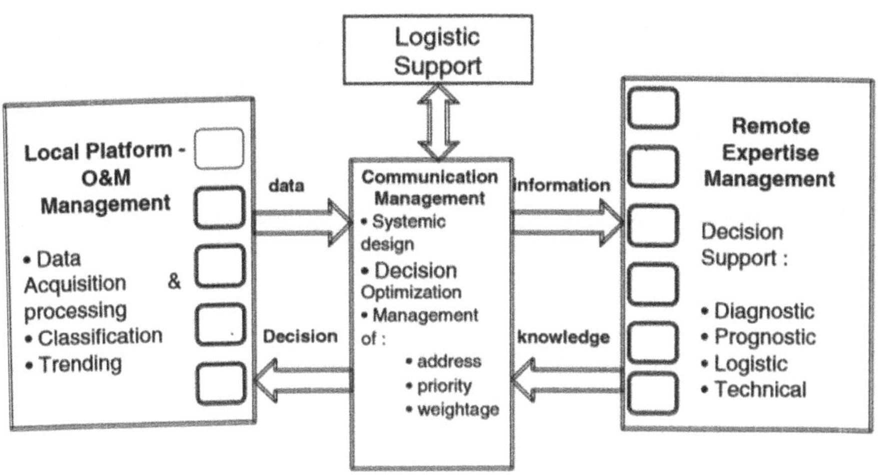

Fig. 3 Framework for e-maintenance [4]

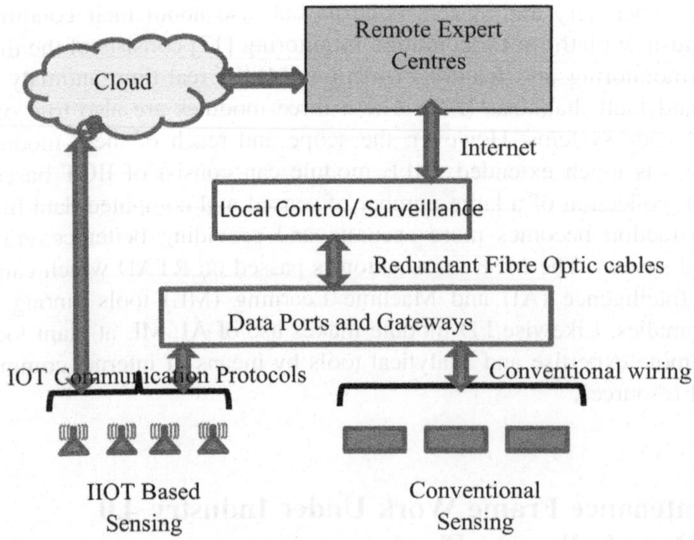

Fig. 4 A scheme for maintenance framework under Industry 4.0

extensions of the local platform capabilities. The communication management module ensures that the overall e-management effort is successfully executed.

Implementation of data collection systems, data analytics and real-time decision making has paved the way for e-maintenance and helped reduce downtime and uncertainty about the current status of the equipment and possible breakdown in the future. Proper use of available technologies will lead towards smart systems which will reduce uncertainty in the decision making process [18].

The above e-maintenance framework is extended using IOT based sensing, cloud technologies as also with effective use of AI and ML. A scheme for maintenance framework under Industry 4.0 is shown in Fig. 4.

A framework for predictive maintenance in line with the above overall framework for maintenance under Industry 4.0 of plant machinery and systems of a remotely located plant is shown in Fig. 5. Sensor and other acquired real time data, historical data, environmental data and asset design data are used by the system represented by the above framework for presentation of processed information in the following forms:

(a) Indication of Asset performance: Key Performance Indicators (KPIs) are defined as a set of quantifiable and strategic measurements in a Performance Monitoring System (PMS) that reflects the critical success factors of an enterprise [19]. Performance data collected through an array of IOT based sensors as well as classical sensors, wireless or wired, are processed against data from other sources to provide information that indicate the performance of the asset on suitable HMI or recorder. Both the number of equipment and the number of sensors are limited to the bare minimum in legacy systems with

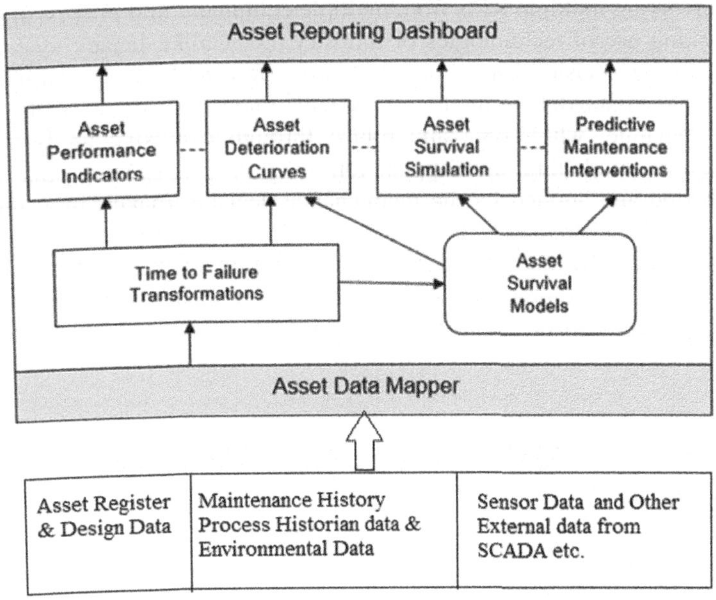

Fig. 5 Schematic diagram of the asset predictive maintenance framework using IIOT

manual modes of monitoring. However, with IOT and wireless based communication of performance parameters, the number of parameters and equipment under coverage can be significantly increased, thereby improving the effectiveness of health monitoring, trending, analysis and predictive maintenance.

(b) Indication of Asset deterioration curves: This feature provides static and dynamic factors that can help explain asset and process failure, and their relative importance. Risk of failure of an asset at given time is also analysed and presented. Feature engineering can be carried out to capture degradation over time by using techniques such as Regression, binary and multi class classification, survival analysis and anomaly detection. In doing so, solution provider will find out the features, which shows degrade pattern with significant predictive power. Technologies of Industry 4.0 such as Artificial Intelligence (AI) and Big Data Analytics can help in making use of all collected data, interpreting suggested trends and continuously improve upon prognosis of failure.

(c) Results of simulation of asset life or survival: The effects of various asset maintenance scenarios, probability of an asset that survive beyond a given time, prediction of failure probability over time are some of the useful simulation that can be undertaken and presented for the maintenance team.

(d) Recommendations for predictive maintenance interventions: The predictive maintenance recommendations are based on collection of a large quantity of data involving multiple KPIs from multiple equipment and processing the same by making use of technologies of Industry 4.0. Unlike legacy systems where maintenance decisions were based on very few KPIs and equipment being monitored in standalone modes, in systems under Industry 4.0, decisions are more dynamic, wholesome and robust through extensive analysis of data, simulation, modelling and application of AI/ML. AI/ML tools if rightly understood and applied by the maintenance decision maker, may unravel the mechanisms or trends behind many of the "sudden/unexpected failures". Further, there is a need for careful determination of the kind of data that is required to be acquired and processed in order to be able to maximise the desired outcomes from the predictive maintenance system.

4 Transition to PdM 4.0 from Legacy Maintenance Systems

A few studies have been conducted regarding the issues connected with the adoption of Industry 4.0 on legacy systems. In one such study, interviewees mentioned that the main adoption challenges are the analysis of data generated, integration of new technologies with available equipment and workforce, and computational limitations as also changes in company's business model through the integration of internal resources with complementary activities of their partners and other cluster companies [20].

One of the features of most remotely located industries is their complexity of systems in terms of their configuration, diversity, size, criticality and inability to provide long breaks in production. This complexity renders transition of legacy maintenance systems of these equipment and systems to Predictive Maintenance under Industry 4.0 (PdM 4.0).

Legacy machine tools are often isolated, not well-equipped with modern communication technologies, and with lack of open Application Programming Interfaces (API). It is therefore difficult to monitor and control the entire production process [21] using legacy systems. It is difficult to easily monitor legacy systems, which can introduce inefficiency and generate higher cost of sensor integration [22]. As a solution, industries can reconfigure their legacy systems into "smart legacy machines". However there are challenges for identifying standard IOT architecture, and establish clearly the benefits of the transition. Industries are also concerned about protection of their data when exposed to a cloud or internet-based architecture [23].

For large plants especially those located in remote areas, progressive conversion of legacy machines to smart legacy machines is a good option, while also moving on to completely Industry 4.0 machines and systems during replacement and modernisation projects. A discussion of a case of requirements for implementing IIOT based predictive maintenance decision support system for a petroleum refinery has been presented later in this chapter. Until a complete change over, there will be a long transition phase having a mix of old and new for which engineers of such remotely locate large plants need to gear up.

5 Customisation of PdM 4.0 Solutions to Remote Process Plants

As mentioned earlier, remotely located large engineering plants be it, process or manufacturing plants, renewable energy farms or ships have unique customised designs, set of equipment, systems and operating philosophies. The set of objectives and constraints including logistics for such industries are also unique. Therefore there is a need to customise PdM 4.0 solution also for such industries. Off the shelf solutions are bound to be sub-optimal due to inadequate coverage and challenges of integration.

5.1 Integration of PdM 4.0 Features During System Design

In order to achieve customisation, the ideal option would be to integrate PdM 4.0 during design of a plant. The plant machinery and systems should be compatible, supportable and integratable with PdM systems. However, this is seldom possible practically since the rates of obsolescence of plant machinery systems and PdM systems are very different. Hence customisation efforts are required to take into consideration a mix of legacy and new equipment for design of PdM 4.0.

5.2 Optimisation of PdM 4.0 Solutions for Remote Process Plants

Major objectives of PdM 4.0 for a remotely located process plant would be maximisation of plant availability, equipment and process reliability, safety and minimisation of costs. However depending on the nature of the plant, there will be several minor objectives as well. All objectives have importance for certain stakeholders at certain times. Many of the objectives would compete with each other resulting in trade offs and not single optimum solution. Similarly there are major and minor constraints in achieving the objectives. Obviously, designing a PdM 4.0 platform for such plants will be a multi-objective, multi-constraint decision optimisation problem. A sample framework for multi-objective Condition Based Predictive Maintenance (CBPM) problem is shown in Fig. 6 [24].

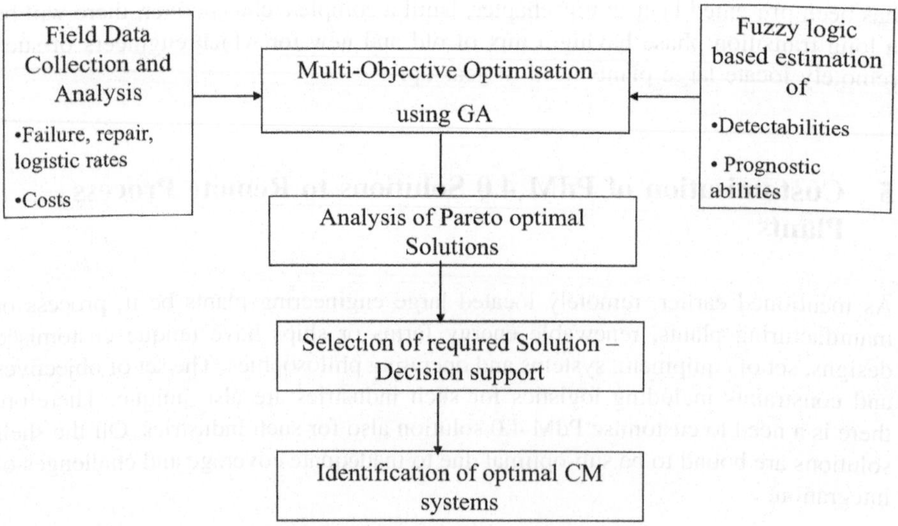

Fig. 6 A framework for multi-objective CBPM problem [24]

As shown in the figure, a Multi-Objective Optimisation Problem (MOOP) is solved with inputs from field data and Detectabilities and Prognostic abilities. The latter two, explained briefly below, are derived based on data or from experts with the help of Fuzzy Logic Framework. Solution to the MOOP consist of a set of optimal points each differing from the other with trade-offs. The decision maker selects that solution which would be the most optimum under the prevailing conditions. This solution corresponds to the set of parameters that would define the CM system that would be selected.

In the process of optimisation, it is essential to identify or create certain parametric metrics for desirable features of the PdM 4.0 system and quantitatively ascertain how they could influence achievement of objectives of PdM 4.0. For instance, the ability of a PdM system to be able to detect the onset of a defect or impending failure and also its capability for prognosis and diagnosis of defects are important requirements for any user. Based on this premise, a study has proposed two metrics, detectabilty and prognostic ability and has undertaken a multi-objective optimisation using these metrics as variables. The influence of these variables on one of the objectives, namely, to minimise unavailability is shown in Fig. 7. It may be observed that the detectability and prognostic ability averaged for the constituent sub systems show an increase with increasing cost.

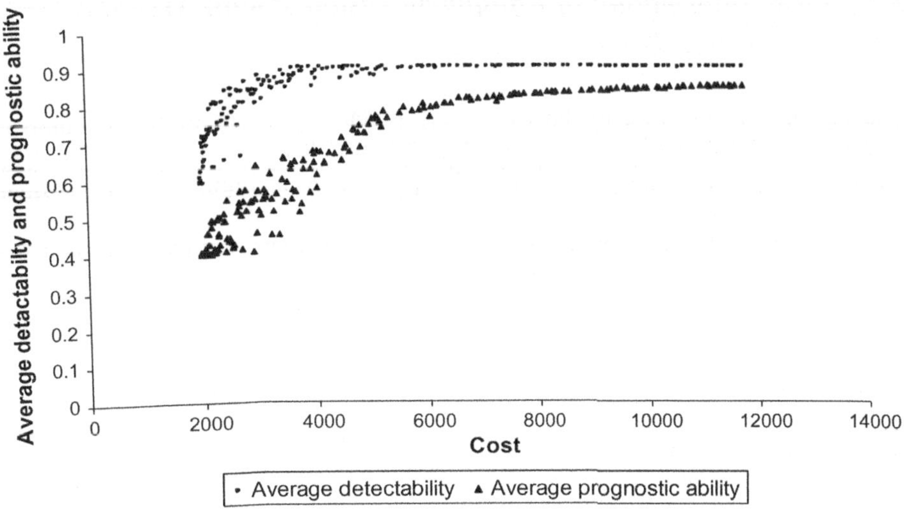

Fig. 7 Cost of PdM versus average detectability and prognostic ability [14]

6 Case Study I: "a Process Critical Equipment of a Petroleum Refinery in North Eastern India Integrated with Industry 4.0 Based Predictive Maintenance System"

6.1 Introduction

Presently online machine monitoring system installed in critical equipment in the refinery is equipped with machine protection system (MPS) apart from diagnostic facility. MPS consist of transducers (vibration, position, temperature, flow, pressure, speed, etc.) with continuous, permanent monitoring instruments installed which are capable of sending alarm and shutdown commands to the machinery control system. MPS thus uses real time measurements and data to automatically shut the machine down when conditions degrade beyond pre-set alarm limits. MPS can be taken off-line and automatic shutdown can be disabled if so desired. Many users would want to disable auto shut down using MPS for the fear of shutdowns due to false alarms.

In this case study, two situations (with and without MPS auto shutdown) encountered in a critical machinery in a major Petroleum Refinery in North Eastern India will be discussed. It will be shown that leveraging global technical expertise and resources pertaining to health monitoring in the realm of Industry 4.0 in case of such critical equipment can provide optimum maintenance decision support to the plant personnel instead of merely depending on an auto shutdown facility.

6.2 *Basic Information of Equipment Installed with Machine Monitoring System*

The equipment discussed in the case study pertains to a set of 03 in numbers Makeup Gas (MUG) compressors (Fig. 8) installed in Hydrocracker Unit (HCU) unit of the refinery. MUG Compressor is a high pressure 3-stage Reciprocating compressor of Hydrogen service. The Makeup H_2 compression section consists of three identical parallel compressor trains, compressing makeup H_2 from the H_2 plant to the reactors. Three compressors are required to run simultaneously at 240% collective load for achieving the maximum load requirement of the HCU Plant.

Equipment Data

Machine: Reciprocating Compressor
Power Rating: 3 MW
Manufacturer: NuovoPignone
Model: 4HF/3
No Cylinders: 4
Service: Make Up Hydrogen

Fig. 8 Picture of one of the cylinders of make up gas compressor with installed sensor of PROGNOST system

No Stages: 3
1st Stage: 19.5/45.6 kg/cm^2 a
2nd Stage: 45/100.8 kg/cm^2 a
3rd Stage: 100/189.1 kg/cm^2 a.

Features of PROGNOST Monitoring System Installed in the Refinery

- Real time data collection.
- Database system.
- Data Management.
- Trend storage.
- Data archive management.
- Analysis of Historical data.
- Data export into MS-EXCEL etc.
- Anomalies banner, that is, a logbook containing errors recorded by the system.
- Thermo dynamical and Mechanical simulations modules.

Basic Reciprocating Compressor Monitoring Parameters

- Suction and Discharge temperature, vibration and pressure of cylinders.
- Vibration and Mechanical looseness.
- Piston Rod run out and drop.
- Temperature of main bearings.
- Crosshead and Crosshead pin parameters.
- Piston rings wear of all stages.
- Rider rings wear of all stages.
- Main packing's temperature.
- Frame vibrations.
- Lube oil system temperature.

The PROGNOST Machine Protection System of MUG Compressor is based on parameters as indicated in Table 1.

When the safety threshold is violated, the automatic shutdown of the machine may be implemented for protection against costly failures. The PROGNOST machine shutdown is based on a voting feature where users during installation can define a group of parameters (Online monitoring recorded data), which when satisfied independent of each other, can trigger a machine shutdown.

The diagnostics system allows comparison of the measured values with the expected values generated by the model. The software processing unit analyses all changes in the thermodynamic condition of the compressors and updates the expected values in real time. Every displayed parameter has four different associated states indicating its condition: Green, Yellow, Orange and Pink. The green state indicates that the value of the measured parameter is normal/good; The Yellow state indicates that the value is at the upper/lower limit of a predefined criticality

Table 1 MUG compressor parameter

Parameter	Description
Crosshead slide vibration	• Threshold monitoring of RMS vibration for 36 segments per revolution • Threshold monitoring of RMS vibration per revolution (1 segment)
Dynamic cylinder pressure	Threshold monitoring of peek to peek dynamic pressure per revolutions
Dynamic Rod drop	Threshold monitoring of peek to peek piston rod vibration for 8 segments per revolutions
Frame vibration	(i) Threshold monitoring of RMS vibration for 36 segments per revolution (ii) Threshold monitoring of RMS vibration per revolution (1 segment)
Alert and shutdown with plausibility check prior to alarm	The plausibility check is a feature that avoids costly false trips that could result from faulty sensors or loosened instrument connections within the loops
Unsafe alarm in case of non-functional loop or system	Any non-functional loop or system feature informs any loosened instrument loops

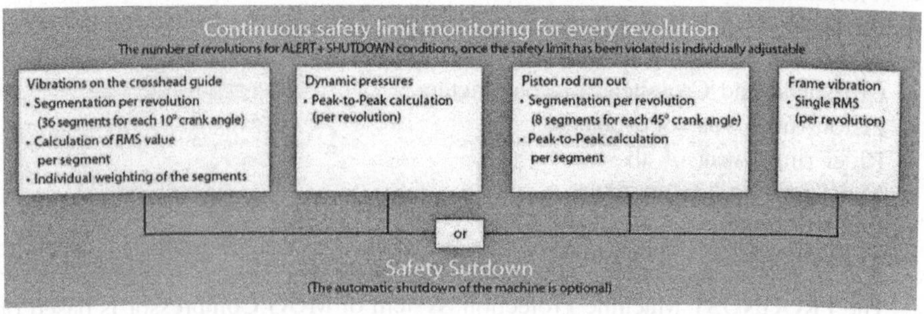

Fig. 9 Schematic diagram relating MPS parameters versus safety shutdown

range; the orange state indicates that the value is at the upper or lower limit of the predefined high criticality range; the Pink state indicates that the value has reached/ crossed the set safety limits of predefined criticality range. With MPS on line, Pink State indicates crossing safety limits and will trigger machine shut down command. The shutdown logic for the compressor is shown schematically in Fig. 9. In the above compressors MPS have been kept offline.

6.3 *Case Study I (Part 1): Compressor System with MPS Auto-Shutdown 'OFF'*

In this case, the MPS Auto-Shutdown for the compressor system was put 'OFF'.

6.3.1 Problem Description and Prognostic System Logging System Alerts

On the 23-Jan-2013, HCU load was at maximum load with the three MUG compressors running with compressor numbers:

- 04-KA-03B at 100% load
- 04-KA-03A at 70–85% load
- 04-KA-03C at 50% load.

On the morning of 23-01-2013, 11 AM, compressor KA-3A was started. The PROGNOST-NT system released several SHUTDOWN alarms approximately 13 min after the start of the compressor. With reference to the Figs. 10 and 11, the

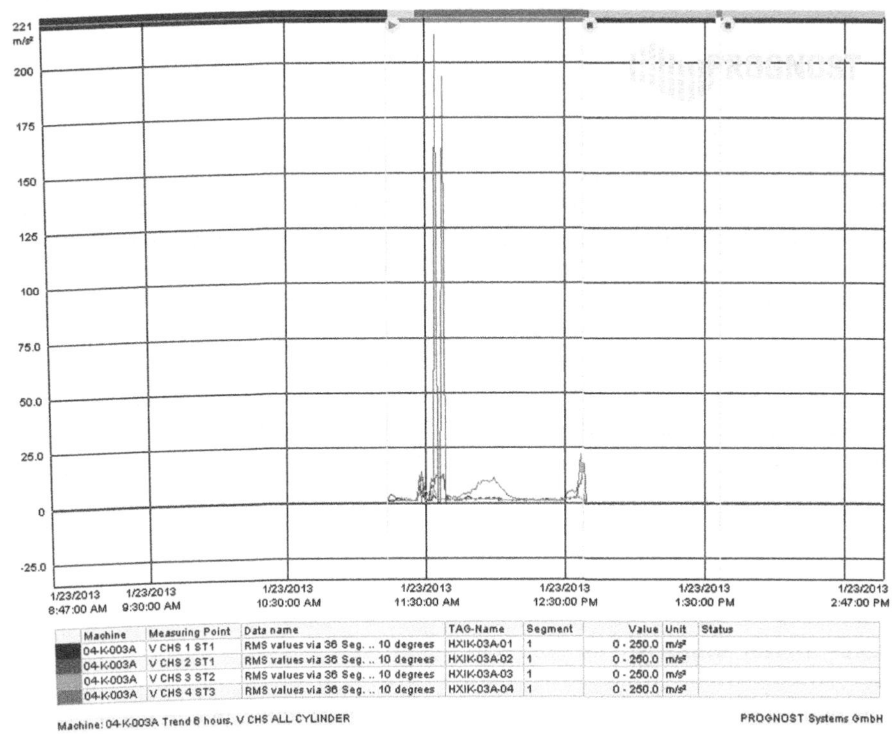

Fig. 10 PROGNOST-NT system all stage crosshead slide vibration trend for the period 23-Jan-2013

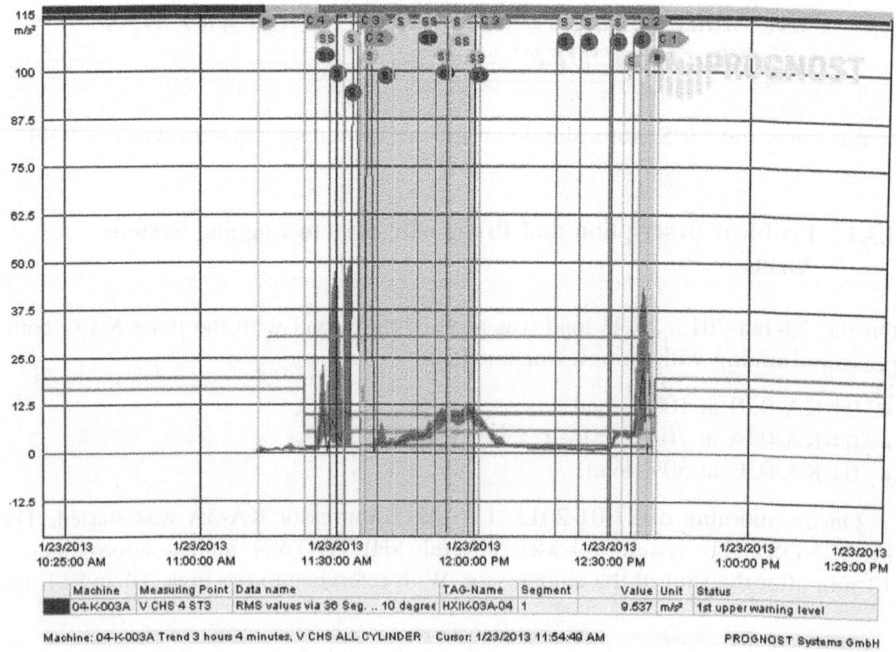

Fig. 11 PROGNOST-NT system 3rd stage crosshead slide vibration RMS trend for the period 23-Jan-2013

crosshead slide vibration on cylinder 4 (third stage) were on a high level and much higher than all other cylinders. Even on other cylinders, e.g. on cylinder 3 high crosshead vibration were recorded and SHUTDOWN alarms were released as well. However the compressor continued to operate since Auto Shutdown feature of MPS had been switched ‘OFF’.

After few minutes abnormal sound from the compressor was heard. The compressor was stopped and opened for inspection by the maintenance team. Cross-Head side cover was opened and white metal chips were found. The cross head gudgeon pin was found seized and cross head overheated and burnt. Pictures of damaged parts are shown in Fig. 12. MPS generated warning signals around 20 times for stopping the compressor but the same was ignored by operation team and machine was continued running suspecting false alarm and fearing negative impact on production if stopped unnecessarily

6.3.2 Defect Analysis

The burnt condition of the crosshead indicated lack of lubrication in the crosshead gudgeon pin. Rod reversal allows lubrication to both sides of the cross head pin and the present damaged condition indicated lack of rod reversal in the 3rd stage. The

Fig. 12 Parts of MUG Compressor showing marks of damage

absence of rod reversal in 3rd stage (Cyl 4) during startup resulted in damage of crosshead and high vibration alert. The Hydrocom stepless flow control on the crank end side (CE) of cylinder 4 was unloaded while the head end side was loaded from the beginning of the run at 11:12 am until 12:37 pm. Such a condition is unusual and typically both sides of the piston will be equally loaded to avoid such an unbalanced situation.

6.3.3 Discussion

If the MPS system was on line, this catastrophic failure could have been avoided. Remote expertise could have been accessed using cloud or direct internet resources. The above failure resulted in the replacement of the Crosshead, crosshead shoes, small end bushing, and main bearings along with complete overhauling of the machine. Total approximate cost for the refinery in damage repair of components was INR **5788348** and in Machine overhauling was INR **660000**. In addition the machine was down for 30 days due to non-availability of spares resulting in the plant operating at part load.

6.4 *Case Study I (Part 2): Compressor System with MPS Auto-Shutdown 'ON'*

In this case, the MPS Auto-Shutdown for the compressor system was kept ON.

6.4.1 Problem Description

On the 23-May-2014, HCU load was at 95% with the Three MUG compressor running with compressor

04-KA-03B at 100% load
04-KA-03A at 50–70% load
04-KA-03C at 50% load

On the morning of 23-05-2014, 9.45 AM, PROGNOST-NT system released several alerts: 'Drive Train Crosshead/Piston Rod/Piston damage', CYL1 ST1 with correlation: 66, 3% and more" in the Cyl 1 stag 1 of 04-KA-003B. At 12.51 pm, PROGNOST-NT system released several safety shutdown alerts for the crosshead vibration of 1ST Stage Cyl 1 of 04-KA-003B.

The crosshead vibration signal of 1ST Stage Cyl 1 of 04-KA-003B showed an increase vibration value of 23.27 m/s^2 from previous recording of 7 m/s^2 (Fig. 13). Normal RMS value for the Crosshead should be below 15 m/s^2.

The PROGNOST-NT system had released several SHUTDOWN alarms for immediate shutdown of the compressor. The vibration trend indicated a major breakdown in the crosshead of 1st stage Cylinder 1 of 04-KA-03B. To analyse the

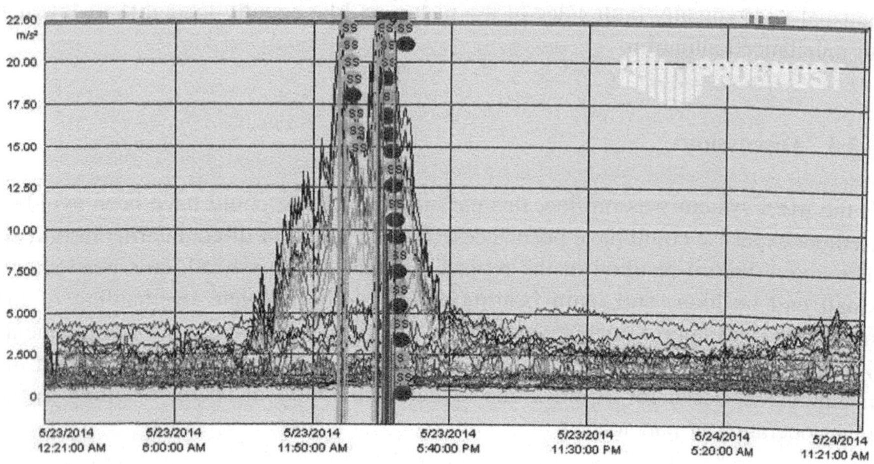

Fig. 13 PROGNOST-NT system 1st stage Cylinder1 crosshead slide vibration RMS trend on 23-May-2014

defect different available trends were analysed from PROGNOST online monitoring system.

The Operation department of NRL reported that no process related abnormality could be recorded in the HCU control room and the trends indicated normalcy in the compressor behavior. No physical abnormality like metallic sound, looseness could be heard. Vibration in Crosshead location in 1st stage Cyl 1 was also checked with offline Vibration Analyzer. The readings were in the range of max 1.8 mm/s and were Normal. The other cylinder readings were also recorded with the Vibration Analyzer and were found to be similar. The ring buffer signals also confirmed that some abnormal vibration signals are being recorded for the VCHS CYL 1 STAGE 1 of 04-KA-03B.

After deliberating with the OEMs of the equipment and PROGNOST, the maintenance team further investigated the issue and the Crank case NDE and DE side vibration signal was trended and analyzed offline. The vibration trend for the period was stable and no abnormal reading was recorded. It was concluded that vibration sensor was generating erroneous data. As anticipated, sensor was found loose and after the tightening of the CHS vibration came down to normal level. There was yet another occasion when a false signal was received due to a faulty sensor. The sensor was replaced and the defect was ruled out after similar analysis.

6.4.2 Discussion

If MPS were online, it would have shut the compressors down resulting in unnecessary production losses. Sudden shutdown of the machine may lead to an emergency situation, if not handled properly may have adverse effect on reactor catalyst leading to catalyst life reduction.

6.4.3 Conclusion

While there are several equipment in a refinery system that may be considered as non-critical and/or having redundancies, there are quite a few process and production critical, large equipment such as the one in this case study. It is essential to prevent catastrophic failures of such equipment that are bound to have large repair costs and time. Hence it may not be prudent to leave preventive shutdown to an automated local safety devise such as MPS discussed in this case study. It was seen in Case Study 1(Part 1) that not having consulted the remote external experts led to the catastrophic failure of the compressor. In the Case Study 1 (Part 2), it was observed that based on consultations with the external domain experts/OEMs, it was found that the fault signals were indeed fault and thus unnecessary interruption of the plant was averted.

Therefore, in the case of process or production critical large equipment, it will be prudent to employ continuous health monitoring by expert agencies although located geographically at distant locations by leveraging the benefits of IIOT and other Industry 4.0 technologies such as Cloud resources and AI. Cost effective solutions can be identified and optimized for such a facility. Any unplanned shutdown of the compressors results in unnecessary production losses. Sudden shutdown of the machine may lead to an emergency situation, if not handled properly may have adverse effect on reactor catalyst leading to catalyst life reduction.

Spurious tripping of 1 no's MUG compressor with three compressor running in NRL Hydrocracker plant would result in plant upset and reduction of load. On the other hand, any spurious tripping of 1 no's MUG compressor with two compressors running in NRL Hydrocracker plant would result in second stage down of the Hydrocracker Unit and in worst case unit shutdown. Hydrogen 2–3 MT (costing around 1.8 Lakhs/MT) in the system will be flared apart from generation of slope and unconverted oil. Hence, continuous monitoring of parameters by external agencies under I4.0 scenario and taking corrective action accordingly will derive value from the technology. Further, organisation will learn from the root cause analysis carried out by external agencies on different conditions.

7 Case Study II: Implementation Requirements for IIOT Based Predictive Maintenance System Under Industry 4.0 in a Remotely Located Petroleum Refinery

In this section, configuring of IIOT based predictive maintenance system as a part of a larger Integrated Business Performance System in a typical remotely located petroleum refinery in North Eastern India will be discussed. The aspects of this system for the refinery would apply for other industries also reasonably well. As suggested earlier, application of Industry 4.0 technology on a business-wide scale is bound to have radical improvements in performance and in redefining the overall business paradigm of remotely located industries. A typical Integrated Business Performance System under Industry 4.0 is shown in Fig. 14.

As it can be seen, the proposed Integrated Business Performance System provides seamless flow of data, practically unrestricted capability for analysis and valuable flow of information to support business operations. With such information integration, dashboards for monitoring and control by senior echelons of the management are also achieved with ease.

The Integrated Business Performance System, as the name suggests, includes all operations of the business enterprise, including business systems and plant systems. Control, Condition Monitoring and Predictive Maintenance will be parts of this overall architecture, taking care of the plant systems.

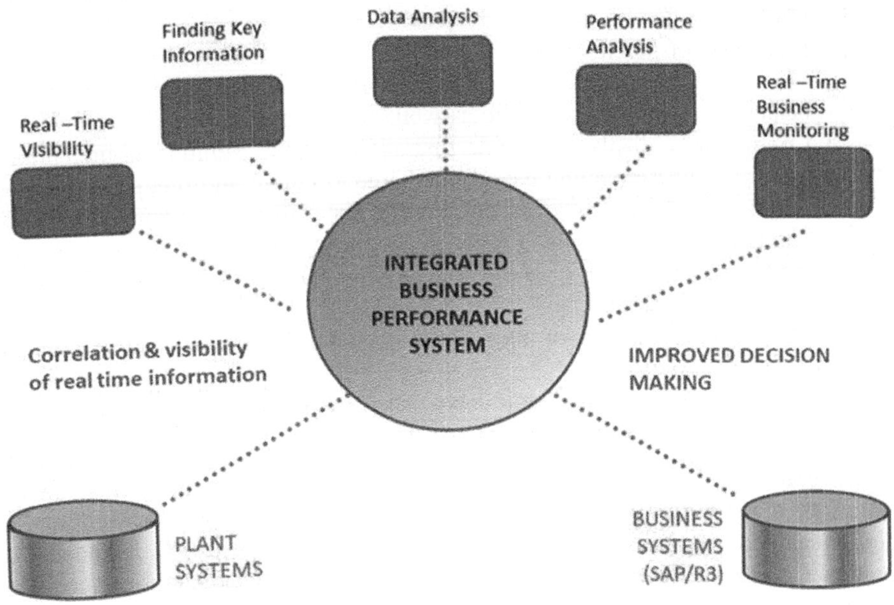

Fig. 14 Integrated business performance system proposed for a typical remotely located petroleum refinery

7.1 *Plant Systems*

A typical petroleum refinery has more than 4000 rotating equipment such as, pumps, motors, turbines, compressors, blowers and mixtures and over 10,000 static equipment such as, Reactors, Columns, Pressure Vessels, Furnaces, Cooling Towers, Valves, Pressure/temperature Safety valves and steam traps. Monitoring the health of these with conventional operator driven preventive and time based inspection is not sufficient. Legacy systems have always had unexpected breakdowns, long logistic delays and repair times leading to poor utility factor. As partial remedial measures these industries had to maintain large inventories and standby workforce. Industry 4.0 technologies now offer a much needed solution to these plants for providing much higher levels of plant RAMS.

7.2 *Predictive Asset Health Monitoring*

Monitoring the health of plant systems is fundamental to Predictive Maintenance. Some of the health monitoring objectives under Industry 4.0 for the refinery are as follows:

(i) Reduce the downtime.
(ii) Prediction of potential failure.
(iii) Dashboard for maintenance and Operation staff for monitoring of asset health accessible from mobile as well as in PC.
(iv) Continuously display real-time operational status, running hours, predictions such as Remaining useful life, Time to failure and predictive alerts as they occur and when the as the running states of the equipment changes.
(v) Automated alert actions triggered by maintenance diagnoses from ML predictive analytical platform and equipment Prediction of failures and performance variations.
(vi) Dashboards, Drill Down, unit wise segregation of sensors and equipment.
(vii) Alerts and events.
(viii) The predictive system should have self-learning features for providing real time signature based diagnosis and prognosis.

Coverage of condition monitoring equipment is an important element in the design of monitoring systems. Coverage has two broad implications: the range of equipment and systems to be covered and the number of signatures or parameters to be covered. Typically coverage of following equipment and parameters are some which are envisaged in a petroleum refinery under Industry 4.0:

(i) Steam traps wireless acoustic sensor for immediate failure detection and steam loss KPI monitoring in Power & Utility offsite.
(ii) Pumps, compressors, Air pre heater, Motor& FD/ID Fan monitoring—Pressure, vibration, temperature, bearing temperature in co-relation with other process historian data and analytics.
(iii) Pressure safety valves for Gas leakage monitoring in Utility Area.
(iv) Cooling towers Pump and Gearbox vibration monitoring and prediction of failure along with cooling tower performance monitoring.

With reference to Fig. 14, the outputs that are expected from the asset predictive maintenance framework will be briefly discussed here.

(a) *Asset Performance Indicators Requirement*: As brought out above, there are several equipment in the refinery and multiple KPIs pertaining to each one of them. With remote sensing based architecture including IIOT for Industry 4.0, the number of parameters and equipment under coverage can be significantly increased as shown in Table 2 as compared with legacy systems. KPIs pertaining to typical pumps in the refinery are shown in the table as a sample. Similar lists of KPIs are drawn up for all equipment and systems.

Industry 4.0 technologies not only make measurement and transmission of KPIs in large numbers feasible but also facilitate processing of such large amount of data using Artificial Intelligence (AI)/Machine Learning (ML)/Data Analytics and generate information immensely useful for predictive maintenance. The role of these technologies will also be leveraged to ensure integrity of data and reliability of information so generated.

Table 2 Comparison of parameters (KPIs) that is available continuously in legacy systems and Industry 4.0 compatible systems in remote sensing based coverage including IIOT in the refinery

Legacy systems with manual recording	Industry 4.0 compatible systems under remote sensing including IIOT
a. Measurements (periodic recording) i. Number of operating hours ii. Pressure (input & output) iii. Temperature (input & output) b. Manual Computations or using hand held equipment i. Delivery head ii. Electrical power	a. Measurements (defined as time series) i. Number of operating hours ii. Motor current iii. Pressure (input & output) iv. Flow (input & output) v. Temperature (input & output) vi. Vibration (for major & critical pumps) b. Direct Computations (defined as time series) i. Delivery head ii. Electrical power c. KPIs average i. Average current during ON stages d. KPIs consumption i. Energy consumption ii. Flow e. KPIs efficiency i. Total efficiency ii. Ratio energy/flow = Energy performance indicator iii. Ratio cost/flow = Financial performance indicator iv. Savings if operated with variable speed drive f. KPIs deviation to design i. Deviation to design power ii. Deviation to nominal flow iii. Deviation to nominal head iv. Deviation to optimal efficiency

Connected with the above requirement of measurement and display of information following features are also included with the aid of technologies such as AI and ML:

(i) Equipment and System Diagnosis
(ii) Dashboard for maintenance engineers.

Typically following are some of the features that will also be included in the dashboard:

- Remaining Useful Life.
- Remaining Useful Life (% of Expected Life) with Age and installation date,
- Time to Failure Prediction within a given time window.
- Survival models for the prediction of failure probability over time.
- MTBF. MTTR, OEE, open work orders.
- Short and detailed report about equipment health to be made accessible from web and mobile devices with maintenance recommendations.

- Figures from Operational metrics such as motor run-hours, no of starts/stops, Ambient temperature and wet bulb temperature and Humidity.
- Spare parts material code and no should be linked with sap current stock availability.
- Pattern recognition or other machine learning techniques for detecting anomalies/predicting failures.
- Performance correlated to a slowly degrading metric—Temperature Bearings Motor windings, Pressure or Delta P of plugging filters in pump, Vibration, amplitude, discharge pressure variation over a time period with same load etc. to be compared in the platform in real time in streaming model.
- All key personnel to be alerted on their mobile device and email about overall Asset health score or developing problems in critical assets when threshold is breached.

(b) *Asset deterioration curves*: Following are some of the features of degradation that are considered in modelling and presentation of results:

- Speed Reduction from Design RPM at rated power
- Design Discharge Pressure versus Actual Discharge Pressure
- Temp Difference between Bearing temp and Ambient Temp in time series
- Difference between (Discharge Pressure—Inlet Pressure) in time series
- Number of times the temp exceeds a threshold value over a number of days
- Peak and RMS value Vibration Pattern over a period (Say In last 10 Days)
- FFT plot reading from Raw feed of vibration sensor in a time period
- Moving Average Voltage and Current drawn over a time series points
- Use feature cross and Static data and frequently updated data
- Remove highly co related and duplicate features in feature engineering
- Degradation pattern in speed, efficiency, pressure, load, heat, noise etc. to be tracked and monitored.

(c) *Asset survival simulation*: Following are some of the simulation that are undertaken as decision support for refinery maintenance team:

- Effects of various asset maintenance scenarios
- Probability of an asset that survive beyond a given time
- Prediction of failure probability in a certain given interval of time.

(d) *Predictive maintenance interventions/recommendations*: The feature provides predictions about the assets at greatest risk of impending failure, so shifting the maintenance regime from fail-and-fix to predict-and-prevent by issuing work order, schedule maintenance rules (alarm, alert or maintenance call). The technical scope of a predictive maintenance decision support platform for a typical refinery under Industry 4.0 envisages a vast domain.

Some of the major technical requirements of such a predictive maintenance decision platform are as follows:

- Fixing and monitoring of KPIs of Throughput, Yield/Production, Profitability, Cost, Target vs Actual Scenario, historical performance, Equipment Overall Efficiency, Health Indices, Probability of failure calculations, Reliability modelling, Failure Reasons, Reliability parameters MTBF/MTTR calculation, trends, spares consumptions, and performance curves of equipment integrated with notifications
- Real Time Performance Based Surveillance based on Stability Modelling for any process and equipment parameter and performance curve
- Root Cause Analysis with integration of relevant data
- Automation of plant-wide repository for all corrective and preventive action taken
- Provision of an analytical platform that can provide a foundation of capability within the refinery for application of Statistics, Data Mining, Reliability, Survival, Discrete Event Simulation, optimisation capabilities to generate in-plant solutions by the utilization of huge amount of data collected
- Provision of tools to track the effectiveness of maintenance
- Provisioning for predictive, prescriptive of cognitive analytics. It should have ANN libraries for image processing, Deep Neural Net, SVM, CNN and all traditional ML algorithms
- Provisioning connectivity to DCS and SCADA systems through open protocols like OPC.

Data set for predictive maintenance of assets: The nature and quantity of data required for predictive maintenance analysis would differ from one application to another. For the case of a refinery, some major categories of data that would be necessary are as follows:-

- Static Data—Equipment Make, Model, Configuration, best practices and OEM recommendations
- Frequently Updated Data & Usage History: Age of asset in days, failure history and the preventive maintenance schedules of assets, KPIs for asset performance tracking
- Maintenance Data: Maintenance/breakdown details, service history
- Time series Data: KPIs that are needed as function of time
- Feature Engineering: Collection of data which are collated in the form of averages, linked data (such as ambient temperature, bearing temperature and vibration) in order to draw better inferences about the health of machinery.

In addition to the above basic structure of Predictive Maintenance Framework, following are some of the major features that are included in a predictive maintenance decision framework for better technology and resource utilisation and enterprise level outcomes:

(i) Asset data mapping and modelling: Mapping of operational, environmental, historical and age related data are done for generating information that point more precisely to equipment/system health. This data mapping can support

Table 3 Sample asset score

Asset Score	Status
9–10	Good
6–8.9	Close monitoring
3–5.9	Action required
0–2.9	Danger

regression models (for predicting RUL, failure time), classification models to predict failure within a given time window, models for flagging anomalous behaviour, Survival models for the prediction of failure probability over time.

(ii) Machine learning based asset health score in dashboard: Using tools of AI and ML, the decision support system can generate certain scores representing the health an equipment or system. This information, when available in the platform for access by maintenance team or on the dashboard for senior management, can serve as a valuable decision making tool.
A sample asset score is shown in Table 3:

(iii) Operational & enterprise reporting automation and dashboards: One of the important requirements of a Predictive Maintenance Decision Support System under Industry 4.0 is a very versatile and rich reporting system. The reports range from direct real time recorded data, data from direct computations, alarms, trends, diagnosis, maintenance reports and dashboard summaries. In order that the benefits of such a decision support system is fully realised, it is necessary to devote adequate attention to the development and configuration of dashboards, HMIs and other reporting mechanisms. For example, a dashboard vibration analysis of typical rotating equipment in a refinery will present the following:

- Time, Frequency Domain Analysis and Phase Analysis
- Bode Plot, Waterfall Plot, Polar Plot, Orbit time base plot
- Enhanced fast Fourier transform
- Constant percentage bandwidth (CPB) and Selective envelope detection (SED)
- Vibration Trend Analysis
- Vibration Spectra and Time waveform Analysis
- Detection of bearing fault harmonics (BPFO/BPFI harmonics) by suitable anomaly detection technique or referring to fault frequency database
- Waterfall full spectrum
- Vibration severity recommendations as per ISO 10816 and ISO 13373.

(iv) Cyber security requirements: The platform is normally developed based on industry-standard, secure software development practices. Implementation of cyber security best practices in accordance to international IT and OT standards (ISO 27001, IEC 62443, NIST SP 800 etc.) are complied with for better inter-operability as well as to ensure integrity and security of data. Necessary safeguards specific to the plant and user department are also built in. However care is to be exercised to ensure that utility of the system and ease of operation is not hampered while instituting security measures.

(v) Safety Requirements: It is needless to state that the system as complex, critical and expensive as the Integrated Business Performance System should have all safeguards against hazardous material in the refinery, electric hazards, fire, vibration and mechanical shock or impact.

7.3 *Suggested IIOT Architecture for a Refinery*

Based on the foregoing discussions, a suggested architecture for a petroleum refinery is shown in Fig. 15. IIOT based sensors as well as conventional sensing equipment pick up signals from operating equipment and systems in two independent distillation units. The signals are transmitted through a set of switches and gateways to dual redundant cabling, firewalls and then to cloud based and remote infrastructure. Control signals are also sent to the distilling plants through the same network.

7.4 *Conclusion*

In modern refining practice where reliability, availability, productivity and foremost safety are the main point of concern, traditional operator driven approach has its draw backs to meet all these criteria fully. So modern approach of IIOT based Industry 4.0 techniques will definitely help to improve the plant's safety and reliability along with other production related parameters. In order to achieve the same, it is necessary to employ Industry 4.0 technologies in a comprehensive, optimal and plant-wide manner as brought out earlier in this chapter and also in the above case study. While it would be desirable to have high-intelligence and complicated-automation in maintenance as brought out in Sect. 2.1, constraints such as intial/recurring costs and human resources/factors need to be borne in mind while finalising the scope for the PdM 4.0 systems.

7.5 *Future Work for Implementation of IIOT Based Industry 4.0*

Knowhow of IIOT based Industry 4.0 is in itself a challenge. Training of management and operators is a first step of the implementation process. To reap the benefits out of the system the detail engineering part can be outsourced to agencies with expertise in data acquisition and processing technologies. Software Solution

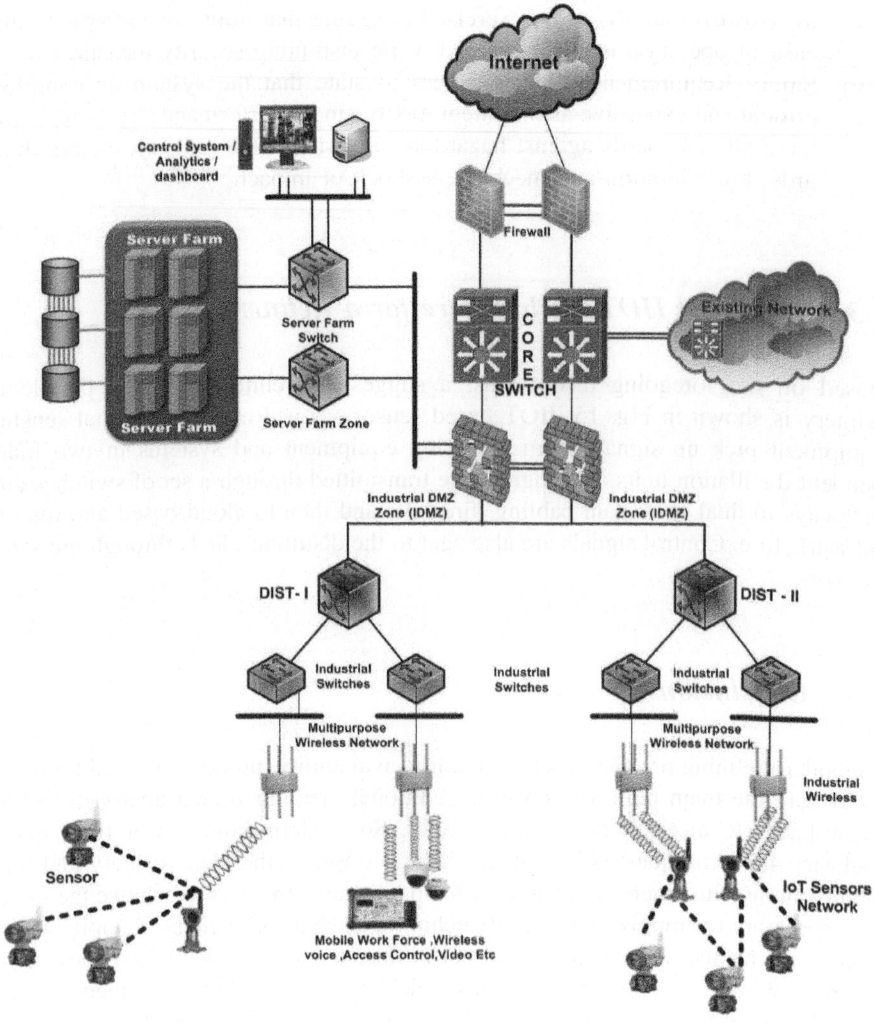

Fig. 15 Suggested IIOT architecture for a refinery

providers in the market can come up with the complete solution and customised packs as per requirement of plant management. The various milestones are described below:

(i) Identification of the requirements- Identification of the critical assets which we want to monitor remotely.
(ii) Choosing the technology-As mentioned earlier, there are service provider companies who can provide various sensors technologies, data acquisition, processing and presentation techniques for IIOT based Industry 4.0.

(iii) The plant management will choose the right technology and train its employees on it.

(iv) Post implementation, it will be necessary to institutionalise maintenance of the system and validation of data as per actual situation.

8 Conclusion

Industry 4.0 is being adopted by manufacturing and process industries in keeping with the global industrial and research trends. Maintenance, as an indispensable element of industrial operations, will also have to align its practices, hardware and frameworks in line with this global trend in order that industries reap the full benefit of Industry 4.0. There is a large dispersion between countries, regions and sectors when it comes to the extent of implementation of Industry 4.0. This dispersions will reduce with greater acceptability and readiness for the constituent technologies on the part of industries worldwide. However, it is amply evident that a comprehensive implementation of Industry 4.0 leveraging all relevant technologies will provide immense mileage that can truly be called the fourth Industrial Revolution. In this chapter, unique considerations for maintenance management of those manufacturing and process plants that are located remotely are discussed. Such remote locations come with an array of challenges which can be effectively handled with the advent of Industry 4.0. Further, application of Industry 4.0 in maintenance is not only desirable but also indispensable for industries to remain competitive. Transition to Predictive Maintenance Under Industry 4.0 (PdM 4.0), especially in respect of legacy systems is a universal challenge and needs to be handled appropriately as also discussed in this chapter. Case studies presented in this chapter highlight the benefits of using technologies of Industry 4.0 such as IIOT, AI, Cloud resources and Automation for better RAMS and hence increased productivity and competitiveness.

References

1. Rohit Robinson (2016) Optimizing remote collaboration, monitoring at process plants. www.controleng.com/articles/optimizing-remote-collaboration-monitoring-at-process-plants
2. Paul J. Galeski When Is Remote Management the Right Move for Your Industrial Application? 05 Oct 19. www.automation.isa.org/when-is-remote-management-monitoring-right-option-factory-plant-industrial
3. Patil SS, Bidkar N, Chavan P, Khamkar A (2018) Industrial plant automation using SCADA. Int J Res Appl Sci Eng Technol (IJRASET) IV
4. Verma AK, Srividya A, Ramesh PG (2010) A systems approach to integrated E-maintenance of large engineering plants. Int J Syst Assur Eng Manag 1(3):239–245
5. Hernabb M, Pentek T, and Otto B (2015) Design Principles for Industrie 4.0 Scenarios: A Literature Review. Working Paper

6. Ravnå R (2016) Industry 4.0 and Maintenance. NorskForening for Vedlikehold
7. Kagermann H, Wahlster W, Helbig H (2013) Recommendations for implementing the strategic initiative INDUSTRIE 4.0. Forschungsunion, acatech
8. Bernata S, Karabag SF (2019) Strategic alignment of technology: organising for technology upgrading in emerging economy firms. Technol Forecast Soc Change. https://doi.org/10.1016/J.TECHFORE.2018.05.009
9. Arbix G, Salerno M, Zancul E, Amaral G, Lins L (2017) Advanced manufacturing: what is to be learnt from Germany, the US, and China. Novos Estud CEBRAP 36:29–49
10. Dalenogare LS, Benitez GB, Ayala NF, Frank AG (2018) The expected contribution of Industry 4.0 technologies for industrial performance. Int J Prod Econ 204:383–394
11. Castelo-Branco I, Cruz-Jesus F, Oliveira T (2019) Assessing Industry 4.0 readiness in manufacturing: evidence for the European Union. Comput Ind 107:22–32
12. Qin J, Liu Y, Grosvenor R (2016) A categorical framework of manufacturing for Industry 4.0 and beyond. Procedia CIRP 52:173–178
13. Moubray J (1997) Reliability-centered maintenance. Industrial Press Inc., New York
14. Masoni R, Ferrise F, Bordegoni M, Gattullo M, Uva AE, Fiorentino M, Carrabba E, Di Donato M (2017) Supporting remote maintenance in Industry 4.0 through augmented reality. Procedia Manuf 11:1296–1302
15. Ceruti A, Marzocca P, Liverani A et al (2019) Maintenance in aeronautics in an Industry 4.0 context: the role of augmented reality and additive manufacturing. J Comput Des Eng. https://doi.org/10.1016/j.jcde.2019.02.001
16. Żabińskia Tomasz et al (2019) Condition monitoring in Industry 4.0 production systems—the idea of computational intelligence methods application. Procedea CIRP 79:63–67
17. Żabiński T, Mączka T, Kluska J (2017) Industrial platform for rapid prototyping of intelligent diagnostic systems. In: Mitkowski W, Kacprzyk J et al (eds) Trends in advanced intelligent control, optimization and automation, KKA 2017, Advances in intelligent systems and computing, vol 577, pp 712–721. Springer, Cham
18. Patwardhan A, Verma AK, Kumar U (2016) A survey on predictive maintenance through big data. In: Kumar U et al (eds) Current trends in reliability, availability, maintainability and safety, Lecture notes in mechanical engineering. Springer International Publishing Switzerland
19. Kang N, Zhao C, Li Jingshan, Horst JA (2016) A Hierarchical structure of key performance indicators for operation management and continuous improvement in production systems. Int J Prod Res 54(21):6333–6350
20. Dalmarco G, Ramalho FR, Barros AC, Soares AL (2019) Providing Industry 4.0 technologies: the case of a production technology cluster. https://doi.org/10.1016/j.hitech.2019.100355
21. Deshpande A, Pieper R (2011) Legacy machine monitoring using power signal analysis. ASME 2011 international manufacturing science and engineering conference MSEC2011, pp 207–214, Corvallis, OR, United States
22. Janak L, Stetina J, Fiala Z, Hadas Z (2016) Quantities and sensors for machine tool spindle condition monitoring. MM Sci J 1648–1653
23. Tedeschi S et al (2018) A cost estimation approach for IOT modular architectures implementation in legacy systems. Procedia Manuf 19:103–110
24. Srivid A, Verma AK, Ramesh PG (2012) Optimal maintenance of large engineering systems–Practical strategies for effective decision making. Narosa Publishing House Pvt. Ltd., N Delhi, India

Dr. P. G. Ramesh is a senior Professor (Mechanical Engineering) and Dean in the Assam Kaziranga University, Assam, India with over 30 years of experience in engineering and management roles in the government, private industries and public and private educational institutions. An alumnus of the National Defence Academy, Naval College of Engineering, University of Pune and IIT, Bombay, he has a wide experience in multiple roles ranging from Naval & Industrial operations, shipbuilding, maintenance management, design & development of engineering products & systems, research and teaching. His areas of interest are futuristic, sustainable and integrated engineering & design, automation, reliability engineering, renewable energy and marine engineering.

Saurav Jyoti Dutta is a Manager (Mechanical Maintenance Department) and Site-in-charge of HCU & H2U units of Numaligarh Refinery Limited, Assam with over 13 years of professional experience. He is an alumnus of Assam Engineering College and is currently pursuing M. Tech in Industrial & Production Engineering from the Assam Kaziranga University, Jorhat, Assam. He has experience in Overhauling and preventive maintenance of various static (Column, Vessel, Furnace, valves etc.), rotary (pumps, compressors, blowers etc.) equipment and Predictive maintenance for reciprocating compressors using online monitoring PROGNOST SYSTEM software.

Subhas Sarma Neog is an Assistant Manager (Mechanical Maintenance Department) and Area-in-charge of Diesel Hydro-treater Unit of Numaligarh Refinery Limited, Assam, India with over 8 years of professional experience. He is an alumnus of Jorhat Engineering College and is and is currently pursuing M. Tech in Industrial & Production Engineering from the Assam Kaziranga University, Jorhat, Assam. He has experience in Overhauling and preventive maintenance of various static (Column, Vessel, Furnace, valves etc.) and rotary (pumps, compressors, blowers etc.) equipment and Predictive maintenance for reciprocating compressors using online monitoring PROGNOST SYSTEM software.

Prandip Baishya is an Assistant Manager (Predictive Maintenance Department) of Numaligarh Refinery Limited, Assam, India with over 12 years of professional experience. He is an alumnus of NIT Bhopal and is currently pursuing M. Tech in Industrial & Production Engineering from the Assam Kaziranga University, Jorhat, Assam. He has experience in Overhauling and preventive maintenance of various static (Column, Vessel, Furnace valves etc.) and rotary (pumps, compressors, blowers, Cranes, DG sets up to 1000 KVA) equipment, Predictive maintenance of rotary equipment using various tools- Vibration analysis, Lube oil wear debris analysis, Motor Current Signature analysis, Ultrasound analysis, Dynamic & in situ balancing.

Indrani Bezbaruah is an Affiliated Faculty (Mechanical Department) and Robotics and Automation lab in-charge of Assam Kaziranga University, Jorhat, Assam, India with professional experience of 2 years. She is an alumnus of Assam Kaziranga University, Jorhat and is trained in industrial Robotics and Automation. Her area of interest is Production engineering, reliability engineering, and automation.

Application of Fuzzy Sets in Reliability and in Optimal Condition Monitoring Technique Selection in Equipment Maintenance

Rajesh S. Prabhu Gaonkar

Abstract In order to improve the design of a system, we need to identify the least reliable component of the system. Unexpected failure of any component of the system may increase the maintenance and down time cost due to unavailability of the system. Though this is easy in simpler systems, it becomes a difficult task as the complexity of the system increases. A methodology using mathematical modelling facility of fuzzy set theory is presented here, which is effective in situations wherein the data available is mostly subjective and it is difficult to get precise quantitative data. After covering basic concepts of various uncertainty modelling theories and fuzzy sets, its application to reliability and fault tree is presented. In the second part of the chapter, multi-attribute decision making methods with application to ranking and optimal condition monitoring technique selection from maintenance engineering domain is presented. These include fuzzy set based Analytic Hierarchy Process (AHP), rating and ranking method, ranking by maximizing and minimizing sets, raking by cardinal utilities and suitability set method.

1 Introduction

In this competitive world, reliability and maintenance of components, sub-systems, systems etc. makes considerable impact on profitability of an enterprise. Reliability, i.e. survival probability of system, was researched world-wide at academic as well as industry level for last several decades. Consequences of inadequate maintenance of equipment was also studied for last many decades. Conventionally, reliability and maintenance modeling is carried out by probabilistic method. Probabilistic methods fared well as product life cycle was long and sufficient data gathering was possible. In present day scenario, product life cycles are short and as a result, there

R. S. Prabhu Gaonkar (✉)
School of Mechanical Sciences, Indian Institute of Technology Goa (IIT Goa), Farmagudi, Ponda, Goa, India
e-mail: rpg@iitgoa.ac.in

D. R. Karanki et al. (eds.), *Advances in RAMS Engineering*,
Springer Series in Reliability Engineering,
https://doi.org/10.1007/978-3-030-36518-9_13

is acute need of newer ways for reliability and maintenance modeling. This chapter presents the application of a different approach i.e. fuzzy set theory approach in reliability and maintenance engineering.

2 Various Uncertainty Modeling Theories

There are two facets of looking at 'uncertainty'. Uncertainty that arises from variability or stochasticity is random. Such uncertainty that is based on randomness is termed as 'aleatory uncertainty'. The other kind of uncertainty is due to lack of information or data, ignorance, subjectivity in expressions, linguistics, ambiguity, etc. This uncertainty is a result of fuzziness and is known as 'epistemic uncertainty'. Aleatory type uncertainty is dealt by Probability Theory that handles randomness. Epistemic uncertainty is dealt by possibility theory or fuzzy set theory.

Probability theory has its roots in the 17th century. Probability is an objective characteristic. It deals with the chance of occurrence of an event such as rolling a dice and getting numbers 1, 2, 3 4, 5 or 6. The conclusions of probability theory are tested by conducting an experiment or by experience. The basics of probability theory lie in set theory and the logic used is that either an element belongs to a set or not. Interval used in probability theory is [0, 1].

Fuzzy set theory was established and developed since the year 1965. Fuzzy set theory is presently a quite well established mathematical theory [1–4]. Fuzzy set theory and its applications developed very comprehensively over the last few decades and has attracted the attention of academicians, researchers, and practitioners world-wide. It considers subjective uncertainty in linguistics e.g. 'beautiful child', 'cold day', 'very hot' etc. The membership grade used here is subjective in nature. This theory focuses primarily on imprecision which is intrinsic in natural languages and is usually associated with the term 'possibilistic'. The term 'variable' is used here in a linguistic sense. Possibility theory also uses [0, 1] interval.

Other prominent theories for uncertainty modeling are Bayesian theory, rough sets theory, vague sets theory, intuitionistic fuzzy sets theory, neutrosophic sets theory, soft sets theory, interval arithmetic theory, evidence theory etc.

Bayesian theory provides a mathematical framework wherein using probability concepts, inferences are drawn and reasoning is performed. It is based on Bayes' theorem where posterior probability is calculated by updating the prior probability. It describes the probability of an event based on prior knowledge of conditions that might be related to the event. Analyst develops a set of initial beliefs and then adjusts them through experimentation and arrives at the Bayesian posterior probability.

Rough set theory proposed by Pawlak in 1982, introduced a new mathematical approach to model imperfect knowledge, i.e. imprecision or vagueness. In this theory, vagueness is expressed by a boundary region of a set. Rough set is defined by means of topological operations, interior and closure, called approximations. This theory is different than that of probability theory in statistics or fuzzy set theory that uses membership grades. This theory does not need any initial information or additional information about data. Interpretation of results is easier in rough set theory and is popular amongst people working in the field of computer engineering.

Intuitionistic fuzzy set and vague set are further extensions or generalizations of fuzzy sets. The definition of intuitionistic fuzzy sets was given by Atanassov in 1983 while vague sets were defined by Gau and Buehrer in 1993. Intuitionistic fuzzy sets are developed prior to vague sets. In Intuitionistic fuzzy sets, there is a membership grade and also a non-membership grade attached with each element. Moreover, there is a constraint that the sum of these two grades, i.e. membership grade and non-membership grade is less than or equal to one. In fuzzy sets, point-based membership values are used while in vague sets, membership values provided are interval based. In vague sets also, an object is characterized by two different membership functions, namely true membership function and false membership function. It is found that interval based membership functions used in vague sets capture data vagueness in a better way and are more expressive. In literature, researchers have considered both intuitionistic fuzzy sets and vague sets as equivalent, i.e. corresponding or similar in form and relations.

Florentin Smarandache in 1995 introduced a theory known as neutrosophic logic and sets. Neutrosophy refers to the knowledge of neutral thought and it represents the main distinction between sets, fuzzy sets and intuitionistic fuzzy sets. In this theory, a different viewpoint is proposed in which each proposition is estimated to have a degree of truth, a degree of indeterminacy and a degree of falsity. A neutrosophic set is a set where each element of the universe has a degree of truth, indeterminacy and falsity. In this type of sets, the indeterminacy is independent of truth and falsity values. All the three degrees i.e. of truth, indeterminacy and falsity, vary within [0, 1].

Another mathematical tool, i.e. theory of soft sets was proposed by Molodtsov in 1999. This theory also deals with uncertainties. Soft set is defined as a parameterized family of subsets of universe set where each element is considered as a set of approximate elements of it, i.e. of soft set. In short, the boundary of the set depends on the parameters. This theory is somewhat a generalization of fuzzy set theory.

Interval arithmetic (also popular as interval mathematics or interval analysis) was proposed by Ramon E. Moore in 1957. This theory characterises each value as a range of possibilities. It considers the computation of both the exact solution and the error term as a single entity i.e. an interval. It is an approach that puts bounds on rounding errors or measurement errors in mathematical computations. It develops numerical methods that yield consistent results. Apart from handling rounding errors and uncertainties, it also helps to get dependable results and definite solutions to equations and also for optimization problems. This concept is effectively used till now in numerous applications in mathematics, computer science and engineering.

Evidence theory (also referred as theory of belief functions) was first introduced by Dempster in 1967 and later extended by Shafer in 1976. This theory is popular as Dempster-Shafer theory. It provides a general framework for reasoning with uncertainty having influence of other theories like probability and possibility theories. Evidence theory is relatively closer to the classical probability theory. Evidence theory considers the evidence from different sources and computes degree of belief i.e. belief function which takes into account all available evidence. The major advantage of this theory is that this requires comparatively less information to describe the phenomenon than the classical probability theory. Evidence theory uses Dempster-Shafer combination rule instead of usual Bayes formula to update the belief function. Four important concepts in evidence theory are—frame of discernment, basic belief assignment, belief function and plausibility function.

3 Fuzzy Set Theory

Human communication and interpretation involves ambiguity and multiple meanings to its expressions. Fuzzy sets have helped researchers and academicians to model the amount of ambiguity and subjectivity. Variety of engineering and management problems involving ambiguous, imprecise, inexact or vague information are well solved using fuzzy sets in literature [1–4]. Fuzzy set theory is suitable to handle problems which lack of adequate data, inconsistency of data, experts differing in their opinions, ignorance, etc.

The originator of fuzzy set and fuzzy logic is Zadeh. In 1965, Zadeh while working in the field of control engineering, introduced fuzzy set concept. He developed the idea of partial belongingness of an object to the set. Traditionally in set theory or in probability theory, an object may belong or not belong to the set. The traditional set is a crisp set. The related logical scheme is like—it may be true or false. The logic of crisp set is extended to fuzzy set wherein the idea of partial truth or partial false is introduced. In practical world, the data values are vague in many of the applications. Fuzzy set theory handles the vagueness by generalizing the idea of membership in a set. An object may be a member of a set to some degree. Fuzzy set theory presents the concept of grades or values or function of membership. Therefore the membership of some object in the given set X may be no membership, full membership or partial membership. Each element of a fuzzy set is linked to a point-value from the unit interval of [0, 1] that is designated as the grade of membership in the fuzzy set.

4 Fuzzy Reliability

Conventional reliability is defined as the probability that a component or a system will perform its intended function adequately/satisfactorily for a specified period of time, understated operating conditions. Conventional reliability concept is based completely on the probability theory. The basic assumption in the conventional reliability is that there are only two states for a component or a system. These are operating or working state and failed state. With the growth in possibility theories in last few decades, there is a vital change in viewing the reliability concepts and assumptions. Reliability theories are modeled using fuzzy sets, mainly with the remarkable development of fuzzy set theory. Although conventional reliability theory cannot be out rightly excluded, the merits of fuzzy reliability must be used in parallel. There are many ideas proposed in the literature in the fuzzy reliability domain. This theory is based on both possibility assumption as well as probability assumption. It also considers binary state assumption and fuzzy-state assumption. Fuzzy reliability concepts use the combination of the above-said assumptions [5–8].

The conventional reliability is PROBIST reliability. It is based on the assumption of probability and the binary-states. PRO stands for probability and BIST stands for binary states. The states are deterministic i.e. crisp and the system will be in one of these two states, operating or failed, at any given point in time. The PROBIST reliability of the component or the system is computed using fuzzy sets which consider it as a fuzzy number. PROBIST reliability can be also in the form of linguistic values. Researchers have considered PROBIST reliability as a fuzzy number with membership functions of triangular, trapezoidal, normal, etc.

PROFUST reliability considers probabilistic assumption and the fuzzy state assumption. The failure time of the system is assumed to follow probability measures. But, in this case, operating and failed states are defined by fuzzy states. The system can be in one of the fuzzy states at any given point in time. As crisp state is one of the states of fuzzy set, PROFUST reliability is one of the cases of PROBIST reliability.

POSBIST reliability is based on possibility assumption and assumes the binary states, while POSFUST reliability is defined based on the concept of possibility theory and fuzzy states.

4.1 *Example of PROBIST Fuzzy Reliability*

PROBIST reliability is considered as Triangular Fuzzy Number (TFN) [9, 10] in this example. Figure 1 shows the reliability block diagram of a system consisting of six components. There are three subsystems configured in series. First subsystem connects two components A and B in parallel. Second subsystem has two components C and D linked in series and third subsystem again has parallel configuration with components E and F. Component reliabilities as symmetric TFN are as below:

$$R_A = [0.65, 0.75, 0.85]$$
$$R_B = [0.70, 0.75, 0.85]$$
$$R_C = [0.70, 0.80, 0.95]$$
$$R_B = [0.85, 0.90, 0.85]$$
$$R_F = [0.60, 0.70, 0.80]$$

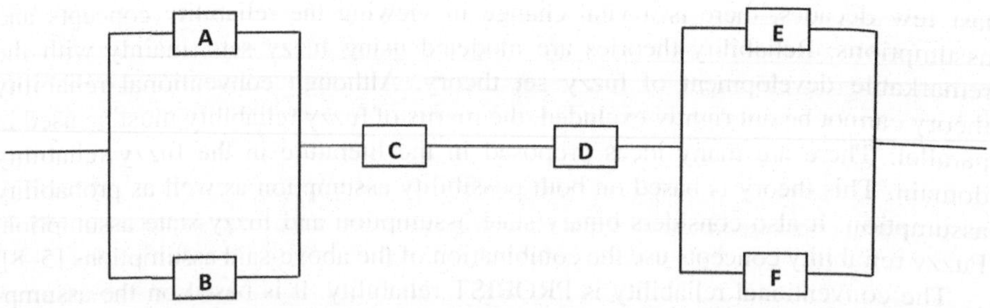

Fig. 1 Reliability block diagram

Series system reliability is given by,

$$[R_{SS}] = \prod_{i=1}^{n} R_i$$

Parallel system reliability is given by,

$$[R_{PS}] = 1 - \prod_{i=1}^{n} [(1 - R_i)]$$

Taking α-cut [9, 10] and solving for parallel subsystem reliability, we get

$$\begin{aligned}[R_{AB}] &= [1 - [\{1 - (0.65 + 0.1\alpha)\} \cdot \{1 - (0.70 + 0.05\alpha)\}], \\ &\quad 1 - [\{1 - (0.85 - 0.1\alpha)\} \cdot \{1 - (0.85 - 0.1\alpha)\}]] \\ &= [0.8950, 0.9375, 0.9775]\end{aligned}$$

$$\begin{aligned}[R_{EF}] &= [1 - [\{1 - (0.75 + 0.05\alpha)\} \cdot \{1 - (0.60 + 0.1\alpha)\}], \\ &\quad -[\{1 - (0.85 - 0.05\alpha)\} \cdot \{1 - (0.80 - 0.1\alpha)\}]] \\ &= [0.9000, \quad 0.9400, \quad 0.9700]\end{aligned}$$

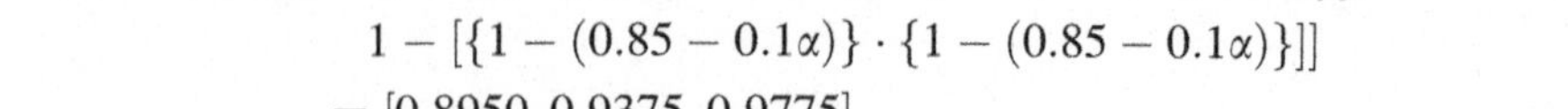

$$\begin{aligned}[R_{AB}] &= [0.8950, \quad 0.9375, \quad 0.9775] \\ [R_C] &= [0.7000, \quad 0.8000, \quad 0.9000] \\ [R_D] &= [0.8500, \quad 0.9000, \quad 0.9500] \\ [R_{EF}] &= [0.9000, \quad 0.9400, \quad 0.9700]\end{aligned}$$

Table 1 PROBIST reliability values

A	R_{SS}(Lower)	R_{SS}(Upper)
0	0.479273	0.81069
0.1	0.493484	0.791743
0.2	0.507978	0.773099
0.3	0.522757	0.754756
0.4	0.537825	0.736709
0.5	0.553186	0.718957
0.6	0.568842	0.701495
0.7	0.584799	0.684322
0.8	0.601058	0.667433
0.9	0.617624	0.650827
1.0	0.6345	0.6345

Taking α- cut and solving for Series system reliability, we get

$$[R_S] = \{(0.8950 + 0.0425\alpha) \cdot (0.7000 + 0.1000\alpha) \cdot (0.8500 + 0.0500\alpha) \cdot (0.9000 + 0.0400\alpha),\\ (0.9775 - 0.04\alpha) \cdot (0.9000 - 0.1000\alpha) \cdot (0.9500 - 0.0500\alpha) \cdot (0.9700 - 0.0300\alpha)\}$$

For various values of α, PROBIST reliability bounds are presented in Table 1.

4.2 *Conceptual Example of PROFUST Fuzzy Reliability*

The illustrated example gives an indication of PROFUST reliability applied to a degrading system [11]. Consider a system consisting of five sub-systems that perform the operation simultaneously. If all five sub-systems work then we say that the system is functioning fully. However, if any of the sub-system fail then the system operates, but in degraded mode. As a result, system performance also degrades. For such a system there are more than two states which can be modeled using fuzzy sets. The solution methodology would involve developing a Markov model for degradable system. Markovian analysis considers the system as being in one of the several states, either in an operating state or failed state. Here the assumptions are that conditional probability of failure during any fixed interval of time is constant and so are the transition rates. Also, the probability that the system would undergo transition from one state to another state depends only on the current state of the system. The transition does not depend on any of the states the system has experienced earlier. Exponential distribution having memoryless property is used to model failure times of the sub-systems, as it satisfies Markovian property.

Let the universe of discourse be $U = \{S_0, S_1, S_2, S_3, S_4, S_5\}$; where fuzzy success states are given by $S = \{S_i,\ \mu_S(S_i)\,;\quad i = 1,\ 2,\ 3,\ 4,\ 5\}$; and corresponding fuzzy failure states are given by $F = \{S_i,\ \mu_F(S_i)\,;\quad i = 1,\ 2,\ 3,\ 4,\ 5\}$. μ is the membership grade. For Markovian analysis, consider the transition from fuzzy success state to fuzzy failure state as $T_{SF} = \{m_{ij}\,,\, \mu_{T_{SF}}(m_{ij})\,;\quad i = 1,\ 2,\ 3, 4,\ 5\}$ where T_{SF} represents and m_{ij} is the transition from state S_i to S_j. The next step is to build the membership functions and derive the equation for the PROFUST reliability using an appropriate method. Using the alpha-cut technique or any other method of fuzzy sets, values of PROFUST reliability can be computed.

4.3 *Fuzzy Fault Tree*

Fault tree analysis is one of the ways to compute probability of failure, and in turn obtain reliability of the system. Traditionally, the failure probabilities of components are obtained from past data. However, several complications such as continuously changing environment and availability of data about components make the calculation of system reliability difficult. If component failure probabilities are not accurate, then the top event probability, i.e. system failure probability will not be accurate as well. Therefore acute need was identified by the system designers to obtain the data from the field experts based on their subjective assessments. In literature, fuzzy sets and its arithmetic have dealt with such situations. The fuzzy sets are extensively used in fault tree analysis that provide a mathematical framework for imprecise and uncertain data situations.

In fuzzy fault tree analysis [7], triangular fuzzy numbers are utilized as possibility distribution of each of the primary event. In quantitative assessment of a fuzzy fault tree, the fuzzy data in the form of triangular fuzzy number is considered for each component at the bottom-most hierarchical level of the fault tree. Usual logic of the trees along with Boolean operations and fuzzy arithmetic is used to provide the failure possibility calculation of all the subsequent higher level events. Finally, possibility distribution for the top-most event i.e. possibility of failure of the system under analysis is determined.

In probabilistic fault tree analysis, exponential time to failure distribution with constant failure rate λ is considered. Based on available data, parameter λ of exponential failure time is a single point estimated value. In fuzzy fault tree analysis it is considered that the uncertainty exists in estimation of parameter λ. This uncertainty is modeled by defining λ as a triangular fuzzy number with triplets λ_1 as lowermost value, λ_2 is middle value and λ_3 as rightmost value i.e. $\lambda = (\lambda_1, \lambda_2, \lambda_3)$. An example is presented here for the fault tree depicted in Fig. 2.

The top event T can be expanded as:

$$\begin{aligned} T &= A_1 \cdot A_2 \\ &= (X_1 + X_2) \cdot (X_3.A_3) \\ &= (X_1 + X_2) \cdot [X_3 \cdot (X_4 + X_5)] \end{aligned}$$

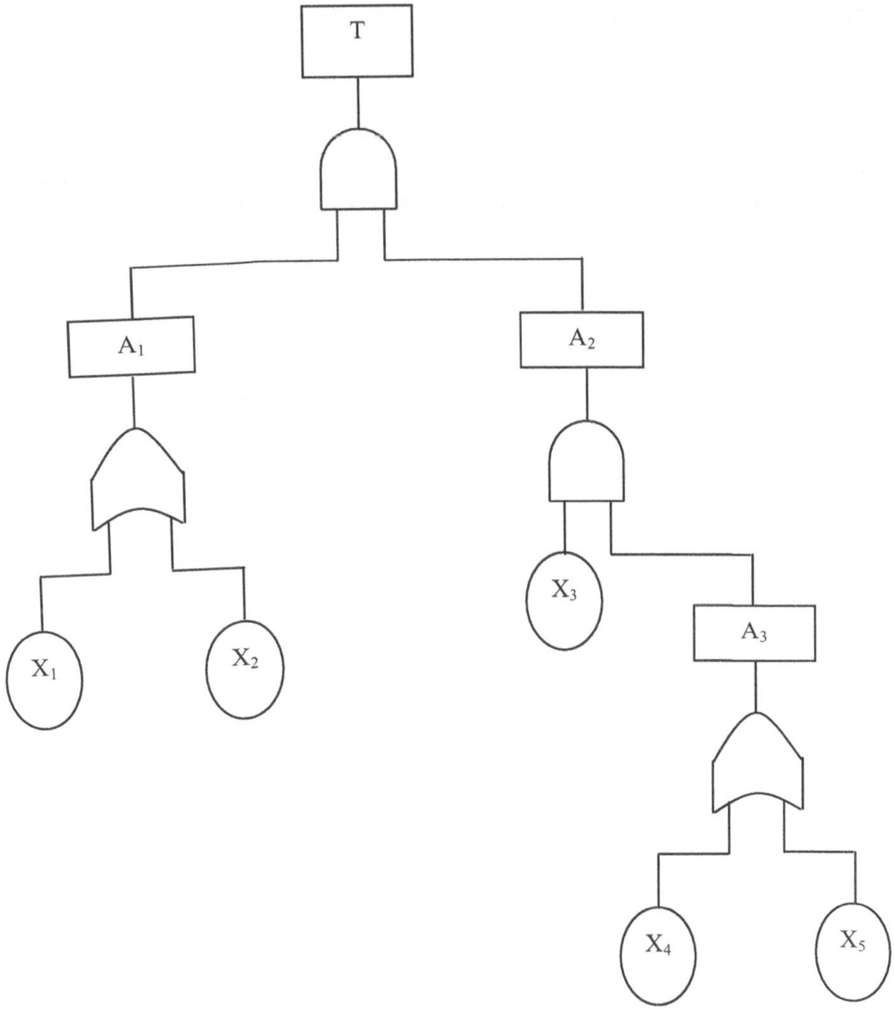

Fig. 2 Fault tree diagram

Given p_i, the failure probability of X_i, the failure probability of top event T is,

$$prob(T) = [1 - (1 - p_1)(1 - p_2)] \cdot p_3 \cdot [1 - (1 - p_4)(1 - p_5)]$$

The exponential time to failure distribution is assumed for all basic events with fuzzy failure rate λ_i i.e. in the form of TFN with triplets $\lambda_{i1}, \lambda_{i2}\, and\, \lambda_{i3}$. The intervals of confidence for the fuzzy failure probability p_i, for the level of presumption α, have been estimated using following equation:

Table 2 Fuzzy probability of the basic events

Basic events	Failure rate λ (crisp value)	Triangular fuzzy number					
		Failure rate			Probability		
X_1	0.0004	0.0001	0.0004	0.0006	0.09516	0.32968	0.45119
X_2	0.0007	0.0004	0.0007	0.00085	0.32968	0.50342	0.57259
X_3	0.0001	0.00005	0.0001	0.00025	0.04877	0.09516	0.22120
X_4	0.00045	0.0002	0.00045	0.0008	0.18127	0.36237	0.55067
X_5	0.00064	0.0005	0.00064	0.00075	0.39347	0.47271	0.52763

Table 3 Fuzzy Probability of the Top Event

Presumption level	Fuzzy probability of the top event	
	Lower bound	Upper bound
0.0	0.009660	0.133376
0.1	0.011807	0.122477
0.2	0.014195	0.111961
0.3	0.016830	0.101835
0.4	0.019711	0.092101
0.5	0.022840	0.082766
0.6	0.026216	0.073831
0.7	0.029837	0.065299
0.8	0.033701	0.057172
0.9	0.037804	0.049453
1.0	0.042141	0.042141

$$[F(t)]_\alpha = \left[1 - \frac{1}{e^{\lambda_{1t} + \alpha(\lambda_2 - \lambda_1)t}}, 1 - \frac{1}{e^{\lambda_{3t} - \alpha(\lambda_3 - \lambda_2)t}}\right]$$

$\alpha = 0$, gives p_{i1} as the lower bound and p_{i3} as the upper bound and $\alpha = 1$ gives p_{i2} which is the middle value. These are given in Table 2.

Table 3 shows the intervals of confidence at different levels of presumption for the top event probability with assumed $t = 1000$.

5 Fuzzy Multi Attribute Decision Making

Decision making in maintenance engineering is a real challenge for decision makers, i.e. engineer/manager as he/she faces a problem involving information and data uncertainty. Much of the decision-making in this field happens in an environment in which the objectives and the constraints are not accurately known. The major source of imprecision in maintenance decision making processes is fuzziness i.e. a kind of imprecision that is associated with fuzzy sets. Although many models have been developed over the years, its discussion is beyond the scope of this chapter. Here, we discuss Multi Attribute Decision Making (MADM) problems involving fuzziness that are solved using various fuzzy set based methods.

The basic MADM problem is to choose between or rank a set of alternatives, given some decision attributes (also known as criteria). Let $A = \{a_i\};\ \ i = 1,\ 2,\ 3, \ldots, n$ be the set of decision alternatives and $C = \{c_j\};\ \ j = 1,\ 2,\ 3, \ldots, m$ be the set of attributes according to which the desirability of an alternative is to be judged. The aim here is to obtain the optimal alternative with highest degree of desirability with respect to all relevant attributes [12–15].

5.1 *Selection or Ranking of Condition Monitoring Techniques—Problem*

The running example [16–21] explained here is related to the maintenance of a turbine. The purpose is to rank or select optimal alternative out of the three alternative techniques namely Vibration analysis (A_1), Lube oil/debris analysis (A_2), Endoscopic examination (A_3). Nine decision attributes are considered, namely, investment cost (C_1), operating cost (C_2), accuracy of the technique (C_3), repeatability of the instruments used (C_4), ease of use (C_5), environmental restrictions (C_6), technical expertise requirement (C_7), ease of maintenance (C_8), ease of mounting (C_9). This problem is solved using fuzzy AHP and other fuzzy set based methods.

5.2 *Fuzzy Analytic Hierarchy Process (FAHP)*

The Analytic Hierarchy Process (AHP), also known as priority theory, is a commanding and flexible decision making procedure that helps people set priorities and make the best decision when both qualitative and quantitative aspects of a decision are to be considered. The AHP engages decision makers in structuring a decision into smaller parts, proceeding from the goal to objectives to sub-objectives down to the alternative courses of action. Decision makers then make simple pair wise comparison judgments throughout the hierarchy to arrive at the overall priorities for the alternatives. The analytic hierarchy process allows users to assess the relative weight of multiple attributes in an intuitive manner. Saaty, in 1980, invented AHP methodology that established a consistent way of converting pair wise comparisons into a set of numbers representing the relative priority of each of the attributes. Many fuzzy set based AHP have been suggested by researchers [14, 15, 17].

The fuzzy set scale for intensity of importance in the form of TFN used in this problem is given in Table 4.

Table 4 Fuzzy Set scale for intensity of importance

Intensity of importance (TFN)	Definition	Explanation
[1, 1, 1]	Equal importance	Two activities contribute equally to the objective
[1, 2.5, 4]	Weak importance of one over another	Experience and judgment slightly favor one activity over another
[3, 4.5, 6]	Essential or strong importance	Experience and judgment strongly favor one activity over another
[5, 6.5, 8]	Demonstrated importance	An activity is strongly favored and its dominance demonstrated in practice
[7, 8.5, 10]	Absolute importance	The evidence favoring one activity over another is of highest possible order of affirmation

Table 5 shows a matrix of relative significance of each pair of attributes. Let r_{ij} be the numerical value assigned to the relative significance i.e. importance of attributes C_i and C_j. Fuzzy set scales for intensity of importance are available in the literature which may also be used. If both C_i and C_j, are equally important then $r_{ij} = 1$; if C_i is more important than C_j then $r_{ij} > 1$ and if C_i is less important than C_j then $r_{ij} < 1$. r_{ij} is a reciprocal matrix and it has positive entries throughout satisfying the reciprocal property i.e. $r_{ji} = \frac{1}{r_{ij}}$. Here, r_{ij} s are in TFN form and its reciprocal is obtained using inverse operation on TFN. Table 6 gives normalized average weights (priorities) matrix of the attributes. It is shown by researchers that normalized column and row weights are adequate as normalized Eigen vectors as used in the case of single value (crisp) intensity of importance, and as such the average of the row and column is taken as the final weight.

The three maintenance alternatives are then compared in pair wise manner under each criterion. These matrices are given in Tables 7, 8, 9, 10, 11, 12, 13, 14 and 15.

The final score of each alternative is then calculated by performing fuzzy multiplication of priority of attributes and priority of condition monitoring technique and then adding them up for each condition monitoring technique. This is given in Table 16. The final scores of techniques in TFN form are: Vibration analysis = [0.139, 0.294, 0.624], Lube oil/debris analysis = [0.104, 0.252, 0.659] and Endoscopic examination = [0.164, 0.371, 0.823]. Ranking of all three techniques is plotted in Fig. 3 in TFN form.

5.3 Linguistic Scales and Input from the Experts

A few more ranking methods are explained in subsequent sub- sections. For this, the ratings and weights have been expressed in linguistic terms. They are considered as linguistic variables, represented by the fuzzy sets. The grade membership for both the variables are considered as TFN's. Figures 4 and 5 show TFN representations of linguistic variables related to rating and weights on the scale [0, 1].

Table 5 Matrix of relative significance of decision attributes

Attributes	C_1	C_2	C_3	C_4	C_5
C_1	1	[0.13, 0.15, 0.2]	[0.1, 0.12, 0.14]	[0.1, 0.12, 0.14]	[0.13, 0.15, 0.2]
C_2	[5, 6.5, 8]	1	[0.17, 0.22, 0.33]	[0.17, 0.22, 0.33]	[0.25, 0.4, 1]
C_3	[7, 8.5, 10]	[3, 4.5, 6]	1	[0.17, 0.22, 0.33]	[5, 6.5, 8]
C_4	[7, 8.5, 10]	[3, 4.5, 6]	[3, 4.5, 6]	1	[5, 6.5, 8]
C_5	[5, 6.5, 8]	[1, 2.5, 4]	[0.13, 0.15, 0.2]	[0.13, 0.15, 0.2]	1
C_6	[5, 6.5, 8]	[1, 2.5, 4]	[0.17, 0.22, 0.33]	[0.17, 0.22, 0.33]	[0.17, 0.22, 0.33]
C_7	[7, 8.5, 10]	[3, 4.5, 6]	[3, 4.5, 6]	[3, 4.5, 6]	[1, 2.5, 4]
C_8	[5, 6.5, 8]	[0.25, 0.4, 1]	[0.17, 0.22, 0.33]	[0.13, 0.15, 0.2]	[0.17, 0.22, 0.33]
C_9	[0.17, 0.22, 0.33]	[0.25, 0.4, 1]	[0.13, 0.15, 0.2]	[0.1, 0.12, 0.14]	[0.13, 0.15, 0.2]
CS	[42.17, 52.72, 63.33]	[12.63, 20.45, 29.2]	[7.87, 11.08, 14.53]	[4.97, 6.7, 8.67]	[12.85, 17.64, 23.06]
Attributes	C_6	C_7	C_8	C_9	RS
C_1	[0.13, 0.15, 0.2]	[0.1, 0.12, 0.14]	[0.13, 0.15, 0.2]	[3, 4.5, 6]	[4.82, 6.46, 8.22]
C_2	[0.25, 0.4, 1]	[0.17, 0.22, 0.33]	[1, 2.5, 4]	[1, 2.5, 4]	[9.01, 13.96, 19.99]
C_3	[3, 4.5, 6]	[0.17, 0.22, 0.33]	[3, 4.5, 6]	[5, 6.5, 8]	[27.34, 36.44, 45.66]
C_4	[3, 4.5, 6]	[0.17, 0.22, 0.33]	[5, 6.5, 8]	[7, 8.5, 10]	[34.17, 44.72, 55.33]
C_5	[3, 4.5, 6]	[0.25, 0.4, 1]	[3, 4.5, 6]	[5, 6.5, 8]	[18.51, 26.2, 34.4]
C_6	1	[0.13, 0.15, 0.2]	[3, 4.5, 6]	[5, 6.5, 8]	[15.64, 21.81, 28.19]
C_7	[5, 6.5, 8]	1	[5, 6.5, 8]	[7, 8.5, 10]	[35, 47, 59]
C_8	[0.17, 0.22, 0.33]	[0.13, 0.15, 0.2]	1	[3, 4.5, 6]	[10.02, 13.36, 17.39]
C_9	[0.13, 0.15, 0.2]	[0.1, 0.12, 0.14]	[0.17, 0.22, 0.33]	1	[2.18, 2.53, 3.54]
CS	[15.68, 21.92, 28.73]	[2.22, 2.6, 3.67]	[21.3, 30.37, 39.53]	[37, 49, 61]	[156.69, 212.48, 271.72]#

RS Row Sum; *CS* Column Sum; # Total of Row Sum

Table 6 Average priority TFN's of attributes

Attributes	C_1	C_2	C_3	C_4	C_5
RS	[4.82, 6.46, 8.22]	[9.01, 13.96, 19.99]	[27.34, 36.44, 45.66]	[34.17, 44.72, 55.33]	[18.51, 26.2, 34.4]
CS	[42.17, 52.72, 63.33]	[12.63, 20.45, 29.2]	[7.87, 11.08, 14.53]	[4.97, 6.7, 8.67]	[12.85, 17.64, 23.06]
N	[0.017, 0.03, 0.052]	[0.033, 0.065, 0.13]	[0.10, 0.17, 0.29]	[0.12, 0.21, 0.35]	[0.068, 0.12, 0.22]
IN	[0.016, 0.019, 0.023]	[0.034, 0.049, 0.079]	[0.069, 0.090, 0.13]	[0.12, 0.15, 0.20]	[0.04, 0.06, 0.08]
Average priority of attributes	[0.017, 0.025, 0.075]	[0.034, 0.057, 0.105]	[0.085, 0.13, 0.21]	[0.12, 0.18, 0.275]	[0.054, 0.09, 0.15]

Attributes	C_6	C_7	C_8	C_9
RS	[15.64, 21.81, 28.19]	[35, 47, 59]	[10.02, 13.36, 17.39]	[2.18, 2.53, 3.54]
CS	[15.68, 21.92, 28.73]	[2.22, 2.6, 3.67]	[21.3, 30.37, 39.53]	[37, 49, 61]
N	[0.058, 0.103, 0.18]	[0.13, 0.22, 0.38]	[0.037, 0.063, 0.11]	[0.008, 0.012, 0.023]
IN	[0.035, 0.046, 0.064]	[0.27, 0.38, 0.45]	[0.025, 0.033, 0.047]	[0.016, 0.020, 0.027]
Average priority of attributes	[0.047, 0.075, 0.122]	[0.2, 0.3, 0.42]	[0.031, 0.048, 0.079]	[0.012, 0.016, 0.025]

N Normalized (Row); *IN* Inverted, Normalized (Column)

Usually, inputs are always taken from individual experts and are then combined as a part of group decision making. Means of combining expert's judgment include averaging method, averaging method with feedback, and combining method for symmetric triangular fuzzy sets. The combined resulting fuzzy set is then required to be interpreted in terms of linguistic variable. To get this, resultant fuzzy set is to be 'mapped' on to the nearest pre-defined fuzzy set. Some mapping methods are—distance measures such as Hamming distance, Euclidean distance, credibility score method, etc. The input data for ratings and weights are given in Tables 17 and 18 respectively.

5.4 Rating and Ranking Method

As per this method (developed by Kanhe), the rating of the alternative i with respect to decision attribute j is r_{ij}; $i = 1, 2, 3, \ldots, n$; $j = 1, 2, 3, \ldots, m$. The relative

Table 7 Investment cost (C_1)

Techniques	A_1	A_2	A_3	RS	N	Average priority
A_1	1	[0.17, 0.22, 0.33]	[1, 1, 1]	[2.17, 2.22, 2.33]	[0.12, 0.15, 0.21]	[0.125, 0.15, 0.21]
A_2	[3, 4.5, 6]	1	[3, 4.5, 6]	[7, 10, 13]	[0.4, 0.69, 1.15]	[0.5, 0.69, 0.95]
A_3	[1, 1, 1]	[0.17, 0.22, 0.33]	1	[2.17, 2.22, 2.33]	[0.12, 0.15, 0.21]	[0.125, 0.15, 0.21]
CS	[5, 6.5, 8]	[1.34, 1.44, 1.66]	[5, 6.5, 8]	[11.34, 14.44, 17.66][#]		
IN	[0.13, 0.15, 0.2]	[0.6, 0.69, 0.75]	[0.13, 0.15, 0.2]			

Table 8 Operating costs (C_2)

Techniques	A_1	A_2	A_3	RS	N	Average priority
A_1	1	[3, 4.5, 6]	[5, 6.5, 8]	[9, 12, 15]	[0.35, 0.57, 0.91]	[0.5, 0.65, 0.84]
A_2	[0.17, 0.22, 0.33]	1	[5, 6.5, 8]	[6.17, 7.72, 9.33]	[0.15, 0.37, 0.57]	[0.15, 0.28, 0.41]
A_3	[0.13, 0.15, 0.2]	[0.13, 0.15, 0.2]	1	[1.26, 1.3, 1.4]	[0.05, 0.06, 0.09]	[0.06, 0.07, 0.09]
CS	[1.3, 1.37, 1.53]	[4.13, 5.65, 7.2]	[11, 14, 17]	[16.43, 21.02, 25.73]#		
IN	[0.65, 0.73, 0.77]	[0.14, 0.18, 0.24]	[0.06, 0.07, 0.09]			

Table 9 Accuracy of the technique (C_3)

Techniques	A_1	A_2	A_3	RS	N	Average priority
A_1	1	[0.17, 0.22, 0.33]	[0.13, 0.15, 0.2]	[1.3, 1.37, 1.53]	[0.05, 0.07, 0.11]	[0.59, 0.77, 0.11]
A_2	[3, 4.5, 6]	1	[0.17, 0.22, 0.33]	[4.17, 5.72, 7.33]	[0.17, 0.3, 0.51]	[0.16, 0.24, 0.38]
A_3	[5, 6.5, 8]	[3, 4.5, 6]	1	[9, 12, 15]	[0.38, 0.63, 1.04]	[0.52, 0.68, 0.91]
CS	[9, 12, 15]	[4.17, 5.72, 7.33]	[1.3, 1.37, 1.53]	[14.47, 19.09, 23.86]#		
IN	[0.067, 0.083, 0.111]	[0.14, 0.17, 0.24]	[0.65, 0.73, 0.77]			

importance of decision attribute j is called weight and is denoted as ω_j. The ranking of the alternatives is performed according to their rank:

$$R_i = \frac{\sum_{j=1}^{m} \omega_j \cdot r_{ij}}{\sum_j \omega_j} \quad ; i = 1, 2, \ldots, n$$

The optimal alternative is the one for which value of R_i is maximum. The method is further extended by Bass and Kwakernaak [22] who proposed following two phases for the method:

Phase 1: Determination of ratings of alternatives: $A = \{a_i\}; \quad i = 1, 2, 3, \ldots, n$ is the set of decision alternatives and $C = \{\tilde{c}_j\}; \quad j = 1, 2, 3, \ldots\ldots m$ are the attributes that are fuzzy in nature. Let $\tilde{R}_{ij}$ be the fuzzy rating of alternative i with respect

Table 10 Repeatability of the instruments used (C_4)

Techniques	A_1	A_2	A_3	RS	N	Average priority
A_1	1	[0.17, 0.22, 0.33]	[0.13, 0.15, 0.2]	[1.3, 1.37, 1.53]	[0.06, 0.08, 0.12]	[0.064, 0.082, 0.12]
A_2	[3, 4.5, 6]	1	[0.25, 0.4, 1]	[4.25, 5.9, 8]	[0.19, 0.34, 0.64]	[0.19, 0.31, 0.55]
A_3	[5, 6.5, 8]	[1, 2.5, 4]	1	[7, 10, 13]	[0.31, 0.58, 1.04]	[0.38, 0.62, 0.88]
CS	[9, 12, 15]	[2.17, 3.72, 5.33]	[1.38, 1.55, 2.2]	[12.55, 17.27, 22.53]#		
IN	[0.067, 0.093, 0.111]	[0.19, 0.27, 0.46]	[0.45, 0.65, 0.72]			

Table 11 Ease of use (C_5)

Techniques	A_1	A_2	A_3	RS	N	Average priority
A_1	1	[0.17, 0.22, 0.33]	[0.13, 0.15, 0.2]	[1.3, 1.37, 1.53]	[0.06, 0.08, 0.12]	[0.064, 0.082, 0.12]
A_2	[3, 4.5, 6]	1	[0.25, 0.4, 1]	[4.25, 5.9, 8]	[0.19, 0.34, 0.64]	[0.19, 0.31, 0.55]
A_3	[5, 6.5, 8]	[1, 2.5, 4]	1	[7, 10, 13]	[0.31, 0.58, 1.04]	[0.38, 0.62, 0.88]
CS	[9, 12, 15]	[2.17, 3.72, 5.33]	[1.38, 1.55, 2.2]	[12.55, 17.27, 22.53]#		
IN	[0.067, 0.093, 0.111]	[0.19, 0.27, 0.46]	[0.45, 0.65, 0.72]			

to j and $\tilde{W}_j \in R$ be the weight or importance of attribute j. The rating of alternative i with respect to attributes j is fuzzy and is given by the grade membership function $\mu_{R_{ij}}(r_{ij})$. The relative importance (weight) of attributes j is given by a fuzzy set $\tilde{W}_j$ with grade membership function $\mu_{W_j}(\omega_j)$.

The evaluation of alternative a_i is a fuzzy set and is computed using $\tilde{R}_{ij}$ and $\tilde{W}_j$:

$$g(z) = g(\omega_1, \ldots \omega_n,\ r_{i1}, \ldots r_{in}) = \frac{\sum_{j=1}^{m} \omega_j \cdot r_{ij}}{\sum_{j=1}^{m} \omega_j}$$

Table 12 Environmental restrictions (C_6)

Techniques	A_1	A_2	A_3	RS	N	Average priority
A_1	1	[0.13, 0.15, 0.2]	[0.1, 0.12, 0.14]	[1.23, 1.27, 1.34]	[0.044, 0.055, 0.07]	[0.047, 0.058, 0.075]
A_2	[5, 6.5, 8]	1	[0.17, 0.22, 0.33]	[6.17, 7.72, 9.33]	[0.22, 0.34, 0.51]	[0.18, 0.26, 0.38]
A_3	[7, 8.5, 10]	[3, 4.5, 6]	1	[11, 14, 17]	[0.4, 0.61, 0.92]	[0.54, 0.69, 0.86]
CS	[13, 16, 19]	[4.13, 5.65, 7.2]	[1.27, 1.34, 1.47]	[18.4, 22.99, 27.67]#		
IN	[0.05, 0.06, 0.08]	[0.14, 0.18, 0.24]	[0.68, 0.77, 0.79]			

Table 13 Technical expertise requirement (C_7)

Techniques	A_1	A_2	A_3	RS	N	Average priority
A_1	1	[3, 4.5, 6]	[5, 6.5, 8]	[9, 12, 15]	[0.38, 0.63, 1.04]	[0.52, 0.68, 0.91]
A_2	[0.17, 0.22, 0.33]	1	[3, 4.5, 6]	[4.17, 5.72, 7.33]	[0.17, 0.30, 0.51]	[0.16, 0.24, 0.38]
A_3	[0.13, 0.15, 0.2]	[0.17, 0.22, 0.33]	1	[1.3, 1.37, 1.53]	[0.054, 0.07, 0.11]	[0.061, 0.077, 0.111]
CS	[1.3, 1.37, 1.53]	[4.17, 5.72, 7.33]	[9, 12, 15]	[14.47, 19.09, 23.86]#		
IN	[0.65, 0.73, 0.77]	[0.14, 0.17, 0.24]	[0.067, 0.083, 0.111]			

Membership function of μ_{zi} is defined as:

$$\mu_{z_i}(z) = min\ \{\overset{n}{\underset{j=1}{min}}\,(\mu_{W_j}(\omega_j)),\ \ \overset{n}{\underset{k=1}{min}}\,(\mu_{R_{ik}}(r_{ik}))\}$$

The final rating is therefore $\tilde{R}_i = \{(r,\ \mu_{R_i})\}$ given by a membership function $\mu_{R_i}(r) = sup_r\ \mu_{z_i}(z)$.

Phase 2: Ranking: Once final ratings are obtained, ranking or rank ordering is carried out in this phase. A measure is established to distinguish the 'preferable alternatives' from each other and rank them. If ratings $\tilde{R}_i$ are fuzzy, then preference set $\tilde{P}_i = \{(p,\ \mu_{P_i}(p)\}$, is obtained using:

$$\mu_{P_i}(p) = \sup_r\ \mu_R(r_1, \ldots, r_m)$$

Table 14 Ease of maintenance (C_8)

Techniques	A_1	A_2	A_3	RS	N	Average priority
A_1	1	[3, 4.5, 6]	[0.25, 0.4, 1]	[4.25, 5.9, 8]	[0.22, 0.44, 0.92]	[0.21, 0.36, 0.69]
A_2	[0.17, 0.22, 0.33]	1	[0.25, 0.4, 1]	[1.42, 1.62, 2.33]	[0.073, 0.12, 0.27]	[0.082, 0.13, 0.24]
A_3	[1, 2.5, 4]	[1, 2.5, 4]	1	[3, 6, 9]	[0.16, 0.44, 1.04]	[0.25, 0.5, 0.85]
CS	[2.17, 3.72, 5.33]	[5, 8, 11]	[1.5, 1.8, 3]	[8.67, 13.52, 19.33]#		
IN	[0.19, 0.27, 0.46]	[0.09, 0.13, 0.2]	[0.33, 0.55, 0.66]			

Table 15 Ease of mounting (C_9)

Techniques	A_1	A_2	A_3	RS	N	Average Priority
A_1	1	[0.17, 0.22, 0.33]	[0.13, 0.15, 0.2]	[1.3, 1.37, 1.53]	[0.05, 0.07, 0.11]	[0.59, 0.77, 0.11]
A_2	[3, 4.5, 6]	1	[0.17, 0.22, 0.33]	[4.17, 5.72, 7.33]	[0.17, 0.3, 0.51]	[0.16, 0.24, 0.38]
A_3	[5, 6.5, 8]	[3, 4.5, 6]	1	[9, 12, 15]	[0.38, 0.63, 1.04]	[0.52, 0.68, 0.91]
CS	[9, 12, 15]	[4.17, 5.72, 7.33]	[1.3, 1.37, 1.53]	[14.47, 19.09, 23.86]#		
IN	[0.067, 0.083, 0.111]	[0.14, 0.17, 0.24]	[0.65, 0.73, 0.77]			

The above fuzzy set is effectively used to judge the degree of preferability of an alternative over all the other alternatives.

Arithmetic operations such as addition, multiplication and division of fuzzy sets are involved in the computation. There are two ways in which this can be performed: by considering only the left, middle and right side values of TFNs or with the help of interval of confidence at each presumption level α (or popularly called as α-cut). R_1 i.e. rating of Vibration Monitoring is [0.465217, 0.661905, 0.86], R_2 i.e. rating of Lube Oil Analysis is [0.473913, 0.646032, 0.8275] and R_3 i.e. rating of Endoscopic examination is [0.445652, 0.633333, 0.82125]. The membership functions of the final ratings are shown in Fig. 6.

It is evident from Fig. 6 that vibration analysis technique is slightly dominant over other two techniques. Nonetheless, one needs to investigate the case further to obtain the optimal technique. To do this, another decisive factor is used which is able to differentiate between 'preferable alternatives'. The preference set is obtained to investigate preferability of vibration monitoring alternative over the others. The preference set thus obtained is [−0.359158, 0.0222225, 0.4002175] and is plotted in Fig. 7. It is observed from Fig. 7 that vibration monitoring technique is the optimal technique under fuzzy decision attributes considered in the example.

Table 16 Computation of scores of condition monitoring techniques

Attributes		C_1	C_2	C_3	C_4	C_5
Average priority of attributes		[0.017, 0.025, 0.075]	[0.034, 0.057, 0.105]	[0.085, 0.13, 0.21]	[0.12, 0.18, 0.275]	[0.054, 0.09, 0.15]
Average priority of techniques	A_1	[0.125, 0.15, 0.21]	[0.5, 0.65, 0.84]	[0.59, 0.77, 0.11]	[0.064, 0.082, 0.12]	[0.064, 0.082, 0.12]
	A_2	[0.5, 0.69, 0.95]	[0.15, 0.28, 0.41]	[0.16, 0.24, 0.38]	[0.19, 0.31, 0.55]	[0.19, 0.31, 0.55]
	A_3	[0.125, 0.15, 0.21]	[0.06, 0.07, 0.09]	[0.52, 0.68, 0.91]	[0.38, 0.62, 0.88]	[0.38, 0.62, 0.88]
Scores of Techniques	A_1	[0.002, 0.003, 0.016]	[0.017, 0.037, 0.088]	[0.005, 0.01, 0.023]	[0.007, 0.015, 0.033]	[0, 0.007, 0.018]
	A_2	[0.008, 0.02, 0.07]	[0.005, 0.016, 0.043]	[0.014, 0.03, 0.08]	[0.023, 0.056, 0.151]	[0.01, 0.028, 0.083]
	A_3	[0.002, 0.003, 0.016]	[0.002, 0.004, 0.009]	[0.044, 0.088, 0.191]	[0.046, 0.112, 0.242]	[0.02, 0.056, 0.132]

Attributes		C_6	C_7	C_8	C_9
Average priority of attributes		[0.047, 0.075, 0.122]	[0.2, 0.3, 0.42]	[0.031, 0.048, 0.079]	[0.012, 0.016, 0.025]
Average priority of techniques	A_1	[0.047, 0.058, 0.075]	[0.52, 0.68, 0.91]	[0.21, 0.36, 0.69]	[0.59, 0.77, 0.11]
	A_2	[0.18, 0.26, 0.38]	[0.16, 0.24, 0.38]	[0.082, 0.13, 0.24]	[0.16, 0.24, 0.38]
	A_3	[0.54, 0.69, 0.86]	[0.061, 0.077, 0.111]	[0.25, 0.5, 0.85]	[0.52, 0.68, 0.91]
Scores of techniques	A_1	[0.002, 0.004, 0.008]	[0.1, 0.2, 0.38]	[0.006, 0.017, 0.055]	[0, 0.001, 0.003]
	A_2	[0.008, 0.02, 0.043]	[0.032, 0.072, 0.16]	[0.002, 0.006, 0.019]	[0.002, 0.004, 0.01]
	A_3	[0.025, 0.05, 0.096]	[0.012, 0.023, 0.047]	[0.007, 0.024, 0.067]	[0.006, 0.011, 0.023]

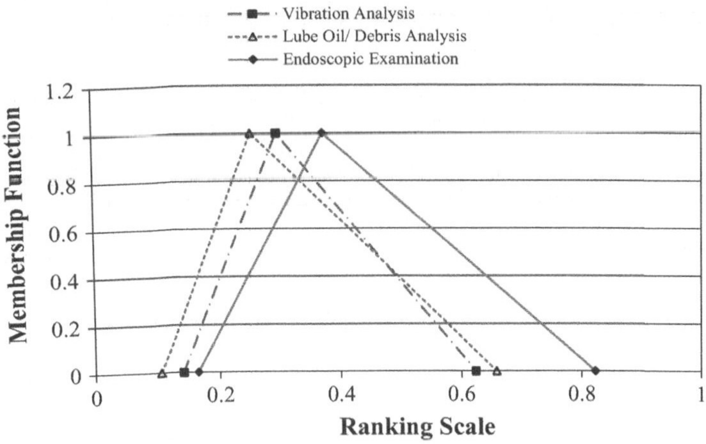

Fig. 3 Final ranking of techniques

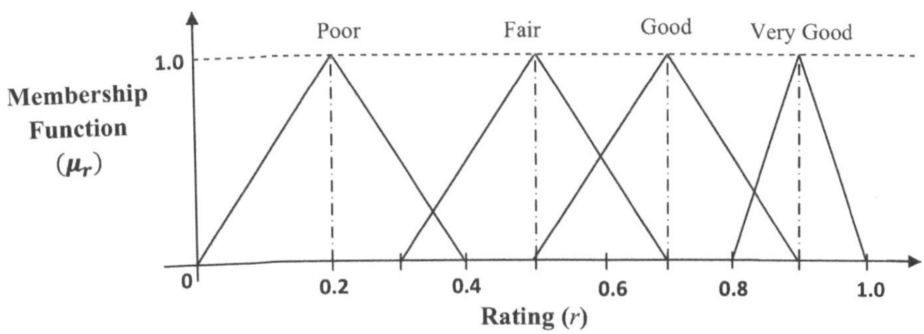

Fig. 4 Fuzzy sets representing ratings in linguistic terms/variables

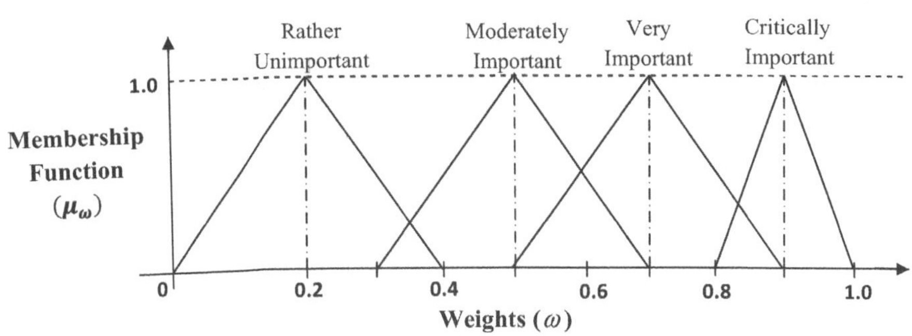

Fig. 5 Fuzzy sets representing weights in linguistic terms/variables

Table 17 Rating of the techniques with respect to decision attributes

Decision attributes	Alternatives/Techniques		
	Vibration analysis (A_1)	Lube oil/debris analysis (A_2)	Endoscopic examination (A_3)
Investment cost (C_1)	Fair	Fair	Good
Operating costs (C_2)	Good	Good	Good
Accuracy of the technique (C_3)	Good	Very Good	Good
Repeatability of the instruments used (C_4)	Good	Good	Very Good
Ease of use (C_5)	Good	Fair	Fair
Environmental restrictions (C_6)	Good	Fair	Fair
Technical expertise requirement (C_7)	Good	Good	Good
Ease of maintenance (C_8)	Good	Fair	Fair
Ease of mounting (C_9)	Fair	Good	Fair

Table 18 Weights of decision attributes

Decision attributes	Weights
Investment cost (C_1)	Moderately important
Operating costs (C_2)	Very important
Accuracy of the technique (C_3)	Critically important
Repeatability of the instruments used (C_4)	Very important
Ease of use (C_5)	Very important
Environmental restrictions (C_6)	Very important
Technical expertise requirement (C_7)	Very important
Ease of maintenance (C_8)	Very important
Ease of mounting (C_9)	Very important

5.5 *Ranking Fuzzy Sets Using* 'Cardinal Utilities'

Baldwin and Guild [23] proposed a relation $\widetilde{P_{ij}} = \{(r_i, r_j),\ \mu_{P_{ij}}(r_i, r_j)\};\ i \neq j$ with membership function as $\mu_{P_{ij}}(r_i, r_j) = f(r_i, r_j)$. This function expresses the 'difference' between the ratings of two fuzzy sets. Such a set is defined as $\tilde{O}(x_i) = \{x_i,\ \mu_O(x_i)\}$ with membership function $\mu_O(x_i) = sup_{r_i, r_j}\ min\ \{\mu_{R_i}(r_i),\ \mu_{R_j}(r_j), \mu_{P_{ij}}(r_i, r_j)\}$. The equation expresses the degree to which alternative x_i is preferable to its best rival alternative. $\tilde{O}(x_i)$ corresponds to *max-min composition* of $\widetilde{R_i}$, $\widetilde{R_j}$ and $\widetilde{P_{ij}}$.

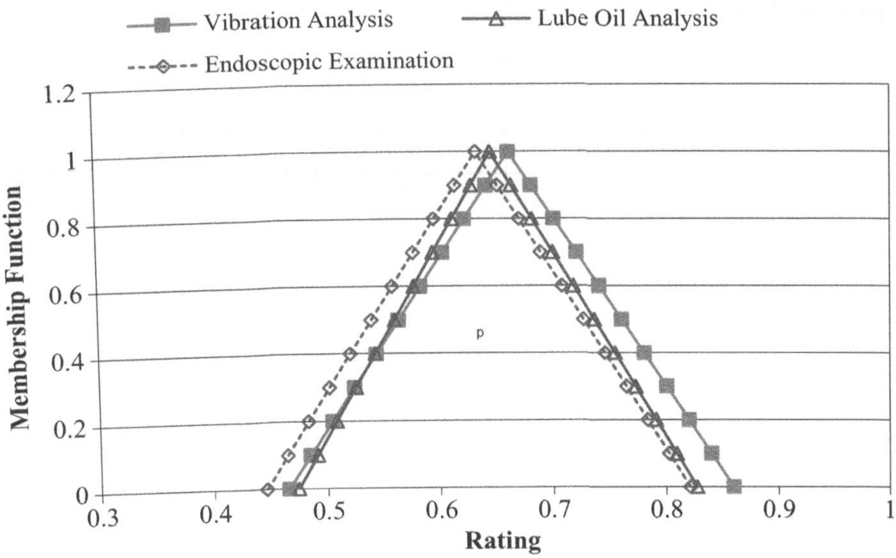

Fig. 6 Final ratings of techniques

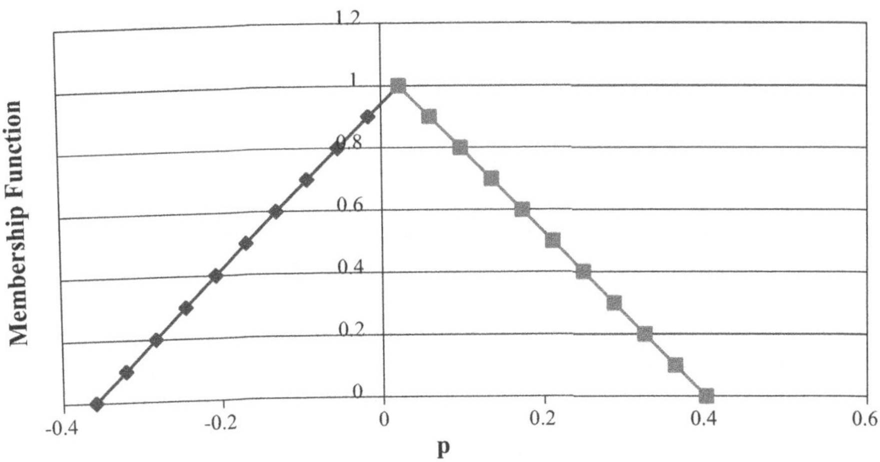

Fig. 7 Membership function of preferability of Technique 1 over Techniques 2 and 3

Without going into the specifics of mathematical analysis, the solution of the problem is presented as:

$$\mu_O(x_i) = \min_j \{\widetilde{\mu}_j\} = \min_j \left[\frac{\delta - \alpha}{1 + (\delta - \gamma) + (\beta - \alpha)}\right]$$

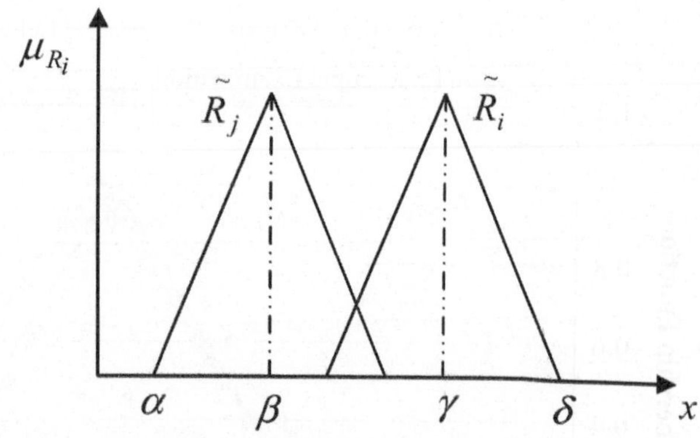

Fig. 8 Membership functions of ratings $\tilde{R}_i$, $\tilde{R}_j$

Table 19 Values of parameters obtained by cardinal utilities method

Parameters of the equation	R_1 and R_2		R_1 and R_3		R_2 and R_3	
	R_1	R_2	R_1	R_3	R_2	R_3
α	0.473913	0.465217	0.445652	0.465217	0.445652	0.473913
β	0.646032	0.661905	0.633333	0.661905	0.633333	0.646032
γ	0.661905	0.646032	0.661905	0.633333	0.646032	0.633333
δ	0.86	0.8275	0.86	0.82125	0.8275	0.82125
μ_j	0.28177	0.26288	0.29900	0.25714	0.27889	0.25539

The parameters of the above equations are depicted in the Fig. 8. Optimal maintenance technique is the one that has maximum $\tilde{\mu}_j$ value obtained through above equation.

All the three maintenance techniques i.e. vibration analysis (with rating R_1), lube oil analysis (with rating R_2) and endoscopic examination (with rating R_3) are compared considering two techniques at a time. From Table 19 one can observe that $\tilde{\mu}_j$ of technique 1 is more than $\tilde{\mu}_j$ of technique 2 and 3, i.e. rating of R_1 is greater rating of R_2 and R_3. This means vibration analysis technique outperforms other two techniques as far as optimality is concerned.

5.6 *Ranking Fuzzy Sets by* 'Maximizing and Minimizing Sets'

This is a modification of ranking approach which is developed by Jain and is further modified by Chen [24] for better discrimination of the ratings. This method utilizes 'maximizing set ($\tilde{M}$)' and 'minimizing set ($\tilde{N}$)' having membership functions defined as:

$$\mu_M(r) = \left[\frac{r - r_{min}}{r_{max} - r_{min}}\right]^n$$

$$\mu_N(r) = \left[\frac{r_{\max} - r}{r_{\max} - r_{\min}}\right]^n$$

where $r \in [r_{min}, r_{\max}]$ is the real interval and $n = 1$ for linear, $n = 2$ for risk prone and $n = 0.5$ for risk averse membership functions.

To rank and get the fuzzy optimal alternative, $O(x_i)$ is obtained using

$$O(x_i) = \frac{R(x_i) + 1 - L(x_i)}{2}$$

where $R(x_i) = sup_r \; min\,\{\mu_{R_i}(r), \mu_M(r)\}$ and $L(x_i) = sup_r \; min\,\{\mu_{R_i}(r), \mu_N(r)\}$.

In this method the analysis is carried out by considering two techniques at a time. The membership functions of the three fuzzy sets (techniques) are obtained as below:

$$\begin{aligned}
\mu_{R1}(r) &= \frac{r - 0.465217}{0.196688} && ;0.465217 \le r \le 0.661905\\
&= \frac{0.86 - r}{0.198095} && ;0.661905 < r \le 0.86\\
&= 0 && ;otherwise
\end{aligned}$$

$$\begin{aligned}
\mu_{R_2}(r) &= \frac{r - 0.473913}{0.172119} && ;0.473913 \le r \le 0.646032\\
&= \frac{0.8275 - r}{0.181468} && ;0.646032 < r \le 0.8275\\
&= 0 && ;otherwise
\end{aligned}$$

$$\begin{aligned}
\mu_{R_3}(r) &= \frac{r - 0.445652}{0.187681} && ;0.445652 \le r \le 0.633333\\
&= \frac{0.82125 - r}{0.187917} && ;0.633333 < r \le 0.82125\\
&= 0 && ;otherwise
\end{aligned}$$

Computation of maximizing and minimizing sets for $n = 1$ is as below:

(1) Considering R_1 and R_2:

$$\mu_M(r) = \frac{r - 0.465217}{0.394783} \quad ; 0.465217 \leq r \leq 0.86$$
$$= 0 \quad ; \textit{otherwise}$$

$$\mu_N(r) = \frac{0.86 - r}{0.394783} \quad ; 0.465217 \leq r \leq 0.86$$
$$= 0 \quad ; \textit{otherwise}$$

(2) Considering R_1 and R_3:

$$\mu_M(r) = \frac{r - 0.445652}{0.414348} \quad ; 0.445652 \leq r \leq 0..86$$
$$= 0 \quad ; \textit{otherwise}$$

$$\mu_N(r) = \frac{0.86 - r}{0.414348} \quad ; 0.445652 \leq r \leq 0.86$$
$$= 0 \quad ; \textit{otherwise}$$

(3) Considering R_2 and R_3:

$$\mu_M(r) = \frac{r - 0.445652}{0.381848} \quad ; 0.445652 \leq r \leq 0.8275$$
$$= 0 \quad ; \textit{otherwise}$$

$$\mu_N(r) = \frac{0.8275 - r}{0.381848} \quad ; 0.445652 \leq r \leq 0.8275$$
$$= 0 \quad ; \textit{otherwise}$$

The ratings of the three techniques obtained by this method and maximizing and minimizing sets are plotted in Figs. 9, 10 and 11.

Ranking values for all three techniques are given in Table 20. When R_1 and R_2 combination is considered, optimal ranking value $(O(x_i))$ of R_1 is more than R_2 and in R_1 and R_3 combination also $(O(x_i))$ of R_1 is more than R_3. Optimal ranking value of R_2 is more than R_3 in R_2 and R_3 combination. Therefore, alternative 1 (vibration analysis) is an optimal technique as per this method.

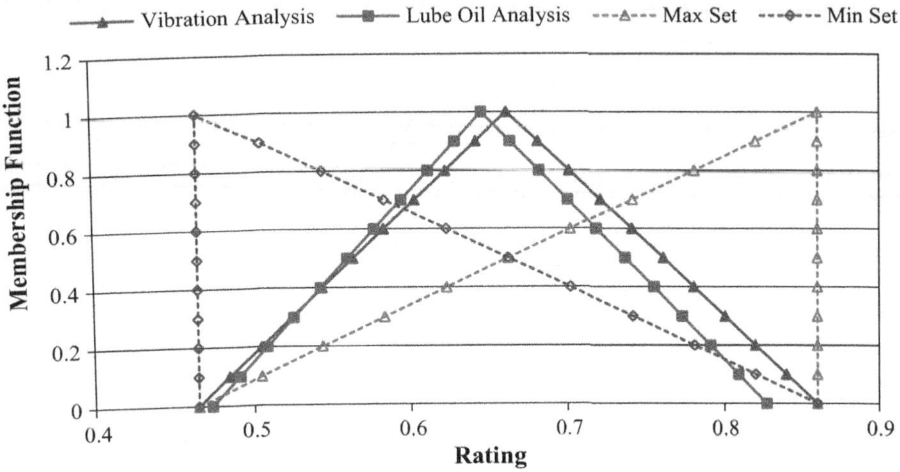

Fig. 9 Final ratings of Technique 1 and 2 (Chen Method)

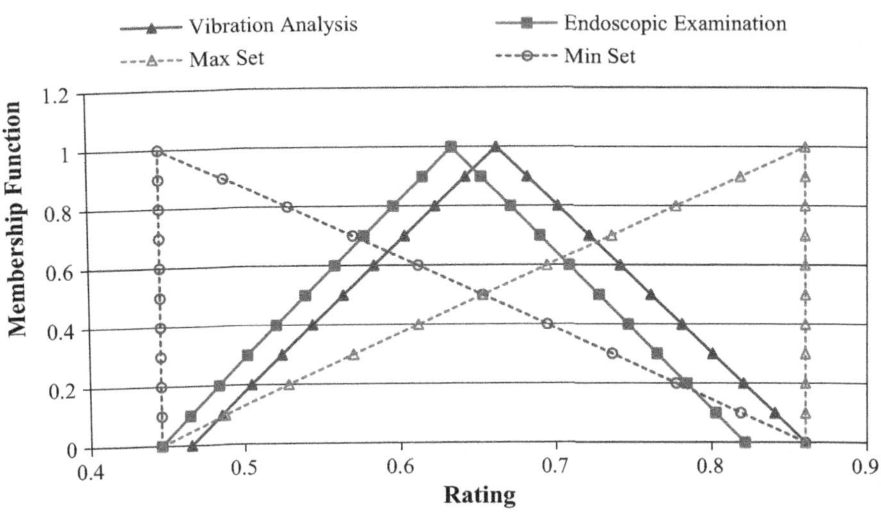

Fig. 10 Final ratings of Technique 1 and 3 (Chen Method)

5.7 *Suitability Set and Dominance Relation*

In this method, rating matrix and weights for attributes are determined followed by computation of suitability set for each alternative [21]. The suitability set for each alternative '*i*' is assumed as fuzzy weighted sum of ratings and is considered as an appropriate measure of suitability:

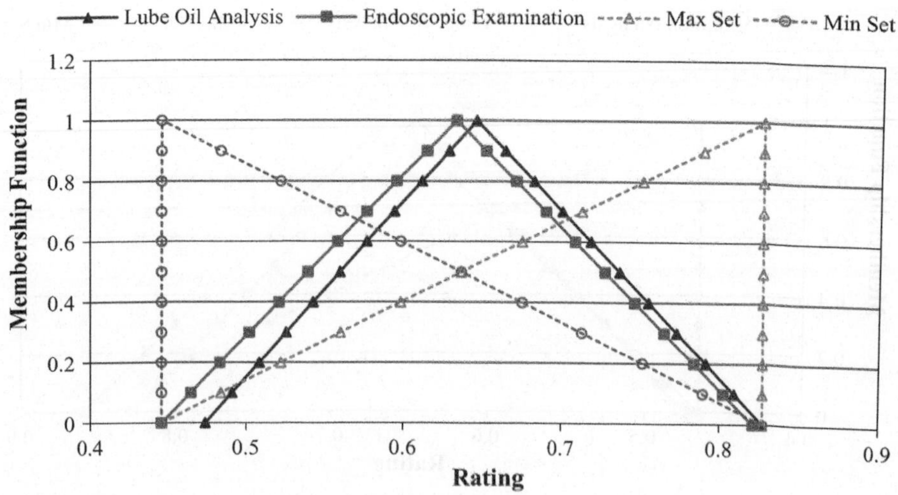

Fig. 11 Final ratings of Techniques 2 and 3 (Chen Method)

Table 20 Values of Parameters Obtained by Chen Method

Parameters of the equation	R_1 and R_2		R_1 and R_3		R_2 and R_3	
	R_1	R_2	R_1	R_3	R_2	R_3
$R(x_i)$	0.66635	0.63017	0.68149	0.62395	0.67919	0.6612
$L(x_i)$	0.670	0.68139	0.64967	0.68919	0.64031	0.67173
$O(x_i)$	0.498175	0.47439	0.51591	0.46738	0.51944	0.494735

$$S_i = \sum_j \omega_j \cdot r_{ij}$$

For choosing an alternative one need to know the concept of dominance. Dominance (δ) of the normal convex fuzzy set A over the normal convex fuzzy set B is defined by:

$\delta(A, B) = \vee_x(\mu_{\leq A}(x) \wedge \mu_B(x))$; where $\vee$ denotes maximum operation, $\wedge$ denotes the minimum operation and $\leq A$ is the fuzzy set '*less than or equal to A*' formed from A by setting:

$$\mu_{\leq A}(x) = 1.0 \quad ; x < x^* \\ = \mu_A(x) \quad ; x \geq x^*$$

where x^* is the leftmost (lowest) value of x for which $\mu_A(x) = 1.0$.

Though one can get an alternative using dominance relation, it is not sufficient to make the choice as it does not take into account the shapes of suitability sets. To include the shapes of membership functions as well and gain confirmation about the optimal technique, 'difference function' is defined as:

$$g_k(S_1, S_{2,}, \ldots, S_n) = S_k - \frac{\sum_{\substack{i=1 \\ i \neq k}}^{n} v(x_i) \cdot S_i}{\sum_{\substack{i=1 \\ i \neq k}}^{n} v(x_i)}$$

where vector, $v(x_i) = min_j\{DR(i,j)\}$ and k corresponds to a position in $v(x_k) = 1$. The set obtained using the above difference function gives the degree of preference of the chosen alternative over the other alternatives. This procedure is particularly useful when two or more suitability sets, i.e. alternatives dominate the other sets/alternatives.

The suitability sets for three alternative techniques in the form of TFN are: $S_1 = [2.14 \quad 4.17 \quad 6.88], S_2 = [2.18 \quad 4.07 \quad 6.62]$ and $S_3 = [2.05 \quad 3.99 \; 6.57]$ and are shown in Fig. 12.

The next step is to get the dominance relation $\{DR(i,j)\}_\delta$, which is the dominance of S_i over S_j. $\{DR\}_\delta$ for this example is:

$$\{DR\}_\delta = \begin{bmatrix} 1 & 1 & 1 \\ 0.98 & 1 & 1 \\ 0.96 & 0.985 & 1 \end{bmatrix}$$

The first row of the $\{DR\}_\delta$ matrix has all entries as 1. This means vibration analysis (with suitability set S_1) dominates over other two alternatives, with weights

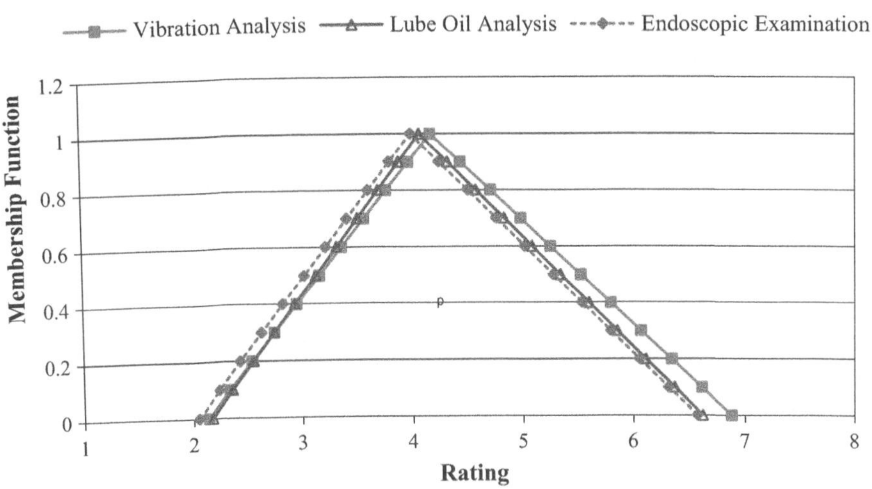

Fig. 12 Final ratings of techniques (suitability set method)

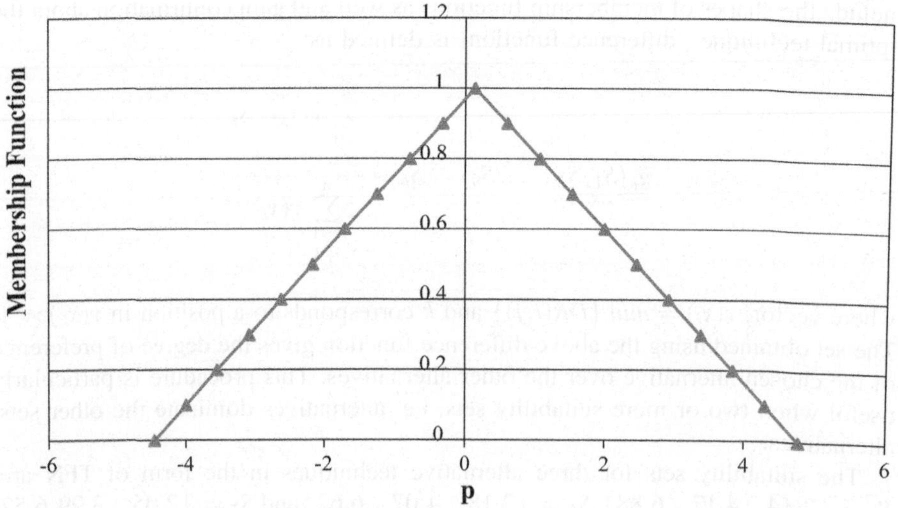

Fig. 13 Preferability of vibration analysis over lube oil analysis and endoscopic examination (suitability set method)

being [1, 0.98, 0.96]. Next, difference function is determined and degree of preference is obtained as [−4.455257732, 0.139587629, 4.764329897]. Plot of preferability of vibration analysis over lube oil analysis and endoscopic examination is shown in Fig. 13.

6 Scope for Hybrid Methods

Reliability, Availability Maintainability and Safety (RAMS) are closely interlinked areas and fuzzy sets application is attempted for several past decades. Readers may study the literature related to dynamic reliability models where fuzzy sets has been successfully applied by researchers. Fuzzy simulation of parameters of RAMS is yet another area that carries good potential for research investigations.

Hybrid approach which considers some uncertain parameters as probabilistic and some others as fuzzy numbers are attempted in the past. Methods for reliability or availability analysis consider all uncertain parameters to be either completely random or entirely fuzzy. However, if both uncertainties are present, as it often happens in real life scenario, there is a need to develop a right hybrid method. For those problems which involve some parameters that are justifiably represented by probability density function and other parameters which are considered to be more effectively represented by fuzzy numbers, the methods such as possibility-probability transformations, belief functions, etc. have been attempted previously. However this domain has lot of scope for further research investigations.

References

1. Zimmermann H-Z (1985) Fuzzy set theory and its applications. Kluwer-Nijhoff Publishing, Hingham
2. Zimmermann H-J (1987) Fuzzy sets, decision making and expert systems. Kluwer Academic Publishers, Norwell
3. Machol RE (Editor in Chief) (1984) Fuzzy sets and decision analysis, Elsevier Science Publishers B.V., Amsterdam
4. Terano T, Asai K, Sugeno M (1992) Fuzzy systems theory and its applications. Academic Press Inc., London
5. Cai KY, Wen CY, Zhang ML (1993) Fuzzy states as a basis for a theory of fuzzy reliability. Microelectron Reliab 33(15):2253–2263
6. Cai KY (1996) Introduction to fuzzy reliability. Kluwer Academic Publishers
7. Verma AK, Srividya A, Prabhu Gaonkar RS (2007) Fuzzy-reliability engineering—concepts and applications. Narosa Publishing House, India
8. Verma AK, Srividya A, Prabhu Gaonkar, RS (2006) An investigation of upcoming approaches in fuzzy reliability. In: Proceedings of the 9th international conference on dependability and quality management (DQM 2006), Belgrade (Serbia), pp 3–9
9. Kaufmann A, Gupta MM (1985) Introduction to fuzzy arithmetic—theory and applications. Van Nostrand Reinhold Company, New York
10. Kaufmann A, Gupta MM (1988) Fuzzy mathematical models in engineering and management science. Elsevier Science Publishers B.V, Amsterdam
11. Verma AK, Srividya A, Prabhu Gaonkar RS (2004) Profust reliability evaluation: an application to degrading diesel engine power plant. In: Proceedings of the international conference on sustainable habitat for cold climates, Leh, Ladakh, pp 114–121
12. Prabhu Gaonkar RS, Verma AK, Srividya A (2008) Exploring fuzzy set concept in priority theory for maintenance strategy selection problem. Int J Appl Manag Technol 6(3):131–142
13. Mechefske CK, Wang Z (2001) Using fuzzy linguistics to select optimum maintenance and condition monitoring strategies. Mech Syst Signal Process 15(6):1129–1140
14. Srividya A, Verma AK, Prabhu Gaonkar RS (2006) Fuzzy Multi-attributes maintenance decision models: a review. In: Proceedings of the 9th international conference on dependability and quality management (DQM 2006), Belgrade (Serbia), pp 10–18
15. Verma AK, Srividya A, Prabhu Gaonkar RS, On the use of fuzzy set concept in maintenance related models. Commun Dependability Qual Manag Int J (Special Issue on Reliability, Maintainability and Safety of Engineering Systems) 9(4):99–109
16. Prabhu Gaonkar RS, Verma AK, Srividya A, Fuzzy set based ranking of condition monitoring methods: a turbine maintenance case study. Commun Dependability Qual Manag Int J (Special Issue on Reliability, Maintainability and Safety of Engineering Systems) 9(4):79–98
17. Verma AK, Srividya A, Prabhu Gaonkar RS (2007) Fuzzy set solutions for optimal maintenance strategy selection. OPSEARCH J 44(3):261–276
18. Prabhu Gaonkar RS, Xie M, Verma AK, Peng R (2010) Using evidential reasoning approach for ship turbine's condition monitoring techniques ranking. In: Proceedings of the IEEE international conference on industrial engineering and engineering management (IEEM 2010), Macau, pp 2398–2402
19. Prabhu Gaonkar RS, Srividya A, Phadke U, Manohar, BS, Naik A (2009) Fuzzy AHP For ranking condition monitoring techniques. In: Proceedings of the international conference on operations research applications in engineering and management (ICOREM 2009), Anna University Tiruchirappalli (AU-T), pp 2265–2281
20. Verma AK, Srividya A, Prabhu Gaonkar RS (2005) Selecting optimal condition monitoring technique using fuzzy multi-attributes decision making methods: a case study. In: proceedings of the 3rd international conference on reliability, safety and hazard (ICRESH 2005), Mumbai, pp 749–756

21. Verma AK, Srividya A, Prabhu Gaonkar RS (2005) Optimal maintenance strategy selection using suitability set and dominance relation. In: Proceedings of the 8th international conference on dependability and quality management (DQM 2005), Belgrade, Serbia, pp 55–62
22. Bass SM, Kwakernaak H (1977) Rating and ranking of multiple-aspect alternatives using fuzzy set theory. Automatica 13:47–58
23. Baldwin JF, Guild NCF (1979) Comparison of fuzzy sets on the same decision space. Fuzzy Sets Syst 2:213–231
24. Chen S-H (1985) Ranking fuzzy numbers with maximizing set and minimizing set. Fuzzy Sets Syst 17:113–129

Dr. Rajesh S. Prabhu Gaonkar is presently Professor on Deputation at School of Mechanical Sciences, Indian Institute of Technology Goa (IIT Goa). He is B.E. (Mechanical Engineering, 1993), M.E. (Industrial Engineering, 1997), P.G. D.O.M. (I.G.N.O.U., 2000), Ph.D. (I.I.T. Bombay, 2007) and a Research Fellow at National University of Singapore (2010–2011). He is Former Professor and Head of the Mechanical Engineering Department at Goa College of Engineering and Former Dean of Faculty of Engineering at Goa University. He is Associate Editor of the International Journal of System Assurance Engineering and Management (a quarterly Journal published by SPRINGER). Dr. Rajesh S. Prabhu Gaonkar is first recipient (awardee) of the Goa State Award for Meritorious Teacher in Technical Education for the year 2015. This is awarded in public recognition of valuable services to the community as a teacher of outstanding merit. He is also recipient of prestigious Fellowship of Indian Institution of Industrial Engineering (IIIE) for his immense contribution in the field of industrial engineering in the state of Goa and recipient of Dr. J. M. Mahajan Award for the year 2016–17 by Indian Institution of Industrial Engineering for contributing substantially to the spread/ propagation of Industrial Engineering in Education i.e. in Institutes and in University. He was awarded "Diploma of Excellence" for the paper presentation at DQM International Reliability Summer School by D.Q.M. Research Center in Serbia in 2005. He has published over 60 research papers in various International Conferences and Journals; co-authored a book on "Fuzzy Reliability—Concepts and Applications" (2007) and co-authored a monograph "Six-Sigma: A Key to Enterprise Excellence" (2008). He is on Reviewer on Panel of about 10 International Journals. He is involved in organization of about 12 International Conferences in various capacities. He has total 25 years of teaching and research experience. Dr. Rajesh S. Prabhu Gaonkar is Founder Chairman of Mechanical Engineering Students Association (MESA) formed to organize technical events in Mechanical Engineering and related fields at Goa College of Engineering and founder Chairman of Indian Institution of Industrial Engineering (IIIE) Goa Chapter. He conceived and implemented MESSERGY (Mechanical Engineering Students' Social and Education Responsibility and Green Initiates) voluntary sustainable project for Mechanical Engineering Students of Goa College of Engineering. His teaching areas include Industrial Engineering, Operations Research, Advanced Optimization, Reliability Engineering/Reliability based Design, CAD/CAM, Maintenance Engineering and Management, Operations Management, System Modeling and Simulation. His research areas include Fuzzy Set Applications in Reliability and Maintenance Engineering and Multi-Attribute Decision Making.

Fuzzy Logic Based Analysis of Dissolved Decay Contents in Transformer Oil

Nitika Ghosh, Vikas Singh Bhadoria, Ram Naresh Sharma and Vivek Shrivastava

Mineral oil in transformer serves as an effective insulator and effective coolant which plays a major role in determining the health of the transformer. It is the decisive factor of a transformer health condition as it contains 70% of the diagnostic information. A fuzzy logic methodology based on statistical techniques has been presented in this paper to monitor, diagnose and predict the health index of the transformer using furanic contents. Furthermore, the proposed model is an effective, yet reliable monitoring technique to address transformer condition, ensuring good health, safety of operation and maintenance.

1 Introduction

Power transformers are not only the costliest, but also one of the most important device used in power sector. It transforms power from one circuit to another circuit, keeping frequency unchanged. Any deformation in its design may affect its per-

N. Ghosh · V. S. Bhadoria
Electrical and Electronics Engineering Department, ABES Engineering College, Ghaziabad, India
e-mail: nitikaghosh21@gmail.com

V. S. Bhadoria
e-mail: vikasbhadoria@gmail.com

R. N. Sharma
Electrical Engineering Department, National Institute of Technology, Hamirpur, Himachal Pradesh, India
e-mail: rnaresh@nith.ac.in

V. Shrivastava (✉)
Electrical Engineering Department, National Institute of Technology, Delhi, India
e-mail: shvivek@gmail.com

D. R. Karanki et al. (eds.), *Advances in RAMS Engineering*,
Springer Series in Reliability Engineering,
https://doi.org/10.1007/978-3-030-36518-9_14

formance, not only efficiency will be lowered but the ageing and oxidation process may lead to the permanent failure of supply. The transformer undergoing replacement or repairing may not only increase the cost, but the large amount of time taken may also hamper the production activities [1]. For this problem to overcome, some effective method is to be devised which optimises capacity management, keeping in mind the optimisation of cost constraints. This is due to the fact that a power transformer makes up to 70% of the total cost in power utility sectors. Therefore a successful monitoring tool is to be devised which indicates the time to time health index of a power transformer, thus saving additional maintenance cost and unnecessary outages [2].

In recent times, studies have been carried out in regard to the overall life cycle of a transformer and the related methods of condition monitoring. The methods will not only makes the system cost effective but also eliminate the chances of sudden power outages resulting due to transformer failure, resulting in the increased nominal life of a power transformer [3]. No matter whether a transformer is slightly damaged or heavily damaged, its capacity to handle shorts circuit current reduces to great extent, which needs to be corrected at earliest. For this, a reliable methodology is to be devised as visual inspection might not be helpful each and every time in this regard. Therefore, in this chapter a strategy for condition monitoring of power transformer by assessing the oil condition has been discussed. Transformer oil acts as an insulator and as a coolant. Assessing oil conditions on some reliable parameters such as acidity, density, interfacial tension, and water content may lead in the reliable monitoring of the device. So various diagnostics tests have to be carried out, which ensure guaranteed high level of performance and effective schedule of maintenance.

The process of transformer ageing is a permanent process and cannot be reversed back. Ageing process affects the life span of the transformer. This chapter relates the oil parameters with the overall health index of the transformers. In this approach, factors like acidity, density and interfacial tension will result in the calculation of dissolved decay contents of the transformer oil. The dissolved decay content, considered with several other parameters determine the overall health of the transformer.

2 Assessment of Transformer Oil Using Dissolved Decay Contents

The oil plays vital part in determining the working condition of power transformer, as it not only acts as a coolant, but also as an effective insulating material. As an impact of various chemical, thermal and mechanical stresses, the insulating oil undergoes an oxidation process [4]. It might be a slow and steady process but its impact is hazardous to the overall performance of the system, resulting in premature ageing of insulation material and associated deterioration. Undoubtedly, the assessment of transformer oil can be the decisive factor in determining the serviceability of device. Any failure, for sure, may be prevented by a successful condition monitoring technique of the operational unit and necessary preventive measures can be taken accordingly [5]. The variations in different oil characteristics may therefore be used to identify/detect the type of incipient failure in the transformer.

Various research studies have shown that oxidation process of oil results in the hydrocarbon chain breakdown, resulting in the generation of soluble gases and also some colloidal suspensions as shown in Fig. 1. Theses invisible colloidal particles results transformer deterioration and encourages further oxidation process. The responsiveness of oxygen further aggravates the process to a worse condition, resulting by products in the form of insolvable sludge and dissolvable gases, which is detrimental to the solid insulation. These products hampers the heat dissipation process in transformers as the produced sludge acts as a barrier to the flow of heat from the fluid placed in cooling unit and from the winding coils to the insulating oil.

Insulating materials used in transformer are subjected to mechanical, electrical, thermal and environmental stress occurring in various parts of the devices. The oxidation processes deteriorates the dielectric strength and the quality of the insulation material used, due to which failure may occur. The durability of the insulation system depends on the magnitude of the stresses applied on it and also the environment in which it is placed. The insulating paper, placed in between the windings cant be retained unless it is overhauled unlike the transformer oil insulation, which can be taken out for maintenance at any point of time. It is the life of the paper insulation that actually determines the life span of the transformer. Therefore, it becomes very important to assess the condition of the paper insulation in ultra voltage transformers.

3 Transformer Oil Ageing Factors

Ageing process in transformer is a permanent and irreversible process and gets accelerated when the insulating material undergoes various thermal, chemical and electrical stresses. Due to the impact of various stresses produced in the transformer and its related by-products, the ageing process becomes a four-dimensional problem. The operating environment has a major impact too. The oxidation process results in the formation of moisture, acids and some colloidal suspended particle of miscible or partially immiscible nature both.

Over a period of time, the problem of ageing escalates to such a level that the performance degrades and ultimately the failure occurs. Here, failure infers to complete breakdown of the solid insulation [6]. The process fastens primarily because of moisture content and increased temperature value. The oxygen level markedly increases by 3–10 times the normal rated value.

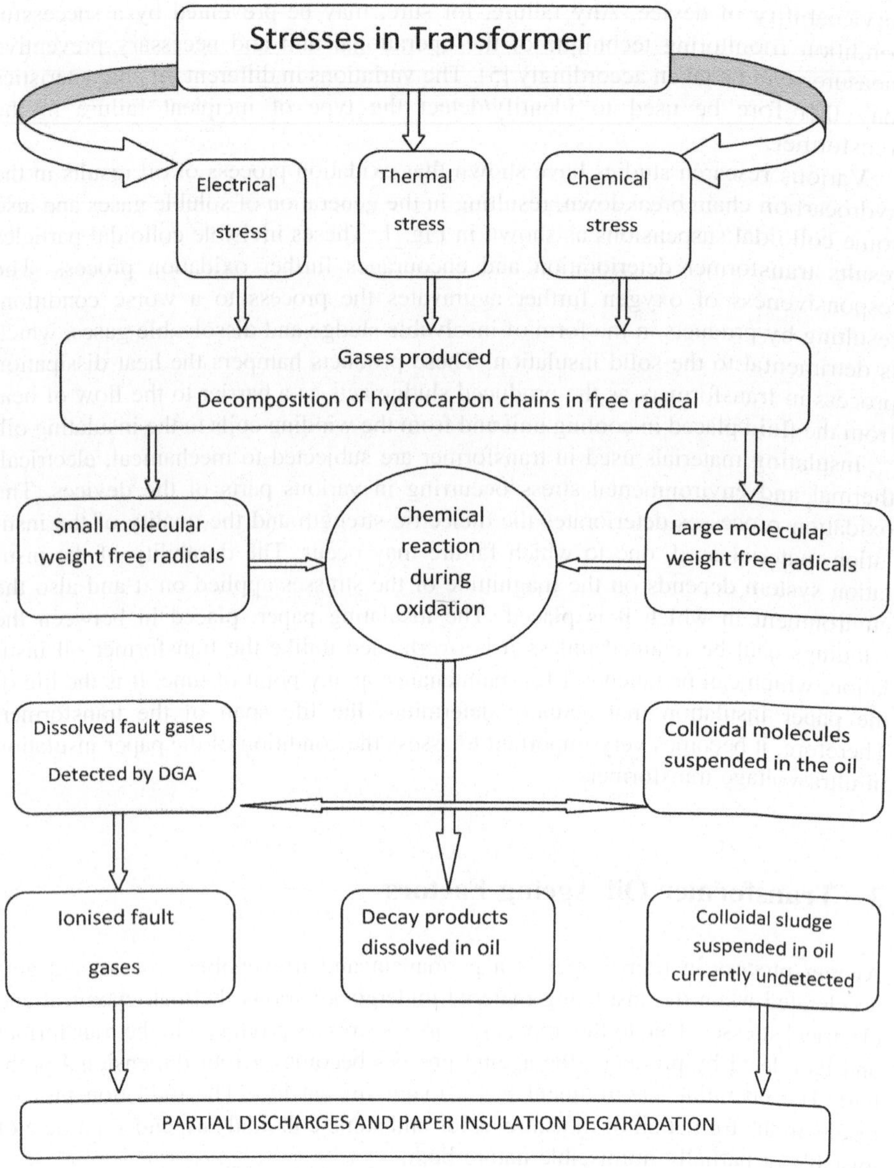

Fig. 1 Sketch of insulating oil decaying properties

3.1 Effect of Heat on Solid and Liquid Insulation

The losses generated in the transformer generate the internal heat, resulting because of the non-dissipation of heat. This not only elevates the process of ageing of non-metallic component of the device and shortens the life span of insulation material. A well designed, the solid insulation can last up to 50 years or even more, provided temperature and other physical conditions remain favourable [7]. The induced faults developed in power transformer leads to the formation of solvable gas bubbles, thus increasing the risk of sudden outages [8]. It is basically the insulation which decides the thermal degradation, together with its decomposition. Cellulose has highest thermal vulnerability. The process of oxidation depends not only on loading conditions, but also on the quality of paper insulation used, water content present in the oil and the acidity level of the oil sample. The electric fields speed up ageing, possibly due to increased precipitation level of acid [8].

3.2 *Reaction of Oxygen in the Oxidation Process*

The moisture content in solid insulation is expressed as the percentage of the weight of the moisture divided by the weight of dry healthy paper insulation. For voltage levels above 120 kV, the permissible moisture content for reliable operation is 4–5%, and for extra high voltage transformers, the limit is 2%. The moisture content present can be found in four states: free water in capillaries, water present in the form of vapour, free imbibed water and the water adsorbed to surfaces [8].

The effect of moisture is detrimental to the transformer insulation and the ageing process associated to it. Moisture content, under sharp temperature transients result in free water that may result in sharp electrical breakdown of the dielectric [8]. This process may lead to thermal imbalance at elevated temperatures where transfer of moisture happens from paper to oil. The moisture content increases in case of oil-conservator type or open-breather type transformer, and are also the by product of the thermal decomposition process. The moisture content may also result by mishandling during storage or transport, either at the time of installation, during minor repairs regarding partial or complete drainage of oil, or as a result of faults in the transformer breather. The increase in the conductivity in the insulation increases the dissipation factor and as a result the oil in service gets contaminated. The contaminants may include hazardous gases, fibres, moisture, debris and various solvable and insolvable colloidal particles. Moisture, being a by product of the degradation, act as an ageing catalyst [8].

3.3 *Effect of Oxygen on Oil and Paper Ageing*

In the whole life span of a transformer, the oil is exposed to different levels of air, moisture and temperature [9]. When the transformer oil comes in contact with any metal, say steel of core and tank, windings made of copper and aluminium ultimately results in oxidation process, which is chemical reaction between unstable hydrocarbons and oxygen. In this action, increased temperature and moisture

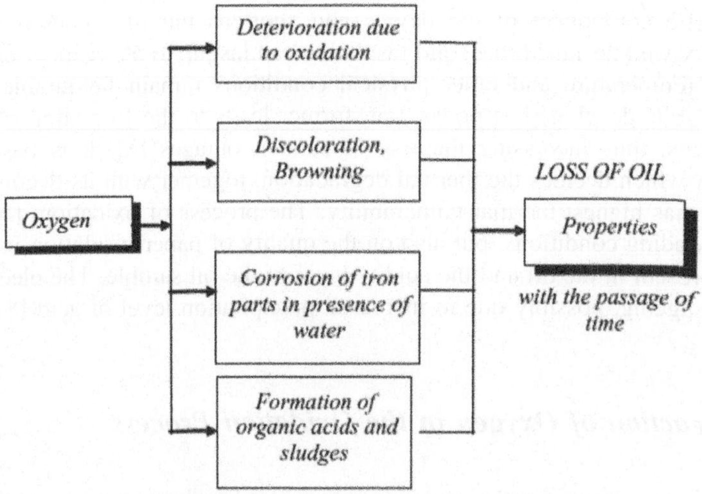

Fig. 2 Effect of oxygen on the insulating oil [9]

content acts as a catalysts and the stresses accelerates this process. The product of oxidative reaction produces sludge and acids as by-products which produces interference in the insulating oil. Chemically speaking, the acid produced will worsen the tensile strength of the insulating paper. Also, in presence of moisture, oxygen aids in the process of corrosion and decomposition as shown in Eq. (1)

$$2Fe + 3H_2O \rightarrow Fe_2O_3 + 3H_2 \quad (1)$$

When water accumulates outside in the outlet valve, the oxidation in the form of galvanic action starts to occur in the metallic valve or the iron wall of the tank.

However the detrimental effect of oxygen on the insulation reduces if the copper metal is present, which in turn enhances oil oxidation and increase the level of oxygen as shown in Fig. 2.

3.4 Furan Contents and Degree of Polymerization (DOP)

The life expectancy of a transformer is dependent on various factors such as electrical and mechanical stresses, increased temperature conditions, over-heating, faults and presence of moisture as shown in Fig. 3 [10]. The health index of a transformer is estimated by a life expectancy of a solid insulation, particularly paper. The state of its working condition depends on its Degree of Polymerization (DOP) and its mechanical strength [10]. The classical method used to determine the DOP requires actual paper samples and is a tedious and cumbersome task to perform. However, the approach presented in this chapter, also known as indirect

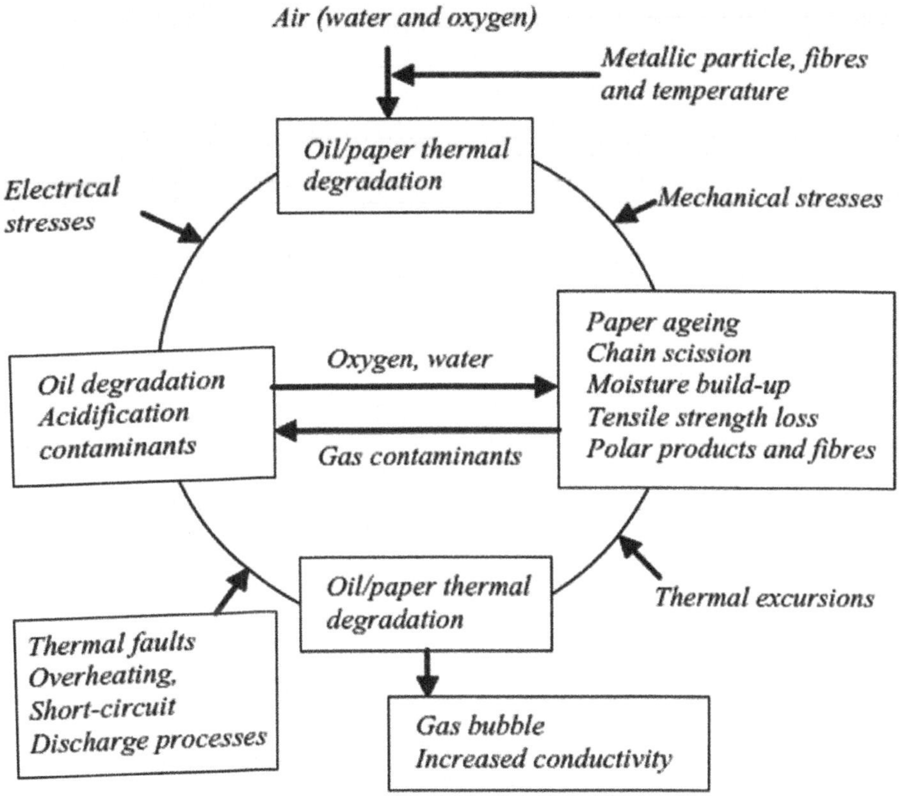

Fig. 3 Factors accelerating the oxidation process in oil [10]

method, estimates the DOP using furan contents, produced in the transformer oil during oxidation process. This method is slightly inaccurate; as the inaccuracy increases with age as more external factors come into account. This is more or less an approximate method. The cases where high accuracy is needed, one should calculate the degree of polymerization using direct approach only [10].

4 Experimental Tests for Calculation of 2-FAL

The furan contents are the by-products of the transformer ageing and deteriorating paper insulation. By carrying out certain tests on transformer oil, we can determine the levels of furan contents and assess the condition of paper insulation due to ageing. In this section, the required tests are discussed in detail.

Table 1 UV-VIS variation with density of oil sample

Density (29.5 °C g/cm)	UV-VIS area
0.860	231.64
0.858	239.38
0.859	216.53
0.818	34.21
0.819	35.16

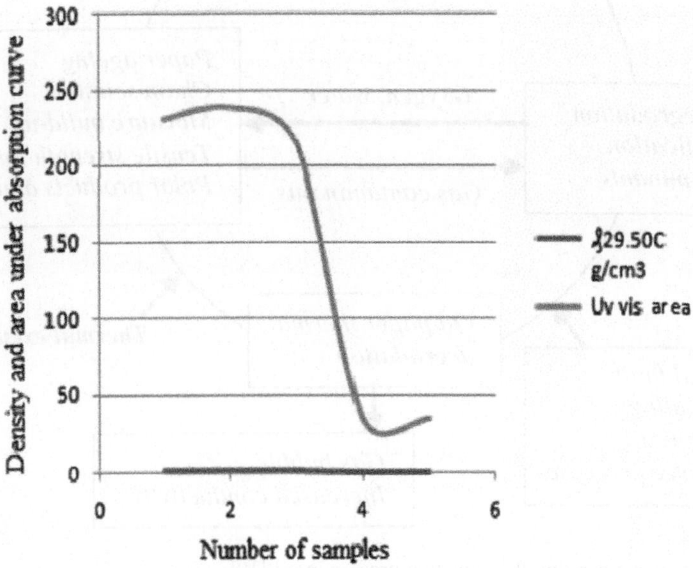

Fig. 4 Plot of density and UV-VIS area

4.1 Density

The density test is of great importance when the transformer is operating under low temperature conditions. The upper density value 30° which ensures that the moisture present in the form of ice restricts to the lower bottom and do not float on the upper surface till the temperature reaches to −13 °C. The oil samples are tested in TIFFAC-CORE at NIT Hamirpur where the densities of oils are determined and the corresponding UV-VIS area is calculated using UV-VIS Photo spectrometer. The data is as shown in Table 1. The plot of UV-VIS area with density is shown in Fig. 4.

Table 2 UV-VIS variation with acidity of oil sample

Acidity (ppm)	UV-VIS area
0.05	231.64
0.17	239.36
0.09	216.53
0.04	34.21
0.05	35.16

4.2 *Overall Acidity Content*

Acidity often referred as the Total Acid Number (TAN) which signifies the overall acid content, measured by the amount of KOH in mg in order to neutralize the hydrogen ions in per gram of oil. There are several methods by which we can estimate the level of acidity. One of the methods is potentiometric Titration Method, where propanol and Toluene are dissolved with small amount of water, along with the sample with aqueous form of NaOH [11]. The oil samples are tested in TIFFAC-CORE at NIT Hamirpur where the acidities of oils are determined and the corresponding UV-VIS area is calculated using UV-VIS Photo spectrometer. The data is as shown in Table 2.

In this method, a glass electrode and other electrode, known as reference electrode are immersed in the oil sample and are connected to potentiometer. Automatic potentiometric titration is used for acidity measurement.

4.3 *Interfacial Tension (IFT)*

IFT is one of the key indicators of transformer oil oxidation. The insulating oils are compounds of carbon and hydrogen, and thus hydrophobic in nature. So, when the oil undergoes oxidation, oxygenated compounds such as carboxylic acids are formed, which in turn are hydrophilic in nature. Because of this process, the whole oil sample becomes hydrophilic. The more the oil becomes hydrophilic; the surface tension of the oil becomes weaker. However, if there is contamination with some other hydrophobic substance, there will be reduction in interfacial tension without corresponding increase in the acid number. The oil samples are tested in

Table 3 UV-VIS variation with IFT

IFT (Mn/m)	UV-VIS area
7.8	230.78
8.1	238.83
7.2	217.35
29.9	33.21
35.5	34.61

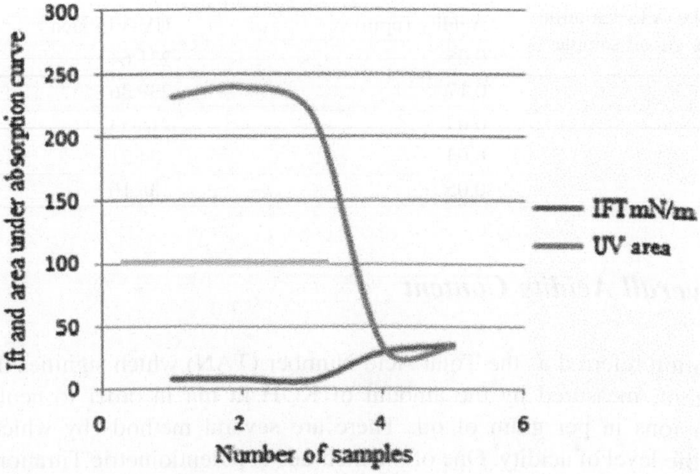

Fig. 5 Plot of IFT versus UV-VIS area by UV-photo-spectrometer

TIFFAC-CORE at NIT Hamirpur where the interfacial tensions of oils are determined and the corresponding UV-VIS area is calculated using UV-VIS Photo spectrometer. The data is as shown in Table 3. The plot of IFT Vs UV-VIS Area is shown in Fig. 5.

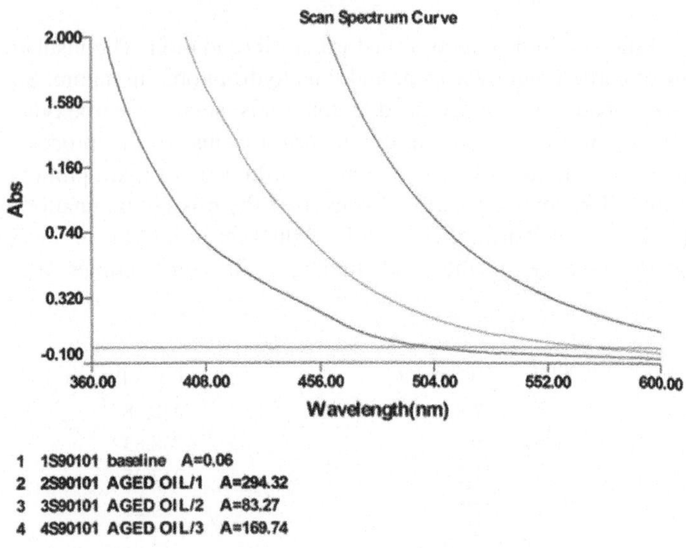

Fig. 6 Scan spectrum curve of test sample by UV-photo-spectrometer

Table 4 Analysis record of samples by UV-spectrophotometer

S. no.	Density (p 29.5 °C g/cm)	Acidity (mg KOH/g oil)	IFT (mN/m)	UV/VI area
1.	0.811	0.01	42.5	29.9
2.	0.814	0.01	41.2	31.5
3.	0.818	0.01	39.1	38.4
4.	0.818	0.02	38	44.0
5.	0.821	0.02	36.4	48.9
6.	0.824	0.03	35.3	61.1
7.	0.828	0.05	31.0	84.8
8.	0.831	0.06	27	99.7
9.	0.836	0.08	24.6	124.3
10.	0.845	0.17	18	186.1
11.	0.858	0.36	8.8	234.1
12.	0.859	0.38	7.6	239.4
13.	0.860	0.41	6.7	243.8

4.4 Dissolved Decay Contents

To calculate the concentration of dissolved decay contents in transformer oil, an offline test is being performed to measure its relative level. UV-VIS photo spectrometer method is used. In this process, the test samples are scanned and graph is plotted between absorbance and wavelength. The area under the curve gives the overall furan contents in the oil sample. The not-so aged oils have an approximate area under the curve less than that of 23 abs X mm of absorption Vs wavelength curve value. If the absorbance curve shifts towards higher wavelength, it indicates that the dissolved decay contents have increased in the concentration, whereas if the curve shifts towards smaller wavelengths, it indicates that the oil has been reclaimed and a sufficient amount of dissolved decay products has been removed by replacement or reclamation.

The variation of area for one of the oil sample is shown below in Fig. 6 and its relationship with acidity, density, and interfacial tension. The testing on the oil sample is carried out at TIFFAC–CORE laboratory, indicating that the breakdown voltage and dissolved gases do not tend to increase the UV-VIS area and their further analysis is not necessary.

4.5 Testing Details

Five oil samples are considered correlating the results of the test and the area under UV-VIS area, with the effect of furan contents, water content and other chemical reactions. The UV-VIS area relates the furan contents with the IFT, density and

Table 5 Range of input and output membership function

Input parameters	Fuzzy input	Low value	Medium value	High value
Input	Density	0.70–0.818	0.816–0.83	0.828–0.858
	Acid level	0.01–0.08	0.06–0.13	0.12–0.27
	IFT	7–20	19–30	29–43
Output	Area under UV VIS curve	27–89	83–237	235–256

acidity of the oil sample taken into consideration. The test is performed at NIT Hamirpur, TIFFAC-CORE Laboratory, using UV-VIS spectrophotometer and the following data is obtained as shown in Table 4.

5 Proposed Fuzzy Based Approach

Fuzzy logic system is a tool in determining the reliability of the transformer health index. If-Then rule is used for mapping the input—output using the FIS Editor GUI tool in MATLAB. While simulating the input-output, the following fuzzy rules have been developed. The three input parameters-density; acidity and IFT have triangular membership function in the range High, Medium and low. The corresponding UV VIS area output is also a triangular membership function. Likewise, UV-VIS area is also classified as High, Medium and Low as shown in Table 5. Low area under UV-VIS curve corresponds to low value of dissolved decay contents and vice versa.

5.1 Rules Defined

For the analysis and interpretation of result, twenty seven expert rules are developed and based on these rules output membership function is obtained and hence corresponding UV-VIS area is determined as shown in Fig. 7. A 3-D plot mapping the UV-VIS area Vs acidity and density is shown in Fig. 8. Rules defined are as under:

(1) If density is low, IFT is low, acidity is low, then UV-VIS area is high.
(2) If density is low, IFT is low, acidity is medium, then UV-VIS area is medium.
(3) If density is low, IFT is low, acidity is high, then UV-VIS area is low.
(4) If density is low, IFT is medium, acidity is low, then UV-VIS area is low.
(5) If density is low, IFT is medium, acidity is medium, then UV-VIS area is medium.

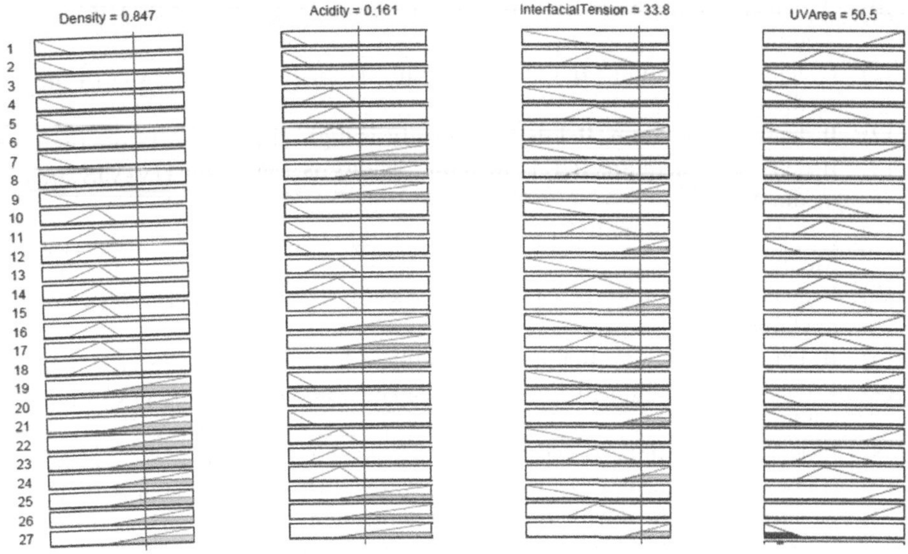

Fig. 7 The calculated VIS area output for lower value of acidity and higher value of IFT

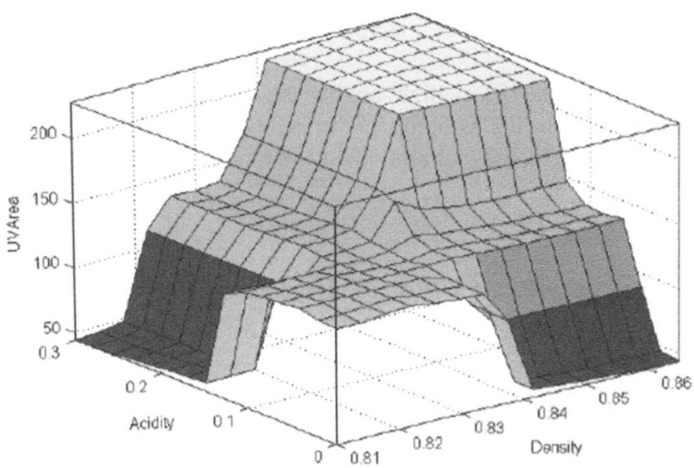

Fig. 8 3-D Plot of UV-VIS area versus acidity and density

(6) If density is low, IFT is medium, acidity is high, then UV-VIS area is low.
(7) If density is low, IFT is high, acidity is low, then UV-VIS area is high.
(8) If density is low, IFT is high, acidity is medium, then UV-VIS area is low.
(9) If density is low, IFT is high, acidity is high, then UV-VIS area is low.

(10) If density is medium, IFT is low, acidity is low, then UV-VIS area is medium.
(11) If density is medium, IFT is low, acidity is medium, then UV-VIS area is medium.
(12) If density is medium, IFT is low, acidity is high, then UV-VIS area is low.
(13) If density is medium, IFT is medium, acidity is low, then UV-VIS area is medium
(14) If density is medium, IFT is medium, acidity is medium, then UV-VIS area is Medium.
(15) If density is medium, IFT is medium, acidity is high, then UV-VIS area is medium.
(16) If density is medium, IFT is high, acidity is low, then UV-VIS area is high.
(17) If density is medium, IFT is high, acidity is medium, then UV-VIS area is medium.
(18) If density is medium, IFT is high, acidity is high, then UV-VIS area is high.
(19) If density is high, IFT is low, acidity is low, then UV-VIS area is high.
(20) If density is high, IFT is low, acidity is medium, then UV-VIS area is low.
(21) If density is high, IFT is low, acidity is high, then UV-VIS area is low.
(22) If density is high, IFT is medium, acidity is low, then UV-VIS area is high.
(23) If density is high, IFT is medium, acidity is medium, then UV-VIS area is medium
(24) If density is high, IFT is medium, acidity is high, then UV-VIS area is medium
(25) If density is high, IFT is high, acidity is low, then UV-VIS area is high.
(26) If density is high, IFT is high, acidity is medium, then UV-VIS area is high.
(27) If density is high, IFT is high, acidity is high, then UV-VIS area is low.

5.2 *Physical Implementation of Model*

The assessment of transformer oil can be the decisive factor in determining the serviceability of device. The Fuzzy Logic provides not only an effective approach but also an acceptable reasoning. The classical method of calculating Degree of Polymerisation is a tedious task as it involves the original insulation samples. The method proposed in this chapter gives fairly reliable results. If we take physical implementation into concern, a fuzzy logic based wireless sensor can be used in this regard. The above mentioned process may employ two sensors. The first sensor will indicate the concentration of furan contents in the insulating material and the second sensor will indicate the overall health index of the transformer. By this estimation, we can determine whether the transformer is demanding maintenance or permanent replacement.

5.3 *Health Index of a Transformer*

One of the most important tools for extracting the information about the current status of the power transformer is the health index. It is considered as a powerful tool for providing one and only index to represent the health of the power transformer. The health index does not reflect the unhealthy condition of any particular part of the device; it also does not indicate any long term degradation, a condition not usually diagnosed by routine or visual inspection. Monitoring a transformer health and giving the transformer an index represents its overall health condition, for planning routine maintenance strategies [12].

Various sets of tests were being performed in the transformer insulating oil and the corresponding results were obtained in terms of acidity level, moisture content, breakdown voltage, dissipation factor dissolved combustible gases and 2-Furfuraldehyde. The gases released during oxidation are indicated by Dissolved Combustible Gases (DCG). This method is particularly of great importance as the individual gas is suited to diagnose the particular type of fault occurred in the transformer. But here in this chapter, we are mainly concerned with the overall health index, so we will collectively consider these gases in terms of DCG.

5.4 *Design of the Membership Functions*

Membership functions were mainly designed on IEEE standards, the context and the experience of the on-field operators. The voltages of the transformers taken into consideration are 65 kV or less.

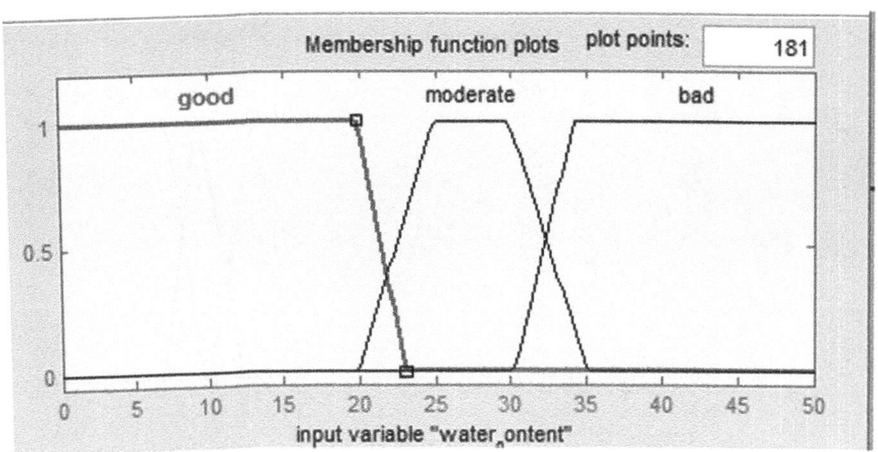

Fig. 9 Membership function for moisture content

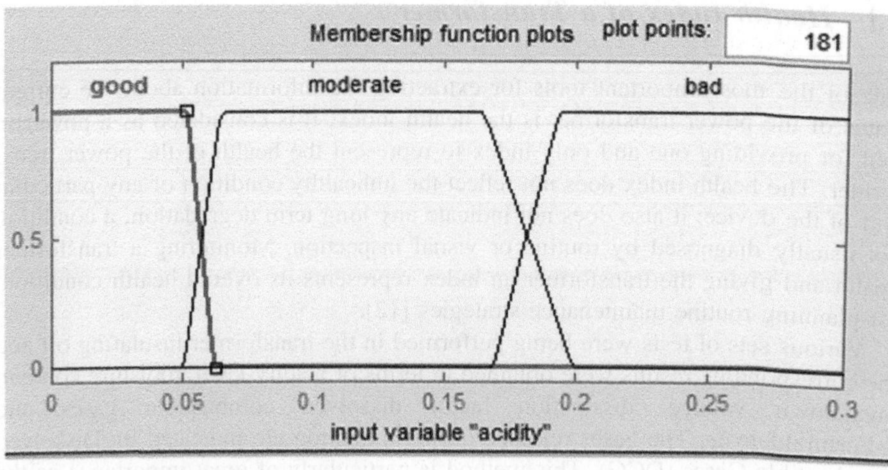

Fig. 10 Membership function for acidity

(A) ***Membership function for moisture content***:

Large *numbers of ageing transformers are* affected by moisture up to certain degree. The presence of moisture accelerates the rate of ageing, reduces the dielectric strength of the transformer oil, produces high saturation in transformer oil and creates various problems caused by the bubbling occurring at high temperatures [12]. The moisture content in transformer oil is a good indicator of the oil condition and oxidation of the oil also adds up to some moisture value, the presence of the water content reflects the condition of the

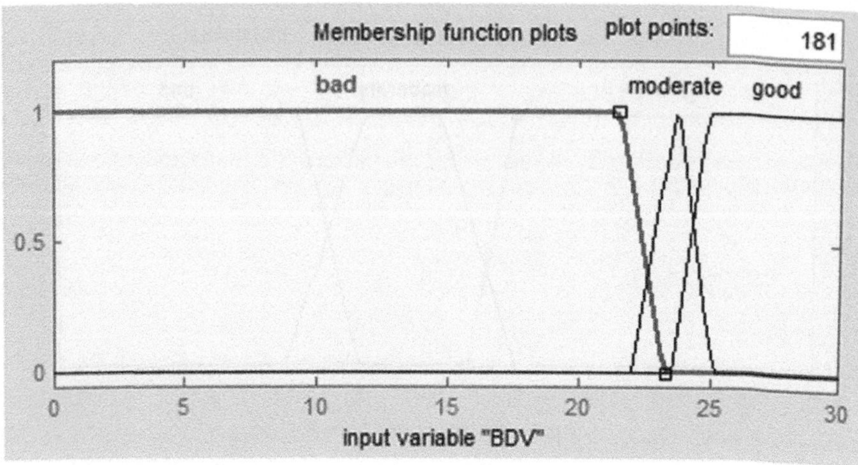

Fig. 11 Membership function for BDV

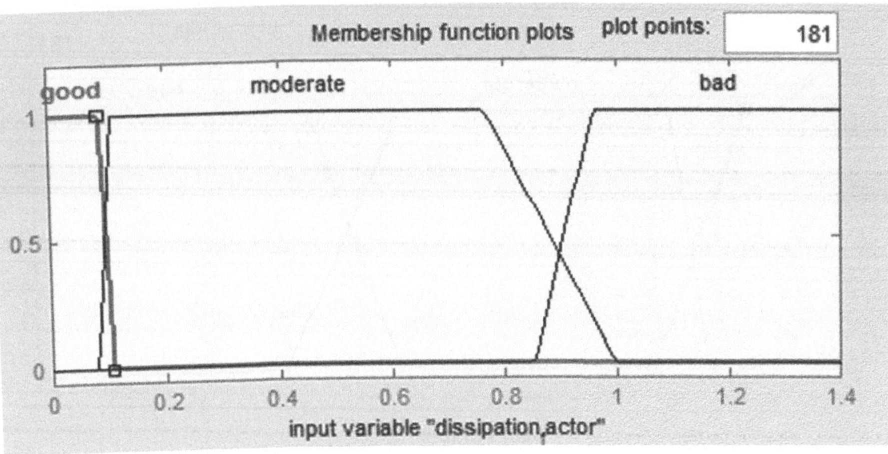

Fig. 12 Membership functions for dissipation factor

insulating material used. The membership functions of moisture content are divided into three linguistic levels: bad, moderate and good, which therefore is used to model the moisture content level as shown in Fig. 9.

(B) ***Membership functions for acidity***:

The total acidity level in the oil considered is measured in mg KOH/g. It is an indication of deteriorating condition of transformer insulation used in the transformer. The overall acidity level of oil increases because of the oxidation processes, with extended periods of service and is a major indicator of transformer health index [4]. The process of oxidation also adds up to the acidity level of the transformer oil. Like the moisture content, the acidity level is also modelled into three linguistic functions namely Bad, Moderate and Good as shown in Fig. 10.

(C) ***Membership functions for oil Breakdown Voltage (BDV)***:

The breakdown voltage of the transformer oil represents its ability to withstand the voltage stresses and is undoubtedly the most reliable feature in determining the oil quality [12]. The oil having lesser BDV value may have increased sparks and partial discharges that contribute to the aging of the transformer insulation. The condition of the transformer insulation is again divided into three linguistic variables; Bad, Moderate and Good as shown in Fig. 11.

(D) ***Membership functions for dissipation factor***:

The power loss in the transformer oil during its operation is given by dissipation factor and is not only used to determine the electric properties of oil, but also its contamination level. As the deterioration of the oil increases, the dissipation factor tend to increase which will ultimately increase the overall temperature of the transformer oil. It contributes in accelerating the aging

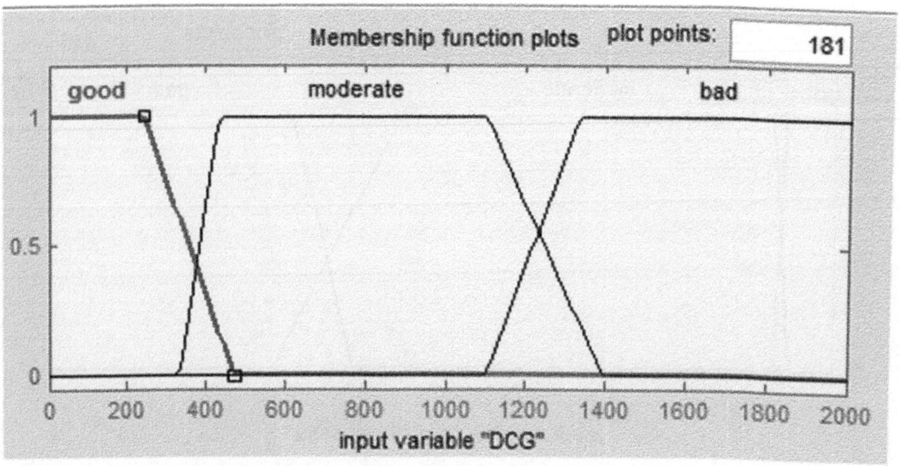

Fig. 13 Membership functions for DCG

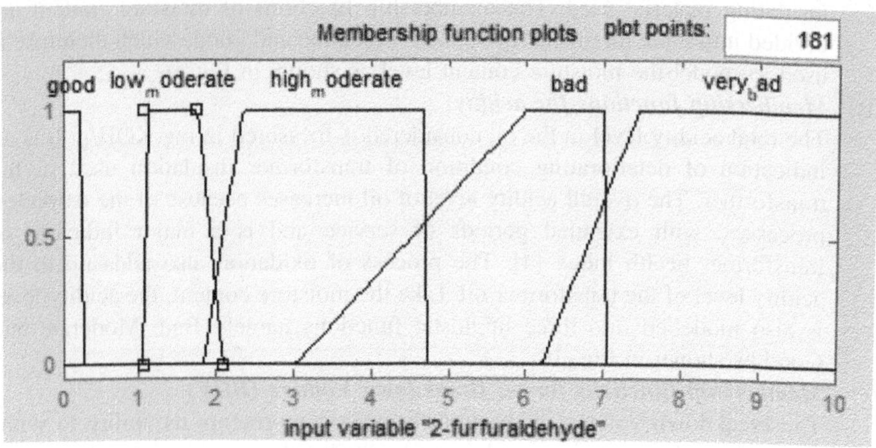

Fig. 14 Membership functions for 2-furfuraldehyde

process. It is further classified as Bad, Moderate and good, in estimating the overall health index of the transformer as shown in Fig. 12.

(E) ***Membership functions for Dissolved Combustible Gases (DCG)***:
At normal temperatures, dissolved combustible gases are produced by all the transformer varieties. However in two cases the concentration of gases increases: (1) If any incipient fault occurs inside a transformer, and (2) If the ageing in transformer takes place. Any predefined critical levels of key gases, produced during oxidation process, can be calculated by one of the ratio methodologies, such as the/Rogers or Dornenberg ratio [11]. These methods

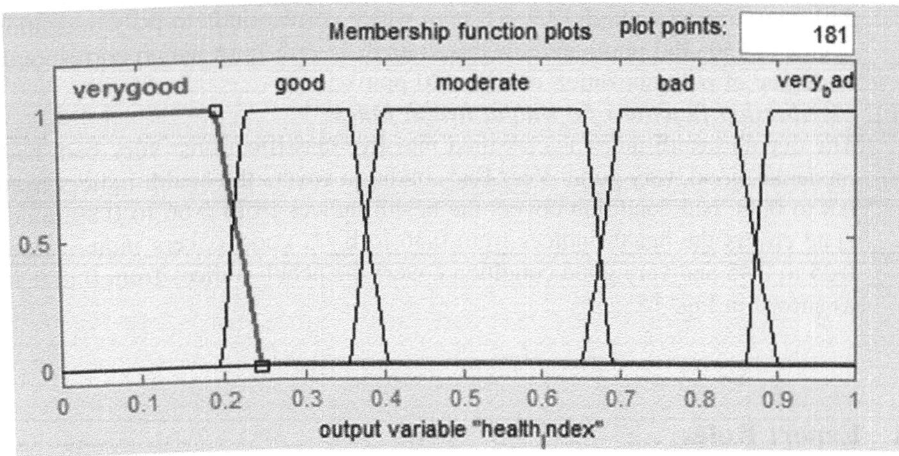

Fig. 15 Membership functions for health index

can determine the type of fault in the power transformer. However, in this proposed methodology, DGA is used to access the overall health index of the power transformer under consideration, not for determining the type of fault. The overall health index of the transformer is therefore accessed using the sum total of the DCG gases obtained. Again, the dissolved combustible gases available are divided into three linguistic variables: Bad, Moderate and good as shown in Fig. 13.

(E) ***Membership functions for 2-Furfuraldehyde***:
One of the most important factors in accessing the health of the transformer is the dissolved decay content or 2-Furfuraldehyde content, because the health of the transformer is directly accessed by it, particularly the paper insulation. It is the major reason for the decline of life of the power transformer [12]. Presence of the water content in poor insulation and increased temperature in the internal winding is the main reason for the production of furan contents in transformer oil. Due to its reliability in the measurement, the overall health index of the test transformer with respect to the dissolved decay content are accessed according to the following conditions: very bad, bad, high moderate, low moderate and good as shown in Fig. 14. The condition of the solid insulation is healthy if the degree of polymerization is greater than 750 units and the corresponding 2-Furfuraldehyde level is about 0.15 ppm. The paper insulation reaches in a very bad condition if the degree of polymerization reduces to 250 units for which the corresponding 2-Furfuraldehyde level is about 7.5 ppm. Three levels are considered between good and very bad: Bad (2.5–7.5 ppm corresponding to the degree of polymerization of 350–

230 ppm), moderate high (0.9–3.8 ppm which corresponds to polymerization degree of 550–300 ppm) and low moderate (0.15–1.5 ppm which corresponds to degree of polymerization of 710–420 ppm).

(F) ***Membership functions for output health index***:
The overall health index is classified into five classifications: very bad, bad, moderate, good, very good. Very Bad condition covers the health indices from 0.8 to 0.98, bad condition covers the health indices from 0.66 to 0.89, moderate covers the health indices from 0.36 to 0.71, good covers indices from 0.25 to 0.45 and very good condition covers the health indices from 0 to 0.30 as shown in Fig. 15.

6 Expert Rules

A set of 33 rules are used for calculating the health index of the transformer. The furan content is set at the highest priority because of its reliable assessment in the health index of the transformer. The overall dissolved combustible gases are set at second priority and the other parameters such as moisture content, acidity level, Breakdown voltage and loss factor are given the same and lower priority levels. The rules are as under:

1. If 2-furfuraldehyde is very bad, then the health index is very bad.
2. If 2-furfuraldehyde is bad and DCG is not bad, then the health index is bad.
3. If 2-furfuraldehyde is bad and DCG is bad, then the health index is very bad.
4. If 2-furfuraldehyde is high moderate and DCG is bad, then the health index is bad.
5. If 2-furfuraldehyde is high moderate and DCG is not bad and water is bad and acidity is bad and (BDV is bad or the dissipation factor is bad), then the health index is bad.
6. If 2-furfuraldehyde is high moderate and DCG is not bad and BDV is bad and dissipation factor is bad and (water is bad or acidity is bad), then the health index is bad.
7. If 2-furfuraldehyde is high moderate and DCG is not bad and water is not bad and acidity is not bad, then the health index is moderate.
8. If 2-furfuraldehyde is high moderate and DCG is not bad and water is not bad and BDV is not bad, then the health index is moderate.
9. If 2-furfuraldehyde is high moderate and DCG is not bad and water is not bad and the dissipation factor is not bad, then the health index is moderate.
10. If 2-furfuraldehyde is high moderate and DCG is not bad and BDV is not bad and dissipation factor is not bad, then the health index is moderate.
11. If 2-furfuraldehyde is high moderate and DCG is not bad and BDV is not bad and acidity is not bad, then the health index is moderate.

12. If 2-furfuraldehyde is high moderate and DCG is not bad and acidity is not bad and dissipation factor is not bad, then the health index is moderate.
13. If (2-furfuraldehyde is good or 2-furfuraldehyde is low moderate) and DCG is bad and water is not bad and acidity is not bad, then the health index is moderate.
14. If (2-furfuraldehyde is good or 2- furfuraldehyde is low moderate) and DCG is bad and water is not bad and BDV is not bad, then the health index is moderate.
15. If (2-furfuraldehyde is good or 2-furfuraldehyde is low moderate) and DCG is bad and water is not bad and dissipation factor is not bad, then the health index is moderate.
16. If (2-furfuraldehyde is good or 2-furfuraldehyde is low moderate) and DCG is bad and acidity is not bad and BDV is not bad, then the health index is moderate.
17. If (2-furfuraldehyde is good or 2-furfuraldehyde is low moderate) and DCG is bad and acidity is not bad and dissipation factor is not bad, then the health index is moderate.
18. If (2-furfuraldehyde is good or 2-furfuraldehyde is low moderate) and DCG is bad and BDV is not bad and dissipation factor is not bad, then the health index is moderate.
19. If (2-furfuraldehyde is good or 2-furfuraldehyde is low moderate) and DCG is bad and water is bad and acidity is bad and (BDV is bad or dissipation factor is bad), then the health index is bad.
20. If (2-furfuraldehyde is good or 2-furfuraldehyde is low moderate) and DCG is bad and BDV is bad and the dissipation factor is bad and (the water is bad or acidity is bad), then the health index is bad.
21. If (2-furfuraldehyde is good or 2-furfuraldehyde is low moderate) and DCG is moderate and (water is bad or acidity is bad or BDV is bad or dissipation factor is bad), then the health index is moderate.
22. If (2-furfuraldehyde is good or 2-furfuraldehyde is low moderate) and DCG is moderate and (water is not bad or acidity is not bad or BDV is not bad or dissipation factor is not bad), then the health index is good.
23. If (2-furfuraldehyde is good or 2-furfuraldehyde is low moderate) and DCG is good and water is bad or acidity is bad and (BDV is bad or dissipation factor is bad), then the health index is moderate.
24. If (2-furfuraldehyde is good or 2-furfuraldehyde is low moderate) and DCG is good and BDV is bad and the dissipation factor is bad and (water is bad or acidity is bad), then the health index is moderate.
25. If (2-furfuraldehyde is good or 2-furfuraldehyde is low moderate) and DCG is good and water is not bad and acidity is not bad and BDV is not bad and dissipation factor is not good, then the health index is good.
26. If (2-furfuraldehyde is good or 2-furfuraldehyde is low moderate) and DCG is good and water is not bad and acidity is not bad and dissipation factor is not bad and BDV is not good, then the health index is good.

27. If (2-furfuraldehyde is good or 2-furfuraldehyde is low moderate) and DCG is good and water is not bad and BDV is not bad and dissipation factor is not bad and acidity is not good, then the health index is good.
28. If (2-furfuraldehyde is good or 2-furfuraldehyde is low moderate) and DCG is good and BDV is not bad and acidity is not bad and dissipation factor is not bad and acidity is not good, then the health index is good.
29. If (2-furfuraldehyde is good or 2-furfuraldehyde is low moderate) and DCG is good and BDV is bad and (water is bad or acidity is bad or dissipation factor is bad), then the health index is moderate.
30. If (2-furfuraldehyde is good or 2-furfuraldehyde is low moderate) and DCG is good and (water is bad or acidity is bad) and (BDV is not bad and the dissipation factor is not bad), then the health index is good.
31. If (2-furfuraldehyde is good or 2-furfuraldehyde is low moderate) and DCG is good and (water is bad or the dissipation factor is bad) and (the BDV is not bad or acidity is not bad), then the health index is good.
32. (2-furfuraldehyde is good or 2-furfuraldehyde is low moderate) and DCG is good and (acidity is bad or the dissipation factor is bad) and (the BDV is not bad and water is not bad), then the health index is good.
33. (2-furfuraldehyde is good or 2-furfuraldehyde is low moderate) and DCG is good and water is good and acidity is good and BDV is good and the dissipation factor is good, then the health index is very good.

7 Result and Discussion

The assessment of the health index of the transformer requires actual paper insulation samples, which at times may not be feasible to obtain. The proposed methodology in this chapter uses transformer oil parameters to obtain the furanic content, which is a major indication of the transformer ageing. The proposed approach shows accuracy up to 82–87%, which makes it a reliable tool to use. Despite of the fact that the fuzzy logic does not work on crisp threshold, the proposed method is reliable enough to indicate whether the transformer is demanding maintenance or permanent replacement. The first fuzzy-set relates the oil acidity, density and interfacial tension with the furan contents of the transformer oil. The furan content is one of the key indicators for transformer ageing. This fuzzy set provides accuracy of around 85%. We have further used this furan content as a decisive factor in estimating the overall health index. The furan contents (2-FAL), along with five more inputs—Dissolved combustible gases (DCG), moisture content, Breakdown Voltage, Dissipation factor and Acidity are mapped as inputs to estimate the health index of the transformer. The fuzzy (Mamdani-model) is used for the same. This fuzzy model provides the health index in the accuracy of about 87%. The above method can be physically implemented in the transformers by using wireless fuzzy sensors. The output of these sensors can predict whether the transformer is demanding maintenance or permanent replacement.

8 Conclusion

The process of the fuzzy inference is formulated for mapping the three input values (acidity, density and interfacial tension) to obtain the dissolved decay content in the transformer. The obtained output is the input along with the five other parameters to determine the health index of the transformer. For this proposed approach, Mamdani minimum-maximum interfacing method is used. The first step in the proposed methodology is to covert the numerical values of three oil parameters into corresponding linguistic variables, the output of which gives the level of dissolved decay content in the transformer oil. In the next step, the output of the first model is taken as the input for the second fuzzy model along with the five other parameters in order to estimate the overall health index of the transformer. The proposed Mamdani model truncates the resulted function which is health index in this case when each rule is fired at minimum membership values for all antecedents. The above mentioned method can be physically implemented by using Fuzzy Logic sensors.

Acknowledgements The authors are thankful to TIFFAC-CORE LAB (NIT HAMIRPUR) for the testing data of the oil samples. Also, we are thankful to MATERIAL SCIENCE LAB, NIT HAMIRPUR for conduction of test in UV-VIS Photo spectrometer in estimating the furan contents in transformer oil.

References

1. Chowdary ML, Singh SK, Jarial RK (2014) Analysis of oil quality index and dielectric dissipation factor. IJEIT 4(4)
2. Lai SP, Abu-Siada A, Islam SM, Lenco G (2008) Correlation between UV-V is spectral response and furan measurement of transformer oil. In: 2008 International conference on condition monitoring and diagnosis, Beijing, China, April 21–24
3. Kachler AJ, Hohlein I (2005) Aging of cellulose at transformer service temperatures. Part 1: influence of type of oil and air on the degree of polymerization of pressboard, dissolved gases, and furanic compounds in oil. IEEE Electr Insul Mag 21(2):15–21
4. Arvind D, Khushdeep S, Deepak K (2008) Condition monitoring of power transformer: a review. In: Proceedings of the IEEE power engineering society transmission and distribution conference and exposition. Chicago, IL, pp 1–6
5. ASTM (2004) Standard test method for interfacial tension of oil against water by the ring method, ASTM D971-99(a)
6. Koreh O, Torkos K, Bashir Mahara M, Boressay J, Izvewkov V (1998) Study of water clusters in insulating oils by fourier transform infrared spectroscopy. IEEE Trans Dielectr Electr Insul 5:896–902

7. Mehta AK, Sharma RN, Chauhan S (2012) Condition monitoring of insulation system in power transformers. IJCEE 4(2)
8. Mehta AK, Sharma RN, Chauhan S, Parminder K (2013) Relative content of dissolved decay and correlation with DGA, IFT and density of transformer oil. ICCPCT
9. MATLAB 7.6 Manual T90/T90+ UV/VIS Spectrophotometer [PG Instrument] TIFAC-CORE, NIT-HAMIRPUR, India
10. Arshad M, Islam SM (2007) Power transformer condition assessment using oil UV-spectrophotometry. In: Annual report at conference on electrical insulation and dielectric phenomena
11. Hasmat Malik SM, Jarial RK, Sood YR (2011) Application and implementation of artificial intelligence in electrical system. In: International conference on advances in computing & communication (ICACC-2011), pp. 499–505 (ISBN 978-81-920874-0-5, Sponsored by IEEE MTTS)
12. Bouaf'chal A, Farzaneh M, Sabau J, Bussie'res D, Robertson EB (2010) Decay products in the liquid insulation of power transformers. IET Electric Power Appl 4(3):177–184

Ms. Nitika Ghosh received her B.Tech. degree from College of Engineering Roorkee in 2012 and M.Tech. from National Institute of Technology, Hamirpur in condition monitoring control and protection in 2016. Currently, she is working as Assistant Professor in Electrical and Electronics Engineering Department at ABES Engineering College Ghaziabad. She is having 2 years of academic experience. She has authored many national and international journals of repute and has qualified GATE exam in Electrical Engineering five times. Her areas of interest include condition monitoring control, electric machines & drives, fuzzy systems, and optimization techniques. She can be reached at nitikaghosh21@gmail.com.

Dr. Vikas Singh Bhadoria is currently working as Associate Professor in Electrical and Electronics Engineering Department, ABES Engineering College, Ghaziabad. He received his Ph.D. in Power System from Gautam Buddha University, Greater Noida, in 2018, and M.E. (Electrical Engineering) and B.E. (Electrical and Electronics Engineering) degrees from RGPV, Bhopal in 2009 and 2005, respectively. He has published more than 20 research papers in reputed international journals and conferences. His research interests include power system restructuring, power system planning, and power system reliability. He can be reached at vikasbhadoria@gmail.com.

Dr. Ram Naresh Sharma received his Bachelor of Engineering (Electrical) from Thapar Institute of Engineering and Technology, in 1987; Masters of Engineering in Power Systems from Punjab Engineering College, Chandigarh, in 1990; and Ph.D. from the University of Roorkee, India, in 1999, respectively. He has joined REC, Hamirpur in 1989. In January 2000, he joined the Department of Electrical Engineering, National Institute of Technology Hamirpur, as an Assistant Professor. Now currently working as a Professor in EED, National Institute of Technology, Hamirpur, Himachal Pradesh, India. He has published a number of research papers in national and international journals of repute. He has been providing the consultancy services to electric power industry. His fields of research interests are artificial intelligence applications to power system optimization problems, evolutionary computation, neural networks, and fuzzy systems. He can be reached at rnaresh@nith.ac.in.

Dr. Vivek Shrivastava has more than 15 years of diversified experience of scholarship of teaching and learning, accreditation, research, industrial, and academic leadership in India, China, and USA. Presently he is a Associate Professor in the Department of Electrical and Electronics Engineering at National Institute of Technology Delhi. Previously he was Associate Professor in the Department of Electrical Engineering at Rajasthan Technical University Kota, Associate Professor at NIT Delhi, and Assistant Professor at Gautam Buddha University Greater Noida, UP India. Prior to his academic assignments, he has worked as System Reliability Engineer at SanDisk Semiconductors Shanghai China and USA.

Dr. Shrivastava has published over 75 journal articles and presented papers at conferences, and has published several chapters in books. He has supervised 4 Ph.D. and 14 master's students, and is currently supervising several Ph.D. students. His diversified research interests are in the areas of reliability engineering, renewable energy, and conventional power systems, which include wind, photovoltaics (PV), hybrid power systems, distributed generation, grid integration of renewable energy, power systems analysis, and smart grid. Dr. Shrivastava is an editor/associate editor of the journals, International Journal of Swarm Intelligence (IJSI) and International Journal of System Assurance Engineering and Management. He is a Fellow of the Institution of Engineers (India) and a Senior Member of the Institute of Electrical and Electronics Engineers (IEEE). He can be reached at shvivek@gmail.com.

Safety Engineering

Probabilistic Safety Assessment in Nuclear and Non-nuclear Facilities: In a Glimpse

Gopika Vinod

1 Probabilistic Safety Assessment of NPPs

1.1 Introduction to Nuclear Safety

The main goal of nuclear safety is to keep the radiation exposure of the public and workers from nuclear facilities as low as reasonably achievable both during normal operational states and in the event of an accident. NPP safety can be assessed both by deterministically and probabilistically. In the deterministic safety analysis, design basis accidents are considered and it is shown that the engineered safety features provided to counter act such accidents result in radiation doses/risks that are acceptable. In contrast, probabilistic safety assessment (PSA) includes all possible accident scenarios and their quantification in terms of plant damage frequency and consequences. PSA is one of the probabilistic tools available to deal the safety of NPPs probabilistically. This chapter mainly focuses on PSA aspects of NPPs. Probabilistic Safety assessment (PSA) provides a comprehensive and structured approach in identifying failure scenarios that pose potential hazards to the plant, its personnel, environment and the public at large. PSA provides valuable insight into potential weak links in the defence in depth concept. PSA is associated with models which predict offsite radiological releases resulting from potential accidents. It deals mainly with the identification of accident sequence applicable to the design of a reactor and further relegating the risk to various process and safety systems and down to the components and operator actions. In fact it is possible to rank the systems and components in terms of their risk significance. In short PSA is an analytical technique for integrating diverse aspect of design and operation in order to assess the risk of a particular nuclear power plant and to develop an information base for analyzing plant specific and generic issues.

G. Vinod (✉)
Reactor Safety Division, Bhabha Atomic Research Center, Trombay, India
e-mail: vgopika@barc.gov.in

D. R. Karanki et al. (eds.), *Advances in RAMS Engineering*,
Springer Series in Reliability Engineering,
https://doi.org/10.1007/978-3-030-36518-9_15

The risk in general for Nuclear Power Plant (NPP) [1] is defined as:

$$Risk = Likelihood\ of\ occurrence\ of\ undesirable\ event\ (an\ accident)$$
$$\times\ Its\ consequences\ in\ terms\ of\ exposure\ to\ radioactive\ material\ release$$

Therefore the objectives of the nuclear safety are:

- Reduce the likelihood of occurrence an accident
- Minimize the release of radio-active material if accidents happen
- Minimize the population exposure if radio-active materials are released.

The first report on Nuclear Power Plant (NPP) accidents WASH 740 [2] was issued in 1957. The consequences predicted were unacceptable at that time. The importance of this report felt only after Three Mile Island NPP accident, since that accident was predicted in this report. Now PSA study of Nuclear Power Plants are carried out in many countries and are mandatory requirement by their regulatory bodies.

1.2 *Probabilistic Safety Assessment for NPP*

For carrying out PSA of NPP, three levels of PSA are used [3].

1.2.1 Level 1 PSA

Level 1 PSA deals with the assessment of plant failures leading to the determination of core damage frequency. It provides insights into design weaknesses and into ways of preventing core damage, which in most cases is the precursor of accidents leading to major radioactive releases with potential health and environmental consequences. In order to reduce the likelihood of occurrence of an accident one needs to design NPP systems for reliable operation and reliability can be engineered in NPP systems by:

- Selection of reliable parts/components
- Use of redundancy techniques with due consideration for minimizing common cause failures
- Design techniques such as derating so that the operating stress is below the specified strength of the parts/components
- Controlling the environmental conditions in which the NPP systems are operating
- Selection of passive systems where the actuation of systems depends upon physical phenomenon rather than electrical/pneumatic signals and devices
- Balance between the automation and operator interaction with the NPP systems for reducing operator errors.

1.2.2 Level 2 PSA

Level 2 PSA addresses the containment system and phenomenological responses, leading, together with Level 1 results, to the determination of containment release frequencies. It provides additional insights into the relative importance of accident sequences leading to core damage in terms of the severity of the radioactive releases they might cause, and insight into weaknesses in (and ways of improving) the mitigation and management of core damage accidents (e.g. severe accident management). To minimize the release of radioactive material, if accidents happen, the NPP design should ensure that

- NPP is established and maintained in safe sub-critical state
- Residual heat from reactor coolant system is removed
- The integrity of reactor coolant system and adequate supply of reactor coolant is maintained
- Containment integrity is maintained
- Reactor building pressure is maintained below the atmospheric pressure with the help of containment ventilation system
- All the opening of the containment are blocked with the help of containment isolation system.

The reliable operation of these systems is to minimize the release of radioactive material.

1.2.3 Level 3 PSA

Level 3 PSA addresses the off-site consequences, leading, together with the results of Level 2 analysis, to estimates of public injuries. It provides insights into the relative importance of accident prevention and mitigation measures expressed in terms of the adverse consequences for the health of the public, and the contamination of land, air, water and food provisions. In addition, it provides insights into the relative effectiveness of aspects of accident management related to emergency response planning. In order to minimize the population exposure if radioactive material is released the NPP design should take into consideration

- Construction of NPP in thin population area or in no population zone
- Procedures for emergency planning and preparedness
- Reliable communication and transportation facilities for speedy evacuation.

In constructing NPP in population zone efforts are required to minimize likelihood of occurrence of an accident and minimize the release of radioactive material.

PSA is performed for both internally initiated events and externally initiated events. In this section, Level 1 PSA will be discussed in detail for internally generated initiating events.

1.3 Level 1 PSA of NPPs

The first step in the Level-1 PSA studies is to identify the scope with respect to source of radioactivity, type of initiating events (IE) and operational state of the reactor. Once the Scope of PSA is identified the various tasks of PSA are mentioned below:

(i) Collection of information on design and operation of plant.
(ii) Initiating Event Identification and Grouping
(iii) Event Tree Analysis
(iv) System Modeling-Involves the following activities

  (a) Fault tree Development
  (b) Data Development and Parameter Estimation
  (c) Common Cause Failure Analysis
  (d) Human Reliability Analysis

(v) Core Damage Quantification and Accident Sequence Analysis
(vi) Uncertainty, Sensitivity and Importance Analyses.

Some of these steps are explained in the following subsections in detail. The plant logic diagram depicting the above process is shown in Fig. 1.

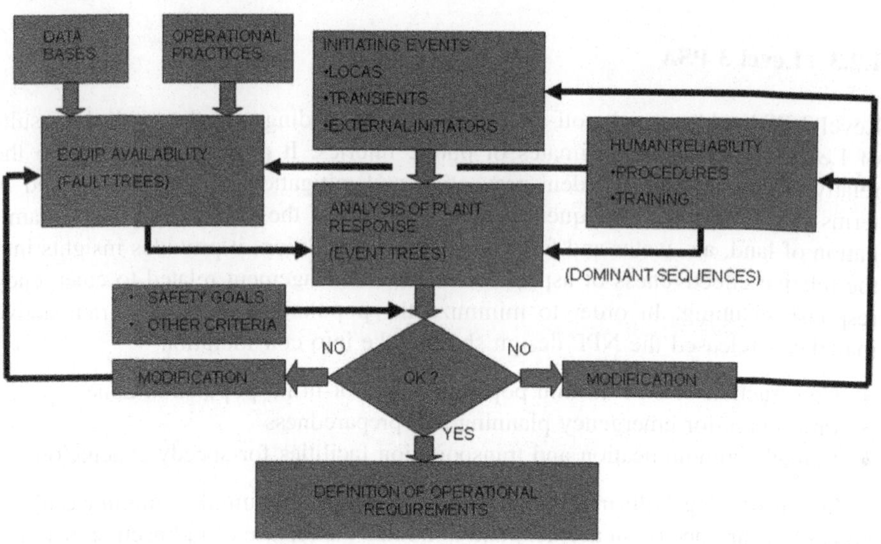

Fig. 1 Plant logic diagram

1.3.1 Identification of Initiating Events

An Initiating Event (IE) is an event that creates a disturbance in the plant and has the potential to lead to core damage depending on the successful operation or otherwise of the various mitigating systems in the plant

- There are two broad categories of Initiating Events:
  - *Loss of coolant accidents* (*LOCA*): Loss of coolant accidents is those accidents that may lead to draining of primary coolant because of isolation failures.
  - *Transient initiators*: Transients initiators are those initiators which affect the removal of decay heat and long term reactivity control.

The initial task of this analysis is to gather information from plant on design aspects and operation practices, which formed the basis for the development of plant models. There are several approaches that are followed for preparing the list of IEs, each approach having its own strengths and limitations. Engineering evaluation, Use of operational experience, Reference to previous lists etc. are few of them.

Once the Initiating events are identified then reactor shutdown and maintaining long term reactor sub-criticality, and decay heat removal are the safety functions necessary after their occurrence. The systems required for the proper functioning of the safety functions and those required for proper functioning of these front line systems are identified. Initiating Events are then grouped in such a way that all IEs in the same group essentially call for similar plant response and the same success criteria on front line systems. Some of the end-states in the event trees are fuel damages resulting from the postulated failures, for this the entire spectrum of fuel damage accidents has to be analysed and categorised (as per the performance of safety systems).

1.3.2 Plant Response Modeling: Event Tree Analysis

Event tree analysis is generally performed as follows: Event sequence modeling considers three basic functions that have to occur in succession for safe termination of the IE viz. reactor shut down, decay heat removal and long term reactivity control. To have a synthesised view of event sequences, small event tree, and large fault tree concept is used for event sequence modeling for various initiating events. In this concept the usual approach followed is that the front line system fault tree models include relevant support systems with suitable boundaries and human actions. Appropriate success criteria and boundary conditions are identified for various modes of front line system operation, which form the basis for their system modeling by fault tree method. For example, a simplified event tree for large Loss of Coolant Accident (LOCA) in a typical Pressurised Heavy Water Reactor (PHWR) is shown in Fig. 2. It has three pivotal events, viz.

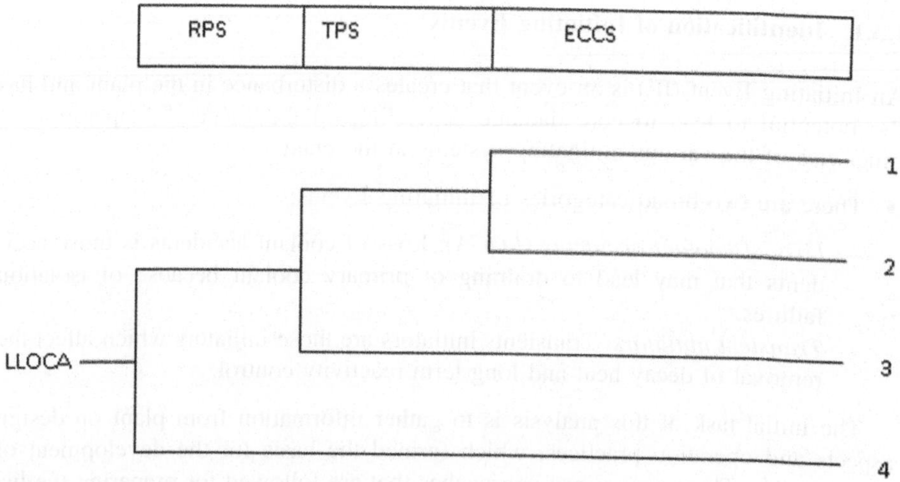

Fig. 2 A simplified event tree for large LOCA in a typical PHWR

- Reactor protection system.
- Total power supply system.
- Emergency core cooling system.

Theoretically event tree should have 8 paths, but it is having only 4 paths. This is because:

- RPS failure will directly have significant consequence irrespective of other events and
- ECCS is dependent on power supply.

The right side of event tree represents the end state that results from that sequence through the event tree. The end state can have healthy or accident consequences. To determine such consequence, through understanding of the system, operating experience, analyses of accidents (like thermal hydraulic or chemical reaction studies) is required.

1.3.3 System Reliability Modelling

Front line and support system modeling has been carried out using fault tree method. Salient features of these Fault Trees are immediate cause concept, detailed component modeling, use of combination of plant specific and generic component failure data for reliability parameter estimation, common cause failure analysis and incorporation of appropriate human error probabilities etc. The fault trees are deduced on the basis of design inputs available from sources like Design Manuals, Operating flow sheets including electrical, control and process drawings, single line diagrams, instrumentation schematics etc. To see the effect of any component failure in the overall risk, it is necessary that all the components and their logical relationships be available in the PSA model. Going down to the last component also allows effective treatment of common cause vulnerabilities.

The basic data required for a component appearing as a Basic Event in any system model is its unavailability which could be due to maintenance, test related problems or random failures of components, human errors etc. Reliability Parameters (for numerical values) considered are failure rate, repair time, test intervals, mission time, common cause component group, Human Error Probability (HEP), etc.

1.3.4 Common Cause Failure Analysis

Reliability of the system is enhanced by better design practice and selecting reliable components and further enhancement of reliability is achieved using redundancy technique. Redundancy means increase in volume, weight, cost and reduced maintainability. Moreover, 100% reliability cannot be achieved using redundancy technique. The limitation is dependent failures or common cause failures (CCF). Common Cause Failures can be defined as multiple failures which are a direct result of a common or shared root cause. The root cause may be design faults existing in redundant components catering to similar function, extreme environmental conditions (fire, flood, earthquake, lightning, etc.), or a human error (miscalibration, incorrect maintenance etc.).

For all other types of dependencies arising out of common design, manufacture, operation and maintenance and environment. Common Cause Failure analysis needs to be carried out with standard approaches such as Alpha factor model, Beta factor model etc. [4]. Most commonly used, beta factor model is a single parameter model; that is, it uses one parameter in addition to the total component failure probability to calculate the CCF probabilities. This model assumes that a constant fraction β of the component failure rate can be associated with common cause events shared by other components in that group. Another assumption is that whenever a common cause event occurs, all components within that common cause component group, are assumed to fail.

$$Q_{It} = (1 - \beta)Q_t$$
$$Q_m = \beta Q_t$$

This implies that

$$\beta = \frac{Q_m}{Q_{It} + Q_m}$$

where,

- Q_t, is the total failure probability of one component ($Q_t = Q_{It} + Q_m$)
- Q_{It}, is the independent failure probability of the single component
- Q_m, is the probability of basic event failure involving m specific components, m is the maximum number of components in a common cause group.

To generalize the equation, it can be written for m components involving failure of k components ($k \leq m$),

$$Q_{It} = \begin{cases} (1-\beta)Q_t & k=1 \\ 0 & 2 \leq k < m \\ \beta Q_t & k = m \end{cases}$$

where,
Q_k is the probability of basic event involving k specific components.

Case study: Shutdown system based on shutoff rods as well as poison injection philosophy. For poison injection system, there are three poison tanks. For the successful functioning of the system, any two out of three poison tanks should be available. Find the system unavailability of Liquid Poison Injection System, if the poison tank unavailability (Q_t) is 1E−3 and β is taken as 10% for common cause failure model for liquid poison tanks in shutdown system.

For poison tank,

$$Q_m = \beta Q_t$$

$$Q_{It} = 9\text{E}{-4},\ Q_m = 1\text{E}{-4}$$

Considering 2/3 redundancy, as shown in Fig. 3, Liquid Poison Injection System unavailability from independent failures = 2.43E−6.

Total Liquid Poison Injection System unavailability = 1.02E−4.

1.3.5 Human Reliability Analysis

HRA has become an essential part of every Probabilistic Safety Assessment (PSA) and is used to identify human errors, quantify their likelihood in terms of Human Error Probabilities (HEPs) and correctly incorporate the HEPs in the assessment of risk. HRA in a PSA involves:

- Identifying the critical human interactions in the system and how they can fail.
- Quantifying their probabilities of failure in terms of HEPs.

Categorization of Human Interactions in PSA

Three categories of interactions can be defined to facilitate the incorporation of HRA into the PSA structure. The three categories are as follows:

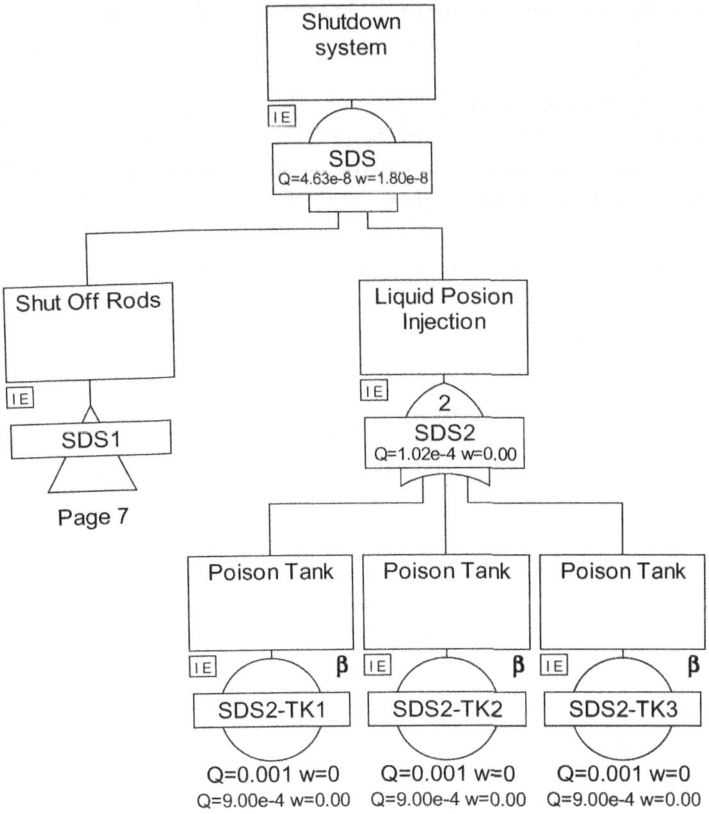

Fig. 3 Fault tree for shutdown system

- **Category A: Pre-Initiators**: These are activities prior to an Initiating Event (IE), related to maintenance, testing and calibration and intended to be positive. Any errors made in these interactions can lead to equipment or systems becoming unavailable when required post-fault. Pre-initiators consist of those actions associated with maintenance and testing, that degrade system availability. They may cause failure of a component/component group of may leave components in an inoperable condition. This unavailability is added to other failure contributions for components or systems, at the fault tree level. Recovery action for such human errors could follow error alarm, post–maintenance testing or post-maintenance inspection checks and may be modelled as applicable at the quantification stage. *These are independent of time and stress.*
- **Category B: Initiators**: Initiators are interactions carried out during normal operation, e.g. control room actions and normal maintenance, test or calibration actions. *Errors in these actions, either by themselves or in combination with other failures (other than human errors) can cause initiating events.* Most important are errors that not only precipitate an accident sequence but which

also concurrently cause failure of safety related systems, either front-line safety systems or support systems. Such 'common cause initiators' need to be specially emphasised.

- **Category C: Post-Initiators**: These are post-incident activities comprising the actions after an initiating event, performed with the intent to bring the plant back to normal safe and stable state. Errors in these interactions can occur while carrying out safety actions or there could be actions/errors that exacerbate the fault sequence. These human errors are associated with detection of the failure situation, diagnostics and subsequent actions for mitigation of Initiating Event. *These are the human actions required to be carried out in a limited time under high/moderate stress.*

Humans make errors due to a variety of causes. Some of the causes are internal to the individual (e.g. not knowing how to operate a particular equipment) and some are external to the individual (e.g. equipment controls not easily and comfortably accessible). The human's performance in a work situation is influenced by factors called Performance Shaping Factors (PSFs). The set of PSFs present in the work situation can greatly affect how safely or otherwise a system is operated. PSFs include factors like work environment quality of the Human Machine Interface (HMI) and quality of procedures and the training imparted. The identification of potential human errors is an important step in HRA. Errors, which alone or in conjunction with hardware/software failures, can lead to degraded system state are to be identified.

Human Error Quantification

After representing the human-error potential, the next step is to quantify the likelihood of errors involved and determine the overall effect of human error on system safety or reliability. Human reliability quantification techniques quantify human errors in terms of Human Error Probabilities (HEPs). HEP, which is measured as the ratio of the number of errors that occurred to the number of opportunities for the error to occur, is the metric of human reliability assessment. Recorded HEPs are relatively few in number. And also there is difficulty in estimating the opportunity for error in many tasks.

When HEP data are scarce, HRA resorts to quantification based on expert judgement or a combination of data and models that evaluate the effects of influences on human performance. The development of techniques of human reliability quantification has always been an area of significant activity. Important techniques are Technique for Human Error Rate Prediction (THERP) [5], Accident Sequence Evaluation Programme (ASEP) [6], Human Cognitive Reliability (HCR) [7], and newer techniques like Cognitive Reliability and Error Analysis Method (CREAM) and A Technique for Human Error Analysis (ATHEANA) [8]. CREAM and AHTEANA are second generation methods and are still not widely used in PSAs.

Human errors are modeled at Fault Tree level for pre initiators and for post initiators they are incorporated at Event Tree level in the Large FT and small ET approach.

1.3.6 Accident Sequence Analysis

Risk analysis is preferred due to its capability for quantification and thereby bringing out dominant risk contributors. In order to quantify, fault tree analysis and event tree analysis techniques are typically used for estimation of likelihood of selected incidents and evaluating frequencies. The process starts with development of event trees for these postulated events identified. Event tree modeling considers the procedure available to prevent the undesirable consequence in case an event happens, considering the role of safety functions in design. This activity comprises of identifying the systems involved in safety functions and probable human actions involved in event mitigation.

For PSA of NPP, major task is quantification of accident sequences that consist of an initiating event, along with mitigating system failures. Accident Sequences are identified using event tree methodology. Event tree analysis generates a large number of event sequences leading to different degrees of fuel damage. In view of the "defense in depth" approach applied in reactor design, an accident situation occurs when an initiating event is coupled with the unavailability of one or more safety systems. Unavailability can be obtained from fault trees for the particular systems. This leads to the evaluation of overall Core Damage Frequency, which in turn gives an indication of associated risk. Here risk is defined as follows:

$$Risk = \sum Accident\,Sequence\,Frequency \times Consequences\,All\,accident\,sequences$$

$$\begin{aligned} &Accident\,Sequence\,Frequency\,(ASF) = Initiating\,Event\,(IE)\,Frequency \\ &\times Unavailability\,of\,one\,or\,more\,Engineered\,Safety\,systems/features\,(ESF) \\ &associated\,with\,the\,IE\,for\,mitigation. \end{aligned}$$

Identifying accident sequence and estimating frequency of accident which include IE frequency estimation and Safety system unavailability estimation. For IE frequency estimation, either generic, plant specific data is used or fault tree analysis technique is used. For estimation of safety system unavailability, fault tree analysis technique is used. For estimating IE frequency, system unavailability and CDF, the following data is required

- Component failure data
- Human error data
- Common Cause Failure data
- Component maintenance data, etc.

In addition, vast amount of design information about reactor process and safety system is also required. The confidence in PSA results depend upon the input data.

In the figure below, on Class IV failure, plant shutdown function is performed by Shutdown system. On success of it, decay heat is removed by shutdown cooling system, which comes on auto. In case of shutdown system failure, fire water system can be injected manually. This is Type C, post initiator human action. Hence it is modeled in event tree, as shown in Fig. 4. Accordingly, consequences are identified based on safety analysis.

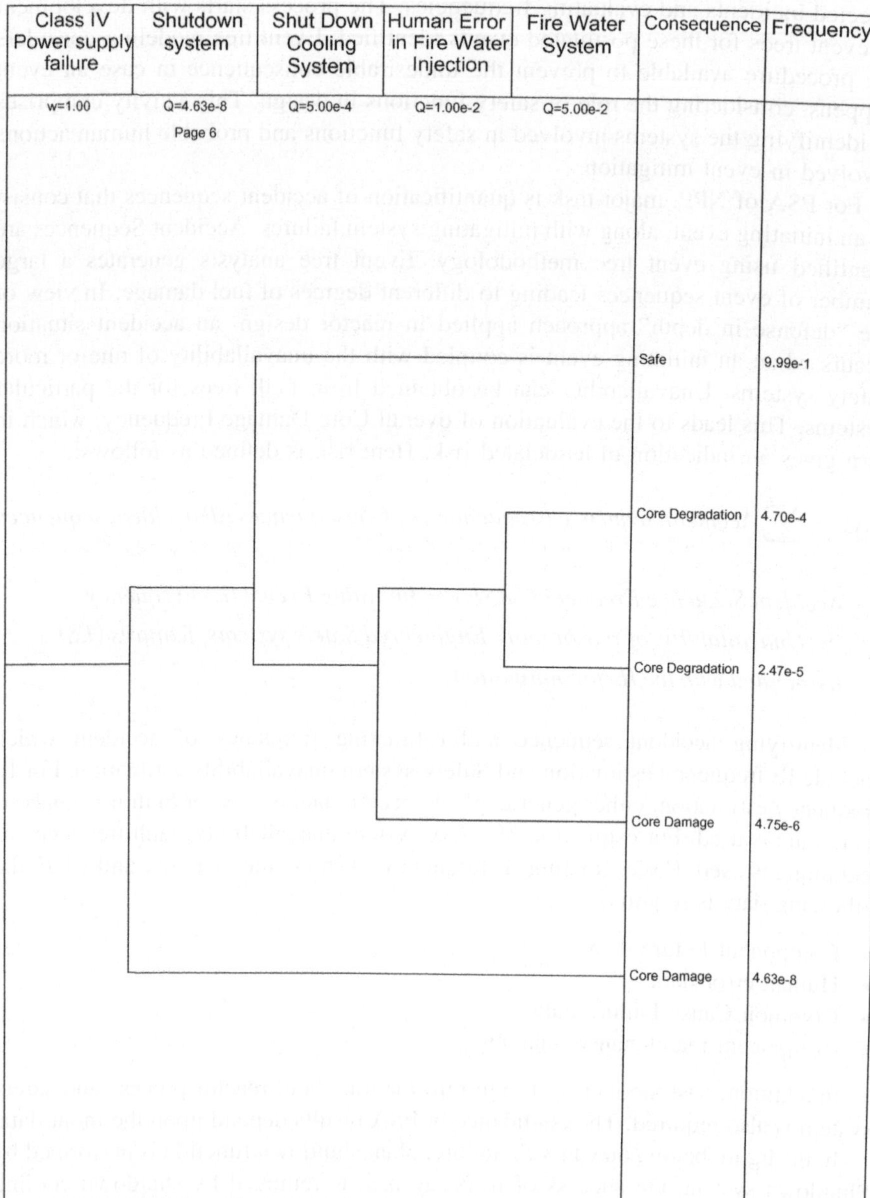

Fig. 4 Event tree for Class IV failure

In order to quantify the event tree, frequency of initiating event and unavailability of safety actions needs to be estimated. Class IV failure frequency is found from site/plant experience data. System availability of shutdown system, Shutdown cooling system and fire water system are found from fault tree analysis. For human error probability estimation, which is a high stress, time dependent action, models such as HCR, THERP, etc. are used.

1.3.7 Uncertainty Analysis

Parameter uncertainties are associated with the fundamental reliability parameters used in the PSA model. Model uncertainties are associated with incomplete understanding of certain processes or phenomena, introducing subjectivity in formulating modelling. Completeness uncertainties are not in itself an uncertainty, but a reflection of scope, limitations etc. and reflects an unanalyzed contribution, it is difficult (if not impossible) to estimate its magnitude.

The Parameter uncertainty, which arises from the quantification of the frequencies and probabilities of the individual Basic Events needs to be addressed in the PSA studies. Importance and Sensitivity Analysis are carried out for all basic events, CCF Groups, various frontline and support systems and for all dominant event sequences for ranking the components to prioritise corrective measures.

1.4 PSA Application in Safety Issues

Risk Informed technology is being extensively applied in various areas as a support tool for routine as well as in critical decision making. Risk Informed techniques are structured to improve the testing and maintenance of highly safety significant components and to reduce unnecessary testing resource allocation towards low-safety significant components. PSA results provide a technical basis for ranking components/systems with respect to their contribution towards Risk. Such ranking can be effectively employed for obtaining solutions for various issues encountered by Regulatory bodies as well as by Nuclear Power Plant Operators. Some important applications of PSA towards Safety Issues are discussed in forthcoming sections.

1.4.1 Probabilistic Precursor Analysis

PSA based analysis of operational events or of precursor analysis answers two basic questions: (a) How could a precursor event have degenerated into an accident with more serious consequences? (b) Is it possible to determine and measure what separates a precursor event from a potential accident with more serious consequences? Thus, the analysis contains a qualitative and a quantitative element. The minimal cutsets provide the Qualitative element to Precursor by identifying the combination of component failures required for the core damage type of accident to happen. ***In addition, PSA helps in providing a Quantitative element to the precursor analysis,*** **viz**., measuring the severity of the event.

Some of the event severity measures considered are—The conditional core damage probability (CCDP). For an initiating event, the CCDP is the conditional probability of core damage given the event. For a condition event, the CCDP is the increase of the core damage probability due to the event (increase in ICDF multiplied by the duration of the condition).

The instantaneous core damage frequency (ICDF), which applies only to condition-type events, is used as an intermediate step in the calculation of the CCDP.

The conditional probability that an operational event would progress to accidents with unacceptable consequences is more widely accepted severity measure. Based on this information, events can be ranked according to their risk significance. Moreover it can be used to prioritize which weaknesses should be handled first, and to assess the level of safety of the plant.

Basically there are two types of precursor events:

(i) The precursor event represents a transient which interrupts normal operation of the plant, thus there is a real effect on plant operation. In this case the event can be easily related to an initiating event of the PSA (if modelled) and the accident scenarios affected by the event are those developing from this initiating event.
When an IE has occurred, the core damage frequency f_{IE} is calculated from the event tree corresponding to the IE, and the CCDP is calculated as

$$CCDP = f_{IE}/\lambda_{IE},$$

with λ_{IE} the frequency of the IE.

(ii) The precursor event involves the unavailability or a degradation of equipment or systems without an immediate impact on plant operation. If the precursor event is related to one (or several) safety functions, a systematic survey of the principal scenarios on which the precursor event impacts needs to be done. First, all the initiators which require the affected safety function(s) need to be identified. In the event scenarios or sequences developing from these initiating events (Event Trees) only the scenarios which entail the precursor event are retained.

Preferably the computerized PSA database is used for this purpose to ensure that the search and identification process is exhaustive.

Precursor events which entail both, an initiating event and equipment or system unavailability, are also possible and both types of impacts need to be included in the subsequent analysis in a combined manner.

The primary result is the ***conditional probability for core damage***, given that the precursor event has happened.

The CCDP value is calculated as:

$$CCDP = T_{\text{event}} \times (CDF_{\text{event}} - CDF_{\text{base}})/A$$

with A the duration of power operation per year, T_{event} the duration of the operational event (h), CDF_{event} the core melt frequency during the event (1/y), and CDF_{base} the base value of core frequency during power operation (1/y).

The main results of precursor investigations are the conditional probabilities for Core Damage from PSA model given that operational event has happened. As a numerical threshold for judging the significance of operational events based on a conservative estimate of the conditional core damage probability a value of 10^{-6} is widely accepted and used. Multiplying the conditional probability of the precursor event j with the frequency, i.e. one event within the observation time in reactor years, and summing up all precursor events within the observation time yields:

$$\lambda = \sum_{j} \frac{CCDP_j}{Observation\,time} \tag{3}$$

where j represents the operational events identified for precursor analysis.

λ is an estimator for the unacceptable consequences, typically either core damage frequency or beyond design basis frequency. The estimator is called ***core damage index***, beyond design basis index, or simply safety or risk index.

The major advantages of this approach are the strong potential for augmenting event analysis which is currently carried out purely on deterministic basis. From the observations it is found that there is slight discrepancy between CCDP values and INES scale associated to an event. Also, the risk index gives an indication about the safety culture followed in plant and can be used as a metric for comparing between various plants.

1.4.2 Probabilistic Vital Area Identification

Identification of vital areas in a facility involves assessing the facility and the locations, whose sabotage can result in undesirable (radiological) consequences. Probabilistic Safety Assessment (PSA) technique can find the component failures leading to core damage (a surrogate for radiological consequence) in a systematic manner, which can be extended to identification of vital areas. The procedure for the generation of location sets (set of locations whose sabotage can lead to possible core damage) and protection sets (set of locations that must be protected to prevent possible core damage). In addition, measures such as ***vulnerability and protectability*** have been introduced, which can be used to rank location sets and protection sets [9].

Vital area identification is helpful, only if, analysis can come up with protection sets. A protection sets represent locations which, when protected, will prevent an adversary from accomplishing sabotage. In order to find the protection set, it is required to construct the location fault tree from the minimal location sets. Following logics are employed to convert location set to location fault tree.

(a) All cut sets are combined using AND gate
(b) Within a cutset, the locations are connected using OR gate.

Protection sets are the minimal cutsets obtained from location fault tree.

Thus the sabotage logic model, (For e.g. the core damage logic model for Nuclear Power Plants, large toxic release model for chemical plants, etc.) is then analyzed to identify the target sets or vital area sets (combinations of areas the adversary must visit to cause radiological sabotage) and the candidate protection sets (combinations of areas that must be protected against adversary access to prevent radiological sabotage).

1.5 Summary

PSA presents an Integrated Picture of the Safety of NPP which encompasses Design, Operational Practices, Component Reliability, Dependencies and Human Reliability. It also helps in identifying predominant contributors to possible Severe Core Damage in terms of Component Failures and Human actions. Also, it identifies any weak-links or imbalances affecting the Safety of the plant with reference to Components/Human actions, which could be improved. In short, PSA of nuclear power plant is an effective technique to minimise accident frequency and consequence by better design technique in power plant system and containment, to develop operational maintenance and accident management procedure and emergency preparedness procedure.

2 Probabilistic Safety Assessment of Non-reactor Nuclear Facilities

2.1 Introduction

The operation of Non Reactor Nuclear Facility (NRNF) differs significantly from Nuclear Power Plants (NPPs). These facilities employ a greater diversity of technologies and processes. Fissile materials, wastes, radiation sources are handled, processed, treated and stored throughout the nuclear installations, in contrast to reactors, where the bulk of the nuclear materials are located in the reactor core and fuel storage areas. Greater reliance is put on the operators, not only to run the facilities during normal operation, but also to respond to fault and accident conditions. Chemical processes, if not managed properly, may lead to inadvertent release of toxic chemicals or radioactive substances. Similar to NPP, facilities also needs to keep the radiation exposure from nuclear facilities to members of the public and workers as low as reasonably achievable (ALARA) during normal operational states (certainly below the limits set by the regulatory bodies) and in the event of accident. Another type of NRNF is Accelerator facilities, which have emerged as powerful tools for research in physics, chemical sciences, material sciences, etc. They are associated with hazards from radiation sources (bremsstrahlung radiation and neutrons), energy sources, hazardous materials etc. In order to adhere to the safety goal, carrying out safety analysis has become almost mandatory for all nuclear facilities.

2.2 *Steps in PSA for Nuclear Facility*

The major steps of PSA for nuclear facility are mentioned below:

- System description
- Hazard identification
- Incident enumeration
- Accident Sequence Quantification
- Risk estimation.

These steps are explained in forthcoming subsections.

2.2.1 System Description

System description involves collection of information on design and operation of plant, i.e., compilation of all technical and human information needed for the analysis (including reliability data). Information was assembled using sources such as safety analysis reports, design manuals, operating practices etc.

2.2.2 Hazard Identification

This is a critical step in risk analysis, a hazard omitted at this stage is a hazard which is not analysed. Typically nuclear facilities, such as accelerator houses potential hazards, like

- Radiation
- Heat load
- Electrical
- Ozone and
- Fire.

For electron accelerator, hazards associated with ionizing radiation arise during several aspects of operation: loss of electrons from the beam at various stages of acceleration; loss of electrons from the beam circulating in the storage ring; and synchrotron radiation emanating from bending magnets and insertion devices located around the storage ring.

2.2.3 Incident Enumeration

Preparation of list of postulated events is a very important task in risk analysis, which needs completeness in identification and tabulation of all incidents without any relevance to their importance or to the initiating event. The approach used for preparation of postulated events is based on:

- Precursor review
- Engineering evaluation
- Use of operational experience.

Since there are no standards/documents listing postulated events from an accelerator and they can vary with design, these approaches may not be conclusive. However, elaborate discussions needs to be undertaken to consider events to the extent possible.

For accelerator, typically initiating events from beam lines are analysed which leads to undesirable consequences such as high radiation due to inadvertent beam dump, Personal exposure in experimental hutch due to failure in safety barriers, etc. [10]. Various events such as 'Loss of target cooling', 'Vacuum degradation due to sputter ion pump failure' can result in beam dump. Similarly, Personal exposure can happen due to Inadvertent entry during experiment, Spurious opening of safety shutter, etc.

2.2.4 Accident Sequence Quantification

For all Postulated Initiating Events, event progression is modelled using event trees. In case of accelerator, the safety function can result either in tripping beam or closing the safety shutters, hence consequence is analysed in terms of dose received from beam dump or personnel exposure. For all PIEs, consequence estimation is carried out to determine the potential damage from radiation (dose assessment to public, accidental exposure, etc.).

A typical event tree for "Spurious opening of Safety Shutter" (SOSS) is given below. This event can happen in experimental hutch, while preparation of experiment is being carried out. Normally, the safety shutter will be in closed condition. Postulated initiating event considered in this case is 'Spurious opening of Safety Shutter', which can have undesirable consequence, if beam is ON and Area is occupied. Frequency of initiating event can be found using fault tree approach enumerating all the causes and their failure rates. Similarly, Probability of conditions such as 'beam is ON' and 'Area is occupied' can be found from plant experience.

From the event tree shown in Fig. 5, frequency for undesirable consequence such as Personnel exposure can be found. Dose assessment, during this event, are considered in consequence estimation.

2.2.5 Risk Estimation

Once likelihood or frequency of incident is estimated along with the specified consequences, it is required to combine them in a meaningful fashion to provide a measure of risk. Many measures of risk have been proposed and are in use, each providing a different view of a particular situation or aspect. Among these measures, perhaps most commonly used ones are those of individual risk and societal risk.

For accelerator like facility, it is appropriate to devise an F-N curve with consequences expressed in released activity in terms of curies (or Becquerel's) of various radionuclides, health effects like early fatalities and latent cancers, and radiation doses (rems or sieverts). NUREG 1860 [11], has proposed an F-C curve in terms of radiation doses, is popularly used for PSA of NRNF, as shown in Fig. 6.

F-C curve, as per NUREG 1860, is applied for communicating the risk from the events identified from the accelerator. Figure 7 shows the typical F-C curve obtained for electron beam accelerator.

It can be found that 'SOSS' falls in the Acceptable region, which ensures the safe regime of Accelerator Fig. 8.

3 Probabilistic Safety Assessment of Non–nuclear Facilities

3.1 Introduction

PSA in non-nuclear facility such as chemical/industrial facility is commonly referred to as Quantitative Risk Assessment (QRA). The QRA methodology has evolved since the early 1980s from its roots in the nuclear, aerospace and electronics industries. The most extensive use of probabilistic risk analysis has been in the nuclear industry. The QRA has since then moved to other sectors, including the Process Industry. The QRA can be defined as the analytical process with which the hazards are identified and the probability of a damage occurring following occurrence of any of the identified hazards is evaluated. From this analysis, the calculated risk is then assessed with reference to suitable tolerability criteria. QRA has a number of specific features:

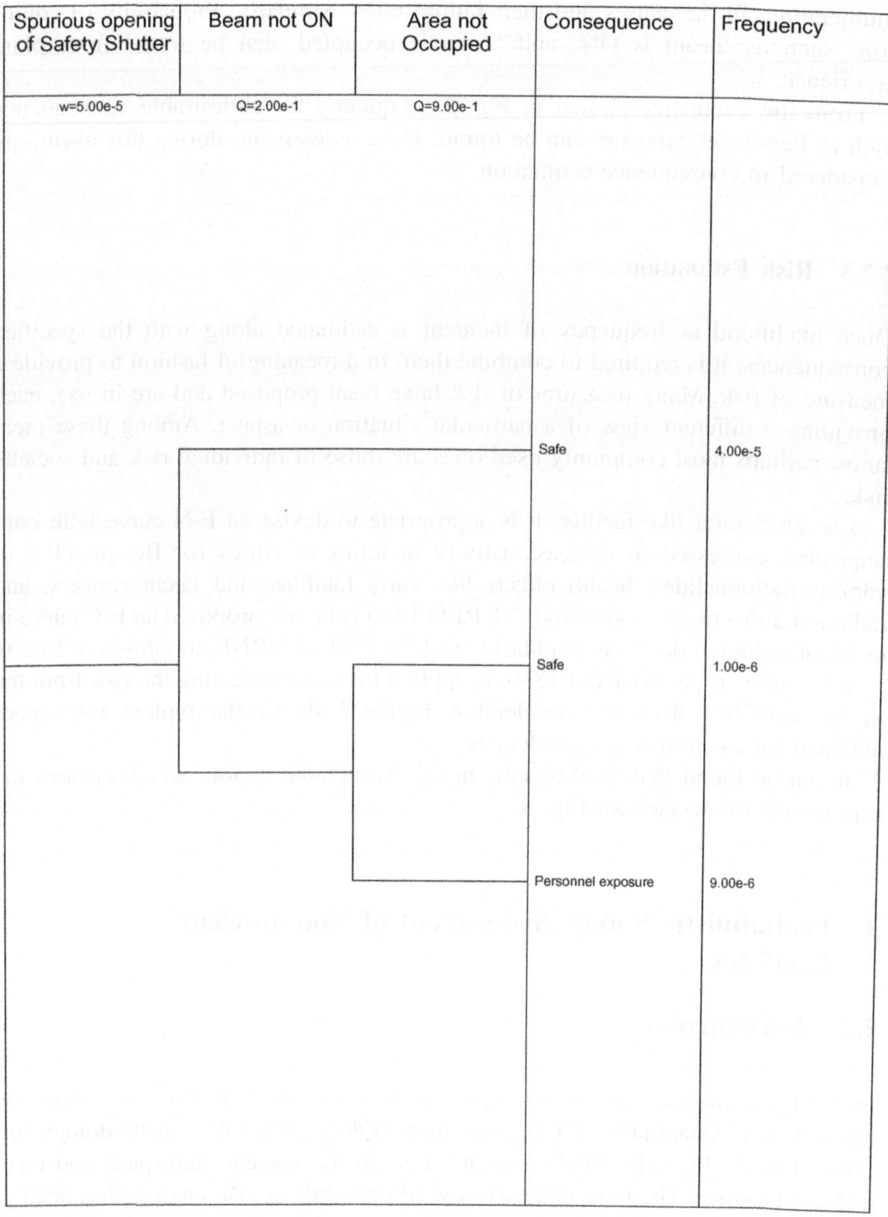

Fig. 5 Event tree for 'spurious opening of safety shutter' (SOSS)

- Chemical reactions may be involved
- Processes are generally not standardized
- Many different chemicals are used
- Material properties may be subject to greater uncertainty

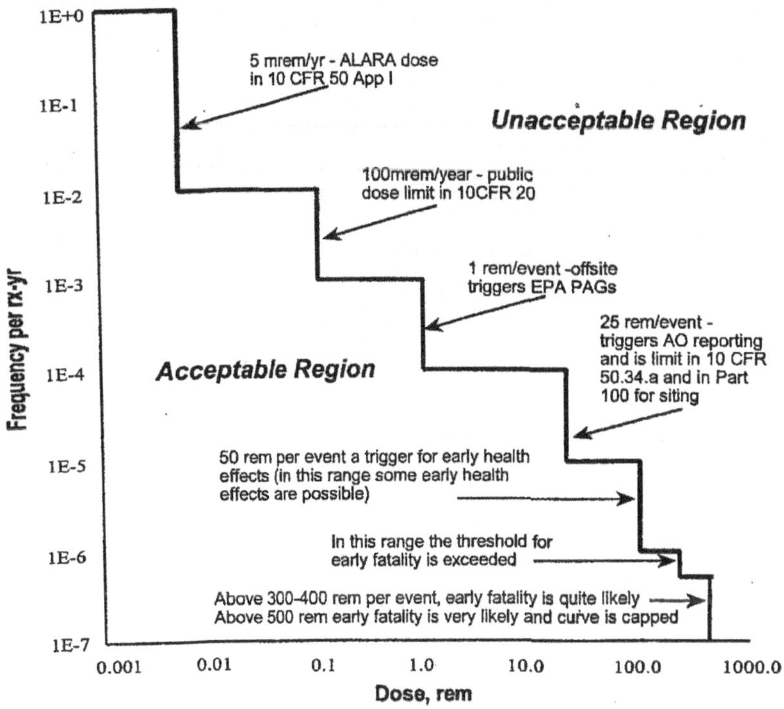

Fig. 6 F-C curve from NUREG 1860

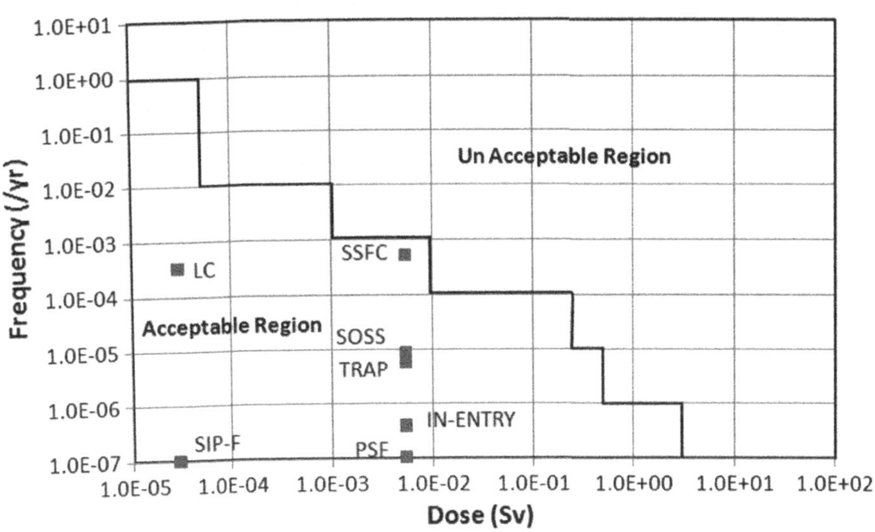

Fig. 7 Frequency versus dose curve for experimental hutch

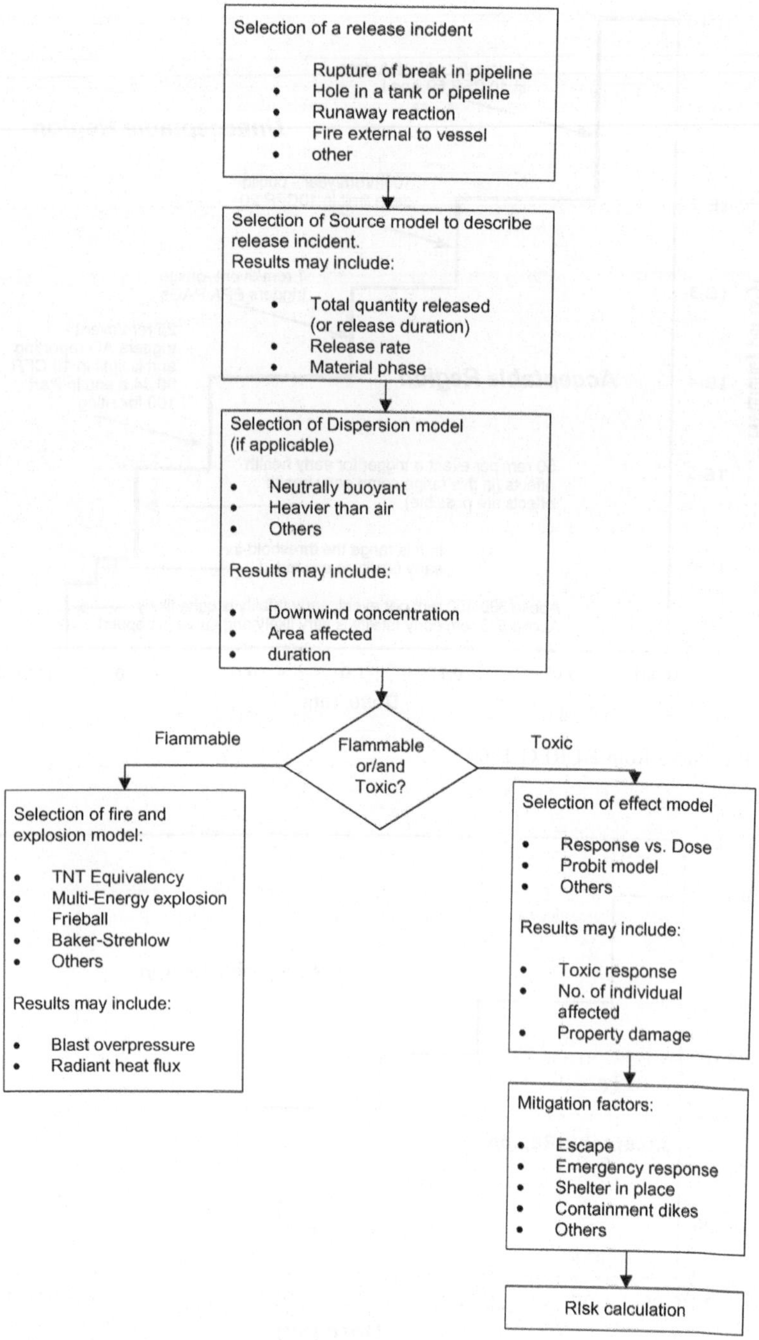

Fig. 8 Consequence analysis from volatile hazardous material release

- Parameters, such as plant type, plant age, location of surrounding population
- Degree of automation and equipment type, vary widely
- Multiple impacts, such as fire, explosion, toxicity, and environmental contamination.

Acute, rather than chronic, hazards are the principal concern of QRA. This places the emphasis on rare but potentially catastrophic events. The risk (*a measure of safety*) in general for industrial facility is defined as:

Risk = Probability of an accident ∗ Consequences in terms of exposure to toxic material

In this session, QRA will be referred as PSA.

3.2 *Steps in PSA for Chemical Facility*

The major steps of PSA for chemical facility are mentioned below:

- Hazard identification
- Incident frequency estimation
- Consequence estimation
- Risk estimation.

These steps are explained in forthcoming subsections.

3.2.1 Hazard Identification

The correct identification of Hazards is the first essential step for a Risk Analysis. Various techniques can be adopted depending on the level of information available. One of the most widely used techniques for Process plant hazard identification is the HAZOP analysis. A HAZOP study is a structured analysis of a system, process or operation, carried out by a multi-disciplinary team. The team proceeds on a line-by-line or stage-by-stage examination of a firm design for the process or operation. The team concentrates on those deviations that could lead to potential hazards to safety, health or the environment. This is done by using a set of guidewords in combination with the system parameters to seek meaningful deviations' from the design intention. A meaningful deviation is one that is physically possible—for example, no flow, high pressure or reverse reaction.

3.2.2 Incident Enumeration

HAZOP forms starting point of incident enumeration. In addition to the identification of hazards, it is common practice for the team to search for potential operating problems. These may concern security, human factors, quality, financial loss or design defects. Where causes of a deviation are found, the team evaluates the consequences using experience and judgment. If the existing safeguards are deemed inadequate, the team recommends an action for change or calls for further investigation of the problem. In summary, approaches used for preparation of postulated events are based on:

(i) HAZOP review
(ii) Engineering evaluation
(iii) Use of operational experience.

Incident enumeration is closely linked with hazard identification and has to be dealt in total. For example, chlorine gas is a 'hazard' while its unplanned emission through a faulty valve is an 'incident'. Once postulated initiating event list is finalized, event tree modeling is performed to understand procedures available to prevent the undesirable consequence in case an event happens. In chemical facility, the major action is focused on preventing toxic exposure. This activity comprises of identifying the systems involved in safety functions and probable human actions involved in event mitigation.

3.2.3 Consequence Estimation

Consequence estimation is the methodology used to determine the potential damage or harm from specific incidents and is closely linked with hazard under consideration. The first step of the Consequence analysis is the calculation of the characteristics of the release of the material: the Source Model. Accidents usually begin with the loss of containment of material from the process. The material has hazardous properties, which might include toxic properties and energy content. Typical incidents might include the rupture or break of a pipeline, a hole in a tank or pipe, runaway reaction, fire external to the vessel, etc. Once the incident is known, source models are selected to describe how materials are discharged from the process. The source model provides a description of the rate of discharge, the total quantity discharged (or total time of discharge), and the state of the discharge—solid, liquid, vapor, or a combination. A dispersion model is subsequently used to describe how the material is transported downwind and dispersed to some concentration levels. For flammable releases, fire and explosion models convert the source model information on the release into energy hazard potentials such as thermal radiation and explosion overpressures.

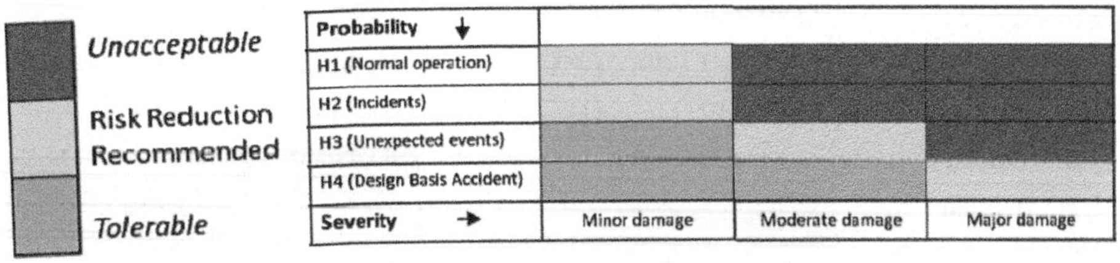

Fig. 9 Qualitative risk matrix

3.2.4 Risk Estimation

Categorization of risk involves assessment of risk taking into consideration the probability and severity of consequence. There are qualitative and quantitative methods available for risk categorization.

Qualitative risk ranking schemes

In typical qualitative risk assessment approach risk matrix is developed as shown in Fig. 9. Three levels of risk are defined; "Unacceptable", "Risk Reduction Recommended" and "Tolerable". Unacceptable risks require risk reducing measures in order for the suggested design to be accepted. Risks Reduction Recommended require a demonstration that the suggested barriers are as effective as reasonably can be achieved considering alternatives and additions. Tolerable risks require no additional barriers, but need to be monitored, for example when design changes, to be kept at a low level.

Quantitative risk ranking schemes

The risk matrix is depiction of the frequency and consequences. Quantitative treatment can also be extended to risk matrix. Typical scheme of risk indexing is shown in Fig. 10. The probability values can be high ($<10^{-1}$ per year), medium (10^{-2} to 10^{-1}), low (10^{-4} to 10^{-2}) or extremely low (10^{-6} to 10^{-4}). The consequences are categorized as high to extremely low based on whether the incident has serious impact on off-site and on-site.

3.3 Risk Based Inspection for Chemical Facility

Last decade saw a trend where life management programmes are globally moving from prescriptive/time-based towards risk- based decision making. Risk analysis finds use/application in decision making, for operation, maintenance and regulatory activities. This methodology has been applied in planning maintenance activities such as testing time, repair time, inspection interval etc. When this is applied to inspection planning, it is termed as Risk based inspection. Risk Based Inspection

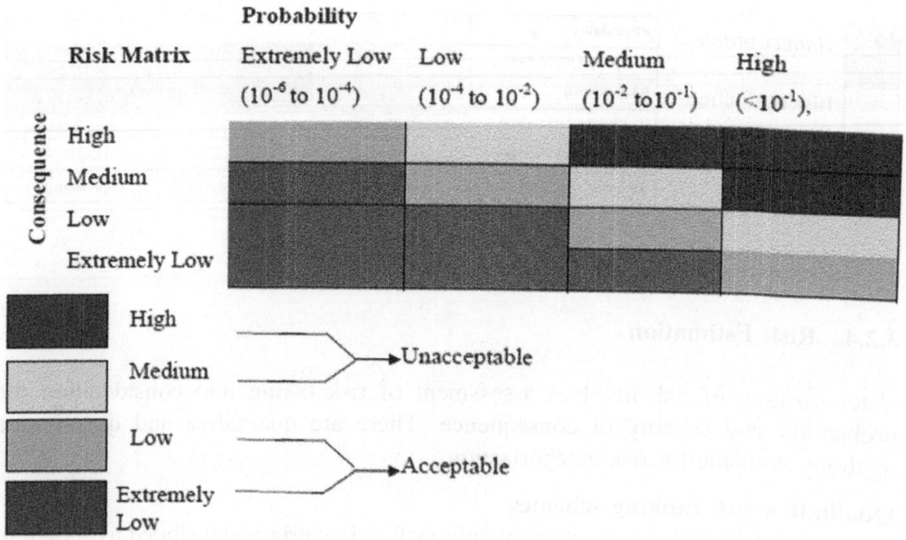

Fig. 10 Risk indexing

(RBI) is a method for using risk as a basis for prioritizing and managing the efforts in an inspection program. Risk Based Inspection focuses the utilization of risk quantification in formulating an In-Service Inspection (ISI) plan thereby emphasizing the importance of surveillance and maintenance activities on plant risk. RBI would be able to establish an effective structural integrity management programme, which reduces plant down time, industry and regulatory burdens, and continue to maintain plant safety. For applying the frame work of Risk Based Inspection, it is required to estimate ***likelihood or probability of failure (PoF)*** of components in process plant and their ***consequence (CoF)***, in terms of damage to the equipment and impact of toxic release to public.

Probability of Failure (PoF) of a component can be estimated from operating experience data/stress-strength models/expert judgment. Service Data Analysis based on operating experience is one of the popularly employed method used for this purpose. Data bases such as OREDA 2002 [12] are the result of various collaborative efforts taken towards methodical collection of operating experience information, which can be termed as generic data base. Consequence of Failure (CoF) estimation, looks into damage as well as toxic impact analysis. Consequence analysis can be performed either in a qualitative manner or detailed quantitative analysis. API 581, Base Resource Document for Risk based Inspection [13], may be used, which provides factors for damage and toxic consequence. After determining the PoF and CoF category, it needs to be applied to Risk matrix to establish the inspection category [14].

4 Concluding Remarks

While doing Probabilistic Safety Assessment, major challenge is in ensuring the correctness and completeness of initiating events possible from the industry. Probability of these events are depend on industry specific, since it needs to consider the environment, test and maintenance practices, safety culture prevailing etc. However, while conducting such analysis at the design, falls back on available information from generic source. It is recommended to have a proper data collection process to improve the predicted failure rates to be made realistic to the extent possible. The consequence analysis finally gives us the individual and collective exposures to hazardous material, number of health effects (both fatal and non-fatal), and costs of disruption. These consequences can be mitigated by countermeasures and the effect of countermeasures on the exposure also accounted in the consequence analysis codes. The analysis tool will help decision makers to estimate numbers of people and areas affected by emergency countermeasures. The probabilistic approach will give statistical results like minimum, maximum, xth percentile number of people exposed to a threshold dose/hazards or areas to be evacuated. Hence Probabilistic Safety Assessment is an essential tool for ensuring the safety of any type of facility.

References

1. Fullwood R, Hall RE (1988) Probabilistic risk assessment in the nuclear power plant industry: fundamental application. Pergamon Press, Brookhaven National Laboratory, New York, USA
2. Reactor Safety Study, WASH-1400, NUREG-75/014, United States Nuclear Regulatory Commission (1975)
3. NUREG-2300, PSA Procedures Guide (1983)
4. EPRI NP-5613, Procedures for treating common cause failures in safety and reliability studies, vol. 1, NUREG/CR-4780 (1988)
5. Swain D, Guttmann HE (1983) Handbook of human reliability analysis with emphasis on nuclear power plant applications. NUREG/CR-1278. US Nuclear Regulatory Commission, Washington DC
6. Swain AD (1987) Accident sequence evaluation program human reliability analysis procedure. NUREG/CR-4772. US Nuclear Regulatory Commission, Washington DC
7. Case study on the use of PSA methods: human reliability analysis. IAEA-TECDOC-592 (1991)
8. Forester J et al (2007) ATHEANA User's Guide. NUREG-1880. US Nuclear Regulatory Commission, Washington DC
9. Vinod G, Sanyasi Rao VVS, Ghosh AK, Kushwaha HS (2012) PSA based vulnerability and protectability analysis for NPPs. Ann Nucl Energy 50:232–237
10. Sanyasi Rao VVS, Vinod G, Sarkar PK, Vaze KK (2012) Application of PSA techniques to synchrotron radiation sources. Indian J Pure Appl Phys 50:776–781
11. NUREG 1860 Feasibility studies for a risk-informed and performance-based regulatory structure for future plant licensing
12. OREDA (2002) Offshore reliability data base

13. API (1998) API Publication 581, Base Resource Documentation—Risk-Based Inspection. American Petroleum Institute
14. Vinod G, Sharma PK, Santosh TV, Hari Prasad M, Vaze KK (2014) New approach for risk based inspection of H2S based process plants. Ann Nucl Energy 66:13–19

Dr. Gopika Vinod joined Reactor Safety Division of Bhabha Atomic Research Centre as a Scientific Officer from 37th batch of Training School after completing her graduation in Computer Engineering. She received her doctoral degree in Reliability Engineering from Indian Institute of Technology, Bombay and also been Post-Doctoral Fellow at Steinbies Advanced Risk Technologies, Germany. She is a recipient of DAE Young Engineer Award 2007. She has been to visiting scientist at Brookhaven National Laboratory. Currently she is heading the Probabilistic Safety Section of Reactor Safety Division of BARC. She also holds the position of Faculty with Homi Bhabha National Institute. She has been actively involved in reliability, safety, and risk analysis of Indian nuclear power plants. She has worked on the development of reliability-based operator support systems such as risk monitor, symptom-based diagnostic system, for Indian nuclear power plants. Her other areas of research activities include risk-informed in-service inspection, reliability of computer-based systems, dynamic reliability analysis, etc.

Passive System Reliability Assessment and Its Integration Into PSA

R. B. Solanki

1 Introduction

Passive systems are defined as 'Either a system which is composed entirely of passive components and structures or a system which uses active components in a very limited way to initiate subsequent passive operation' [1]. Most of the traditional reactor safety systems are 'active' in the sense that they involve electrical or mechanical operation when demanded to actuate. Some engineered systems operate passively, e.g. pressure relief valves, check valves etc. Passive, as its name suggests that something will happen on its own, depending on laws of physics, such as natural convection, gravity driven flows or pressure differentials. Inherent or full passive safety design depends only on physical phenomena such as conduction, convection, gravity or pressure difference, not on functioning of engineered active components, which requires external motive power. All reactors have some elements of inherent safety, but in some of the recent designs, the passive safety systems/features are being introduced as a back up to the active systems to enhance the safety further.

The objective of introducing the Passive systems in Nuclear Power Plants are simplify the design of the safety systems and enhancing the of plant safety by relying on driving forces, which do not require external motive power for the system function. The 'Active' safety systems failures are observed mainly due to two of the attributes such as 'Operator error' and 'malfunction of active component'. These attributes do not contribute in failure of 'Passive' safety systems and hence, Passive safety systems are considered to be 'more reliable' than active systems. This claim is not proven yet due to inadequate operating experience gathered so far. Figure 1 shows a schematic of a typical natural convection based passive safety system.

R. B. Solanki (✉)
Atomic Energy Regulatory Board (AERB), Mumbai, India
e-mail: rbrajsolanki@aerb.gov.in

D. R. Karanki et al. (eds.), *Advances in RAMS Engineering*,
Springer Series in Reliability Engineering,
https://doi.org/10.1007/978-3-030-36518-9_16

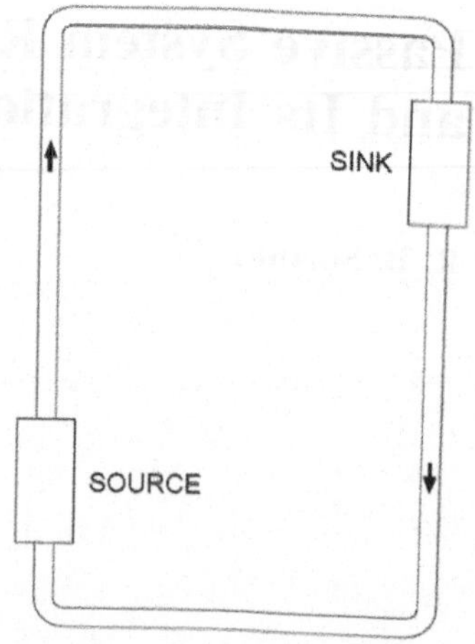

Fig. 1 Schematic of a typical natural convection based passive safety system

The primary function of such system is to transfer heat from a source to a sink. In general, a heat source, a heat sink and the pipes connecting them form the essential hardware of a typical natural circulation system. The pipes are connected to the source and sink in such a way that it forms a continuous circulation path. When the flow path is filled with a working fluid, a natural circulation system is ready where fluid circulation can set in automatically following the activation of the heat source under the influence of a body force field like gravity. With both the source and sink conditions maintained constant, a steady circulation is expected to be achieved, which can continue indefinitely if, the integrity of the closed loop is maintained. The fluid circulation is the result of buoyancy forces, which in turn is the result of the density differences thermally induced by the transport of heat from the source to the sink. Usually, the heat sink is located above the heat source to promote and sustain natural circulation. Such loops in which the fluid circulation is caused by the thermally induced buoyancy force are also known as natural circulation loops, thermo-syphon loops or natural convection loops [2].

The main advantage is that the heat transfer is achieved without the aid of any fluid moving machinery. The absence of moving/rotating parts to generate the motive force for flow makes it less prone to failures reducing the maintenance and operating costs.

2 Reliability Assessment for Passive Safety Systems

The driving forces in passive safety systems are relatively weaker than those of the active safety systems, the counter-forces such as flow resistance, friction, pressure drops etc. becomes of comparable magnitude of that of driving forces, which cannot be ignored as normally done for active systems. The associated uncertainties are also required to be considered. The specific plant conditions and configurations that

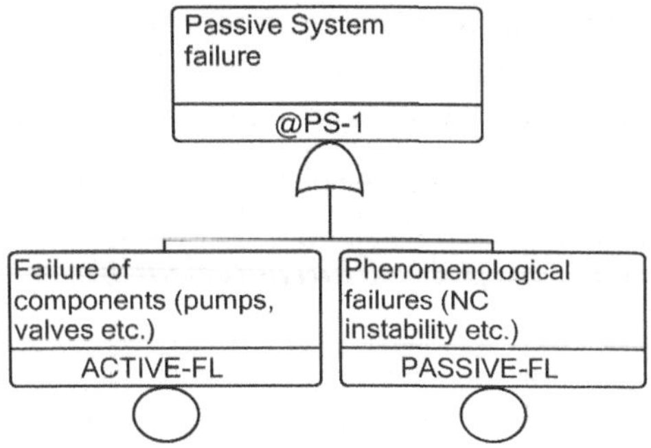

Fig. 2 Fault tree for a typical passive safety system

exists at the time a system is invoked to perform its intended safety function governs the magnitude of such natural driving forces. These uncertainties affect the Thermal-Hydraulic performances of passive safety systems. For this reason, careful design and analysis methods must be employed to assure that the safety systems perform their intended functions satisfactorily.

The Reliability Analysis is carried out using a system Fault Tree. In a case of passive systems, the fault tree would be very simple, consisting of several basic events, representing failure of physical phenomena (e.g. natural circulation) and failure of activating valve or other means of initial system activation, connected together by "OR" gate as given in Fig. 2.

The contribution from the 'Active Component' part of the fault tree of the passive system reliability can be evaluated in a classical fault tree method using the available component failure database. However, different methods are required to evaluate the contributions from the 'phenomenological' failures. The research is ongoing worldwide on the reliability assessment of the passive safety systems and their integration into PSA, however consensus is not yet reached. Some of the widely used approaches are presented in next section of the chapter.

3 Reliability Evaluation of Passive Safety Systems (REPAS) Method

The Reliability Evaluation of Passive safety systems (REPAS) Method was developed in late 1990s and it is one of the initial efforts that were made for assessment of passive system reliability. The method was cooperatively by ENEA, the University of Pisa, the Polytechnic of Milan and the University of Rome, that was later incorporated in the EU (European Union) RMPS (Reliability Methods for Passive Systems) project. The REPAS methodology characterizes the performance of passive systems in an analytical way using the uncertainty propagation approach. This methodology is based on the evaluation of a failure probability of a system to carry out the desired function from the epistemic uncertainties of those physical and geometric parameters which can cause a failure of the system [3]. Major steps involved in the REPAS method is described here, for detail explanation, reader may refer a technical report prepared by NEA [4].

3.1 Step 1—Characterization of Passive Safety System

The mission of the system and relevant phenomenology involved in the system should be identified. The design parameters (like pressure, level, temperature etc.) and critical parameters (like presence of non-condensable gases, heat losses in piping etc.), which govern the system should be identified. The full characterization of safety systems requires large number of parameters. Experts' judgment or other statistical techniques such as Analytical Hierarchy Process (AHP) could be used to identify the optimum system parameters.

3.2 Step 2—Development of Analytical Model

The experimental database for the operation of the passive safety systems is sparse. Hence, for performance evaluation of the system, one has to rely upon the numerical modeling through Best-estimate computer codes. The system analysis to be carried out using validated computer codes and appropriate analytical model that represent the system in the simulation. The analytical model developed should be validated through experimental results or through code-to-code comparison.

3.3 Step 3—Assigning Probabilistic Distributions

The nominal values and the range for the design and critical parameters must be identified as a part of system characterization. The probability distributions for the occurrence of that value of the parameters are assigned in this task. This can be done through experts' judgment, possibly taking into account available data on operation and maintenance of the T-H passive systems. As a general rule, a central pivot is to be identified, and then the range is to be extended to higher and lower values. The pivot value represents the nominal condition for the parameter. Once the discrete ranges have been set up, discrete probability distributions may be applied to represent the probabilities of occurrence of the values. The higher probability of occurrence corresponds to the nominal value for the parameter. Then lower probabilities may be assigned to the other values, as much low the probability as much wide the distance from the nominal value [5].

3.4 Step 4—Performance Evaluation of Passive System

Once the ranges and associated probabilities are fixed, a stochastic selection of a limited number of system configurations is performed through Monte-Carlo procedure. Through stochastic selection of the system configurations, a limited set of accident and operational transients are analysed with Best-estimate computer code with different initial values for the design and critical parameters. The data set of the system configurations and corresponding output parameter (i.e. performance indicator) represents the general physical behavior of the passive safety system.

3.5 Step 5—Development of Response Surface

Having obtained the representative data of the system behavior, a mathematical relationship between the input parameters and the output parameter is established using statistical appropriate methods. This mathematical relationship is known as 'Response Surface'. This response surface is used to predict the parameter of interest for 'new' set of system configurations for large number of system configurations. The accuracy of the prediction depends on the accuracy with which these coefficients are predicted, which in turn depends on the adequacy of the sample size of input parameters and the capability of the analytical model.

3.6 Step 6—Assessment of Passive Safety System Reliability

The system reliability is estimated in terms of failure probability using direct Monte-Carlo method using the data obtained in step no 4 and 5. In order to evaluate the system performance, 'Failure Criteria' needs to be derived from the knowledge of the mission of the system under various postulated accident scenarios. The Failure Criteria are generally defined considering the fuel cooling and thereby maintaining its integrity. The specific fuel related safety limits could only be established with detailed plant-specific design data. In absence of such details, more generic criteria may also be adopted for preliminary assessment during the conceptual design stage.

The system failure probability is estimated with the following formula:

$$System\,Failure\,Probability = \frac{No.\,of\,Times\,the\,Failure\,Criteria\,is\,Reached}{Total\,No.\,of\,Simulations}$$

REPAS methodology is based on estimating the reliability through parametric uncertainty analysis. To address the issue of large computation time required for performance evaluation using computer codes, it relies on the Wilk's theory of finding out the optimum number of computer code runs for a given confidence level. Using the results of these limited code runs, a Response Surface (mathematical model) is generated, which is used to predict the parameter of interest for different system configurations. The Response Surface is derived through multivariate linear regression of the limited computer code results; the failure probability does not really change substantially by performing large number of simulations through identified Response Surface to predict system performance.

4 Assessment of Passive System ReliAbility (APSRA) Method

Assessment of Passive System ReliAbility (APSRA) Method [6] was developed to overcome the issues associated with assigning the probability density functions to the input parameters. The APSRA method is not based on uncertainty propagation approach followed in REPAS Method. Initial few steps of APSRA method is similar to REPAS method (Step 1 and Step 2). Other major steps of APSRA is described in this section.

4.1 Generation of Failure Surface

The deviation of the critical parameters from their nominal value is considered and system behavior is predicted using a Best estimate code. The system behavior in terms of success/failure is represented in a parametric space and a failure surface demarcating the failure and the success regions is generated.

4.2 Root Diagnosis

After establishing the domain of failure, the next task is to find out the cause of deviation of key parameters which eventually result in the failure of the system. Through root diagnosis approach, the deviations of the critical parameters are identified and the mechanical components in the system are identified, the failure of which could be attributed for such deviations during the mission time.

4.3 *Reliability Assessment of Passive System*

The system unavailability is estimated using the classical fault tree approach. The component failure data are obtained from the generic failure database for active components. The phenomenological failures are modeled as 'Basic Event' corresponding to the critical parameters attaining values such that the system does not fulfil its intended function. The failure probability for such basic events are subjectively assumed. The system reliability is estimated in terms of the system failure probability.

5 Use of Artificial Neural Networks (ANN) in Passive System Reliability Assessment

The Artificial Neural Networks (ANN) have found many applications in a wide range of disciplines due to its ability to reproduce and model nonlinear processes efficiently and effectively [7]. The ANN contains an input layer to accept the input, an output layer to calculate the final results with several numbers of hidden/ intermediate layers in between to process the information. In this structure, each neuron output is connected to every other neuron in the subsequent layers connected in cascade with no connection between neurons in the same layer [8]. Figure 3 shows a Typical ANN with multi-layered perceptron structure.

Input layer represents the raw information that is fed into the ANN network. This part of network does not change its values. Every single input to the network is duplicated and send down to the nodes in hidden layer. The hidden Layer accepts

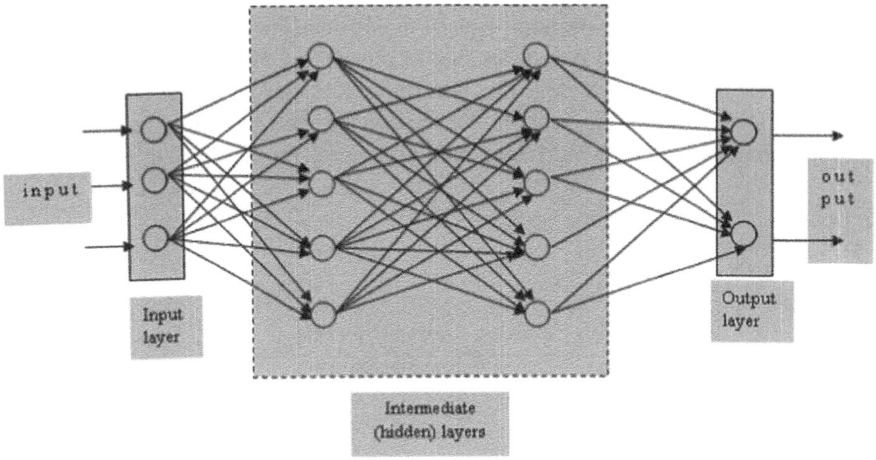

Fig. 3 A typical artificial neural network structure

data from the input layer. It uses input values and modifies them using some weight values, these new values are than send to the output layer but they also undergo modification through weights from connection between hidden layers and output layer. Output layer process information received from the hidden layers and produces an output. This output is than further processed by activation function.

The ANNs are adjusted, based on a comparison of the predicted output and the 'desired' target, until the network output matches the desired target. Typically, many such input/target pairs are needed to train a network. This process is generally known as 'Supervised Learning'. Figure 4 illustrates the 'Supervised' learning process of ANN.

In Supervised Learning process, a 'Teacher' is assumed to be present during the learning process, when a comparison is made between networks' computed output and correct expected output, to determine the error. This error can then be used to change the network parameters, which results in improvement in the performance of ANN in predicting the output accurately.

In this way, each input is related to the output parameter through a relationship based on the weights adjusted through training. The ANN develops non-linear Response Surface without relying on the subjective assumptions about the relationship between input parameters and output parameter. This would eliminate the need for simplified or subjective assumptions about the relationship between the input parameters and the output parameter for circumstances in which underlying relationship is 'unknown'.

The use of simplified assumptions in development of 'Response Surface' are one of the shortcomings identified in the existing REPAS methodology for passive system reliability assessment. REPAS assumes linear relationship between input parameters and output parameter, which may not be true in all cases. The subjective assignment of probability distributions for characterizing the input parameter uncertainty was addressed by APSRA methodology. However, APSRA methodology estimates the passive system reliability using the classical fault tree approach, in which the subjective assumption is used to consider the probability of attaining a specific value of the input parameter due to physical phenomena.

The ANN based method develops non-linear Response Surface without relying on the subjective assumptions about the relationship between input parameters and output parameter. This would eliminate the need for simplified or subjective

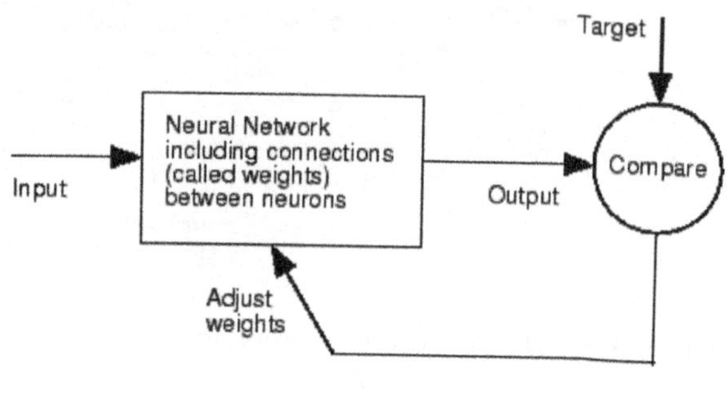

Fig. 4 Supervised learning process of ANN

assumptions about the relationship between the input parameters and the output parameter.

The ANN performance depends on several considerations such as design of network, adequacy of training data etc. The design considerations include the number of input and output nodes, number of hidden layers and number of nodes in each of these layers. The training considerations include determination of input and output parameters, selection of size of training data sets, initialization of network weights, selection of learning rate and stopping criteria for training. According to the nature of the problem, various working structures with or without feedback loops can be used. The most suitable structure for linear/non-linear modelling is multi-layered perceptron structure.

The ANN model can be trained to generalize well within the range of inputs for which they have been trained. However, they do not have the ability to accurately extrapolate beyond this range, so it is important that the training data span the full range of the input space. Further, in multivariate models (more than one independent variable), the variation of output parameter due to 'combined' effect of more than two parameters is also important to be included in the training dataset as the ANN does not have the ability to comprehend the interdependency among the input parameters and their combined effect on the desired output.

In view of these, if the number of parameters are more, the design space for ANN would be very large. Hence, some design space optimization technique should be used as a pre-requisite to use of ANN model for passive system reliability assessment so that ANN performance in prediction improves without using a large number of T-H code runs to generate the training data.

6 Integration of Passive System Reliability Into Probabilistic Safety Analysis (PSA)

The integration of passive systems in the Probabilistic Safety Assessment (PSA) models is a difficult and challenging task. No commonly accepted practices exist so far on how to estimate reliability of passive systems. The main challenge arises from the nature of passive systems for which the main operating principles are based not on active components, but on physical phenomena [9]. There are number of different ways how to integrate passive system reliability model into the whole plant PSA model. It could be done directly in the event tree of relevant accident sequence as a single basic event, or a separate fault tree could be developed. The most commonly used approach is to develop a fault tree for each safety system. The system fault tree in a case of passive systems would be very simple, consisting of several basic events, representing failure of physical phenomena (e.g. natural circulation) and failure of activating valve or other means of initial system activation, connected together by "OR" gate as given in Fig. 2.

The component part of a fault tree typically models reliability of active or passive components of a passive system. This part of a fault tree is strongly related to specific design of a system, but the usual design includes some active components responsible mostly only for actuation of the system, the later operation of the system is dependent on passive means. This includes valves, along with logics of actuation signals or manual actuation through human intervention. The modeling should include different failure modes of the components during start-up and during operation of a system. The examples of failure modes could be 'Failure to Open', 'Failure to Close', 'Failure to Run', 'Failure to Start', 'Failure to Remain Open', 'Failure to Remain Closed' etc.

Reliability of passive components (e.g. piping, tanks, heat exchangers etc.) could also be modeled in this part of the fault tree, but usually omitted due to high reliability of these components. In general, the component part of a fault tree is modeled by conventional methods, applied also to other systems in conventional PSA studies.

Another part of a passive system fault tree deals with modeling of physical phenomena. There is no common approach so far how to model this type of failures. The RMPS project [9] proposed to derive reliability estimate of physical phenomena from T-H modeling calculation. In this approach, the fault tree would contain a single basic event, showing unavailability of the process to complete its function when demanded.

Integrating such passive system into the PSA model requires the determination of plant demands for the specific passive system given specific initiating events and then to develop relevant accident sequences. Depending on the particular initiating event, initial and boundary conditions could be different and this usually impacts on the reliability of the system. The accident sequence modeling techniques by event trees are not different for active or passive systems.

The new element in the probabilistic modeling of the passive system is the methodology to quantify reliability of the physical process, represented as a single basic event. Methodology for estimation of passive system reliability is evolving since its inception. Researchers worldwide devoted efforts to address various open issues involved in reliability assessment. However, the research efforts carried out so far in order to establish an 'acceptable' method for passive system reliability assessment has not been conclusive yet. Some of the widely used methods are described in this chapter. This section provides the details of the case study [10] performed for integrating a typical passive system into PSA.

6.1 Description of the Isolation Condenser System (ICS)

A case study is performed for a typical Isolation Condenser System (ICS), deployed in pressure-tube type Boiling Water Reactors (BWRs) and working on the natural circulation principle. There may be two or more identical, redundant trains of ICS depending on the design of NPP. Typically, each ICS train consists of seam drum,

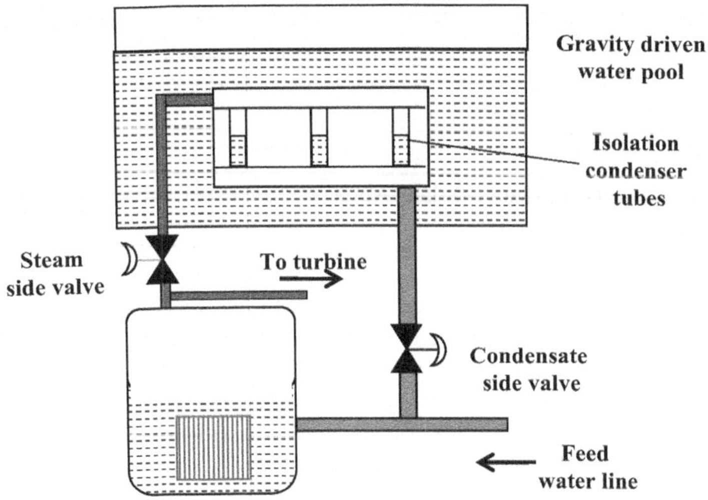

Fig. 5 Schematic of isolation condenser system [11]

isolation condenser submerged into elevated water pool, 'normally open' steam supply valves, 'normally closed' condensate return valves and associated piping. Figure 5 provides a schematic of a typical ICS loop deployed in advanced NPP.

During the normal operation of reactor, steam produced in the steam drum is fed to the steam turbine through steam lines. The main condenser reject the heat from to atmosphere through condenser cooling system and feed the condensate back to steam drum through series of feed water heaters and feed pumps. Tap off connections are made from the main steam lines to ICS loops. When reactor is shutdown, the normal heat removal system is isolated and ICS is actuated to remove the decay heat produced in the reactor. Normally steam side valve is kept open and condensate side valves gets open on actuation signal. The steam rises into isolation condensers submerged into water pool generally located at a higher elevation for creating density difference in the loop, which a driving force of the operation of ICS. The steam gets condensed into submerged isolation condenser and the condensate is returned to steam drum due to gravity when the return valves opens when ICS is actuated.

6.2 *Estimation of ICS Reliability*

Two-layer feed forward ANN network with 10 numbers of 'sigmoid hidden layer neurons' and 'liner output layer neurons' is developed in MATLAB. The 93 sets of six numbers of input parameters were provided in ANN model as columns in a matrix. 93 sets of output vectors (i.e. Integral Power Ratios obtained from

Thermal-Hydraulic analysis of ICS) were also provided in another matrix. The 'Levenburg-Marquardt' back propagation algorithm is used as training function for the ANN model. Inside ANN model, the input vectors and target vectors is randomly divided into three sets as follows [12]:

- 70% (65 data) are used for training
- 20% (18 data) is used to validate ANN
- 10% (10 data) is used as a completely independent test of network generalization.

The trained ANN model was then used to estimate the output parameter values for the 'new' set of input parameters. The results obtained through ANN model were validated against the corresponding results obtained by Thermal-Hydraulic performance analysis carried out using the RELAP5 computer code. In the present study, Isolation Condenser ratio (ICR) was used as performance indicator parameter for the ICS. The ICR is defined as follows:

$$ICR = \int_0^t W_i dt / \int_0^t \dot{W}_{Base} dt \quad (1)$$

where, W_i and W_{Base} are the cumulative heat rejected into GDWP during the specific system configuration and cumulative heat rejected into GDWP during the Base case for the specified mission time.

If the estimated value of ICR falls below 0.6, the ICS is considered to have failed in its mission to successfully remove the decay heat from the NPP. This way the ICR values were obtained using the trained ANN model for new set of input parameters for large number of configurations using Monte-Carlo simulation approach. Total 20 sets of Monte-Carlo simulations with the size of 10,000 were carried out (total nos. of simulations = 2,00,000) for estimating the passive system failure probability. Using the above-mentioned failure criteria (ICR less than equal to 0.6), the failure probability of ICS was estimated to be 7.08E−03 ± 0.79E−03 [10].

6.3 Integration of Passive System Reliability Into PSA

For the purpose of demonstration of approach of integrating passive system reliability into Probabilistic Safety Assessment (PSA), a hypothetical reactor design based on general engineering design concepts being deployed in NPPs is considered. For simplification, in this study, only those systems, which are required to function under the operational transient are described. These includes Reactor Protection System (RPS), Normal Heat Removal System (NHRS) and passive Isolation Condenser System (ICS).

The reactor is operating at the rated power with all systems working normal. One of the process parameters changed due to small deviation in the operating condition, the disturbance is beyond the capability of the control system and reactor trips on process parameter reaching a pre-defined set point. The RPS actuates and brings reactor to shutdown state. If RPS fails to trip the reactor, core damage occurs.

Subsequent to reactor trip, decay heat is to be removed from the core to achieve the safe state. The NHRS continues to operate and takes the heat from the reactor core and transfer it to ultimate heat sink. If NHRS fails, the standby passive ICS actuates and provide the necessary cooling to the reactor coolant system. If both active as well as passive decay heat removal systems (i.e. NHRS and ICS) fail, then 'core damage' occurs. The event tree model is developed using RiskSpectrum® PSA software tool [13] is shown in Fig. 6.

The reliability analysis of ICS has been estimated as per the methodology given in Sect. 6.2. The reliability for Reactor Protection System (RPS) and Normal Heat Removal System (NHRS) is estimated using Fault Tree Analysis method [14] using RiskSpectrum® PSA software tool. The generic failure data [15, 16] are used for estimation of component failure rates and probabilities. Common Cause Failures are considered appropriately using alpha factor method [17]. For Human Reliability Analysis, only latent human errors are considered, human failure probabilities for latent actions are estimated using THERP method [18].

The RPS consists of two independent shutdown systems. The first system (SYS-1) consists of mechanical shutoff rods, which are parked outside the reactor and drops into reactor under gravity upon actuation. The second system (SYS-2) provides back-up to the first system, which consists of shut-off tubes in which liquid neutron absorber solution is added to provide negative reactivity. If both SYS-1 and SYS-2 fails to function, RPS fails to trip the reactor. The broad level fault tree models for RPS is shown in Fig. 7 for illustration purpose.

The Fault Tree top gate with description 'Reactor Protection System failure' is AND gate with two input nodes namely; SYS-1 failure and SYS-2 failure. This means that Reactor Protection System (RPS) fails only and only if both SYS-1 and SYS-2 fails to function.

SYS-1 failure is a voting gate with 'failure criteria' of more than or equal to 2 rods failure having 14 input nodes, each representing the failure of shut-off rods to get inserted into the reactor upon actuation. Each of this is further decomposed into detailed fault tree models up to Basic Events, which are the smallest element of the fault tree. Due to brevity, the details of the fault tree models are not included here.

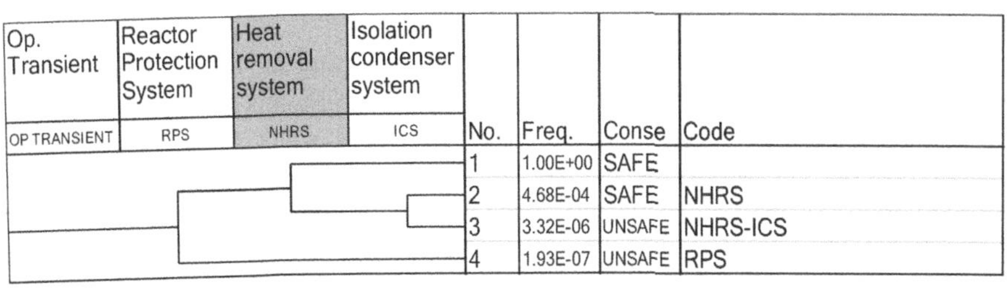

Fig. 6 Event tree model for operational transient [10]

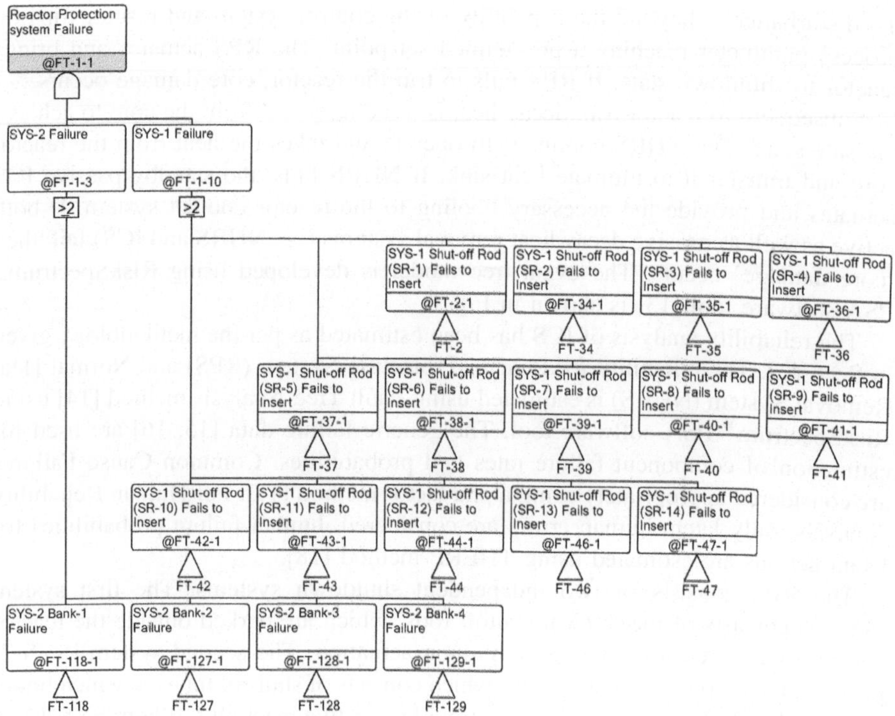

Fig. 7 Broad level fault tree model for RPS [10]

The NHRS supplies feed water to Steam Drum during normal operation of the reactor. When reactor trips, the steam from the Steam Drum is dumped into Main Condenser through Steam Dump Valves. The steam from the turbine exhaust is condensed in the Main Condenser. The condensate water is supplied to the deaerator storage tank through series of feed water heaters, Condensate Extraction Pumps (CEPs) and the Feed Water Pumps (FWPs). Broad level fault tree model is shown in Fig. 8 for illustration purpose.

The Fault tree top gate with description 'Normal Decay Heat Removal System failure' is OR gate with four input nodes namely; Feed Water not available from Steam Drum-1 to Steam Drum-4. This means that NHRS fails even if feed water is not available in any of the Steam Drum.

The immediate cause concept is adopted in developing the detailed system fault tree. According to this concept, the feed water to Steam Drums are supplied through feed water discharge header, which in turn gets feed water from the feed water supply header. The supply header gets the feed water from the Main Condenser through series of feed water heaters, Feed Water Pumps (FWPs) and Condensate Extraction Pumps (CEPs). The steam from the Steam Drum is fed through Steam Discharge Valves into Main Condenser during the shutdown condition for normal decay heat removal.

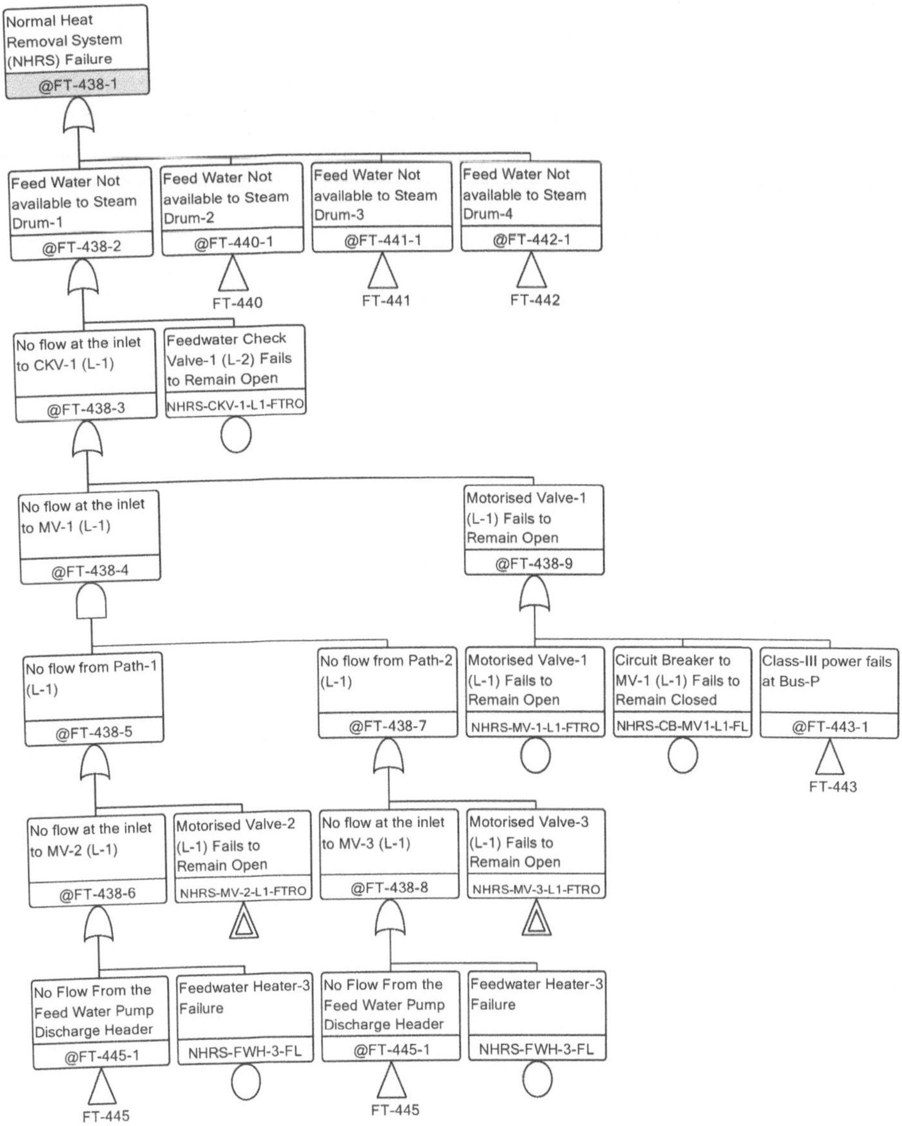

Fig. 8 Broad level fault tree model for NHRS [10]

The reliability is estimated in terms of unavailability in this case study, which is shown in Table 1.

Having estimating the reliability of RPS, ICS and NHRS, the next step is to estimate the accident sequence frequencies and eventually Core Damage Frequency (CDF) induced due to an operational event considered in the case study. Each accident sequence is categorized with end state category as 'Safe' or 'Unsafe'. The

Table 1 Results of system reliability analysis

System	Unavailability
RPS	1.93E−07
NHRS	4.68E−04

sequences leading to 'Unsafe' failures are considered to eventually results into 'Core damage'. For estimating Core Damage Frequency induced due to operational transient (CDFOT), the frequencies of all event sequences depicted in Fig. 6, which results into the end-state category of 'Unsafe' Failure are summed together. With this approach, the CDFOT works out to be 3.51E−06/year.

7 Conclusion

The passive systems reliability assessment involves active component failures as well as phenomenological failures. Different approaches available to estimate the passive system reliability is provided in this chapter.

The reliability estimates of the passive systems can be integrated into PSA with different approaches. One of such possible approach to integrate passive system reliability in a PSA have been demonstrated with a case study in this chapter.

The input parameters may be different if the active component of ICS operate at different time during the accident progression or the component operates in desired position but not exactly as intended in the design (i.e. partial opening of valve), which would affect the Thermal-Hydraulic Response of the passive systems. This would in turn affect the phenomenological failure estimates. Such dynamic interactions are not included in the case study. Further research is required to address this issue.

References

1. International Atomic Energy Agency, 1991, IAEA-TECDOC-629, Safety related terms for advanced nuclear power plants
2. International Atomic Energy Agency, 2005, IAEA-TECDOC-1474, Natural circulation in water cooled nuclear power plants
3. Burgazzi L (2011) Addressing the challenges posed by advanced reactor passive system performance assessment. Nucl Eng Des 241:1834–1841
4. Bianchi F, Burgazzi L, D'Auria F, Ricotti ME, NEA/CSNI/R(2002)10, The REPAS approach to the evaluation of passive safety systems reliability. In: Proceedings of an international workshop hosted by the Commissariat à l'Energie Atomique (CEA) held in Cadarache, France, 4th–6th March 2002
5. Burgazzi L (2007) Addressing the uncertainties related to passive system reliability. Prog Nucl Energy 49:93–102

6. Nayak AK, Vikas Jain MR, Gartia HP, Anthony A, Bhatia SK, Sinha RK (2009) Reliability assessment of passive isolation condenser system of AHWR using APSRA methodology. Reliab Eng Syst Saf 94:1064–1075
7. Beale MH, Hagan MT, Demuth HB, Neural network ToolboxTM, getting started guide, R2014b, The MathWorks, Inc., 3, Apple Hill Drive, Natick, MA 01760-2098, United States
8. Meireles Magali R G, Almeida Paulo E M, Simoes MG (2003) A comprehensive review for industrial applicability of artificial neural networks. IEEE Trans Ind Electron 50(3):585–601
9. Kopustinskas V, EC-JRC-IE, Burgazzi L, ENEA, Ivanov I, Hristova R, Kalchev B, Marinov M, Hadjiev V, TUS, Marquès M, CEA (2003) Reliability methods for passive safety function, deliverable 7. Approaches for introducing passive system unreliability in accidental Sequence, EVOL—RMPS—D07
10. Solanki RB, Singh S, Verma AK, Varde PV, Estimating passive system reliability and integration into probabilistic safety assessment. In: Proceedings of 4th international conference on reliability, safety and hazard (ICRESH-2019)
11. Solanki RB, Kulakarni HD, Singh S, Verma AK, Varde PV (2018) Optimization of regression model using principal component regression method in passive system reliability assessment. Prog Nucl Energy 103:126–134
12. Solanki RB, Kulakarni HD, Singh S, Varde PV, Verma AK, Artificial neural network (ANN) based response surface approach for passive system reliability assessment. In: Proceedings of 1st international and 4th national conference on reliability and safety engineering (INCRS-2018), PDPM IIITDM Jabalpur, February 26–28, 2018
13. RiskSpectrum® PSA, Advanced fault tree and event tree software tool. A member of the Lloyd's Register Group
14. Atomic Energy Regulatory Board, Probabilistic safety assessment guidelines, AERB Safety Manual, AERB/NF/SM-1, June 2008
15. International Atomic Energy Agency (1998) Component reliability data for use in probabilistic safety assessment, IAEA-TECDOC-478
16. Atomic Energy Regulatory Board (2006) Compendium of standard generic reliability database for probabilistic safety assessment of nuclear power plants, AERB/NPP/TD/O
17. NUREG/CR-5801 (1993) Procedures for treating common cause failures in safety and reliability studies
18. NUREG\CR. 1278, Handbook of HRA with emphasis on nuclear power plant application, final report, Sandia National Laboratories, August 1989

Dr. R. B. Solanki is an Electrical Engineer, presently working as Scientific Officer in Atomic Energy Regulatory Board (AERB), India. He has completed his M.Tech. (Energy Science) from Indian Institute of Technology (IIT), Bombay in February 2000. Subsequently, he has completed his Ph.D. in the field of Passive System Reliability Analysis and its integration into PSA at IIT, Bombay in the August 2019.

Since his joining to AERB, he has been involved in review and assessment of probabilistic safety assessment (PSA) studies of the utility organizations for about 19 years. He was deputed to United States Nuclear Regulatory Commission (USNRC) for 1 year during the year 2006–2007, where he was assigned to the Office of Nuclear regulatory research (RES), the Office of Nuclear Reactor Regulation (NRR), and to the Office of New Reactors (NRO) for PRA-related activities of USNRC. Shri Solanki has been involved in IAEA-CANDU PSA working group during 2010–2013 and contributed technical papers on open issues of PSA for CANDU reactors.

He has been involved in review of Level-1 PSA (internal events, external events, full power, shutdown) and Level-2 PSA reports of all operating NPPs in India, as member of AERB PSA committee. Currently, he is also involved in developing regulations for AERB on different types of reactor designs such as light water reactors, pressurized heavy water reactors, and sodium-cooled fast reactors.

Project Stage Considerations for an Inherently Safe and Reliable Chemical Plant

Raghavendra Pai, Suraj Rane and Anusha Pai

Abstract Safety, operability and reliability considerations are vital components of a chemical manufacturing unit. It is a well-accepted fact that the operating team inherits a facility that is already built and its performance is dependent on how robust or otherwise, the installed facility is. Therefore it is important to ensure that safety and reliability factors are given due consideration right from the inception and project execution stage. This work examines the factors that influence eventual safety-operability-reliability of the unit, right at the inception and execution stage. The work also looks at relevant frameworks, practices and tools to ensure adequate coverage. Safety, operability and reliability considerations during inception and project stage are presented by conducting detailed analysis of characteristics of chemical manufacturing units, recommending best practices and concepts used/can be used. The relevant tools and practices used for the study are Engineering Project Process, Process Risk Assessment and management tools like Hazard and Operability (HAZOP) Methodology, Layers of Protection Analysis (LOPA), Safety Integrity Level (SIL) and Management of change. Finally, reliability review, design and project implementation is focused on, in building a more reliable plant with respect to downtime, high on-stream hours etc. Different tools and best practices adopted to improve long term reliability are discussed. The work on developing a process (including templates, methodology) to identify factors that influence long term reliability is a novel attempt based on the experience and analysis of the authors.

R. Pai
Project Management Office Lead (Asia Pacific), Syngenta,
Corlim, Ilhas 403110, Goa, India
e-mail: ragi_pai@yahoo.co.in

S. Rane (✉)
Mechanical Engineering Department, Goa College of Engineering,
Farmagudi, Ponda 403401, Goa, India
e-mail: ssr@gec.ac.in

A. Pai
Computer Engineering Department, Padre Conceicao College of Engineering,
Verna 403722, Goa, India
e-mail: anusha.pai@gmail.com

D. R. Karanki et al. (eds.), *Advances in RAMS Engineering*,
Springer Series in Reliability Engineering,
https://doi.org/10.1007/978-3-030-36518-9_17

Keywords Process safety · HAZOP · Risk assessment matrix · LOPA · SIL · Reliability · Process risk assessment

1 Introduction

Over the years there is mass awareness among infrastructure developers on the issue of accident prevention in large-scale critical systems. This is more prevalent in organizations which operate at global levels. With stringent safety norms and shorter deadline to develop infrastructure like chemical plants, there is a need to build safety and reliability at the design stage. This requires thorough study and planning of activities considering the likelihood of failure events.

This chapter strives to look into the safety, operability and reliability considerations at the project stages of chemical plants. Although many of the aspects could be extrapolated to other similar facilities such as assembly lines, power plants, this paper has chemical plants as the primary focus. It is a well-accepted fact that the operating team inherits a facility the way it has been built and the operating performance is significantly dependent on how robust or otherwise, the installed facility is. This paper examines the factors that influence safety-reliability-operability of the unit at the inception and project execution stage, inhibiting factors that lead to inadequate considerations and recommendations on tools and processes that could help build a robust plant. It must be noted that the work analyses the various facets at a conceptual and qualitative level. To clarify the context, the chemical plant under discussion is presumed to consist of both batch processing units and continuous units. Use of techniques like HAZOP, LOPA and SIL is discussed in the context of safe and reliable chemical plant. As most of the chemical manufacturing units have hazards associated, be it on account of toxic materials or fire and explosion, process safety assumes significant importance to ensure safe operation. Continuous operation on 24 $\times$ 7 basis is the other feature of chemical plants that makes reliability and on-stream performance very important.

The chapter is organized as follows: Sect. 2 describes role of reliability in engineering projects. Project management process is explained in Sect. 3. Safety, operability and reliability considerations during Inception and project stage is discussed in Sect. 4. Section 5 deliberates process risk management techniques. Qualitative analysis of general reliability is explained in Sect. 6. Conclusion of the chapter is presented in Sect. 7.

2 Engineering Projects and the Role in Reliability

It must be recognized that any investment is a business decision and to a great extent guided by the returns. So the quantum of investment is a crucial consideration and there is a natural tendency to reduce the project costs to improve the return on investment. The total cost of installation is only one element, and increasingly now life cycle cost is recognized as against just the one time installed project cost. It makes tremendous business sense to invest adequately at the inception stage and ensure selection of right technology, products and scheme that consider safety and reliability. Adopting the right-first-time approach means one has to deal with only incremental cost as against making changes at a later stage. A poorly conceptualized plant can simply add perineal cost arising out of reduced availability, frequent failures and interruption. The moot question however is how to identify the most optimal scheme that looks at long term benefits. Adopting "Gold standard" always may not make a great business sense but a fit for purpose approach makes sound commercial sense.

The following sections look at different processes, best practices, tools, templates which can be adopted to cover safety and reliability issues right at the project stage.

3 Project Management Process

Projects all though characterized as unique endeavors, there are mature project processes that have lent a framework to enable flawless execution. Project Management Body of Knowledge [1] is one such popular standard by the Project Management Institute, USA the other notable one being, Projects in Controlled Environment (PRINCE2) from AXELOS, UK. Although these elaborate standards consider various knowledge areas and stages of project management focus on safety, operability and reliability are quite generic. This is quite legitimate as the project process is intended to provide a framework for efficient project delivery and it is tedious to include technical consideration in such a framework. Project management process has developed as a separate domain with primary focus on delivering to scope, on time and within the promised schedule which often are referred to as triple constraints. Modern project methodologies do focus significantly on project safety and quality but for long term operability and reliability issues are largely dependent on user requirement specification or the scope agreed upfront. These factors therefore depend on how much analysis has gone into defining the requirement and how detailed the scope is. It is important also to look at factors that can inhibit focus on reliability and operability aspects. Some of the factors are listed below.

1. Too much focus on initial project cost: As discussed in the earlier section, to improve the financials of the business case, there is a tendency to reduce project cost which can inhibit sufficient investment in reliability, e.g. Selection of lower grade steel rather than higher grade steel for fabrication; Congested layout; No provision for redundant instrumentations even for critical service.
2. Lack of a standardized process for evaluating reliability: For risk and operability assessment there are excellent qualitative tools such as Hazard and Operability

(HAZOP) review. These will be discussed in the later sections. However there are hardly any systematic processes to evaluate reliability aspects. Some of the tools such as Life Cycle Cost analysis under the value improvement practices are available but not that popular in practice.

3. Insufficient involvement of operating and maintenance executives: It is customary to have a representation from different disciplines such as process, mechanical, instrumentation, electrical and civil disciplines in the project team. But in many cases, due to limited interaction between design teams and operations teams, important inputs do not get properly communicated.
4. Overemphasis on project execution dominates other issues: Reliability being a long term aspect, often, doesn't get the kind of attention it deserves. The success of project is measured by the cost, time and recently construction safety. Quality management is one dimension that has a direct impact on reliability. However the emphasis on quality varies vastly from industry to industry and is not consistent.

A conscious awareness of these factors will help in getting a more reliable plant.

4 Safety, Operability and Reliability Consideration During Inception and Project Stage

In this section, some of the assessments, tools and practices applied during project stage, which can enhance safety, operability and reliability are discussed. To some extent, safety, operability and reliability are intertwined. However the stakes associated with deviation in each varies considerably. In fact safety has no room for any deviation as it is directly associated with the freedom to operate. Operability refers to the ease of operation and how flexible the operational controls are. Although reliability is embedded in the above two factors, often it is seen as the ability to operate for prolonged periods without unexpected disruptions and is critical for continuity of the supply chains and business. Reliability of individual component in turn impacts safety and operability.

Before discussing the tools and other details, it is useful to understand the typical stages a project goes through. PMBOK identifies, initiation, planning, execution, monitoring, controlling and closing stages for a typical waterfall methodology of execution. It is important that activities in each of the groups are carried out in a timely way to ensure progressive detailing so that changes and reworks are minimized [1].

Figure 1 depicts the interaction between each stage. It is important to note that most of the design work should be completed well before execution picks up pace.

Although there is a degree of overlap of individual process groups, some activities can be commenced only after the completion of preceding activity. For instance, if the foundation design is not complete and available, it is not possible to start with civil construction. As rework of design will severely hamper both the

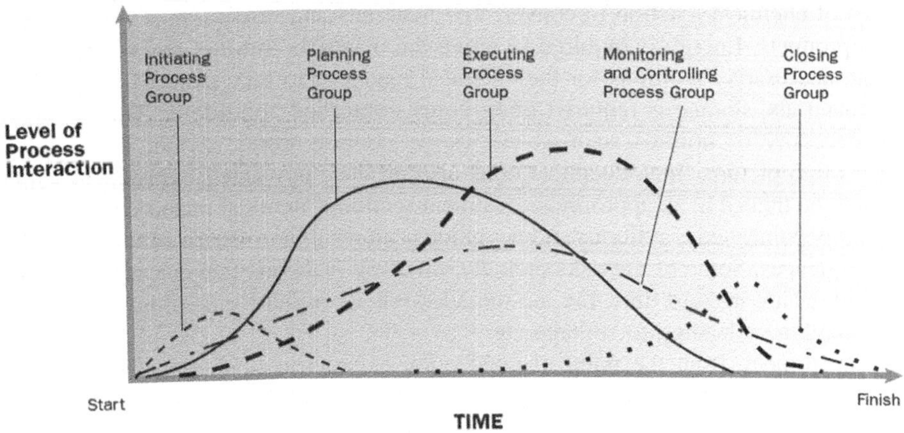

Fig. 1 Interaction between stages [1]

project schedule and costs, validation of design in terms of safety and reliability is carried out timely so that the recommendations can be incorporated in time. A typical flow of different project stages of any project with recommended timelines for activities associated with risk assessment is shown in Fig. 2. Blocks in green depicts important activity stages, corresponding blocks in light blue shows the important design achievements, and orange block shows the design activities.

To appreciate the impact of each of these, characteristics of a chemical manufacturing unit is first discussed along with importance of each of project stages. A typical chemical plant consist of series of unit operations that help carry out

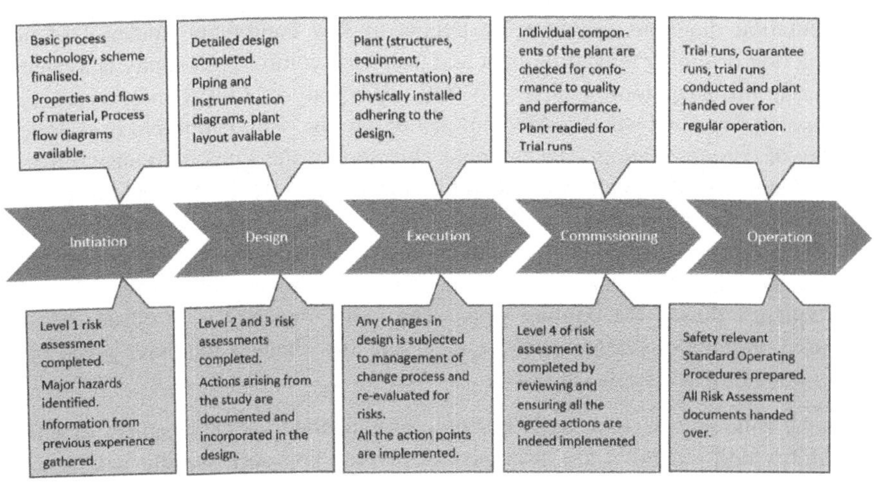

Fig. 2 Flow of project stages

series of chemical reaction to convert raw materials and intermediate chemicals to final products. Energy is required to move the materials through the stages and to create necessary conditions for the chemical reactions to take place. To handle all the materials, storage is required often under controlled conditions. This is where the necessity of utilities such as electricity, cooling water, steam, refrigeration comes into picture. Such chemical units also produce unwanted byproducts which need to be treated in an appropriate treatment section. Chemical reactions take place in highly controlled conditions and deviations can result in mishaps of wide ranging consequences. Some of the risks include, run away reactions that can result in high energy release in short time, fire, un-intended release of fluids into atmosphere, loss of containment leading to spillage etc.

Any project, from the point of conception to maturity undergoes progressive, stage-wise development in terms of technology chosen, engineering and finally execution in the field. To avoid sudden surprises and the necessity of rework on account of safety related issue, risk assessment too follows a progressive path.

Normally, five levels of risk assessments are identified although there is no hard and fast rule to have the exact same number. The five levels of risk assessments are:

- Level 1: This is the first level of risk assessment that is carried out at the initial stage. This is done by assembling a short team and require information such as, size of the project, location, familiarity with the process, corporate memory and broad risks involved. All high level risks that emerge from the discussion are noted and recorded. These will be revisited during subsequent can.
- Level 2: During this stage, all the data related to the chemicals proposed to be used are tabulated along with the properties and toxicity data. The chemicals are then rated for toxicity and this data forms an important input at the time of evaluating risk.
- Level 3: This is the most elaborate level of risk assessment. In addition to the material data, it is expected that process flow diagram, process and instrumentation diagrams, engineering details of key equipment, material of construction of piping and equipment and proposed control philosophy is available. During this stage tools such as HAZOP, SIL Analysis, LOPA analysis are used.
- Level 4: This analysis is carried out just before the plant is put into operation by way of commissioning. The essential objective of the study is to ensure that all the recommendations have been fulfilled in full and the plant is safe to proceed with the commissioning.
- Level 5: Risk assessment is a assessment post commissioning of the plant. Usually this is done after 5 years of plant start up and continued at the same frequency during the coming years. The main objective is to ensure that the initial assumptions continue to be valid and any changes that have taken place do not introduce fresh risks.

In the following section, tools and processes applied for management of safety, operability and reliability are discussed. A layered approach starting with process safety, operability and long term reliability is strongly advocated with common thread of reliability running through each of these. It is important to note that it is the project manager who should account for these while committing the schedule. To ensure that required people are available for each of the meetings, it is essential to organize interactions and plan the activities well in advance. The test of a good project manager therefore rests in the ability to plan well and execute all these activities.

5 Process Risk Assessment Techniques

As mentioned earlier, safety during both the construction and operation of chemical plant is extremely important and non-negotiable. Process Risk Assessment is a methodical way of assessing and managing risks emanating from process hazards, right at the design stage of the project. There are many well established methodologies such as HAZID (Hazard Identification), HAZAN (Hazard Analysis) and HAZOP (Hazard and operability) available for application. HAZID has primary focus of identification of hazards and HAZAN is a quantitative analysis for hazards in a system. HAZOP is a more detailed analysis that looks at both the hazard and operability issues for a given system and is discussed in more detail.

Risk Assessment is generally carried out in stages starting at the beginning of the initiation stage and develops along with the design and detailing. It should be carried out by well trained and experienced specialist who is independent and not easily swayed by others' pressure.

The study is carried out assembling a team of members representing different discipline such as process, mechanical, electrical engineering among others along with the latest sets of documents. The next step is to gather relevant data. The intention is to collect all relevant data into a single dossier that forms a complete assessment. Individuals working separately can begin to complete the forms. Documents could be pertaining to different aspects related to the properties of materials (physical properties, toxicology etc.), reactions and interactions, and equipment information.

In the second stage major hazards such as fire and explosion and toxic release are identified and reviewed. In this stage the following details such as project scope, process flow diagrams, intrinsic process hazards and other issues affecting project scope and implementation are studied.

Once hazards are identified, the team should begin to understand and manage significant risks. It is likely that these meetings will identify data gaps, and that this process will be iterative in nature. In the hierarchy of risk management, elimination of the hazard is the first choice. This strategy seeks to eliminate the hazardous situation and can lead to what is called the inherently safe design e.g. if the catastrophic rupture of a vessel due to very high pressures is a high hazardous scenario, the questions to be evaluated are, whether there is low pressure process available or can the vessel have a design value that exceeds the maximum pressure that the process may generate. Similarly can a highly toxic catalyst be replaced with something inert even if it is at reduced efficacy?

Once all practical ways to eliminate hazards are explored, the next strategy is to prevent conditions using process controls to reduce severity and frequency or both. For this study more detailed design documents such as process flow diagrams, piping and instrumentation diagrams, instrumentation logic etc. are required. The process risk assessor must ensure that the project manager is aware of the critical controls identified during the process risk assessment. The project manager must ensure that the project scope and cost estimate includes critical controls with contingency based on the level of risk assessment completed.

Two types of risk assessment methodologies, namely Hazard and Operability (HAZOP) study and Process Risk Assessment (PRA) are discussed.

5.1 Hazard and Operability Study

HAZOP or Hazard Operability study is one of the universally accepted methodology used for carrying out risk assessment. This can be defined as: "The application of a formal systematic critical examination of the process and engineering intentions of the new facilities, to assess the hazard potential of mal-operation or malfunction of individual items of equipment and the consequent effects on the facility as a whole" [2]. HAZOP seeks to identify potential hazards when there is deviation from the design intent. Multidisciplinary project team is required while carrying out the HAZOP. There has been substantial progress in improving the research aspects in HAZOP studies, some of the recent ones are use of simulation [3], use of graph [4] and use of fuzzy multi-attribute technique [5]. Even the applications have been found in diverse fields like breakthrough behaviors of hydrogen [6], biopharmaceutical industry [7] and evaluate errors related to human aspects in three waste pickers cooperatives [8]. It can be carried out only after the design is completed and the following documents are available.

- Process description.
- Process flow diagrams: Provides main equipment involved in the process and flow of material with the quantities but does not provide full list of piping and instrumentation.
- Piping and instrumentation flow diagrams (P&ID: Diagrammatic representation of piping, instruments, vessels as shown in Fig. 3). It must be noted P&ID do not give the dimension details such as length, location etc.

The HAZOP lead divides the P&ID into many subsystems which are called "nodes" (refer the node marked in Fig. 3). A series of "guide words" tabulated in Table 1, denotes possible deviation to different "process parameter" are applied within the node. "No", "More of", "Less of" are the some of the guidewords used

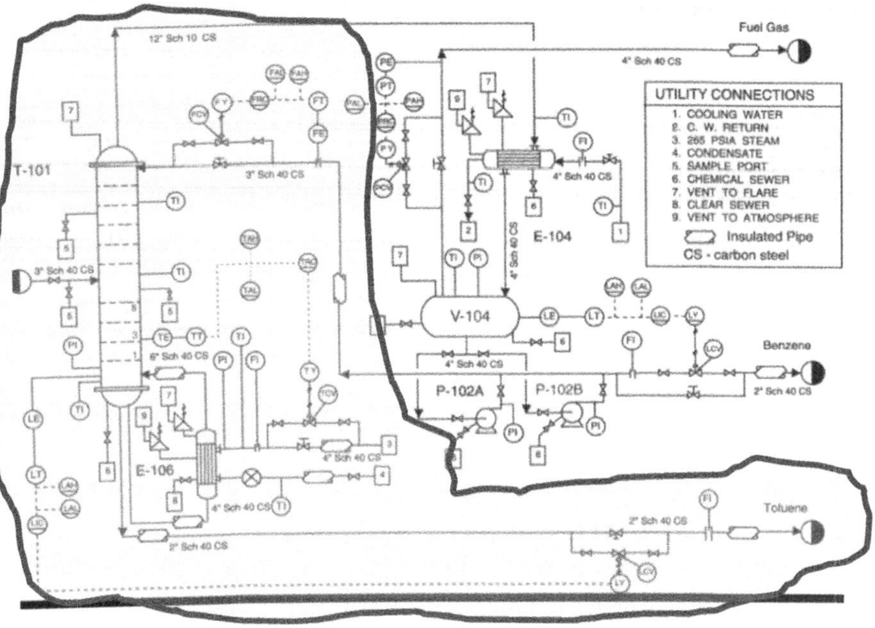

Fig. 3 Piping and Instrumentation flow

Table 1 Guide words in P&ID

Guide words	Meanings	Deviations example
No	Absence of intended	No flow of steam
More	More than intended	Higher temperature
Less	Less than intended	Lower than required pressure
As well as	In addition to the intent	Ingress of foreign material
Part of	Only part of the intent	One of the three reactants does not enter the reactor
Reverse	Opposite of intended	Flow in the reverse direction
Other than	Abnormal situation	Shut down scenario

on process parameters such as "Temperature" "Pressure" "Flow" etc. Table 2 depicts HAZOP worksheet.

In essence, HAZOP study looks at different scenarios that can present itself during the plant operation and enables envisaging the consequences. One HAZOP essentially ensures that full plant is covered in detail by forcing to envisage the consequence and the reason for deviation. It is implicit that the recommendations from the study are fully implemented by the project team.

Table 2 Typical HAZOP worksheet

Plant								
P&ID number								
Node								
Title								
Design Intent								
Sr. No	**Guide word**	**Deviation**	**Causes**	**Conse-quence**	**Prevented by**	**Recomm-endation**	**Action by**	**Due date**

5.2 *Process Risk Assessment Using Matrix*

This method fundamentally is similar to HAZOP and changes slightly in the way the work sheet is formatted. This study too is conducted in a team with the risk assessor identifying possible hazardous event pertaining to a particular part of the system. Main reference document for the study is the Piping and Instrumentation Diagram (P&ID). At first the risk is evaluated by entering details in the Process Risk Assessment (PRA) form discussed in Sect. 5.2.4. Before proceeding further few concepts and tools used to effectively carry out process risk assessment are discussed.

5.2.1 Risk Assessment Matrix

The matrix has severity of consequence on the X-axis and expected frequency of the hazardous event occurring in the Y-axis, as shown in Fig. 4. The frequency indicated can be quantified by assigning values such as number of occurrence in 10000 years (virtually impossible) to say once in a year (very likely). Similarly consequence of low, medium etc. can be further objectified by assigning high severity to consequence such as loss of life, serious business interruption, loss of reputation etc. and low severity to consequence such as minor loss of production, possibility of minor injury. As this is subjective, for the discussion in this work, stratification and quantification of these is omitted. The color coding is in line with the generally accepted denotation, green being safe, amber requiring additional measures and red being unacceptable. At this stage, it is important to understand the engineering controls philosophy, which will enable better appreciation of reliability evaluation.

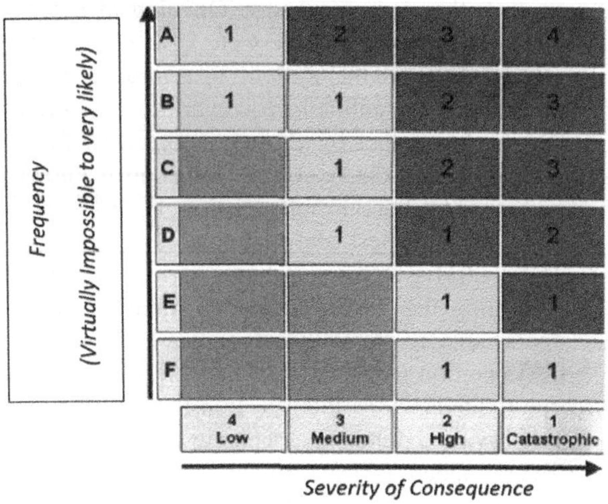

Fig. 4 Risk assessment matrix

5.2.2 Layers of Protection Analysis (LOPA)

The intended mode of operation of any plant is benign and is expected to be safe and as discussed in the HAZOP section, risk can turn into a hazard only when there is deviation from the intended state. LOPA has been applied recently in risk assessment of critical subsea gas compression systems [9] and Hydrogen Sulfide Risk Management [10]. Process plants are provided with the layers of controls as shown in Fig. 5, provided to ensure redundancy or back up. The layers are described as follows:

- Basic Process Control System (BPCS): This is generally based on computer controls such as a PLC (Programmable Logic Controller) or a DCS (Distributed Control System). These are programmed to control the process parameters based on the set points. Field instruments (temperature, pressure, flow sensors) continuously measure different parameters and send the signal to the control system which processes the input and sends an output to the final control elements that

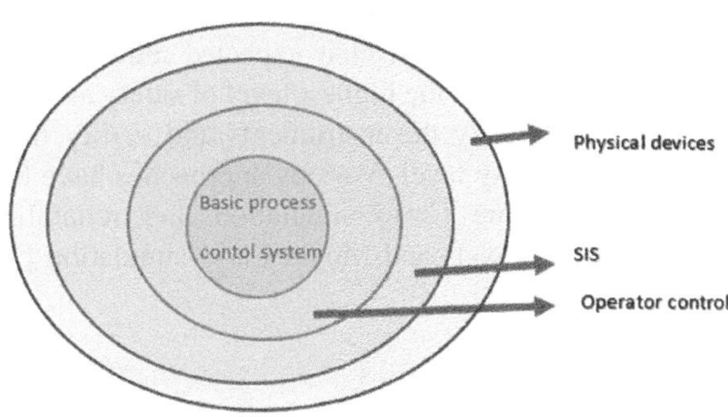

Fig. 5 Layers of control

controls the parameters (e.g. Opening of valves to regulate flows, start/stop of pumps, start/stop of heaters etc.). At least in theory once the correct set points are entered and the system is activated, the process plants are expected to operate as programmed. However this is seldom the case as there are deviations taking place either by choice or forced by limitations.

- This brings to the next level of control known as operator control, which is nothing but human intervention by the operating staff, who continuously monitor parameters and make necessary changes.
- The control systems also have alarm and trip in built in the system. The alarm will bring the deviation to the attention of the operator so that there is an opportunity to manually intervene. There is background program that will stop the operation (trip) and bring the system to a safe state. This however is not a true redundancy as this rides on the back of BPCS.
- To provide redundancy, separate and independent Safety Instrumented System (SIS) is added, which is based on a separate Programmable logic control (PLC). Signal sources for these are independent (redundant) sensors and so are the final control elements. These are not provided for all the parameters but restricted to critical parameters and critical controls to affect an emergency safe shutdown. The objective of this system is not to keep the plant running but to bring it to safe stop.
- Measures listed above will try to avoid undesired state and the next level of protection comes in the form of mitigation of undesired consequence. As an example, consider a case where over-pressurization is an undesired event, measures such as physical safety devices such as rupture discs, safety valves to minimize the impact in case of breach of earlier layers of protection.
- There are further layers but these are mainly aimed at containing the damage rather than preventing a hazardous event. (Bund wall, fire suppression system, fire protection system, emergency plan etc.)

5.2.3 Safety Integrity Level (SIL)

The control system provided as protective measure will be only as good as their reliability of performance when required, generally referred to as probability of failure on demand (lower the value more is the reliability). As of now four levels of integrity are defined and their expected reliability is as shown in the Table 3. As evident, SIL 4 offers the highest level of safety and SIL 1 relatively lower. There are agencies that verify the instruments and certify the instruments to conform to a particular integrity level. Various approaches have been reported on verification of SIL using Monte Carlo simulation and reliability block diagrams [11] and Stochastic Petri Nets and Monte Carlo simulation [12].

Table 3 Safety integrity levels (IEC 61508 standard)

Safety intergrity level	Safety (%)	Probability of failure on demand (%)	Risk reduction factor (%)
SIL 4	>99.99	0.001–0.01	100,000–10,000
SIL 3	99.9–99.99	0.01–0.1	10,000–1,000
SIL 2	99–99.9	0.1–1	1,000–100
SIL 1	90–99	1–10	100–10

5.2.4 Process Risk Assessment Form

After understanding the tools, the process for evaluating the risk is discussed in the remainder of this section. The risk assessment session is a group exercise and is started by selecting a particular section of the plant and the relevant drawing. In the risk assessment form shown in Table 4, all the necessary details of the project are first entered. Following this, the form is progressively filled with rest of the details:

- *Hazardous event*: After discussion with the group, the risk assessor will pick up a possible hazardous event, for example over-pressurization of the vessel.
- *Caused by*: Under this, description of circumstances in terms of deviations that can lead to such an event are entered.
- *Consequences*: In this both immediate and ultimate possible consequences are detailed out.
- *Prevented by*: Under this, measures that exist in the design that will prevent occurrence of the undesired event are entered. At this stage only basic process controls are considered and no credit is taken for the safety controls provided. To continue with the example, vessel designed for pressure more than the maximum pressure of the pumps feeding to the vessel will lead to assigning a "very low" for frequency. This is the basic level and reflects inherent safety state. Descriptions of circumstances in terms of deviations that can lead to such an event are entered.
- *S&F rating (Severity and Frequency)*: These two columns depict the raw risk by assigning the appropriate color coding under the respective columns. The Risk Assessment Matrix in Fig. 4 is referred, to arrive at the color coding and the risk assessor consults the group while assigning the ratings. As an example this could be say D, 1 that falls in Red category.
- *Reducing Severity and Reducing Frequency*: Under these columns, the existing available controls that aid in reducing severity or reducing frequency are described.
- *S&F rating (Severity and Frequency) considering controls*: The process of evaluation is the same as before but now considering other control measures that help to reduce either the Severity or Frequency. The number indicated in the boxes of Risk Assessment Matrix indicates the number of independent protection layers required to make the system safe. The number of red boxes to be

Table 4 Process risk assessment form

Project			Risk assessor					Members					
Date			Reference document										
Ref No.	Hazardous event	Caused by	Consequences (Immediate and ultimate)	Prevented by	S	F	Reducing severity	Reducing frequency	S	F	Action	Action by	Comments
1	A short summary of location, cause and consequence	Specific cause or series of causes in sufficient detail to assess probability and the consequences	A description of consequences with a chronological order of events	The normal control system—items that are used every batch or routinely in normal continuous production			Any additional system that reduces consequence of failure once it has occurred and not normally called upon	Any additional system that prevents the failure if working, but has its own chance of failure			Agreed action resulting from the discussion		
2													

crossed to reach amber state by applying SIS gives the SIL level required. After applying these and evaluating the risk, safety rating is again provided with require color coding. The intention is to ensure that the final ratings are preferably green, amber being acceptable and red unacceptable. If the final rating is red, the project team has to rework the design. The risk assessor may ask the team to have appropriate organizational operating measures for amber final risk scenario. After these assumptions if any, actions agreed are entered in the sheet. This process is repeated for all the different type of risks to cover the entire plant.

A follow up assessment is made before the plant start up. During this study, red risks are reviewed and rated again considering implementation of the agreed actions. A site visit is also made to recheck some of the critical items agreed. In short before the plant is cleared for startup, it is made sure that there are no red risk scenarios and all the actions committed during the course of the risk assessment.

It is not unusual for the designs to undergo changes during the course of the project due to some changes necessitated either due to technical or any other reason. So this can make the risk assessment carried out incomplete as there has been a change in the basic assumption made during risk assessment. To deal with this there is a management process called Management of Change. This is a documented and structured process that mandates change proposals to be sent to a review committee where the impact of the change with respect to the risk profile is carefully studied. If required, the entire risk assessment for the affected portion may have to be carried out again.

6 Qualitative Analysis of General Reliability

The discussion so far demonstrated effective use of processes to incorporate reliability to safety critical operation right at the time of design and project execution. This gives rise to the possibility of exploring processes that can help achieve reliable operations in general not just restricted to safety critical portion.

At the current plant, reliability of any plant in general, is dependent on reliability of the technology, plant detailed design, quality of individual components supplied, quality delivered by the individual contractors working on the project. In practical terms, this depends on competency and the selection criteria set by the team for technology and supplier selections, contractual agreements with suppliers and contractors and Quality Assurance plan. As evident there is no process specifically to try and define the reliability of operation. In the section below an attempt is made to develop a frame work.

It is important to develop and define metrics which can help in quantifying reliability. As a starting point, Overall Equipment Efficiency (OEE) is proposed as the metric to set and track the reliability [13]. OEE for availability in its simplest form is the ratio of expected productive operating time to the total time. Generally a span of one year is considered as the basis. So as an example if annual turnaround of 30 days in a year is considered, OEE will be 100% if the plant remains in productive operation for 335 (365 minus 30) days in a year. The objective will be to attain OEE of 100%, but this being ideal condition, there is an opportunity to identify and analyze reasons which can lower OEE and then further working on how to reduce the down time.

Reliability of any plant is dependent on many factors but for this discussion interruptions unrelated to plant reliability (downtime due to non-availability of raw material or utilities) is not considered. From purely a reliability point of view, overall plant reliability is a function of reliability of individual components and as the adage goes, the strength of the chain depends on the weakest link. If these can be identified and worked on overall reliability can be improved.

6.1 Suggested Process

A brain storming session comprising of representatives from mechanical, instrumentation process and operations is suggested. Matrix such as the one provided below can be used to classify the different equipment into various groups. An equipment can fail due to different reasons such as fatigue failure, improper design, subjecting to conditions other than the equipment is designed for, ageing, and corrosion. To narrow down the analysis to a more meaningful exercise, reasons such as non-conformance to design is not considered. A matrix with a full list of all the equipment either individually or as group where appropriate is prepared. The matrix would also contain dimensions that give rise to vulnerability for failures. For those equipment considered vulnerable, remedial measures are agreed and acted upon. Obviously this can result in new requirements (addition of installed spares, change of metallurgy, change in maintenance philosophy etc.) and consequently the total cost. The final decision will then require careful balance between potential losses due to downtime versus the additional investment required to avoid the downtime. The process can also include assignment of probability of failures before and after the recommendations is very useful in benchmarking. The format of the matrix is provided as Table 5. The details of the items to be included are as below.

- *Equipment name*: Equipment name and Identification number should be entered in this column. For example, Instrument Air compressor.
- *Duty*: Indicate whether the equipment needs to run continuously or intermittently or occasionally. Options could be Continuous/Intermittent/Occasional.
- *Potential down time in case of failure (in days)*: Indicate the potential downtime if the said equipment fails.
- *Type*: Indicate the type of equipment. Options could be Static/Rotary/Electrical/Instrumentation/Structure.
- *Operating Condition*: This gives an idea about the vulnerability of the particular equipment for failure. Options could be high temperature, high pressure, high

Table 5 Reliability analysis

Sr. No	Equipment	Duty Continuous/ Occasional	Type Rotary/ Static/	Potential down time in case of failure (in days)	Operating conditions	Existing protection	Vulnerability to failure. High/ Medium/ Low	Remedial measure / Agreed action/ Responsible	Vulnerability to failure. High/ Medium/ Low (after the remedial measure)	Approximet cost

speed, corrosive environment, material build up over time, possibility of rapid deviation of process condition.
- *Existing protection*: Describe the strategy adopted in the current design. This can be descriptive entry with use of options such as Right material of construction, design per standard, in line spare, spare parts, periodic maintenance etc.
- *Vulnerability*: As a group, vulnerability for failure of the equipment in question is assigned either as High/Low/Medium.
- *Remedial Action*: As a group, remedial action to improve the reliability is agreed for all equipment that are HIGH in vulnerability and those having high downtime in case of failure and MEDIUM vulnerability. The recommendations can range from installing spare equipment, Restricting the source of supplier to the most reliable supplier, Selecting the appropriate supplier, stage-wise inspection, additional instrumentation, maintenance strategy, change in process parameter, additional monitoring devices etc.

Finally the actions are agreed and the responsibility of action assigned. Approximate cost will help in taking a final decision in case the cost of change is disproportionally higher than the expected benefit.

7 Conclusion

This work has tried to look at application of practical approaches to ensure safe and reliable operations. With increased focus on completing projects in time and to cost, long term operations considerations are prone to get a bit obscured. As demonstrated through the above sections, it is possible to incorporate these issues in the project process by creating stage gates which gives the necessary visibility as well as the means to consider these important aspects. It is worthwhile noting here there is sufficient number of processes, standards to address these issues but the problem is that these need to be assembled together so that there is sufficient awareness for effective application. Attempt is also made to develop a frame work for reliability assessment at the project stage and needs more work to define a detailed process for a holistic application.

References

1. PMBOK® Guide (2017) 6th edn. Project Management Institute, Newtown Square PA
2. WHO (1984) Major hazard control: a practical manual. International Labour Office, Geneva
3. Danko M, Janošovský J, Labovský J, Jelemenský U (2019) Integration of process control protection layer into a simulation-based HAZOP tool. J Loss Prev Process Ind 57:291–303
4. Kościelny JM, Syfert M, Fajdek B, Kozak A (2017) The application of a graph of a process in HAZOP analysis in accident prevention system. J Loss Prev Process Ind 50(A):55–66

5. Cheraghi M, Baladeh AE, Khakzad N (2019) A fuzzy multi-attribute HAZOP technique (FMA-HAZOP): application to gas wellhead facilities. Saf Sci 114:12–22
6. Jung W-C, Jung P-K, Moon H-M, Chang M-H, Yun S-H, Lee H-G, Hwang M-W, Woo I-S (2018) Hydrogen breakthrough behaviors for cryogenic adsorption and HAZOP study. Fusion Eng Des 134:123–127
7. de la O Herrera MA, Luna AS, da Costa ACA, Lemes EMB (2015) A structural approach to the HAZOP—hazard and operability technique in the biopharmaceutical industry. J Loss Prev Process Ind 35:1–11
8. Fattor MV, Vieira MGA (2019) Application of human HAZOP technique adapted to identify risks in Brazilian waste pickers' cooperatives. J Environ Manag 246:247–258
9. Ifelebuegu AO, Awotu-Ukiri EO, Theophilus SC, Arewa AO, Bassey E (2018) The application of Bayesian—layer of protection analysis method for risk assessment of critical subsea gas compression systems. Process Saf Environ Prot 113:305–318
10. Yazdi M (2017) The application of bow-tie method in hydrogen sulfide risk management using layer of protection analysis (LOPA). J Fail Anal Prev 17(2):291–303
11. Kaczor G, Młynarski S, Szkoda M (2016) Verification of safety integrity level with the application of Monte Carlo simulation and reliability block diagrams. J Loss Prev Process Ind 41:31–39
12. Zhao X, Malasse O, Buchheit G (2019) Verification of safety integrity level of high demand system based on Stochastic Petri Nets and Monte Carlo simulation. Reliab Eng Syst Saf 184:258–265
13. Gibbons PM (2010) Incorporating six sigma thinking and asset management strategy performance indicators into the overall equipment effectiveness measure. In: European research conference on continuous improvement and lean six sigma, Bristol, UK

Raghavendra Pai is a senior professional with over 27 years of experience in engineering project & portfolio management, process engineering, process safety, operational excellence and operations in the specialty chemical and hydrocarbon processing industry. Currently working for Syngenta as Project Management Office Lead for the Asia Pacific region. He holds a Bachelor's degree in Chemical Engineering from NITK, Surathkal and has a PG Diploma in Management from Goa Institute of Management. He also holds a Post graduate Certificate in Process Business Management from University of Warwick, UK. He is a Certified Project Management Professional (PMP) from the Project Management Institute, USA, Certified Energy Auditor from Bureau of Energy Efficiency, New Delhi and Management of Portfolios (MoP) certification from Axelos UK. He has keen interest in the areas of Project & Portfolio Management, Environmental Engineering, Process Safety, Leadership and High performance teams.

Suraj Rane Professor in the Department of Mechanical Engineering at Goa College of Engineering, Goa, India has a total of 21 years of teaching experience and 1 year of industrial experience. He has completed his B.E (Mechanical Engineering) and M.E (Industrial Engineering) from Goa University, Goa, India and Ph.D. in Reliability Engineering from Indian Institute of Technology (IIT) Bombay. He was the Chairman of Indian Institution of Industrial Engineering (IIIE)—Goa Chapter (2016–19). He was the Chairman Board of Studies in BE (Mechanical Engineering) and M.E (Industrial Engineering) of Goa University. He is member of various Professional bodies like QCI, ISTE, ORSI, SAE-INDIA, IAPQR, SREQOM, SRESA, IISA and ISPS. He is on the reviewer panel of various international journals in the areas of Quality, Reliability and Maintenance Engineering. He was a member of Technical Advisory Committee of Goa State Pollution Control Board. His areas of interest are Quality Engineering, Reliability Engineering, Maintenance Engineering, Six Sigma and Optimization Studies.

Anusha Pai Associate Professor in the department of Computer Engineering at Padre Conceicao College of Engineering, Goa has over 20 years of teaching experience. She has completed her B.E in Information Science and Engineering from Bangalore University and Masters in Information Technology from Goa University. She is currently pursuing her Ph.D. at Visveswaraya Technological University, Belgaum, Karnataka, India in the area of Software Reliability. She is the Board of Studies member in Computer Engineering at Goa University. She is member of professional bodies like ISTE and ACM. Her areas of interest include Data Mining, Information Retrieval, Operation Research and Software Engineering.

Integrated Deterministic and Probabilistic Safety Assessment

Durga Rao Karanki

Deterministic safety analyses as well as probabilistic safety assessments are widely used in risk management of complex engineering systems such as nuclear and process plants. Challenges to these approaches include modeling of dynamic interactions among physical process, safety systems, and operator actions as well as propagation of these model uncertainties. Dynamic Event Tree (DET) analysis allows for integrated deterministic and probabilistic safety assessment (IDPSA) by coupling thermal-hydraulic/process system models with safety system and operator response models. This chapter introduces the concept of IDPSA, highlights the benefits of the approach as well as its limitations. Case study on a medium loss of coolant accident in a nuclear power plant is presented, which focuses on a comparison between IDPSA and traditional approaches considering impact of accident dynamics.

1 Introduction to Integrated Deterministic and Probabilistic Safety Assessment

In the safety analysis of complex engineering systems, we develop accident sequence models to quantify the risk. In this process, it is a challenge to consider dynamic interactions and capture their impact on accident models. Such dynamics (time dependent interactions) can arise due to human interactions, digital control systems, & passive system behavior [1–4]. The main objective is to increase realisam in modeling dynamics while quantifying risk [5]. In this section, an integrated deterministic and probabilistic safety analysis approach (IDPSA) is introduced including its basic elements and their relationships.

D. R. Karanki (✉)
Siemens Mobility AG, Hammerweg 1, 8304 Wallisellen, Switzerland
e-mail: durga.karanki@siemens.com

D. R. Karanki et al. (eds.), *Advances in RAMS Engineering*,
Springer Series in Reliability Engineering,
https://doi.org/10.1007/978-3-030-36518-9_18

1.1 Probabilistic and Deterministic Safety

In the Risk Analysis of Nuclear Power Plants (NPPs), we primarily address these questions: What is the hazard? How likely is it? What are the consequences? How to improve the level of safety? We use a systematic and comprehensive methodology Probabilistic safety assessment (PSA) to evaluate the risk associated with complex engineering systems like NPPs. Here are the high level tasks of PSA: (1) Identify accident initiators (2) developing accident models including sequence delineation (sequence delineation—which sequences lead to core damage) and success criteria definitions. Success criteria—identifying the requirements for success of the safety systems. For example, what is the success criteria of safety equipment. These requirement concern how many systems or pieces of equipment must operate, the latest time by which the operators must intervene, how long the equipment must function. (Typical questions to identify Success Requirements: *How many systems must operate? *How long they must function? *Latest time for operator intervention). (3) Quantifying the risk then corresponds to estimating the likelihood that the requirements are not met and an accident follows.

How is it done with the classical approach?

Risk/Safety assessments are widely performed to evaluate risks associated with complex engineering systems such as aeronautical systems, nuclear power plants, marine systems, etc. This is called PRA in aerospace and PSA in nuclear industry. Figure 1 shows elements of a typical PSA. Methods are quite common. Classical combination of event tree and fault trees are used to build risk models. Quantification of risk models require inputs such as data, simulations, etc. Typical results include risk, contribution of sequence, basic events to risk.

Offline (with predefined boundary conditions) thermal hydraulic/process analysis are performed for determination of sequence outcome and defining success criteria (deterministic safety analysis). Probability safety models (classical combination of event tree and fault tree models) address the probability that these criteria are not met. Figure 2 illustrates high level tasks in the current PSA practice. Accident scenario is simulated with a standalone tool, for instance thermal hydraulic code in case of nuclear power plants. Typically, few sequences with predefined boundary conditions are simulated to investigate if sequences lead to a safe state or an undesirable consequence.

1.2 Issues in Current Approach

How complex engineered system works in normal conditions is well known. What about in an accident scenario? What are the accident dynamics? Process behavior evolves with the time; i.e. the process parameters change with respect to time influencing the response of safety systems. Operator response may influence the

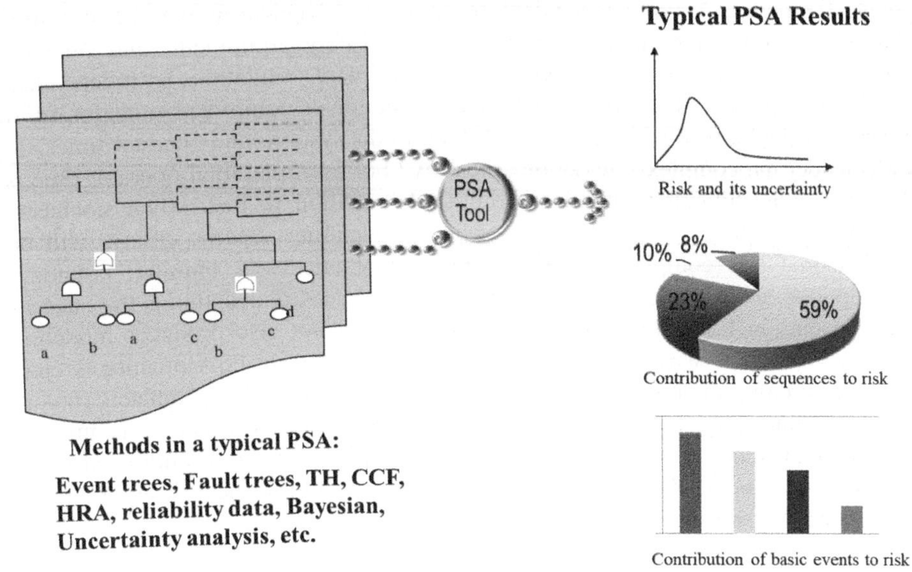

Fig. 1 Elements of a typical PSA/PRA

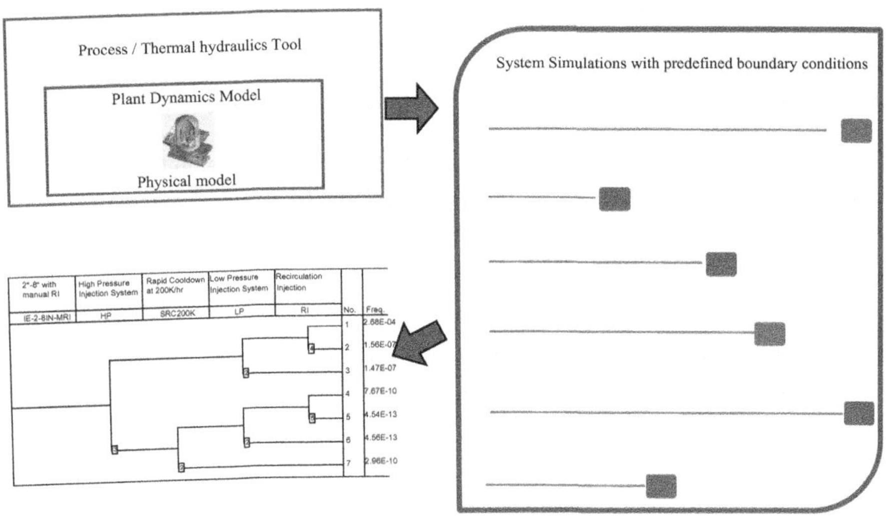

Fig. 2 Flow of high level tasks in a classical PSA

physical process. A typical accident scenario in a Nuclear Power Plant (NPP) involves complex interactions among process, safety equipment, and operator actions. The big question is: In current practice of plant simulations by independent thermal hydraulic codes (which don't have operator or equipment models), or A matrix of calculations) with pre-decided states of the systems and operator times, do we consider the complex interactions properly? For example, total operator time is dynamic, it is difficult to predict it offline. In addition to that, when stochastic variabilities in those responses are considered, defining success criteria will be cumbersome and complex. Also, the detrimental effects of binning; bounding assumptions are necessary while enveloping sequences and defining the success requirements in PSA; bounding is not only difficult but also produces modeling artifacts in certain cases. One of the major issues in current PSA practice is epistemic uncertainty in physical/process models [6–8]. PSA Model parameters (failure rates or failure probabilities) are already accounted, but uncertainties in physical model parameters (e.g. thermal hydraulic models) also needs to be addressed while building risk models. These may impact sequence outcomes, success requirements, subsequently risk estimates and contributors.

1.3 IDPSA Using Dynamic Event Trees

Dynamic Event Tree (DET) analysis provides a framework for integrated accident simulation of physical process, equipment, and operator actions. In other words, DET provide the means to simulate physical system evolutions, the evolution of system states due to stochastic events, and the dynamic interactions between these evolutions (DET simulates the dynamic interactions among physical process, safety system responses, and operator responses). DET models include deterministic (physical) as well as Stochastic Models.

Accident scenario is simulated considering dynamic interactions and stochastic variabilities to generate sequences. The outcome of sequences are labelled based on the values of physical parameters. Risk is estimated considering all the undesired sequences.

In the dynamic event tree: The transient is simulated in deterministic dynamic model and the process parameter values are obtained from plant dynamic model with respect to time. Scheduler has the integrated model of the plant describing the behavior of the various elements as a set of rules. When the process parameter reaches a level, it would fire one of these rules. As a result, event sequences are generated based on the rules (scheduler). When the process parameter demands intervention of safety system or human action, one of the rule in scheduler gets fired, and branching takes place in the DET.

Several DET implementations tools and their applications can be seen in the literature. Interested readers may refer the following works in the literature [9–18].

2 A Simple Example: Water Leaks into a Ship/Vessel

Flooding is one of the important hazards for ship safety and stability. This example (see Fig. 3) is extremely simplified version of that problem. A Ship/vessel begins to accumulate water due to leak, which could be due to valve/pipe. The objective is to estimate the likelihood of vessel reaching the critical level, considering the stochastic variabilities and time dependent interactions. The tank will accumulate to a critical level if the operator does not isolate the leak in time. The operator receives a cue from an alarm caused by the rise in level, in response the operator should close the valve. System failure depends on the failures on demand of alarm and valve, as well as human response. The time dependent element in the problem is the human response time completing with the time taken by the tank to reach the critical level, which depends on leak size and location.

How do we solve this problem with the classical method? In the current practice, offline TH/physical process simulation-based analysis are performed to determine the sequence outcomes and define success criteria in developing accident sequence models. Success criteria analysis involved identifying the requirements on safety systems associated with the success branch for a function. These requirements concern how many components must operate, how long they must function, the latest time by which operator must intervene. The developed accident sequence model is typically an event tree with coupled fault trees, which are quantified to obtain risk. In the process of developing event tree models, bounding assumptions are necessary to envelop the sequences and define the success requirements. The accident dynamics complicate such grouping.

Figure 4 shows a classical event tree modeling of the vessel problem. Event trees are graphical models that order and reflect events according to the requirements for the mitigation of initiating events. Event trees are used to determine the sequences of system failures that can lead to undesired consequences.

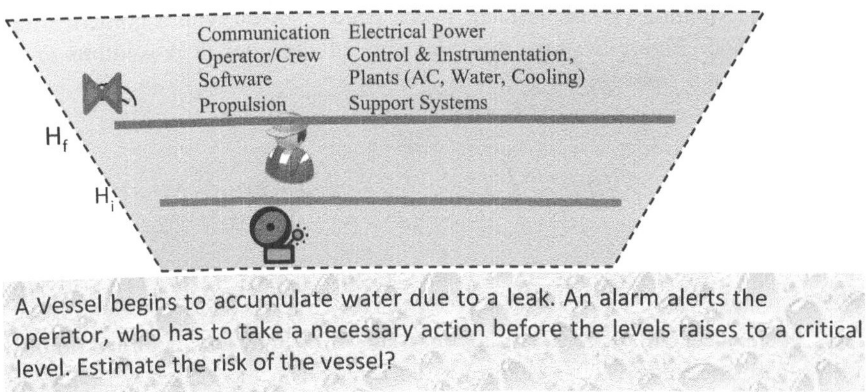

Fig. 3 An accumulating vessel with an initial level H_i and a critical level H_f

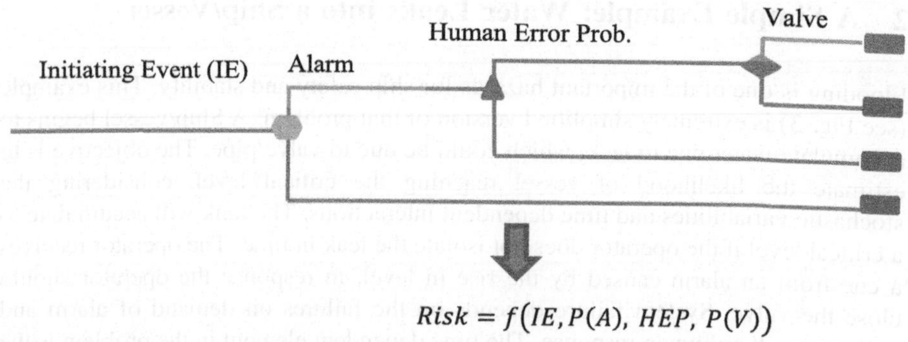

Fig. 4 Classical event tree modeling of vessel problem

Fig. 5 Analytical Solution of the vessel problem

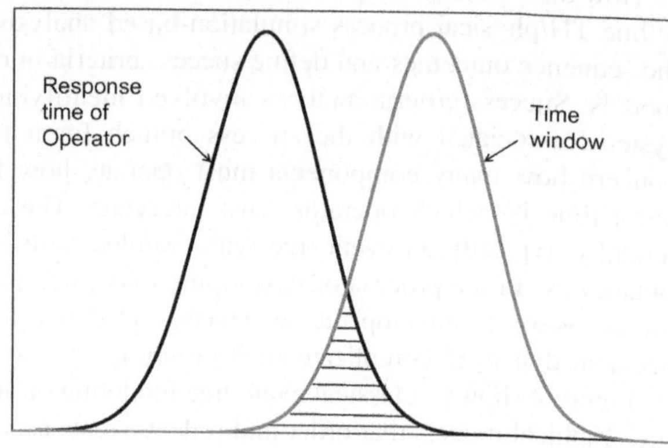

How do we solve this dynamic risk problem? There are several methods available in the literature such as analytical solution, analog Monte Carlo simulation, Dynamic Flowgraph Methodology, and DET, etc. As this is a simple problem, an analytical solution can be derived, which can be solved with a numerical integration method. Figure 5 shows the elements of the analytical solution of this problem, adapted from [6].

$$HEP = \int_{0}^{\infty} f_R(r) \int_{0}^{\left(\frac{r+k_2}{k_1}\right)^2} f_H(h)dh \cdot dr \tag{1}$$

where $k_1 = \frac{A}{aC}\sqrt{\frac{2}{g}}$ and $k_2 = \sqrt{H_f} \times k_1$

As depicted in Fig. 6, DET simulation framework integrates both deterministic and probabilistic models. In response to an accident in the vessel, several safety systems and crew actions are demanded by the process. The transient is simulated in deterministic dynamic model and the process parameter values are obtained from

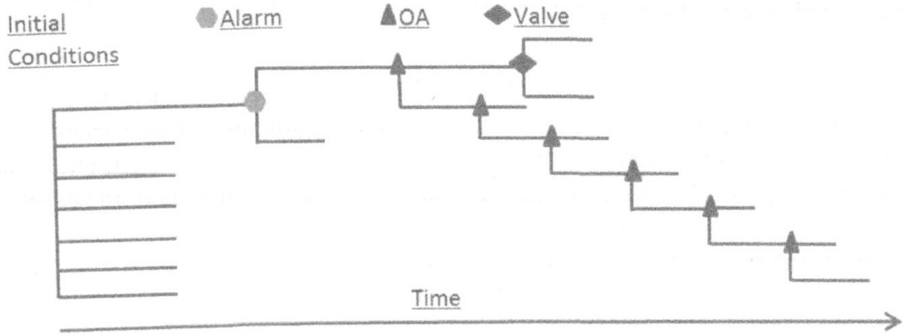

(a) Discrete dynamic event tree with discretization of continuous variables

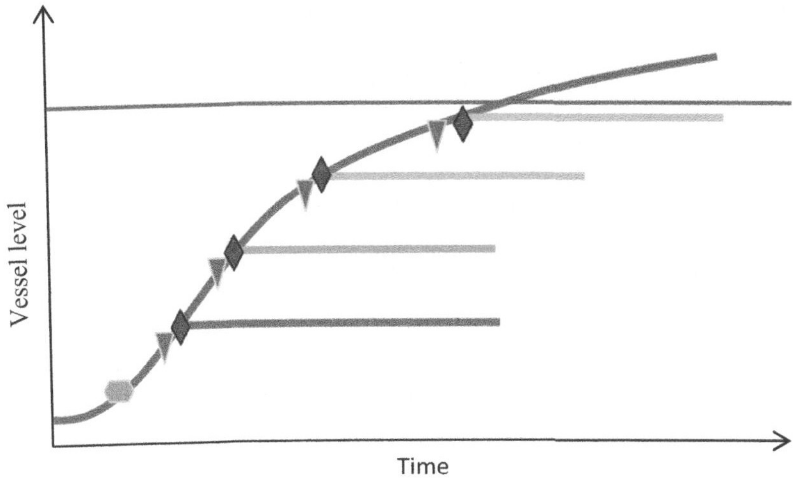

(b) Physical parameter evolution over time for different DET sequences

Fig. 6 DET solution of the vessel problem

vessel dynamic model with respect to time. The branches represent the possible states of the safety systems or/and crew. The sequences and vessel parameters with respect to time are obtained from the DET simulations.

3 IDPSA Methodologies

This section describes two IDPSA methodologies available in the literature, namely, DET Informed PSA [19] and quantified DET [20].

3.1 DET Informed PSA

Figure 7 shows the high level tasks of DET Informed PSA approach, which include DET Modeling, accident simulations with DET simulator, and success criteria analysis to develop event tree models and their evaluation. Primarily DET models consists of physical models of system behavior, stochastic models of the equipment, and operator response models.

Focusing on the detailed tasks involved in this approach, firstly the scope of the overall analysis is defined including the boundary conditions of initiating event, safety functions to be considered and the variabilities to be addressed, end sequence criteria (undesirable consequence), etc. The simulation models for physical process, response of safety functions, and operator responses are developed. The accident scenarios are simulated with Simulators considering the random variabilities. The results from the simulations are analyzed to understand the accident dynamics and identify the evolutions that lead to undesirable consequence. The success criteria are identified by examining the sequences generated by DETs, initially for the individual safety functions and initiating events. Initiators and sequences with similar success criteria for each safety function are then grouped, as a basis for defining one or more event trees to represent the overall variability of initiating event. Finally, the success criteria for these trees are defined. For operator actions, additional DET simulations are used to estimate time windows (TWs) in order to calculate the Human Error Probabilities (HEPs) in consideration of the sequence boundary conditions. The overall risk is quantified and important contributors are also identified.

The quantification of event trees requires various tasks such as fault tree modeling, common cause failure modeling, human reliability analysis, and failure data

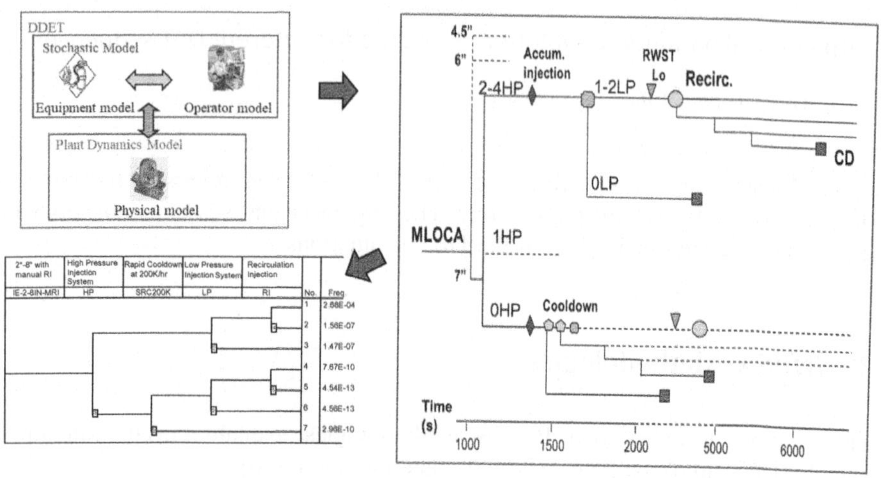

Fig. 7 Macro tasks of DET informed PSA

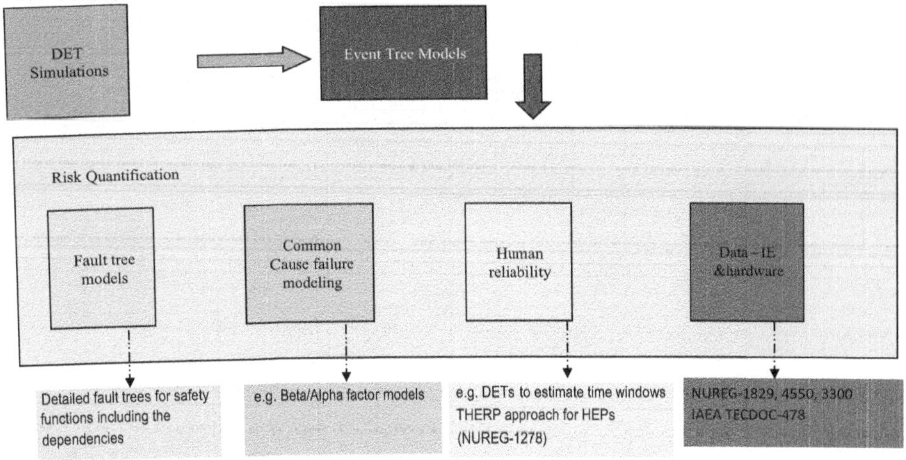

Fig. 8 Flow of tasks in DET informed PSA

(See Fig. 8). Fault trees for safety functions are developed and linked to event trees. The common cause failure modeling accounts for any implicit dependencies among similar safety equipment. DET simulations provide useful information to estimate time windows for operator actions, which are used to estimate human error probabilities. PSA parameters including probabilities and frequencies of the initiating events, hardware, and operator actions are also necessary. Finally, PSA tools are used for quantification of accident sequence models to obtain risk results including point risk estimate and important contributors to risk basic events and sequences.

3.2 *Quantified DET Based IDPSA*

Figure 9 shows the high level tasks of DET quantification approach, which are DET modeling, simulation, and evaluation. DET Modeling and simulation task are quite similar to DET Informed approach. DET Evaluation replaces success criteria and bounded compact event trees modeling in this approach. In DET evaluation, all of the individual sequences that are generated are explicitly quantified. The DET sequences whose outcomes lead to undesirable consequences are identified and the frequencies for each of these failure sequences as well as total risk from the scenario are estimated.

Figure 10 shows a comparison between both IDPSA methodologies. Both DET Informed approach and DET quantification use the same integrated DET simulation tool. Although the DET tool and models are same, the simulations in DET-informed are with bounding while the latter approach perform simulations without bounding.

DET Informed PSA provides an integrated framework to account for complex dynamic interactions and stochastic variabilities among physical process, safety

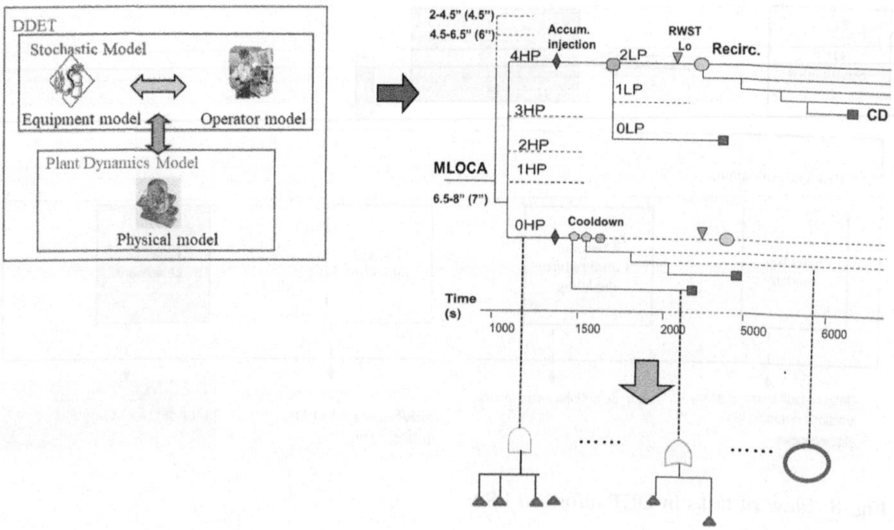

Fig. 9 Macro tasks of DET quantification

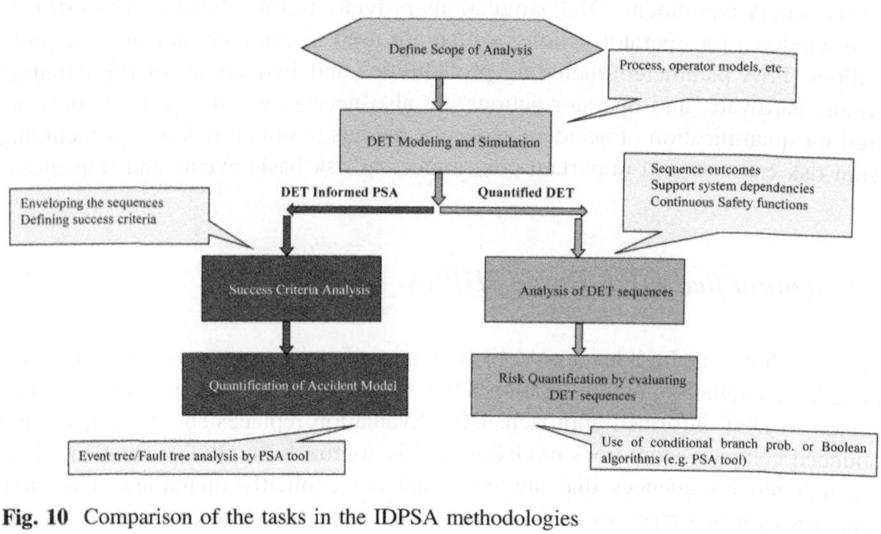

Fig. 10 Comparison of the tasks in the IDPSA methodologies

equipment, and operator actions. DET Informed PSA helps to get Success criteria definitions and to build a compact event tree which is quite practical in large scale PSAs of complex engineering systems. On the flipside, in DET informed PSA, bounding assumptions are inevitable to envelop the sequences and define the success requirements. Some detrimental effects of bounding may arise due to accident dynamics, which could be overlooked. Quantified DET approach does not need any bounding assumptions, thus bypassing issues associated with bounding effects. However additional computations are necessary.

DET-informed approach yields classical event trees that is compatible with the current PSA practice. Usually the classical event trees contain binary branches. On the other hand, Quantified DET approach may generate several branches for safety systems representing various combinations of conditions among the safety functions. Regarding quantification of risk, both approaches differ how safety systems are handled whose response involves a continuous aleatory variable. For example, recovery time of power supply or response time of operator actions. In classical event trees, the human error probability or recovery probability are estimated offline, then used as inputs to quantification of binary branches. In contrast, Quantified DET does not require the time windows to estimate these probabilities.

Quantified DET approach has to deal with the following practical issues: it must evaluate all generated sequences, treat support system dependencies among safety functions, as well as account for safety systems whose response is continuous. Some possible solutions for these issues were proposed in [20].

4 Case Study—MLOCA Scenario

This section presents a case study on Medium Loss of Coolant Accident (MLOCA) scenario. The application of both IDPSA approaches presented in preceding Section are described. The analysis of results primarily focuses on impact of accident dynamics as well as impact of bounding. This accident scenario is particularly chosen because its break range has strong effect on the sequence dynamics and consequently on the requirements on the safety function (success criteria). The DET models and some of the simulations of MLOCA accident scenario are adapted from the references [19, 20].

4.1 Description

The accident scenario is derived from the Zion Pressurized Water Reactor (PWR), a decommissioned 4-loop PWR with a thermal power of 3250 MW and one of the plants addressed in the NUREG-1150 study of severe accident risks [21].

In a PWR, the primary coolant (water) is pumped under high pressure to the reactor core where it is heated by the energy generated by the fission of atoms. The heated water then flows to a steam generator where it transfers its thermal energy to a secondary system where steam is generated and flows to turbines which, in turn, spin an electric generator. This is about the physical process during the normal operation of NPP. In accident conditions, we need safety systems like Emergency Core Cooling System (ECCS) to prevent core damage. When automatic systems don’t function, operator has to intervene. Figure 11 shows a simplified schematic

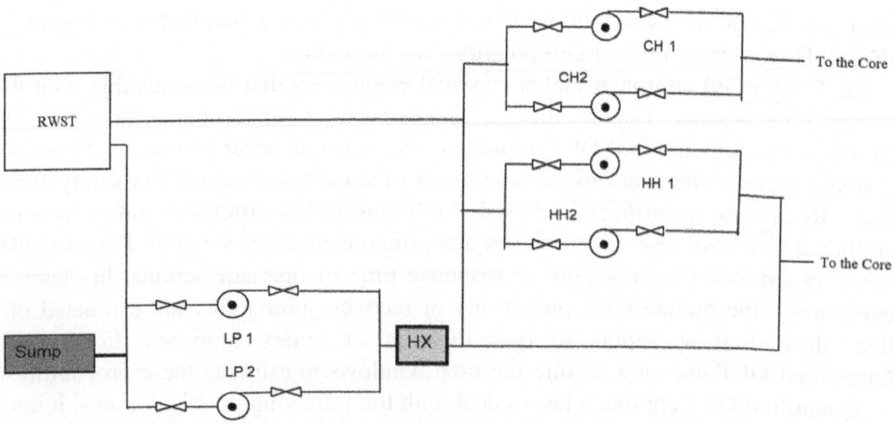

Fig. 11 A schematic diagram of the ECCS of a PWR

diagram of the ECCS of the PWR. ECCS consists of a high-pressure injection (HPI) system, a low-pressure injection (LPI) system, and an accumulator in each loop (not shown); HPI consists of two charging (CH) pumps as well as two high head (HH) injection pumps and theirs associated Valves. LPI includes two Low Pressure (LP) pumps in their injection trains. All these injection trains take suction from Refueling Water Storage Tank (RWST). When the water level in RWST tank reaches low level, an alarm prompts operator in control room to manually switch the injection to recirculation from containment sump.

If HPI fails, operator must perform rapid cooldown to depressurize reactor coolant system to low pressures conditions, which allows LP injection system to prevent core damage. This rapid cooldown is manual action using steam generator dump valves. Figure 12 shows operator actions during this accident sequence, which include reactor trip alerts operator in control room, emergency procedures will be examined, required action to depressurize must follow.

Figure 13 shows an Event Sequence Diagram (ESD) of MLOCA scenario. The sequence of events occurring, and their interactions are shown at a high level. Following reactor trip and turbine trip, the response to a MLOCA scenario begins with a HPI phase. Subsequently, the RCS pressure drops below the set point of the accumulators (40 bars) and finally to the low pressure injection (LPI) set point (15 bars). The summary of events and their potential stochastic variabilities are also depicted in Fig. 13.

Figure 14 presents ESDs during Secondary Rapid Cooldown (SRC) and Recirculation Injection (RI) phases.

The pivotal events involved in MLOCA scenario can be represented with a simplified event tree as shown in Fig. 15. There are four safety functions and seven sequences. Each header will have a fault tree (Fault trees are used to identify all combinations of component failures/events that can lead to system failure). Event tree represents sequences of system failures and associated consequences. The

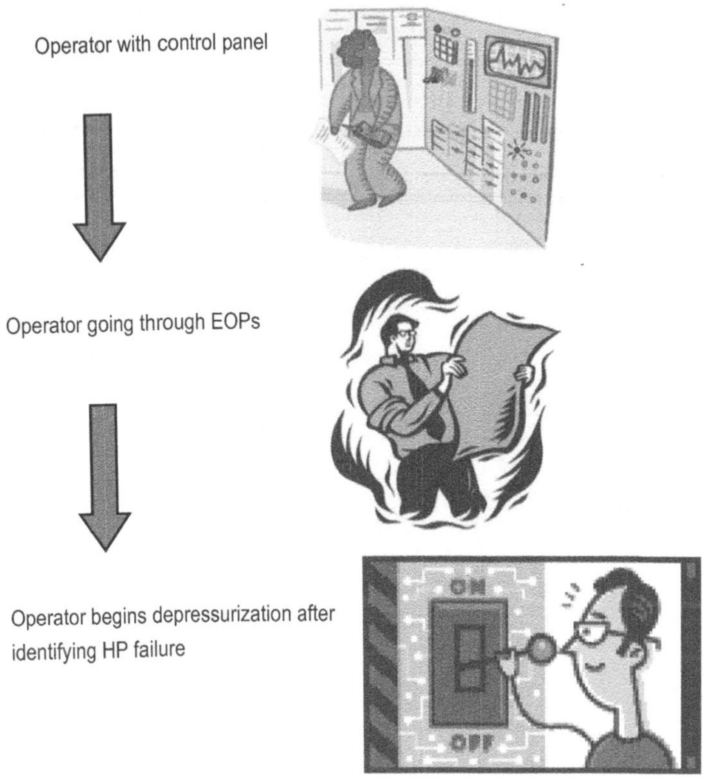

Fig. 12 Operator actions to initiate rapid cooldown

success criteria definitions of safety functions (which decide their respective probabilities) are input to the fault tree (in order to calculate their branch probabilities) modeling. The important elements of success requirements are: number of injection trains, rate and timing of cooldown, timing of recirculation. Irrespective of what method (PSA or IDPSA) we choose, plant simulations are designed (scope is defined) to provide information to know the answers to these questions which give SC definitions.

Figure 16 shows the computational framework for MLOCA scenario. Simulation of accident scenario with different break sizes results in about 1000 sequences. A typical event tree for this scenario may have about 7 sequences as shown in Fig. 15, which is compact and of practical use with cut sets, importance measures, etc. Classical PSA approach usually based on a single limiting break, which may ignore the dynamic interactions, or in principle we could also take hybrid of most challenging criteria (it gives unnecessary conservatism and it does not represent any physical break). To account for dynamics, variabilities, and diversity in success requirements, split break range model can be used, which has about 30 sequences; additional sequences and appropriate success criteria

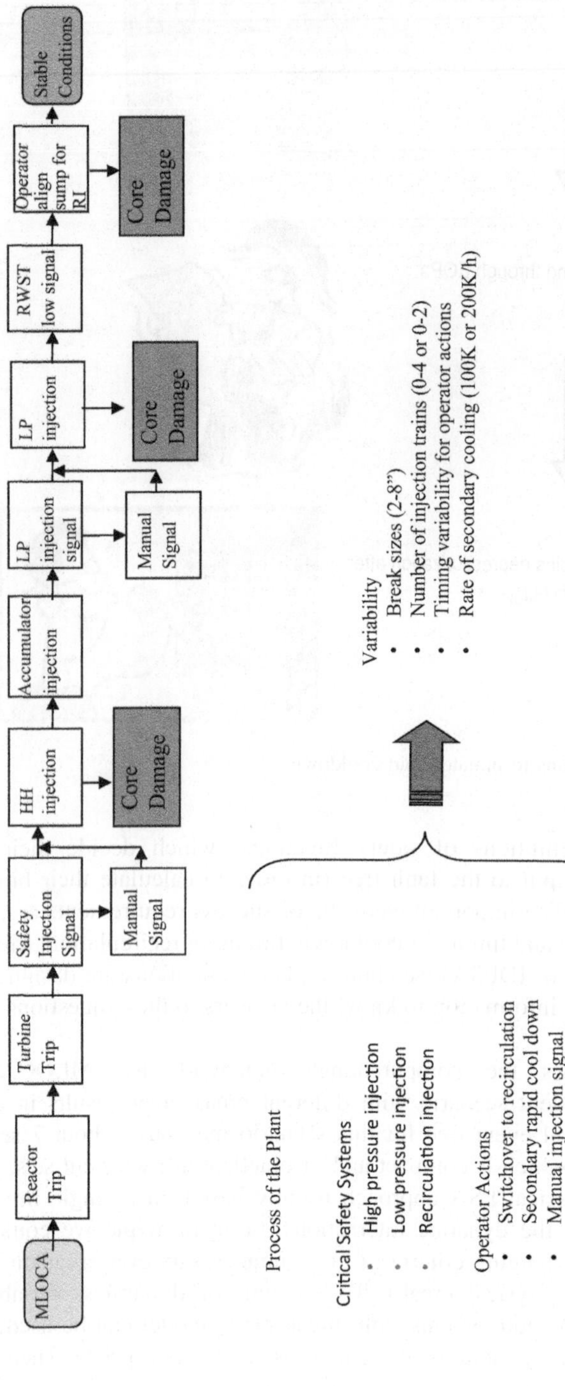

Fig. 13 Events and their interactions in the accident scenario (simplified ESD)

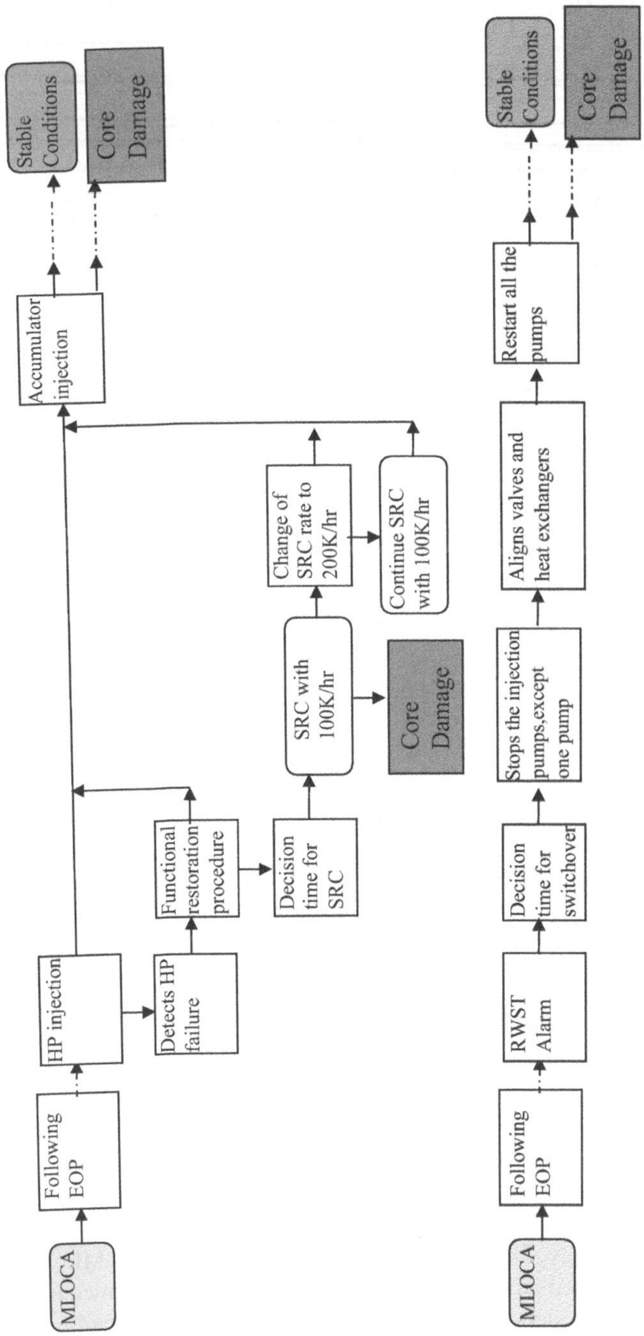

Fig. 14 Detailed events in secondary rapid cooldown and recirculation modes

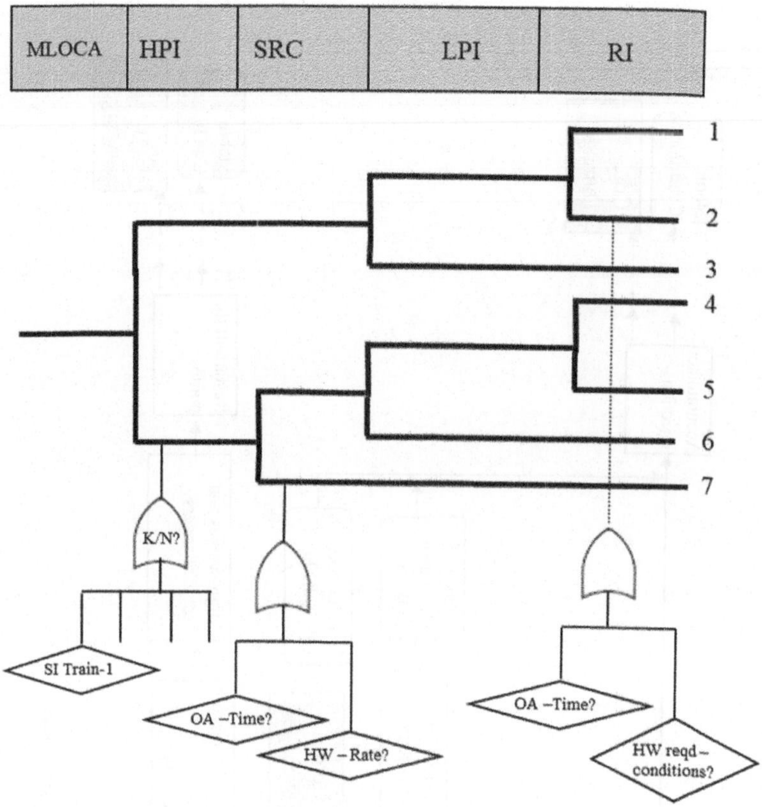

Fig. 15 Qualitative event tree of MLOCA scenario

definitions are used. To bypass issues associated with bounding, we can use DET quantification, meaning quantifying all the 1000 sequences.

4.2 Results—Comparison of IDPSA Approaches with PSA

The results obtained from the three methods, namely, classical PSA, DET informed PSA, and Quantified DET are considered for comparison. The obtained results include conditional core damage probabilities, core damage frequencies, critical sequences, and important basic events.

Figures 17 and 18 show event tree models from classical PSA and DET-informed approaches respectively. 7″ limiting break is assumed to represent MLOCA scenario in PSA approach, based on the results reported in [22]. DET-informed includes three split break range models, 2–4.5″, 4.5–6.5″, and 6.5–8″. Regarding event tree headers, they are different because of different time windows and high or low

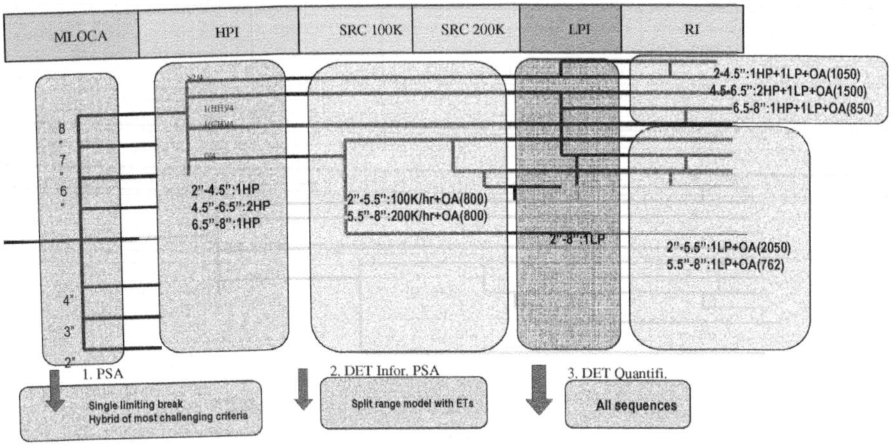

Fig. 16 Computational framework with three approaches

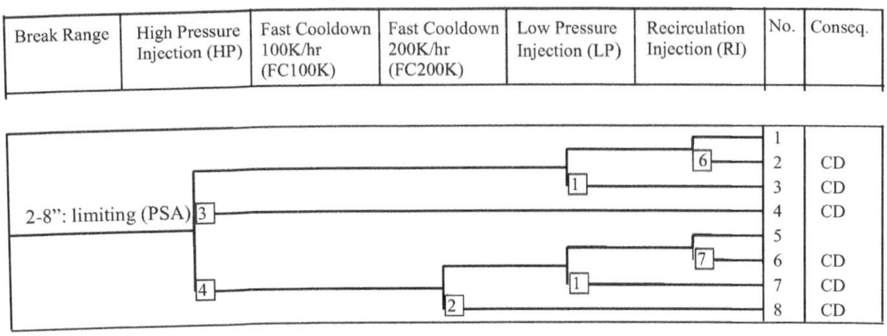

Fig. 17 Event tree of classical PSA approach

pressure conditions leading to different events or probabilities in fault tree. Sequence 4 (1HP) in 4.5–6.5″ event tree leads to core damage (CD) as there is no cue for OA for cooldown. To quantify risk estimates, fault trees of safety systems, CCF models, HRA models, and failure data were plugged into the event trees. In Quantified DET approach, large event trees consisting of 1000 sequences were quantified (a section of DET can be seen in Fig. 9).

Figure 19 shows a comparison of the obtained core damage frequency (CDF) estimates among three methods. The core damage frequency (CDF) indicates the likelihood of an accident that would cause damage to reactor core, which also incorporates the probability of the breaks occurring. The results indicate the classical PSA approach is the most conservative, followed by DET-informed approach. The classical approach is significantly higher than quantified DET by a factor of 14 while DET-informed is higher by almost a factor

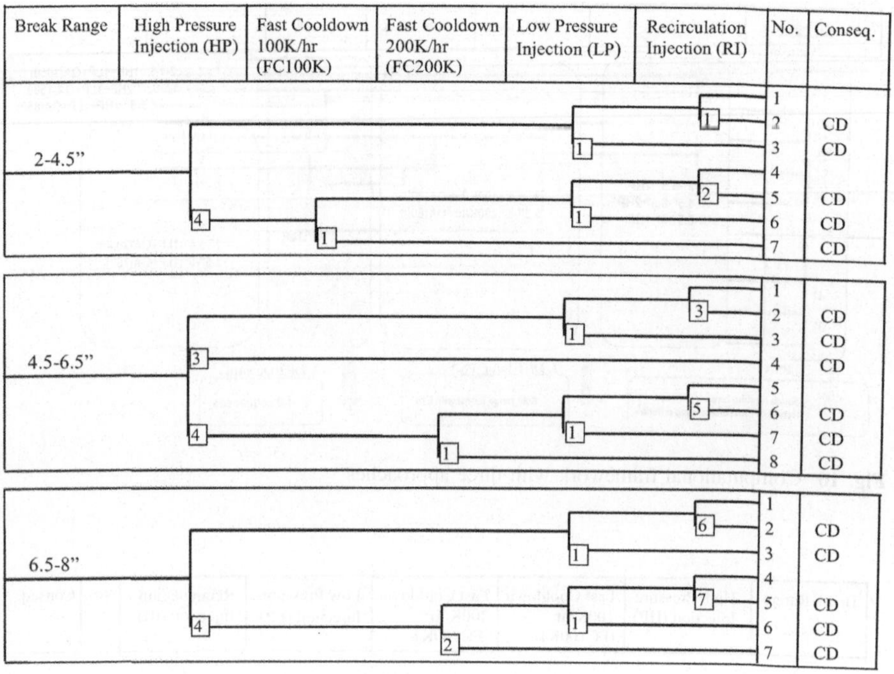

Fig. 18 Event trees of DET informed PSA approach

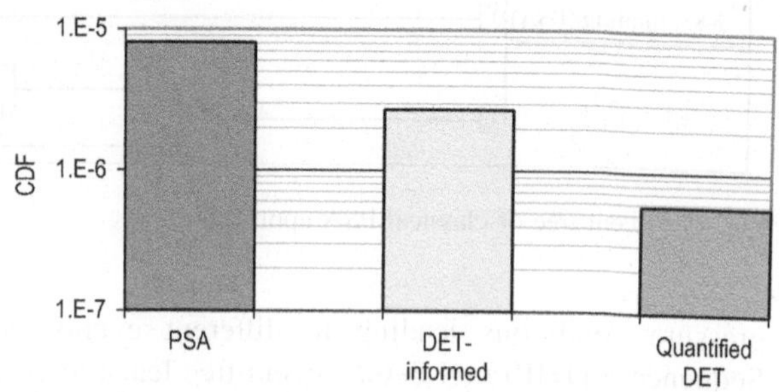

Fig. 19 Comparison of core damage frequencies among three methods

of 5. PSA and DET-informed estimates are higher than quantified DET, due to conservative bounding success criteria (limiting time windows).

Figure 20 shows a comparison of Fussel-Veseley importance measure of basic events. In PSA and DET-informed quantification, operator action during recirculation has the highest importance measure and none of the other basic events has comparable measure. Interestingly in quantified DET results, CCF event of LP injection is as important as operator action. Due to the conservative human error probability in PSA and DET-informed quantification, the importance of hardware events was underestimated. Quantified DET gave more realistic results as the impact of bounding is not present.

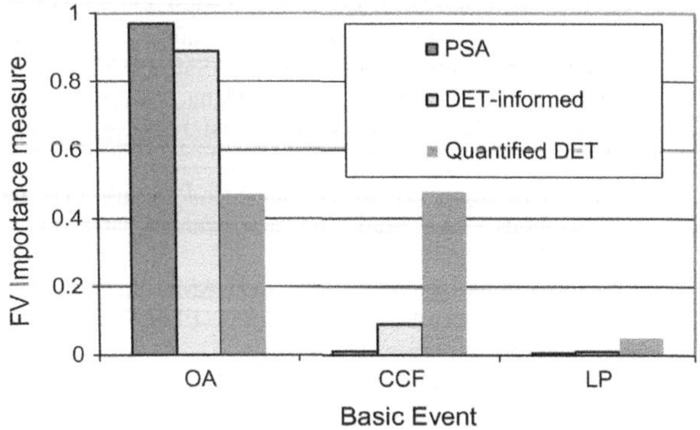

Fig. 20 Comparison of important events

5 Summary

In the current case study, bounding in PSA and DET informed quantification lead to conservative estimate of overall risk (core damage frequency), but it can be sensitive to bounding assumptions and their ability to capture accident dynamics. Also both quantifications underestimated the percentage risk contribution and importance of events due to bounding success criteria. Bounding gets complicated when we consider physical parameter uncertainties and epistemic uncertainties of failure and repair parameters. The IDPSA approach, quantified DET, improves the modeling of accident dynamics, eliminates the effects of bounding, and provides a framework propagate the epistemic uncertainties of the physical model, safety system model, and operator response models. However, quantified DET requires additional computations as well as the discretization of continuous stochastic responses.

Acknowledgements The author is grateful to Vinh N. Dang, Tae-Wan Kim, Saidur Rahman, and Omar Zerkak, for the guidance as well as the collaborative work at Paul Scherrer Institute.

References

1. Siu N (1994) Risk assessment for dynamic systems: an overview. Reliab Eng Syst Saf 43:43–73
2. Hsueh K-S, Mosleh A (1996) The development and application of the accident dynamic simulator for dynamic probabilistic risk assessment of nuclear power plants. Reliab Eng Syst Saf 52:297–314
3. Aldemir T (1989) Quantifying set point drift effects in the failure analysis of process control systems. Reliab Eng Syst Saf 24:33–50
4. Labeau PE, Smidts C, Swaminathan S (2000) Dynamic reliability: towards an integrated platform for probabilistic risk assessment. Reliab Eng Syst Saf 68:219–254
5. Siu N, Marksberry D, Cooper S, Coyne K, Stutzke M (2013) PSA technology challenges revealed by the great East Japan earthquake. In: PSAM topical conference in light of the Fukushima Dai-Ichi accident, Tokyo, Japan, 15–17 Apr 2013

6. Karanki DR, Rahman S, Dang VN, Zerkak O (2017) Epistemic and aleatory uncertainties in integrated deterministic-probabilistic safety assessment: tradeoff between accuracy and accident simulations. Reliab Eng Syst Saf (ISSN: 0951-8320) 162:91–102
7. Rahman S, Karanki DR, Zerkak O, Dang VN (2018) Deterministic sampling for propagating epistemic and aleatory uncertainty in DET analysis of a nuclear power plant. Reliab Eng Syst Saf 175:62–78
8. Karanki DR, Dang VN, MacMillan MT, Podofillini L (2018) A comparison of dynamic event tree methods—case study on a chemical batch reactor. Reliab Eng Syst Saf (ISSN: 0951-8320) 169:542–553
9. Amendola A, Reina G (1984) DYLAM-1, a software package for event sequence and consequence spectrum Methodology. EUR-924, CEC-JRC ISPRA, Commission of the European Communities, Ispra, Italy
10. Cacciabue PC, Amendola A, Cojazzi G (1986) Dynamic logical analytical methodology versus fault tree: the case of auxiliary feedwater system of a nuclear power plant. Nucl Technol 74:195–208
11. Cojazzi G (1996) The DYLAM approach to the dynamic reliability analysis of systems. Reliab Eng Syst Saf 52:279–296
12. Acosta C, Siu N (1993) Dynamic event trees in accident sequence analysis: application to steam generator tube rupture. Reliab Eng Syst Saf 41:135–154
13. Kloos M, Peschke J (2006) MCDET: a probabilistic dynamics method combining Monte Carlo simulation with the discrete dynamic event tree approach. Nucl Sci Eng 153:137–156
14. Hakobyan A et al (2008) Dynamic generation of accident progression event trees. Nucl Eng Des 238:3457–3467
15. Aldemir T (2013) A survey of dynamic methodologies for probabilistic safety assessment of nuclear power plants. Ann Nucl Energy 52:113–124
16. Hsueh KS, Mosleh A (1993) Dynamic accident sequence simulator for probabilistic safety assessment. In: PSA international topical meeting, conference proceedings, Florida, 26–29 Jan 1993
17. Catalyurek U et al (2010) Development of a code-agnostic computational infrastructure for the dynamic generation of accident progression event trees. Reliab Eng Syst Saf 95:278–294
18. Alfonsi A et al (2013) Dynamic event tree analysis through RAVEN. In: ANS PSA 2013 international topical meeting on probabilistic safety assessment and analysis, American Nuclear Society, CD-ROM, Columbia, SC, USA, 22–26 Sep 2013
19. Karanki DR, Kim T-W, Dang VN (2015) A dynamic event tree informed approach to accident sequence modeling: dynamics and variabilities in MLOCA. Reliab Eng Syst Saf (ISSN: 0951-8320) 142:78–91
20. Karanki DR, Dang VN (2016) Quantification of dynamic event trees—a comparison with event trees for MLOCA scenario. Reliab Eng Syst Saf (ISSN: 0951-8320) 147:19–31
21. USNRC (1990) Severe accident risks: an assessment for five US nuclear power plants, NUREG-1150
22. Karanki DR, Dang VN, Kim TW (2012) The impact of dynamics on the MLOCA accident model—an application of dynamic event trees. In: Proceedings of 11th probabilistic safety assessment and management/European safety and reliability (PSAM11/ESREL2012), CD-ROM, Helsinki, Finland, 25–29 June 2012

Dr. Durga Rao Karanki is currently working as a RAMS Manager at Siemens Mobility AG (Switzerland) since 2017. He is responsible for planning, coordination, and performing RAMS activities for development and execution of European Train Control System (ETCS) projects. Previously, he worked as a Scientist at Paul Scherrer Institute (PSI) from 2009 to 2017. His research at PSI primarily focused on dynamic safety assessment and uncertainty management. Prior to joining PSI, he worked as a Scientific Officer (2002–2009) at Bhabha Atomic Research Centre (India), where he conducted research on dynamic fault tree analysis, uncertainty analysis, and risk informed decision making of nuclear power plants. He is also a visiting faculty at several technical institutes. He has actively been involved in Reliability and Safety Assessment research and development for the last 17 years. His work resulted in more than 70 publications including 4 books, 15 first author journal papers, and several conference papers, with more than 900 citations. He received two awards for research Excellency from Society for Reliability Engineering, Quality and Operations Management (SREQOM). He is on the editorial board of three international journals in the area of reliability and risk analysis. He holds B.Tech in Electrical and Electronics Engineering from the Nagarjuna University (India), M.Tech in Reliability Engineering from the Indian Institute of Technology (IIT) Kharagpur and Ph.D. from the IIT Bombay.

The manufacturer's authorised representative in the EU is Springer Nature Customer Service Centre GmbH, Europaplatz 3, 69115 Heidelberg, Germany. If you have any concerns regarding our products, please contact ProductSafety@springernature.com

Printed and bound by CPI Group (UK) Ltd, Croydon, CR0 4YY
30/09/2026
02225237-0017